Cereal Crops: Science and Technology

Cereal Crops: Science and Technology

Edited by Alabaster Jenkins

SYRAWOOD
PUBLISHING HOUSE

New York

Published by Syrawood Publishing House,
750 Third Avenue, 9th Floor,
New York, NY 10017, USA
www.syrawoodpublishinghouse.com

Cereal Crops: Science and Technology
Edited by Alabaster Jenkins

International Standard Book Number: 978-1-68286-380-0 (Hardback)

Cataloging-in-publication Data

Cereal crops : science and technology / edited by Alabaster Jenkins.
    p. cm.
Includes bibliographical references and index.
ISBN 978-1-68286-380-0
1. Grain. 2. Grain--Biotechnology. 3. Grain--Genetic aspects. I. Jenkins, Alabaster.
SB189 .C46 2017
633.1--dc23

Printed in the United States of America.

TABLE OF CONTENTS

Permissions

List of Contributors

Index

PREFACE

Cereals are the most widely consumed food products and they constitute a major part of the world's diet. Cereal production depends on the availability of land, water as well as nutrients. The various studies that are constantly contributing towards advancing technologies and evolution of this field are examined in detail. Different approaches, evaluations, methodologies and advanced studies on crop science have been included in this book. It covers in detail some existent theories and innovative concepts revolving around this discipline. It will provide comprehensive knowledge to the readers. This book is meant for students and experts who are looking for an elaborate reference text on cereal crops, cereal production and cereal crop management.

After months of intensive research and writing, this book is the end result of all who devoted their time and efforts in the initiation and progress of this book. It will surely be a source of reference in enhancing the required knowledge of the new developments in the area. During the course of developing this book, certain measures such as accuracy, authenticity and research focused analytical studies were given preference in order to produce a comprehensive book in the area of study.

This book would not have been possible without the efforts of the authors and the publisher. I extend my sincere thanks to them. Secondly, I express my gratitude to my family and well-wishers. And most importantly, I thank my students for constantly expressing their willingness and curiosity in enhancing their knowledge in the field, which encourages me to take up further research projects for the advancement of the area.

Editor

Biochemical and Molecular Characterization of Potential Phosphate-Solubilizing Bacteria in Acid Sulfate Soils and Their Beneficial Effects on Rice Growth

Qurban Ali Panhwar[1], Umme Aminun Naher[2,3], Shamshuddin Jusop[1]*, Radziah Othman[1,2], Md Abdul Latif[2,3], Mohd Razi Ismail[2]

1 Department of Land Management, Faculty of Agriculture, Universiti Putra Malaysia (UPM), Serdang, Selangor, Malaysia, 2 Institute of Tropical Agriculture, Universiti Putra Malaysia (UPM), Serdang, Selangor, Malaysia, 3 Bangladesh Rice Research Institute, Gazipur, Bangladesh

Abstract

A study was conducted to determine the total microbial population, the occurrence of growth promoting bacteria and their beneficial traits in acid sulfate soils. The mechanisms by which the bacteria enhance rice seedlings grown under high Al and low pH stress were investigated. Soils and rice root samples were randomly collected from four sites in the study area (Kelantan, Malaysia). The topsoil pH and exchangeable Al ranged from 3.3 to 4.7 and 1.24 to 4.25 $cmol_c$ kg^{-1}, respectively, which are considered unsuitable for rice production. Total bacterial and actinomycetes population in the acidic soils were found to be higher than fungal populations. A total of 21 phosphate-solubilizing bacteria (PSB) including 19 N_2-fixing strains were isolated from the acid sulfate soil. Using 16S rRNA gene sequence analysis, three potential PSB strains based on their beneficial characteristics were identified (*Burkholderia thailandensis*, *Sphingomonas pituitosa* and *Burkholderia seminalis*). The isolated strains were capable of producing indoleacetic acid (IAA) and organic acids that were able to reduce Al availability via a chelation process. These PSB isolates solubilized P (43.65%) existing in the growth media within 72 hours of incubation. Seedling of rice variety, MR 219, grown at pH 4, and with different concentrations of Al (0, 50 and 100 µM) was inoculated with these PSB strains. Results showed that the bacteria increased the pH with a concomitant reduction in Al concentration, which translated into better rice growth. The improved root volume and seedling dry weight of the inoculated plants indicated the potential of these isolates to be used in a bio-fertilizer formulation for rice cultivation on acid sulfate soils.

Editor: Luis Herrera-Estrella, Centro de Investigación y de Estudios Avanzados del IPN, Mexico

Funding: The authors acknowledge the financial and technical support given by Universiti Putra Malaysia and the Ministry of Education, Malaysia under the Long-term Research Grant Scheme (LRGS) fund for Food Security. The funders had no role in study design, data collection and analysis, decision to publish, or preparation of the manuscript.

Competing Interests: The authors have declared that no competing interests exist.

* Email: shamshud@upm.edu.my

Introduction

Acid sulfate soils are sporadically spread throughout the coastal plains around the globe. The estimated worldwide extent of acid sulfate soils is about 24 million hectares [1]. In Malaysia, acid sulfate soils mainly occur in the west coastal plains of Peninsular Malaysia and Sarawak (Borneo Island). The soils are characterized by the presence of pyrite (FeS_2), which is associated with high acidity and Al content, resulting from its oxidation upon exposure to the atmosphere [2]. Under acidic soil conditions, Al^{3+} restricts the growth of roots either by inhibiting cell division, cell elongation or both [3]. It is also known that rice grown on acid sulfate soils are subjected to Fe^{2+} toxicity [4].

Acid sulfate soils are normally not suitable for crop production unless they are properly ameliorated and their fertility improved [5]. They usually contain high concentration of Al and Fe which influence the biochemical properties of the soils. There are only a few acid-tolerant plant species and microbes that are able to survive in these soils. The soils contain low amount of total microorganisms, which vary considerably depending on the vegetation type and soil management. Microorganisms perform a major role in nutritional chains that are an important part of the biological balance in soils [6]. Bacteria are important for closing nutrient and geochemical cycles, involving carbon, nitrogen, sulfur and phosphorous. Beside chemical treatment, addition of microbes may improve nutrient availability in soils (especially phosphorus) and reduce Al toxicity.

There are some species of bacteria which have the potential to mineralize and solubilize organic and inorganic phosphorus in soil [7]. Bacterial strains such as *Pseudomonas*, *Bacillus* and *Rhizobium* are the dominant phosphate solubilizers [8] and single novel genus *Sphingomonas* spp. has the capability to fix nitrogen [9]. Moreover, in recent years, a growing number of *Burkholderia* strains and species have also been reported as plant-associated bacteria. *Burkholderia* spp. can be free-living in the rhizosphere as well as epiphytic or endophytic, obligate endosymbionts or phytopathogens [10]. Among the beneficial properties of these bacteria is the production of plant growth promoting phytohormone, polysaccharides and organic acids that are important for rice growth. Indoleacetic acid produced by the bacteria is known to promote an extensive root architecture capable of absorbing nutrient elements efficiently from the surroundings which ultimately improves rice growth [11–12].

PSB produce large amounts of organic acids [13] resulting in Al binding via a chelation process which is a plausible mechanism for reducing Al toxicity to rice roots in acid sulfate soils. Few attempts have been made to isolate and characterize these potential microorganisms living in the rhizosphere of rice plants grown in acid sulfate soils. Hence, the present study was undertaken to isolate and enumerate microbial occurrences in acid sulfate soils, and to identify beneficial isolates such as N_2-fixing and P-solubilizing bacteria that have important role in nutrient cycling, reducing Al toxicity and producing IAA in an acid sulfate soil environment.

Materials and Methods

Sampling location and isolation of microbes

Experimental site and conditions. This experiment was conducted at the Soil Microbiology Laboratory, Department of Land Management, Faculty of Agriculture, Universiti Putra Malaysia, Serdang, Malaysia. Soil samples were randomly collected from four sites in an acid sulfate soil area at Semerak, Kelantan, Malaysia which is located at a latitude of 30.01°N and longitude of 101°.70E.

We are working in a Government agency (University) so there is no need to take any permission for research activities from any one as we are working in a Government Project" Long term Research Grant Scheme" (LRGS) by the Ministry of Higher Education Malaysia. Hence, no specific permissions were required for these locations/activities. It has no any conflict on this issue." The methods to determine the soil physico-chemical analyses are given in Table 1.

Enumeration of the total microbial population from rice cultivated on acid sulfate soil. A series of 10-fold dilutions were prepared up to 10^{-8} for soil and rhizosphere microbial population determinations using the spread plate count method. Total fungal, bacterial and actinomycetes populations were determined using potato dextrose agar (PDA), nutrient agar (NA), and actinomycetes agar plates, respectively in five replicates.

Enumeration of PSB from rice cultivated on acid sulfate soil. Phosphate-solubilizing bacterial (PSB) population was determined from the soil, rhizosphere and endosphere using selective media plates at various pH levels: (i) the National Botanical Research Institute's phosphate growth medium (NBRIP) [17] at pH 5.0 and 6.7; and (ii) PDYA-AlPO$_4$ at pH 3.5 and 5. For the determination of the rhizosphere population, approximately 3 g of rice plant roots with its adhering soil were

transferred into conical flask containing 99 mL of sterile distilled water and the contents were vigorously shaken. A 10-fold dilution series was prepared up to 10^{-8} and 0.1 mL aliquots were spread on selective media and incubated at $28\pm2°C$ in an incubator. For the determination of the endophytic population, fresh roots were taken and surface sterilized with 70% ethanol for 5 minutes and treated with 3% Clorox for 30 seconds [11]. Roots were cut into small pieces and surface sterilized by dipping into 95% ethanol and a flame. Surface sterilized roots were homogenized using a sterilized mortar and pestle. Endophytic PSB populations were determined using the total plate count method in five replicates.

Estimation of diazotrophs from rice cultivated on acid sulfate soil. A series of 10-fold dilutions were prepared up to 10^{-8} using rhizosphere and non-rhizosphere soil and the diazotroph populations were determined using the most-probable number (MPN) method in nitrogen free (Nfb) semi-solid medium in five replicates. The Nfb semi-solid medium [18] contained 5 g malic acid, 0.5 g K_2HPO_4, 0.2 g $MgSO_4$. 7 H_2O, 0.1 g NaCl, 0.02 g $CaCl_2$ and 0.5% bromothymol blue in 0.2 N KOH (2 mL), 1.64% Fe-EDTA solution (4 mL) and 2 g agar.

Determination of the beneficial traits of isolated bacteria

Determination of indoleacetic acid (IAA). The isolates were inoculated in nutrient agar (NB) broth with the addition of tryptophan (2 mg L^{-1}) and incubated at $28\pm2°C$ for 48 hours. The culture was centrifuged at 7000 rpm for 7 minutes and 1.0 mL of the supernatant was mixed with 2 mL of Salkowsky's reagent [19]. The IAA concentration was determined using a spectrophotometer at 535 µm in five replicates.

Phosphate solubilization

Phosphate solubilization on media plates. The phosphate solubilizing activities of the isolates were assayed by spotting 10 µL of the cultures on different phosphate containing media such as NBRIP [17], Pikovaskaya, Christmas Island Rock Phosphate (CIRP) and PDYA-AlP in five replicates. The plates were incubated at 30°C for one week and the phosphate solubilization efficiency was measured [20]:

$$Solubilization\ efficiency = \frac{Solubilization\ diameter}{Growth\ diameter} \times 100 \quad (1)$$

Phosphate solubilization in broth culture. The isolated PSB strains were evaluated for their phosphate-solubilizing activity in broth culture. Three different media containing insoluble

Table 1. Analytical methods used in the study.

S No.	Analysis	Procedure
1	Soil pH and EC	Soil: water (1:2.5) extract using PHM210 standard pH meter and EC meter by glass electrode [14]
2	CEC	1 M NH$_4$OAc solution buffered at pH 7.0 was used [14]
3	Soil texture	Pipette method [15]
4	Total N	Kjeldahl digestion method [16]
5	Exchangeable cations (Ca, Mg, K)	Extracted by 1M NH$_4$OAc solution at pH 7 [14]
6	Exchangeable Al	Extracted by 1 M KCl solution
7	Total carbon	CN analyzer (LECO CR-412)
8	Micronutrients (Cu, Mn, Zn, Fe)	Inductivity coupled plasma - atomic emission spectroscopy (ICP-AES)

ICP-AES = inductively coupled plasma atomic emission spectroscopy, CEC = cation exchange capacity.

phosphate: i) calcium phosphate in NBRIP broth [17] containing (g L^{-1}) $MgCl_2.6H_2O_5$ g, $MgSO_4.H_2O$ 0.25 g, KCl 0.2 g, $(NH_4)_2SO_4$ 0.1 g, $Ca_3(PO_4)_2$ amended with glucose 10 g; ii) CIRP broth modified from NBRIP supplemented with phosphate rock (CIRP) instead of calcium phosphate; and iii) PDYA-AlP broth [21] containing (g L^l): PDA agar 39 g, yeast extract 2 g, sterilized 10% K_2HPO_4, and 10% $AlCl_3$ (100 mL). Each media (200 mL) were inoculated with PSB inoculum and were incubated at 30°C on a Kottermann 4020 shaker at 80 rpm for 3 days. Each treatment was replicated five times.

Determination of the population in broth culture. One milliliter of broth was taken from the respective flasks at different periods (initial & after 72 h) for determination of the bacterial population. A series of 10-fold dilutions were prepared up to 10^{-10}. The population was determined using the drop plate count method according to the method of Somasegaran and Hoben [22].

Determination of phosphorus solubilization in broth cultures. Exactly 2 mL of samples were taken for P determination. The samples were first allowed to sediment for 15 minutes and were then centrifuged at $4000 \times$ g for 5 minutes. The supernatant was filtered through 0.2 µm filter paper and kept at $-20°C$ until analysis. The available P was determined according to the published methods [23].

Nitrogen-fixing activities. The N_2-fixing activity was determined using the Nfb semi-solid liquid medium [18]. The presence of pellicle formation below the surface of media indicated the N_2-fixing activity.

Molecular identification of PSB strains

DNA extraction and primers. The identification of PSB was carried out on the basis of 16S rRNA gene sequencing. The genomic DNA of PSB isolates was extracted by the Qiagen DNeasy Plant Mini Kit (Qiagen, Valencia, CA). Forward primers D1 (5-AGAGTTTGATCCTGGCTCAG-3) and reverse P2 (3-ACGGCTACCTTGTTACGACTT-5) were used for amplification of 16S rRNA gene [24]. Each sample was replicated five times.

PCR protocols and gel electrophoresis. The total PCR reaction mixture was 50.0 µL, comprising 200 µM dNTPs, 50 µM of each primer, 1× PCR buffer, 3 U Taq polymerase and 100 µg genomic DNA. The thermal cycler (MJ Mini personal Thermal Cycler, Bio-Rad, Model- PTC-1148) conditions were as follows: 95°C for 3 minutes, followed by 30 cycles of denaturation at 95°C for 1 minute, annealing at 48°C for 1 minute and primer extension at 72°C for 2 minutes. This was followed by a final extension at 72°C for 10 minutes. The reaction products were separated by running 5 µL of the PCR reaction mixture in a 1.0% (w/v) agarose gel and staining the bands with ethidium bromide.

Strain identification using gene sequencing. Three potential isolates with greater beneficial characteristics were selected for 16S rRNA gene sequencing analysis. Sequence data were aligned and compared with the available standard sequences of bacterial lineage in the National Center for Biotechnology Information GenBank (http://www.ncbi.nlm.nih.gov/) using BLAST [25]. A phylogenetic tree was constructed by the neighbor-joining method using the software MEGA 4 [26]. The obtained sequences were deposited in the European Molecular Biology Laboratory data (accession number NR 074312.1, NR 042635.1 and NR 25363.1).

Principal Components Analysis. Principal Components Analysis (PCA) is an ordination technique. Here we performed an eigen analysis of the covariance matrix. The eigen value is a measure of the strength of an axis, the amount of variation along an axis, and ideally the importance of an ecological gradient. The precise meaning depends on the ordination method used. Two-dimensional graph (two axes of PCA) was constructed using co-variance matrix and find out the variation among the characters of 21 bacterial isolates.

Efficiency of isolates to improve rice seedling growth with high Al and low pH. Three potential phosphate-solubilizing bacterial strains (PSB7, PSB17 and PSB21) were selected on the basis of their beneficial characteristics. Seven-days-old MR 219 rice seedlings were grown in Hoagland solution containing different concentrations of Al (0, 50 and 100 µM) with five replicates. The initial pH of the solution was adjusted to 4.0. Rice seedlings were harvested 21 days after sowing. The bacterial population, plant dry biomass, solution pH and organic acids were determined soon thereafter.

Determination of the microbial population at different Al concentrations. One milliliter of broth was taken from the respective flasks at different time periods (6, 12, 24, and 48 h) for the determination of bacterial growth. A series of 10-fold dilution were prepared up to 10^{-10}. The population was determined using the drop plate count method [22].

Determination of organic acids. About 20 µL of the samples from each treatment were injected into HPLC with a UV detector set at 210 nm. A Rezex ROA-organic acid "H^+" (8%) column (250×4.6 mm) from Phenomenex Co. was used, the mobile phase was 0.005 N H_2SO_4 with a flow rate of 0.17 mL min^{-1}.

Determination of root morphology. The root morphology of the rice plants was determined using a root scanner (model Epson Expression 1680 equipped with root scanning analysis software). Total root length (cm), total surface area (cm^2) and total volume (cm^3) were quantified using a scanner (Expression 1680, Epson) equipped with a 2 cm deep plexiglass tank (20.30 cm) filled with up H_2O [27]. The scanner was connected to a computer and scanned data were processed by Win-Rhizo software (Regent Instruments Inc., Québec, Canada).

Statistical analysis

All data were statistically analyzed using SAS Software (Version 9.2), and treatment means were separated using Tukey's test (P< 0.05).

Results

Properties of the acid sulfate soils

The pH of the acid sulfate soils under study ranged from 3.3 to 4.7 pH with values decreasing with depth (Table 2). The soils were taxonomically classified as Typic Sulfaquepts. Soils of this nature normally contain pyrite in the subsoil, having pH<3.5 below a depth of 50 cm [2]. It is the oxidation of this pyrite that produces acidity and the subsequent release of Al and/or Fe into the environment. The low soil pH is consistent with the presence of high exchangeable Al in the soils. Exchangeable Al in the topsoil at all sampling sites ranged from 1.24 to 4.25 $cmol_c$ kg^{-1}, occurring at a toxic level for rice growth. Total N and exchangeable K were found to be at sufficient levels for rice growth, while available P, exchangeable Ca and Mg were insufficient. The Zn, Cu and Mn contents were low, while extractable Fe was high (124 to 181 mg kg^{-1}) (Table 2). Total C was high in the topsoil with values above 2% and it decreased consistently with depth. The high organic matter in the topsoil would have a profound effect on the availability of Al and Fe, which can be partly fixed (chelated) by it and hence deactivated. Chelated Al and Fe are non-toxic to rice plants in the field.

Table 2. Chemical characteristics of the acid sulfate soils.

Site	Soil depth (cm)	Soil pH	Total C (%)	Total N (%)	Avail. P (mg kg⁻¹)	Exchangeable (cmol_c kg⁻¹)				Extractable (mg kg⁻¹)			
						K	Al	Ca	Mg	Fe	Zn	Cu	Mn
1	0–15	4.0b	2.1c	0.18b	26.3a	0.05b	1.7cd	0.43c	1.0bc	174abc	1.6ab	3.2a	8.1ab
	15–30	4.0b	1.6d	0.14c	19.1b	0.05b	2.1c	0.40c	0.9cd	170abc	1.6ab	2.5b	7.8ab
	30–45	3.8bc	1.2e	0.10d	13.1def	0.04c	4.5b	0.30d	0.6ef	129bcd	0.9d	2.3bc	7.4abc
	45–60	3.6c	0.9f	0.09d	12.9def	0.04c	5.5a	0.25d	0.5ef	124d	0.7d	2.0bcd	6.4bcd
2	0–15	4.7a	2.9a	0.17b	25.2a	0.06a	1.24d	0.57b	0.7de	181a	2.3a	2.4b	8.8a
	15–30	3.6c	1.1e	0.12c	16.6bc	0.04c	1.5cd	0.43c	0.5ef	176ab	1.8a	1.9cde	8.2ab
	30–45	3.4de	1.1e	0.09de	15.5cd	0.04c	1.7cd	0.40c	0.4f	167abcd	1.8a	1.7def	7.9ab
	45–60	3.3e	0.9f	0.08de	11.4f	0.03d	1.9c	0.12e	0.4f	163bcd	1.5b	1.5ef	6.0bcd
3	0–15	3.8bc	2.3b	0.21a	19.2b	0.05b	1.8cd	0.80a	1.3a	180a	2.0a	1.4ef	7.5ab
	15–30	3.5d	1.1e	0.13c	14.8cde	0.04c	1.9c	0.73a	1.2b	178a	1.3bc	1.5def	6.1cd
	30–45	3.4de	0.9f	0.09d	12.2ef	0.04c	1.9c	0.63b	0.94abc	145cd	1.2c	1.5ef	5.3cd
	45–60	3.3e	0.6g	0.06e	11.6f	0.04c	4.3b	0.60b	0.7de	124d	1.1c	1.1f	4.5d
4	0–15	3.8bc	2.2b	0.20a	19.1b	0.05b	1.7cd	0.72a	1.2b	179a	1.97a	1.3ef	7.2ab
	15–30	3.6c	1.15e	0.11c	14.3cde	0.04c	1.8cd	0.70a	1.0bc	170abc	1.2c	1.4ef	6.0cd
	30–45	3.5d	1.01ef	0.08de	11.5f	0.03d	2.03c	0.60b	0.94abc	154c	1.1c	1.4ef	5.2cd
	45–60	3.4de	0.5g	0.05e	10.7fg	0.03d	4.03b	0.56b	0.8de	130d	1.0c	1.2f	4.3d

Data values are means of five replicates. Means followed by the same letter within a column are not significantly different ($P<0.05$).

Total population of microorganisms in the rhizosphere, non-rhizosphere and endosphere

Higher microbial populations were found in the rice rhizosphere compared to the non-rhizosphere. The total bacterial and actinomycetes population were higher than fungal populations (Table 3) and higher PSB populations were observed on NBRIP plates compared to PDYA-AlP media plates (Table 4). However, there were no significant population differences of PSB present in the NBRIP media plates at media pH of 5.0 and 6.7. The highest diazotrophic population was found in the endosphere (25×10^6 cfu root g^{-1}) and the lowest was in the non-rhizosphere (13×10^3 cfu soil g^{-1}) (Table 4).

Biochemical properties of the strains isolated from rice cultivated on acid sulfate soils

The indigenous bacterial isolates were capable of producing indoleacetic acid (IAA). Among all the isolates, PSB21 (14.96 mg L^{-1}), followed by PSB7 (13.16 mg L^{-1}) produced high IAA levels as compared to the other isolates. The lowest amount of IAA was produced by PSB20 (Table 5). A total of 21 phosphate-solubilizing bacterial (PSB) isolates were identified from the acid sulfate soils. However, only 19 strains were able to grow in nitrogen-free medium (Table 5).

Phosphate solubilization in media plates

All the selected 21 PSB isolates were able to solubilize P on NBRIP and Pikovskaya media plates, with P-solubilizing activity indicated by a clear halo zone (Plate S1). The highest P solubilizing activity in NBRIP media plate was contributed by PSB17 (70.23%), followed by PSB7 (57.5%), while the lowest activity (23.23%) was contributed by PSB20 strain (Table 5). On the other hand, on Pikovskaya media plates, the highest activity was contributed by PSB17 (76.03%), followed by PSB21 (70.12%).

Phosphate solubilization in broth culture

The isolates from the soils were able to solubilize P from different forms of inorganic phosphate incorporated into the broth culture after 72 hours of incubation (Table 5). Comparatively, the highest P solubilization activity was found in NBRIP broth with PSB7 (43.65%), followed by PSB21 (43.34%). Most of the strains were able to solubilize P in CIRP broth, but higher P solubilization was observed in samples containing PSB1 and PSB5 compared to other isolates, and lower amounts of soluble P were observed in the PDYA-Al broth.

Principal component analysis among 21 bacterial isolates based on biochemical properties

In order to assess the patterns of variation, PCA was done by considering seven biochemical characters. The first three principal components (PCs) explained 99% of the total variation in 21 bacterial isolates and showed 73.27, 21.63, 4.14% variations, respectively (Table 6). The PC1 and PC2were loaded with all seven characters such as IAA production, Nfb activity, P solubilization in different sources of P media and culture (NBRIP, Pikovskaya, PDYA-AlP and CIRP). All the characters either in PC1 or PC2 showed positive contribution for the differences between the bacterial isolates. The PC1 was strongly responsible for variation of the 21 bacterial isolates.

The PCA scatter plot is shown in Fig. 1. All the bacterial isolates were grouped into seven clusters. The highest number of bacterial isolates from acid sulfate soil was in cluster III (PSB3, PSB4, PSB6, PSB8, PSB9, PSB16 and PSB19), followed by cluster II (PSB10, PSB11, PSB12 and PSB13), cluster I (PSB1, PSB2 and PSB6), cluster VII (PSB7, PSB17 and PSB21) and cluster VII (PSB15 and PSB18). Clusters IV and V each contained one strain. The results show that cluster VII with bacterial isolates PSB17, PSB21 and PSB7 was the most diverged group.

Identification of the potential strains

Molecular analysis by the 16S rDNA identification technique was adopted in this study. These excellent markers for the clarification of bacterial phylogeny are ribosomal ribonucleic acids. In this study, we used gene sequences from the β-subclass of Proteobacteria to determine the phylogenetic relationships among the tested isolates. The neighbor-joining tree was subjected to the numerical re-sampling by bootstrapping, and the resulting bootstrap values were observed at the tree branch nodes. Each value represents the number of times (out of 1000 replicates) that the represented groupings occurred in the re-samplings. The consensus tree showed 98–100% confidence levels between 3 potential isolates from the β-subclass of Proteobacteria PSB7 [*Burkholderia thailandensis*] and PSB21 [*Burkholderia seminalis*], whereas PSB17 [*Sphingomonas pituitosa*] had a 100% confidence level (Fig. 2 & 3).

All of the sequences have higher than 98% identity with the queried sequence. Two PSB7 and PSB21 sequences were identified as *Burkholderia* spp. with accession numbers NR 074312.1 and NR 042635.1, respectively, while one belonged to *Sphingomonas* sp. (NR 25363.1).

Table 3. Total microbial population of microorganisms isolated from the acid sulfate soils.

Site	Bacterial		Fungal		Actinomycetes	
	Rhizosphere	Non-rhizosphere	Rhizosphere	Non-rhizosphere	Rhizosphere	Non-rhizosphere
	------- (*cfu soil g^{-1}) -------					
1	1.5×10^{7b}	7.1×10^{5a}	7.0×10^{3ab}	6.0×10^{3ab}	3.0×10^{4c}	2.1×10^{5a}
2	9.0×10^{6c}	3.5×10^{5b}	10.0×10^{3ab}	3.0×10^{3b}	8.0×10^{4b}	3.0×10^{4c}
3	2.8×10^{7a}	1.8×10^{5c}	7.0×10^{3a}	9.0×10^{3a}	1.2×10^{5ab}	10.0×10^{4ab}
4	1.9×10^{7ab}	1.5×10^{5c}	2.0×10^{3b}	5.0×10^{3ab}	1.2×10^{5ab}	4.0×10^{4c}

*cfu = colony forming unit, Data values are means of five replicates.
Means followed by the same letter within a column are not significantly different (P<0.05).

Table 4. Phosphate-solubilizing bacteria and diazotroph population from rice cultivated on acid sulfate soils.

Site	Phosphate-solubilizing bacterial population												Diazotrophs population		
	Rhizosphere (*cfu soil g⁻¹)				Non-rhizosphere (cfu soil g⁻¹)				Endosphere (cfu root g⁻¹)				Rhizosphere (cfu soil g⁻¹)	Non-rhizosphere (cfu soil g⁻¹)	Endosphere (cfu root g⁻¹)
	†NBRIP		‡PDYA-AIP		NBRIP		PDYA-AIP		NBRIP		PDYA-AIP		------------- N-free media -------------		
	pH 5.0	pH 6.7	pH 3.5	pH 5.0	pH 5.0	pH 6.7	pH 3.5	pH 5.0	pH 5.0	pH 6.7	pH 3.5	pH 5.0			
1	2×10^{7a}	4×10^{4b}	-	-	5.7×10^{4a}	3×10^{4a}	-	-	28×10^{4a}	19×10^{4b}	-	-	23×10^{5b}	24×10^{3b}	25×10^{6a}
2	70×10^{4b}	41×10^{4b}	-	-	2×10^{3b}	3×10^{3b}	-	-	8×10^{4}	4×10^{4a}	-	-	15×10^{6a}	3×10^{4a}	11×10^{5b}
3	57×10^{4b}	60×10^{4b}	-	5×10^{4a}	3×10^{4a}	21×10^{4a}	-	6×10^{4a}	9×10^{4}	9×10^{4a}	-	6×10^{4a}	16×10^{5b}	13×10^{5b}	38×10^{5b}
4	24×10^{4b}	5×10^{4b}	-	2×10^{4a}	2×10^{4a}	21×10^{4a}	-	5×10^{4a}	50×10^{4a}	71×10^{4a}	-	5×10^{4a}	23×10^{5b}	24×10^{3b}	25×10^{6a}

*cfu = colony forming unit,
†NBRIP = National Botanical Research Institute's phosphate growth medium,
‡PDYA-AIP = Potato dextrose Yeast Agar- Aluminium Phosphate respectively.
Data values are means of five replicates. Means followed by the same letter within a column are not significantly different (P<0.05).

The efficacy of PSB strains to reduce Al toxicity

It was observed that high Al concentration severely affected the growth of the rice seedlings. Plant height and dry biomass significantly decreased with increased Al concentrations (Table 7). In comparison, the bacterial inoculated plants were less affected by Al toxicity. High plant height (18 cm) and dry biomass (0.76 g) were observed in the *Burkholderia seminalis* inoculated treatments at 0 µM Al (Table 7). The PSB associated with the plant roots produced organic acids that provided P and chelated the Al toxicity. As shown by the SEM and TEM micrographs (Plate S2), the bacteria existed in association with the rice seedlings.

The morphology of the PSB-inoculated and untreated rice roots

This study found that root length, root surface area and volume varied with the Al concentration and bacterial inoculation. Presence of Al affected rice root development, especially at high concentrations (Table 7). Generally, greater root length, surface area and volume were found in inoculated compared to non-inoculated rice plants. Significantly greatest root length (58.51 cm) was observed at 0 µM Al by PSB21 inoculated rice plant compared those at high Al concentration. The greatest root surface area of 74.12 cm^2 was recorded in PSB17 inoculated rice at 0 µM Al, while the highest root volume was observed in the *Sphingomonas pituitosa* and *Burkholderia seminalis* (6.9 and 6.88 cm^3, respectively) inoculated rice plants at the same Al concentration.

Effects of Al and PSB inoculation on the release of organic acids

It was observed that organic acids released by the rice roots varied with the Al concentration and bacterial inoculation (Table 8). Plants with highest Al concentration was found to secrete the highest amounts of organic acids. The release of organic acids was affected by the Al concentration. Higher amounts of organic acids were released by PSB-inoculated plants compared to the non-inoculated rice seedlings. Among the organic acids released, higher amounts of oxalic and citric acids were observed at 100 µM Al compared to malic acid, while the amount of malic acid was found to be low, particularly at 0 and 50 µM Al concentration.

Effect of Al on the population of PSB strains with or without rice seedlings

It was found that the population of PSB was affected by the Al concentrations in both of the plant and without plant systems. It seemed that a higher Al concentration lowered the population of PSB (Fig. 4a &b). Hence, at 0 µM Al concentration, higher PSB population was observed compared to that observed at higher Al concentrations. Among the inoculated PSB strains, the highest population was recorded by PSB21 at all the Al concentrations without plant system, while in plant system all strains were showing the same response. The population decreased with the increasing Al concentration, and this trend was observed for all the PSB strains in both culture systems. It was clear from the current study that some bacteria strains survived under the condition of low pH and high Al concentration; these strains have the potential to be used for rice cultivated on acid sulfate soil containing high Al concentration.

Table 5. Biochemical properties of the isolated PSB strains.

Isolates	IAA production (mg L⁻¹)	Nfb activity	P solubilization (%) in different media plates		P solubilization in broth culture after 72 hours of incubation					
					*NBRIP		†CIRP		‡PDYA-AlP	
			NBRIP	Pikovskaya	(ppm)	(%)	(ppm)	(%)	(ppm)	(%)
PSB1	10.00bc	+ve	50.56c	45.26d	104c	31.30	8.43bc	2.54	3.21c	0.97
PSB 2	9.60c	+ve	45.23d	55.23c	103c	31.00	5.67de	1.71	4.32b	1.30
PSB 3	9.00c	+ve	40.00e	40.24e	108c	32.51	7.23c	2.18	1.34e	0.40
PSB 4	5.28e	+ve	36.67f	45.65d	104c	31.30	4.32f	1.30	2.87d	0.86
PSB 5	7.32d	+ve	45.23d	50.13c	106c	31.91	10.12a	3.05	4.89ab	1.47
PSB 6	4.32f	+ve	30.22f	45.81d	109c	32.81	3.21g	0.97	3.23c	0.97
PSB 7	13.16a	+ve	57.5b	60.12b	145a	43.65	11.3a	3.40	5.43a	1.63
PSB 8	3.84fg	−ve	42.2e	45.36d	107c	32.21	6.32d	1.90	3.45c	1.04
PSB 9	4.40f	+ve	34.28f	35.93f	106c	31.91	5.64de	1.70	1.76e	0.53
PSB 10	5.44e	+ve	45.65d	50.22c	79de	23.78	6.54d	1.97	2.65d	0.80
PSB 11	9.20c	+ve	51.11c	52.16c	65f	19.57	5.21e	1.57	3.43c	1.03
PSB 12	10.00bc	+ve	55.73b	47.27d	89d	26.79	4.21f	1.27	3.28c	0.99
PSB 13	11.60b	+ve	50.32c	34.44fg	73e	21.97	4.02f	1.21	2.69d	0.81
PSB 14	5.60e	+ve	40.24e	35.38f	85d	25.59	3.41g	1.03	3.38c	1.02
PSB 15	8.16d	+ve	49.41cd	45.42d	142a	42.74	6.03d	1.82	2.76d	0.83
PSB 16	9.63c	+ve	31.11f	40.83e	112bc	33.71	4.07f	1.23	3.48c	1.05
PSB 17	12.16b	+ve	70.23a	76.03a	142a	42.74	9.34b	2.81	5.73a	1.72
PSB 18	11.36b	−ve	45.09d	44.1d	122b	36.72	5.74de	1.73	4.21b	1.27
PSB 19	10.56b	+ve	50.21c	32.5g	107c	32.21	6.24d	1.88	4.89ab	1.47
PSB 20	1.56h	+ve	23.23g	31.02g	124b	37.32	2.02h	0.61	4.1b	1.23
PSB 21	14.96a	+ve	56.65b	70.12a	144a	43.34	9.57b	2.88	5.23a	1.57

*NBRIP = National Botanical Research Institute's phosphate growth medium,
†CIRP = Christmas Island Rock Phosphate and
‡PDYA-AlP = Potato dextrose Yeast Agar- Aluminium Phosphate respectively.
Data values are means of five replicates. Means in each column followed by the same letters are not significantly different according to Tukey's HSD at P≤0.05. Note: (+ve) for N₂ fixing and (−) for not N₂ fixing activities.

Table 6. Principal component analysis of seven biochemical characters and proportion of variation for each component.

Biochemical characters	[a]PC1	PC2	PC3
% Variation	**73.27**	**21.63**	**4.14**
[b]IAA production (mg L^{-1})	0.49485	−0.04136	0.46886
Nfb activity	0.71264	0.4016	−0.03145
P solubilization (%) on [c]NBRIP	0.5499	−2.0133	1.8433
P solubilization (%) on Pikovskaya media plates	0.60905	−1.5349	−2.3093
P solubilization (ppm) in NBRIP broth culture	2.4185	1.2426	0.19441
P solubilization (ppm) in [d]CIRP broth culture	−0.55919	0.18271	−0.0435
P solubilization (ppm) in [e]PDYA-AIP broth culture	−0.63265	0.33844	−0.06201

[a]PC = Principle component,
[b]IAA = Indoleacetic acid,
[c]NBRIP = National Botanical Research Institute's phosphate growth medium,
[d]CIRP = Christmas Island Rock Phosphate and
[e]PDYA-AIP = Potato dextrose Yeast Agar- Aluminium Phosphate respectively.

Effect of PSB inoculation on nutrient solution pH at different Al concentration

There was a clear effect of PSB inoculation found in the nutrient solution pH at different Al concentrations in both of the plant and without plant systems (Fig. 5a &b). The initial solution pH was 4.0 and after 24 h of inoculation, the pH increased to 7.00 for all PSB inoculated treatments (with and without plant system). However,

for the non-inoculated treatment, it remained almost the same as the initial pH (3 to 4.0) in both of the conditions regardless of Al concentrations. Without plant, the highest pH of 7.69 was obtained for the 100 μM Al treatment with PSB 21(Fig. 5a). For the inoculated plant, the solution pH was a bit higher in 0 μM Al compared to the other treatments (Fig. 5b). The lowest solution pH of 2.80 was found in the non-inoculated control at 100 μM of

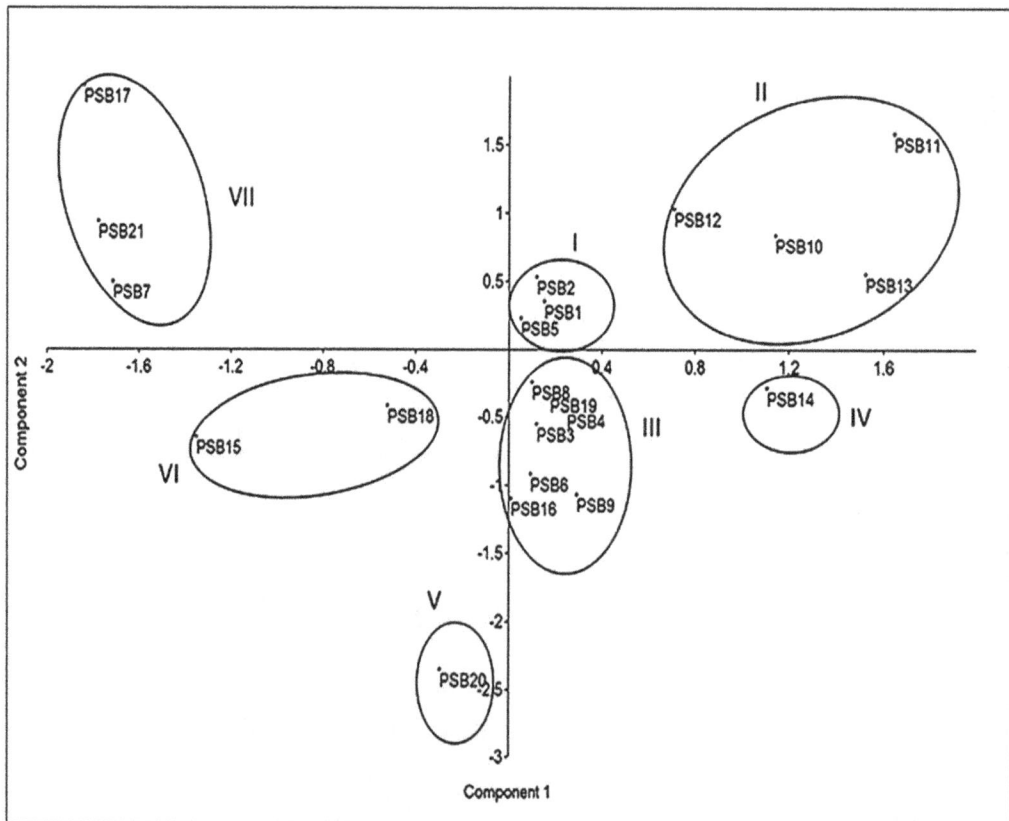

Figure 1. Plotting of two principal axes in principal component analysis showed the variation among 21 bacterial isolates based on seven biochemical characters using co-variance matrix.

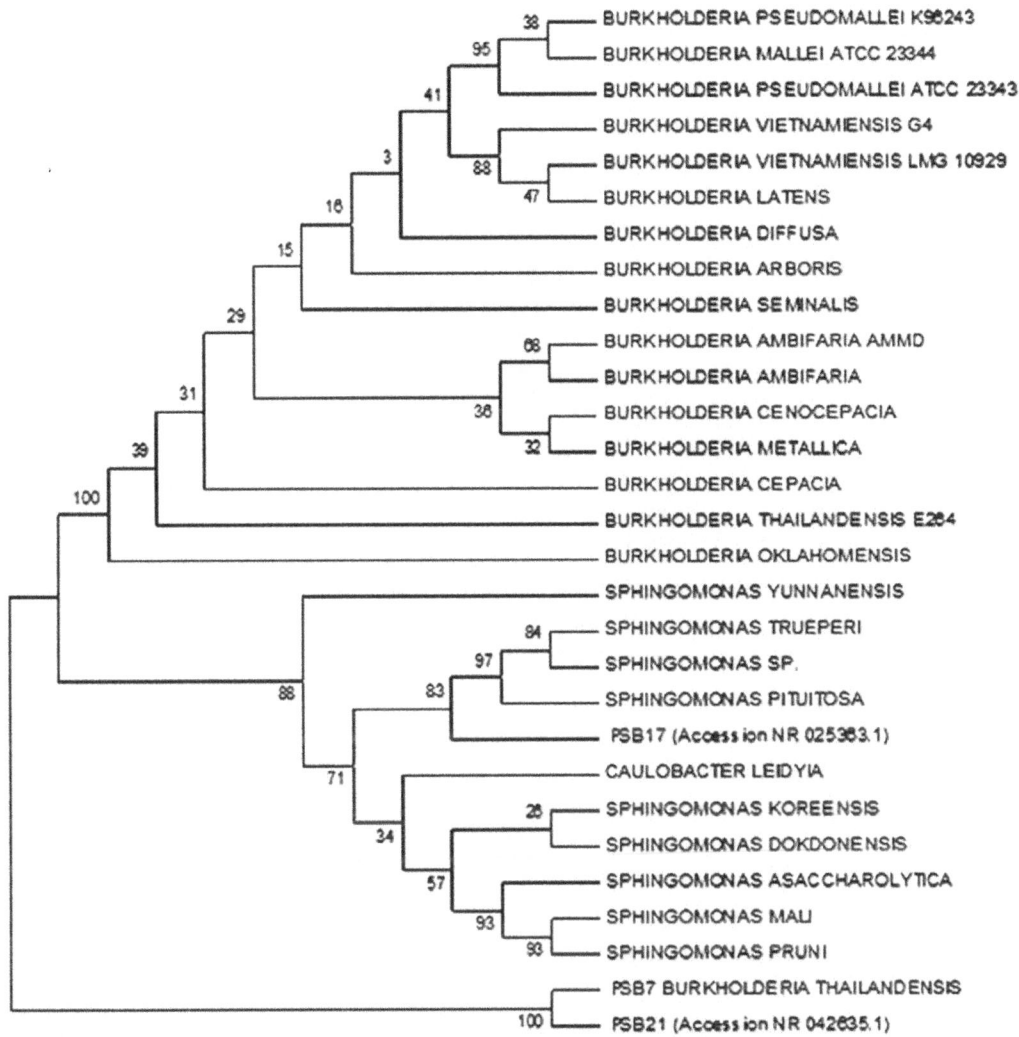

Figure 2. Phylogenetic tree with bootstrap values. Tree constructed using Neighbor-Joining (NJ) method. PSB7 accession NR 074312.1, PSB17 accession NR025363.1, and PSB21 accession NR 042635.1.

Al concentration (with rice seedling). Rice seedlings were affected severely by the low pH. However, PSB inoculation increased the pH of solution that resulted in enhanced growth of rice seedlings (Fig. 6).

Discussion

Acid sulfate soils are dominated by sulfur or oxidizing-reducing bacteria [28]. These soils may also contain some beneficial microorganisms that help improve rice growth. In this study, it was found that the pH of the acid sulfate soils was very low and Al concentration was very high, especially at the depth below 50 cm, proving that these were true acid sulfate soils [29]. Although the soils were very acidic there were still some microorganisms living in them. These microorganisms (bacteria, actinomycetes and fungus) can be potentially beneficial for rice production. It was found that the population of the bacteria changed from place to place and varied according to the native vegetation. Furthermore, the bacterial population was found to be higher at the rhizosphere compared to the non-rhizosphere, indicating bacterial synergism with plant roots [11].

A total of 21 PSB were isolated from the acid sulfate soils in the present study. These PSB isolates have the potential to be used in rice production. Higher PSB populations were observed in the calcium phosphate (NBRIP) medium compared to other P media sources. In this medium, the PSB population was not affected by pH. Panhwar *et al.* [13] reported significantly higher PSB population observed in tricalcium phosphate medium. Further-more, it was reported that most of the strains after an incubation period formed pellicles in N free semisolid-malate media, pointing to their N_2 fixing abilities [30].

The isolated strains have the capability to produce growth hormones (IAA) that could help to the plants to enhance their root and shoot growth. This shows the potential of these bacteria for use in crop production [31]. It is known that rhizosphere microorganisms mediate many soil processes, such as decomposition, nutrient mineralization and nitrogen fixation [32]. Other researchers have reported that bacteria in rice fields have the potential to produce IAA and are able to fix N [11–33]. In the current study, it was found that the isolated strains were able to solubilize phosphate in NBRIP and Pikovskaya media, with P-solubilizing activity indicated by clear halo zone around colonies [34–35]. A similar trend was observed for liquid broth culture

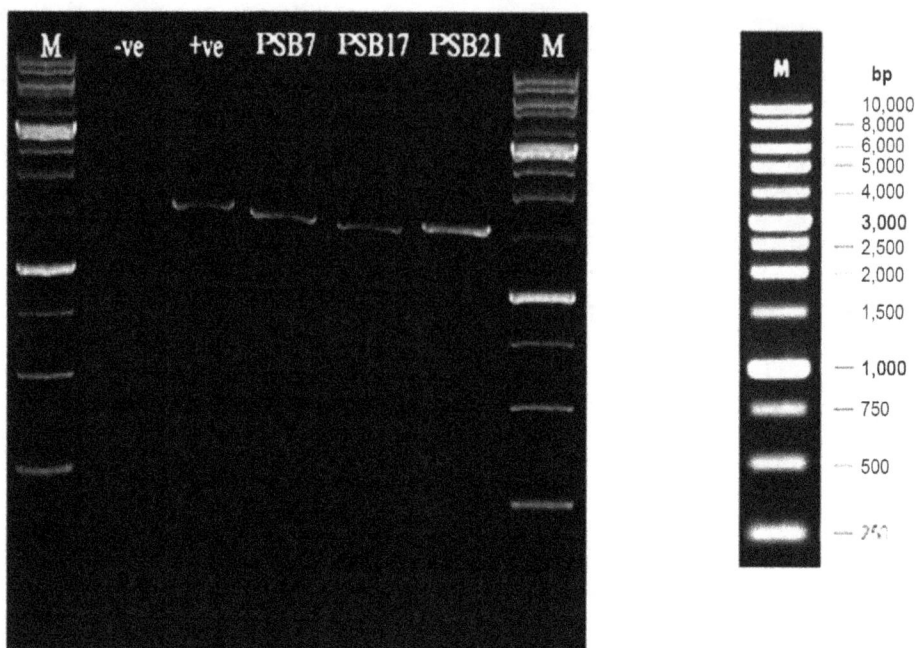

Figure 3. SDS-PAGE of the PSB7 (*Burkholderia thailandensis*), PSB17 (*Sphingomonas pituitosa*) and PSB21 (*Burkholderia seminalis*). (M: DNA ladder; −ve: negative control; +ve: positive control).

using calcium phosphate (NBRIP) and rock phosphate (CIRP). The highest P solubilization occurred in NBRIP broth, while the lowest was in the CIRP and PDYA-AlP broth cultures. This is similar to the findings of Chakraborty *et al.* [36] who reported that isolated PSB strains solubilized P from calcium phosphate to a greater amount than rock phosphate, aluminum phosphate and iron phosphate.

The bacterial population was significantly affected by the different forms of inorganic phosphate incorporated into the broth culture. NBRIP and CIRP broths were found to have the highest bacterial population, while lower bacterial growth was found in the PDYA-AlP broth after 72 hours of inoculation. The low content of solubilized P in the PDYA-A broth may be due to its insolubility although the isolates had the ability to solubilise it. Al in the soils may fix P, even further reducing solubilization. It has been established that phosphate rock has a lower content of soluble P compared to calcium phosphate [37].

The three principal components showed 99% variation of the total variation in 21 bacterial isolates. Similarly, Naher *et al.* [38] found distinct variations in different soil bacterial isolates using principal component analysis. Based on the production of indoleacetic acid and phosphate solubilizing activities, three isolates PSB7, PSB21 and PSB21 were found to have great potential. This noteworthy finding is supported by the PCA scatter plot based on the biochemical properties and these three bacterial isolates which were grouped distantly from the other strains.

The assessment of the bacterial 16S rRNA gene sequence has emerged as a preferred genetic technique as it can better identify weakly described, rarely isolated, or phenotypically aberrant strains [39–40]. In our study, we employed a molecular phylogenetic approach based on 16S rRNA sequences to identify pure potential isolates. Three potential isolates, namely PSB7 (*Burkholderia thailandensis*), PSB21 (*Burkholderia seminalis*), and PSB17 (*Sphingomonas pituitosa*) were identified.

The present isolates demonstrated beneficial traits like N_2 fixation, phosphate solubilization and phytohormones production (IAA). Several *Burkholderia* spp. (*Burkholderia unamae*, *Burkholderia xenovorans*, *Burkholderia silvatlantica*, *Burkholderia tropica*, *Burkholderia tuberum*, *Burkholderia phymatum*, *Burkholderia mimosarum* and *Burkholderia nodosa*) have been known to fix N_2 [41] and are commonly find in rice, maize, sugar cane, sorghum, coffee and tomato [42]. The *Sphingomonas* sp. has beneficial attributes like N_2 fixing [43] and high polysaccharide production. These species are able to synthesize the bacterial exopolysaccharide gellan and related polymers and were shown to possess constitutive gellanase activity [44]. Exopolysaccharides have been known to perform a major role in providing protection to the cell as a boundary layer [45], as well as by chelating heavy metals due to the presence of several active functional groups [46].

The isolated PSB were able to grow well under low pH conditions. This might be due to the deposition of lipopolysaccharide in bacterial cultures at a pH below 4.5. The possibility of lipopolysaccharide incorporation into the cell wall is episodic at acidic pH that would argue against the involvement of these polysaccharides in the acid pH toterant bacteria (*R. tropici* UMR1899) as other outer-membrane components may confer greater pH tolerance to the strain [47]. Exopolysaccharides were partially distinguished by their FTIR spectra indicating the presence of carboxylic acid group and H^- bonded group that might have been the reason for the increase of pH in broth culture [44]. However, in the present study, increase in the pH of the nutrient solution from 4 to 7 (both plant and without plant system) might be the result of bacterial exopolysaccharide production. An increase in the pH of bacterial culture medium was the main factor that governs cell growth and exopolysaccharides production [48]. Moreover, bacteria produce polysaccharides in the presence of carbon source and when the pH of the broth is increased from 5.0 to 7.0 [49].

Table 7. Effects of Al and PSB inoculation on the growth of rice seedlings.

Treatments	Plant height (cm)			Dry weight (g)			Root length (cm)			Root surface area (cm^2)			Root volume (cm^3)		
	Al conc. (µM)			Al conc. (µM)			Al conc. (µM)								
	0	50	100	0	50	100	0	50	100	0	50	100	0	50	100
Control	14.5c	13.3c	11.6c	0.61c	0.59c	0.58c	24.31c	23.69c	18.99d	42.63d	28.29d	25.47c	3.51c	2.62c	1.03c
Burkholderia thailandensis	17.7b	15.6b	14.1a	0.71b	0.69a	0.67a	57.69b	36.32b	33.78c	69.08c	58.43a	42.36a	4.80b	4.19b	3.42a
Burkholderia seminalis	18a	16.4a	14.1a	0.76a	0.69a	0.65b	58.02a	47.28a	36.85b	74.12a	53.02b	43.40a	6.88a	3.76a	2.03b
Sphingomonas pituitosa	17b	15b	13b	0.71b	0.66b	0.65b	58.51a	49.66a	38.01a	72.59b	45.45c	38.52b	6.90a	5.18a	3.81a

Data values are means of five replicates. Means within the same column followed by the same letters are not significantly different at $P<0.05$.

Table 8. Effects of Al and PSB inoculation on the release of organic acids.

Treatments	Oxalic acid (µM)			Citric acid (µM)			Malic acid (µM)		
	Al conc. (µM)								
	0	50	100	0	50	100	0	50	100
Control	72c	82d	94d	13d	40d	52d	3d	4d	55d
Burkholderia thailandensis	78c	131a	265 a	45b	196a	257b	40a	121a	336a
Burkholderia seminalis	98a	123b	114 c	146a	170b	268a	7b	86b	315b
Sphingomonas pituitosa	95ab	94c	185 b	38c	137c	241c	5c	54c	151c

Data values are means of five replicates. Means within the same column followed by the same letters are not significantly different at $P<0.05$.

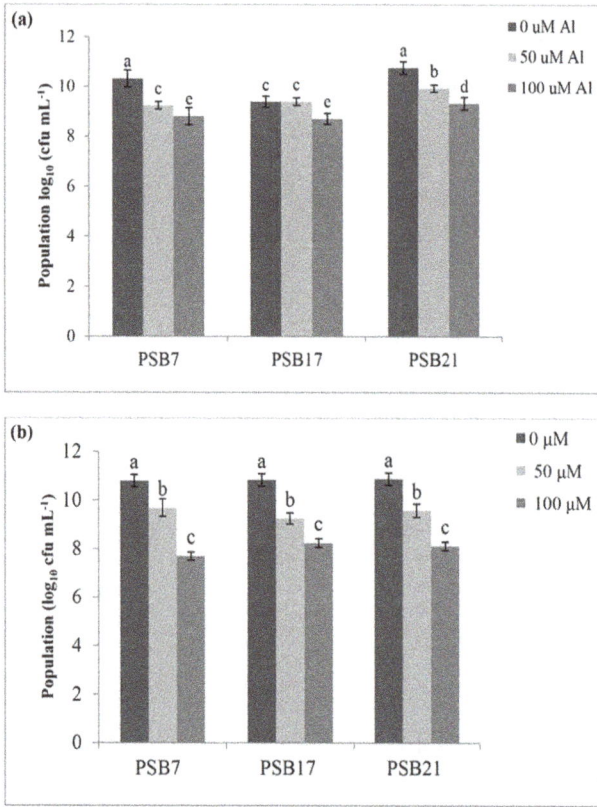

Figure 4. Effect of different Al concentrations on the PSB population (a) without plant, (b) with plant system.

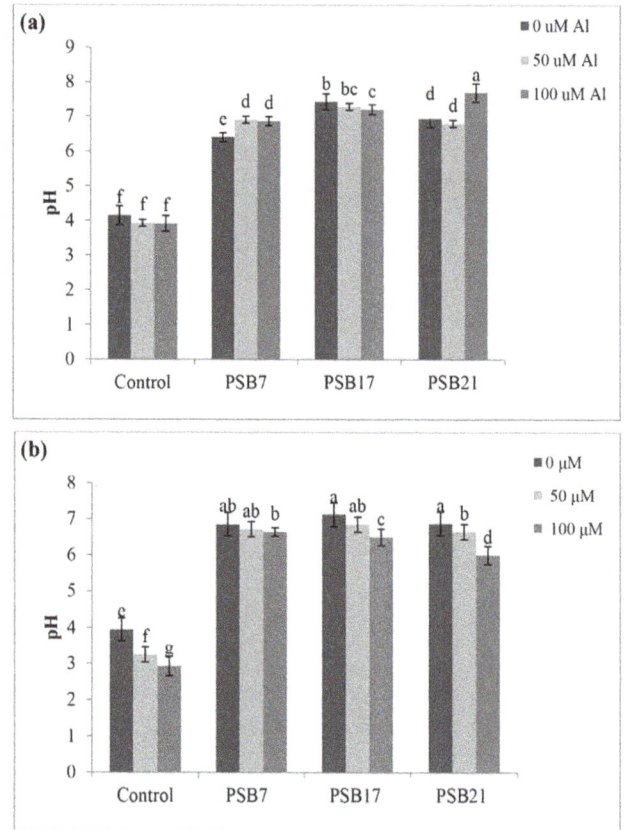

Figure 5. Effect of PSB inoculation on solution pH at different Al concentration a) without plant, b) with plant.

Al toxicity is one of the major constraints to rice root development. In the present study, PSB-inoculated rice seedlings showed better root growth compared to non-inoculated seedlings. The pH of the solution changed from acidic to neutral, providing a favourable environment for plant growth. The better root growth was also promoted by the release of organic acids produced by PSB that chelated Al, rendering it inactive [4]. It is known that rice roots also secrete organic acids (citric, oxalic, and malic acids) and the secretion of these organic acids is localized to the root apex [50]. The inoculated rice seedlings without the presence of Al produced higher plant biomass. This might have been due to the production of phytohormone by the bacteria (such as IAA) that stimulated rice plant growth.

In the present study, multiple reasons might have contributed to a reasonable seedling growth under high Al and low pH conditions. First, the strains were able to produce some polysaccharides that increased the solution pH. Moreover, the strains produced organic acids that helped chelate Al in the solution and consequently reduced Al toxicity. The organic acids performed two main important functions: 1) to detoxify Al; and 2) to make P available [51]. There has been evidence of aluminum resistance in some plant species that has been ascribed to organic acid exudation from roots and by microorganisms that have the ability to chelate Al [52]. However, the most effective organic acids to chelate Al, in decreasing order, are citric, oxalic and tartaric acids [53].

In this study, it is proven that these isolates have the potential to reduce Al toxicity, fix nitrogen, solubilize phosphate, increase soil

pH and produce phytohormone. Therefore, these three bacteria

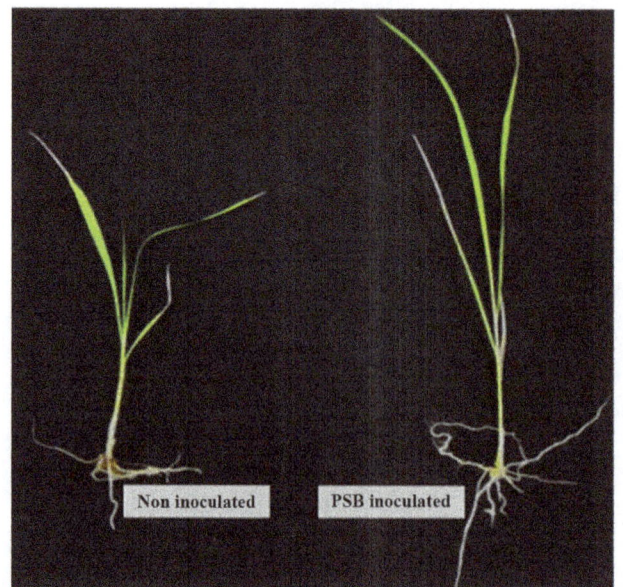

Figure 6. Effect of Al concentration (100 µM) at pH 4.0 on rice seedlings.

strains can be exploited to enhance the productivity of rice planted on acid sulfate soils either in Malaysia or other part of the tropics.

Conclusions

Rice growing on acid sulfate soils suffers from Al^{3+} toxicity and H^+ which affects its growth and eventually the yield. To some extent, rice with the help of PSB is able to reduce Al^{3+} by excreting organic acids (oxalic, citric and malic acids) via its roots at the appropriate time, and these acids are able to detoxify Al^{3+} via a chelation process. The growth of rice can be improved further with the help of PSB. In this study, three locally isolated bacteria; *Burkholderia thailandensis* (PSB7), *Sphingomonas pituitosa* (PSB17) and *Burkholderia seminalis* (PSB21) of these essential microbes were found to exist under adverse conditions (low pH and high Al concentration). These PSB not only helped to solubilize P in the soils, but their activities resulted in increased pH due to the release of exo-polysaccharides. If the water pH increased above 5, Al would have precipitated as inert Al-hydroxides, making it unavailable to the growing rice in the field.

These PSB could also produce growth enhancement phytohormones such as IAA. The potential of these PSB should be exploited further for the sustainable management of acid sulfate soils, especially for rice production. For instance, they can be used as bio-fertilizers to improve the fertility of acid sulfate soils.

Author Contributions

Conceived and designed the experiments: QAP UAN SJ RO. Performed the experiments: QAP UAN. Analyzed the data: QAP UAN MAL. Contributed reagents/materials/analysis tools: RO MRI. Wrote the paper: QAP UAN SJ RO.

References

1. Sullivan S (2004) Sustainable management of acid sulfate soils. Aust J Soil Res 42(5/6): 595–602.
2. Shamsuddin J, Elisa Azura A, Shazana MARS, Fauziah CI, Panhwar QA, et al. (2014) Properties and management of acid sulfate soils in Southeast Asia for sustainable cultivation of rice, oil palm and cocoa. Adv Agron 124: 91–142.
3. Marschner H (1991) Mechanisms of adaptation of plants to acid soils. Plant Soil 134: 1–20.
4. Shamsuddin J, Elisa AA, Shazana MARS, Che Fauziah I (2013) Rice defense mechanisms against the presence of excess amount of Al^{3+} and Fe^{2+} in the water. Aust J Crop Sci 7(3): 314–320.
5. Shamsuddin J, Che Fauziah I (2010) Alleviating acid soil infertility constraints using basalt, ground magnesium limestone and gypsum in a tropical environment. Malaysian J Soil Sci 14: 1–13.
6. Arias ME, Gonzalez-Perez JA, Gonzalez-Vila FJ, Ball AS (2005) Soil health-a new challenge for microbiologists and chemists. Inter Microbiol 8: 13–21.
7. Khiari L, Parent LE (2005) Phosphorus transformations in acid light-textured soils treated with dry swine manure. Can J Soil Sci Soc 85: 75–87.
8. Rodriguez H, Fraga R (1999) Phosphate solubilizing bacteria and their role in plant growth promotion. Biotech Adv 17: 319–339.
9. Takeuchi M, Hamana K, Hiraishi A (2001) Proposal of the genus *Sphingomonas sensu stricto* and three new genera, *Sphingobium*, *Novosphingobium* and *Sphingopyxis*, on the basis of phylogenetic and chemotaxonomic analyses. Inter J Syst Evo Microbiol 51: 1405–1417.
10. Janssen PH (2006) Identifying the dominant soil bacteria taxa in libraries of 16S rRNA and 16S rRNA genes. App Environ Microbiol 72: 1719–1728.
11. Naher UA, Radziah O, Shamsuddin ZH, Halimi MS, Mohd Razi I (2009) Isolation of diazotrophs from different soils of Tanjong Karang Rice growing area in Malaysia. Inter J Agri Biol 11(5): 547–552.
12. Naher UA, Radziah O, Shamsuddin ZH, Halimi MS, Mohd Razi I, et al. (2011) Effect of root exuded specific sugars on biological nitrogen fixation and growth promotion in rice (*Oryza sativa*). Aust J Crop Sci 5(10): 1210–1217.
13. Panhwar QA, Radziah O, Zaharah AR, Sariah M, Mohd Razi I (2012) Isolation and characterization of phosphorus solubilizing bacteria from aerobic rice. African J Biotech 11(11): 2711–2719.
14. Benton J Jr (2001) Laboratory guide for conducting soil tests and plant analysis. CRC Press LLC, New York.
15. Teh SC, Talib J (2006) Soil physics analyses volume 1. Universiti Putra Malaysia Press, Kuala Lumpur, Malaysia.
16. Bremner JM, Mulvaney CS (1982) Nitrogen-Total. In: Methods of soil analysis. Agron. No. 9, Part 2: Chemical and microbiological properties, 2nd edn. Madison, WI, USA, 595–624 pp.
17. Nautiyal CS (1999) An efficient microbiological growth medium for screening phosphate-solubilizing microorganisms. FEMS Microbiology Letters 170: 265–270.
18. Prasad G, James EK, Mathan N, Reddy PM, Reinhold-Hurek B, et al. (2001) Endophytic Colonization of Rice by a Diazotrophic Strain of *Serratia marcescens*. J Bacteriol 183(8): 2634–2645.
19. Gordon AS, Weber RP (1950) Colorometric estimation of indoleacetic acid. Plant Physiol 26: 192–195.
20. Nguyen C, Yan W, Le Tacon F, Lapeyrie F (1992) Genetic variability of phosphate solubilizing activity by monocaryotic and dicaryotic mycelia of the ectomycorrhizal fungus *Laccaria bicolor* (Maire) P.D. Orton. Plant Soil 143: 193–199.
21. Katzenelson H, Bose B (1959) Metabolic activity and phosphate dissolving ability of bacterial isolates from wheat roots rhizosphere and non-rhizosphere soil. Can J Microbiol 5: 79–85.
22. Somasegaran P, Hoben HJ (1985) General Microbiology of Rhizobium. Methods in Legume- *Rhizobium* Technology 39–53.
23. Murphy J, Riley JP (1962) A modified single solution method for the determination of phosphate in natural waters. Anal Chim Acta 27: 31–36.
24. Weisburg WG, Barns SM, Pelletier DA (1991) 16S ribosomal DNA amplification for phylogenetic study. J Bacteriol 173: 697–703.
25. Chen YP, Rekha PD, Arun AB (2006) Phosphate solubilizing bacteria from subtropical soil and their tricalcium phosphate solubilizing abilities. App Soil Ecol 34: 33–41.
26. Chung H, Park M, Madhaiyan M (2005) Isolation and characterization of phosphate solubilizing bacteria from the rhizosphere of crop plants of Korea. Soil Biol Biochem 37: 1970–1974.
27. Hamdy EL Z, Czarnes S, Hallett PD, Alamercery S, Bally R, et al. (2007) Early changes in root characteristics of maize (*Zea mays*) following seed inoculation with the PGPR *Azospirillum lipoferum* CRT1. Plant Soil 291: 109–118.
28. Mathew EK, Panda RK, Nair M (2001) Influence of subsurface drainage on crop production and soil quality in a low-lying acid sulphate soil. Agriculture Water Management 47: 191–209.
29. Shamsuddin J, Muhrizal S, Fauziah I, Van Ranst E (2004) A laboratory study of pyrite oxidation in an acid sulfate soils. Comm Soil Sci Plant Anal 35: 117–129.
30. Azlin CO, Amir HG, Chan LK (2005) Isolation and characterization of diazotrophic rhizobacteria of Oil Palm roots. Malaysian J Microbiol 1: 31–35.
31. Glick BR (2005) Modulation of plant ethylene levels by the enzyme ACC deaminase. FEMS Microbiology Letters 251: 1–7.
32. Pradhan N, Sukla LB (2005) Solubilization of inorganic phosphate by fungi isolated from agriculture soil. African J Biotechnol 5: 850–854.
33. Woo SM, Lee MK, Hong IS, Poonguzhali S, Sa TM (2010) Isolation and characterization of phosphate solubilizing bacteria from Chinese cabbage. 19th World Congress of Soil Science, Soil Solutions for a Changing World 1–6 August 2010, Brisbane, Australia.
34. Kumar A, Bhargava P, Rai LC (2010) Isolation and molecular characterization of phosphate solubilizing Enterobacter and Exiguobacterium species from paddy fields of Eastern Uttar Pradesh, India. African J Microbiol Res 4(9): 820–829.
35. Parasanna A, Deepa V, Balakirashna Murthy P, Deecaraman M, Sridhar R, et al. (2011) Insoluble phosphate solubilization by bacterial strains isolated from rice rhizosphere soils from southern India. Inter J Soil Sci 6(2): 134–141.
36. Chakraborty BN, Chakraborty U, Saha A, Sunar K, Dey PL (2010) Evaluation of Phosphate Solubilizers from Soils of North Bengal and Their Diversity Analysis. World J Agri Sci 6(2): 195–200.
37. Nahas E (1996) Factors determining rock phosphate solubilization by microorganisms isolated from soil. World J Microbiol Biotechnol 12: 567–572.
38. Naher UA, Radziah O, Latif MA, Panhwar QA, Puteri AMA, et al. (2013) Biomolecular Characterization of Diazotrophs Isolated from the Tropical Soil in Malaysia. Inter J Mol Sci 14: 17812–17829.
39. Clarridge JE (2004) Impact of 16S rRNA Gene Sequence Analysis for Identification of Bacteria on Clinical Microbiology and Infectious Diseases. Clinical Microbiol Rev 17(4): 840–862.
40. Ludwig W, Amann R, Martinez-Romero E, Schönhuber W, Bauer S, et al. (1998) rRNA based identification and detection systems for rhizobia and other bacteria. Plant Soil 204: 1–19.

41. Suárez-Moreno ZR, Caballero-Mellado J, Venturi V (2008) The new group of non-pathogenic plant-associated nitrogen-fixing *Burkholderia* spp. shares a conserved quorum-sensing system, which is tightly regulated by the RsaL repressor. Microbiology 154(7): 2048–2059.

42. Caballero-Mellado J, Onofre-Lemus J, Estrada-de los Santos P, Martinez-Aguilar L (2007) The tomato rhizosphere, an environment rich in nitrogen-fixing Burkholderia species with capabilities of interest for agriculture and bioremediation. App Environ Microbiol 73: 5308–5319.

43. Zhang JY, Liu XY, Liu SJ (2010) *Sphingomonas changbaiensis* sp. nov., isolated from forest soil. Inter J Syst Evo Microbiol 60: 790–795.

44. Sunil PT, Amarsinh Bhosale A, Trishala Gawade B, Tejswini Nale R (2013) Isolation, screening and optimization of exopolysaccharide producing bacterium from saline soil. J Microbiol Biotechnol Res 3(3): 24–31.

45. Caiola MG, Billi D, Friedmann EI (1996) Effect of desiccation on envelopes of the *cyanobacterium Chroococcidiopsis* sp. (Chroococcales). European J Phycol 31: 97–105.

46. Kaplan D, Christiaen D, Arad SM (1987) Chelating Properties of Extracellular Polysaccharides from *Chlorella* spp. App Environ Microbiol 53(12): 2953–2956.

47. Graham PH, Draeger KJ, Ferrey ML, Conroy MJ, Hammer BE, et al. (1994) Acid pH tolerance in strains of *Rhizobium, Bradyrhizobium*, and initial studies on the basis for acid tolerance of *Rhizobium tropici* UMR 1899. Can J Microbiol 40: 198–207.

48. Singh S, Das A (2011) Screening, production, optimization and characterization of cyanobacterial polysaccharide World J Microbiol Biotechnol 27: 1971–1980.

49. Bueno SM, Garcia-Cruz CH (2006) Optimization of polysaccharides production by bacteria isolated from soil. Brazilian J Microbiol 37(3): 296–301.

50. Ma JF (2000) Role of organic acids in detoxification of aluminum in higher plants. Plant Cell Physiol 41: 383–390.

51. Haynes RJ, Mokolobate MS (2001) Amelioration of Al toxicity and P deficiency in acid soils by additions of organic residues: a critical review of the phenomenon and the mechanisms involved. Nutr Cycl Agroecosyst 59: 47–63.

52. Klugh KR, Cumming JR (2007) Variations in organic acid exudation and aluminum resistance among arbuscular mycorrhizal species colonizing *Liriodendron tulipifera*. Tree Physiol 27: 1103–1112.

53. Hue NV, Craddock GR, Adams F (1986) Effect of organic acids on aluminium toxicity in subsoils. Soil Sci Soc Am J 50: 28–34.

2

Assessing Reference Genes for Accurate Transcript Normalization Using Quantitative Real-Time PCR in Pearl Millet [*Pennisetum glaucum* (L.) R. Br.]

Prasenjit Saha, Eduardo Blumwald*

Department of Plant Sciences, University of California Davis, Davis, California, United States of America

Abstract

Pearl millet [*Pennisetum glaucum* (L.) R.Br.], a close relative of Panicoideae food crops and bioenergy grasses, offers an ideal system to perform functional genomics studies related to C4 photosynthesis and abiotic stress tolerance. Quantitative real-time reverse transcription polymerase chain reaction (qRT-PCR) provides a sensitive platform to conduct such gene expression analyses. However, the lack of suitable internal control reference genes for accurate transcript normalization during qRT-PCR analysis in pearl millet is the major limitation. Here, we conducted a comprehensive assessment of 18 reference genes on 234 samples which included an array of different developmental tissues, hormone treatments and abiotic stress conditions from three genotypes to determine appropriate reference genes for accurate normalization of qRT-PCR data. Analyses of Ct values using Stability Index, BestKeeper, ΔCt, Normfinder, geNorm and RefFinder programs ranked *PP2A, TIP41, UBC2, UBQ5* and *ACT* as the most reliable reference genes for accurate transcript normalization under different experimental conditions. Furthermore, we validated the specificity of these genes for precise quantification of relative gene expression and provided evidence that a combination of the best reference genes are required to obtain optimal expression patterns for both endogeneous genes as well as transgenes in pearl millet.

Editor: Xianlong Zhang, National Key Laboratory of Crop Genetic Improvement, China

Funding: This work is funded by The United States Agency for International Development (USAID) under the Grant No. APS M/OAA/GRO/EGAS-11-002011. The funders had no role in study design, data collection and analysis, decision to publish, or preparation of the manuscript.

Competing Interests: The authors have declared that no competing interests exist.

* Email: eblumwald@ucdavis.edu

Introduction

Increasing global population has raised the need of both food and fuel production. In addition, the growing use of fossil fuel is contributing to global climate changes due to elevated greenhouse gas emission. Pearl millet [*Pennisetum glaucum* (L.) R. Br., formerly *P. americanum*] is an excellent food and forage crop of arid to semiarid regions of the world [1,2] and a close relative of Panicoideae bioenergy grasses like switchgrass and foxtail millet [3]. It is well adapted to drought, heat, high salinity, poor soil fertility and low pH with an efficient C4 carbon fixation and high yield potential [4]. Thereby, pearl millet provides an ideal crop for functional genomics studies related to C4 photosynthesis and abiotic stress tolerance. Although several genetic engineering studies have been conducted in pearl millet [5,6], functional genomic studies under abiotic stress conditions are scanty [7].

Quantitative real-time polymerase chain reaction (qRT-PCR) provides an important platform for measuring gene expression changes due to its high sensitivity, specificity and wide range of application [8]. However, its accuracy is influenced by the expression stability of the internal control reference genes for reliable transcript normalization of target genes [9,10]. An ideal reference gene should be constitutively and equally expressed across developmental stages and experimental conditions [9]. According to the 'golden rules' [11], identification of the most suitable and highly stable internal reference genes for accurate normalization is one of the prerequisites for qRT-PCR. So far most of the studies published deal with model plant species with known genome sequence, for e.g. Arabidopsis [12], rice [13], brachypodium [14]; however, relatively few studies have been documented in plants with limited or no genome information [15,16]. Thus the lack of suitable reference genes is one of the major limitations for gene expression studies using qRT-PCR in crop plants [16], including pearl millet.

Over the past few years emphasis has been given to identify and validate suitable reference genes from important plant species such as bamboo [17], barley [18], brachypodium [14], cotton [19], foxtail millet [20], mustard [21], peanut [22], wheat [23,24] and switchgrass [25]. The commonly used traditional housekeeping reference genes include *actin* (*ACT*), *elongation factor 1α* (*EF1α*), *glyceraldehyde-3-phosphate dehydrogenase* (*GAPDH*), *tubulin* (*TUB*), *ubiquitin-conjugating enzyme* (*UBC*) and *18S ribosomal RNA* (*18S rRNA*) which are involved in basic cellular processes [26]. Moreover, no single traditional reference gene with stable constant expression across tissues and experimental conditions was found, thus leading to explore additional new reference genes for reliable normalization of qRT-PCR data [26]. Recent reports illustrated that *F-box/kelch-repeat protein* (*F-box*), *phosphoenolpyruvate carboxylase-related kinase* (*PEPKR*), *protein phosphatase 2A*

(*PP2A*) and *TIP41-like family protein* (*TIP41*) genes are superior compared to traditional reference genes [17,21,22,27]. Several statistical algorithms, namely, Stability Index [28], ΔCt [29], BestKeeper [30], geNorm [31], NormFinder [32], and RefFinder [33] have been employed for proper validation and stability ranking of the best reference genes for qRT-PCR data normalization in numerous plant species. However, to the best of our knowledge, no systematic analysis for the selection of suitable reference genes for qRT-PCR analysis in pearl millet has been reported. Therefore, a comprehensive validation of reference genes under different experimental conditions for accurate transcript normalization is needed in pearl millet.

In this work, we evaluated 18 potential candidate reference genes in 234 samples from three important pearl millet genotypes using qRT-PCR. Expression patterns of these genes were monitored in tissue samples under different developmental processes, hormone treatments and abiotic stress conditions. Expression stability of these genes was validated using six statistical algorithms in order to assign appropriate reference genes suitable to each experimental condition for accurate transcript normalization. Our results showed that sets of genes are appropriate for accurate transcript quantification of endogenous genes as well as transgenes from different tissue samples. We further illustrated detailed expression patterns of three essential pearl millet endogenous genes specific to development, hormonal stimuli and abiotic stresses.

Materials and Methods

Plant materials

Pearl millet (*Pennisetum glaucum* [L.] R. Br.) genotypes ICMR01004, IPCI1466 and IP300088 were used in this study. Seeds of ICMR01004 and IPCI1466 were obtained from the International Crops Research Institute for the Semi-Arid-Tropics (ICRISAT), India, while seeds of IP300088 were acquired from the Germplasm Resources Information Network's (GRIN), USA. Seeds were kept in wide mouth polypropylene bottles (VWR) and stored in a seed vault at 9°C with a relative humidity of 50%.

Developmental tissue samples

For developmental tissue samples, three genotypes were grown in 5 liter pots containing agronomy mix (equal parts of redwood compost, sand and peat moss) under greenhouse conditions of 16 h day/8 h night photoperiod at 30±2°C until maturity. Plants were watered every alternate day with tap water and fertilized biweekly. Tissue samples of vegetative and reproductive stages included callus 30DPC (days post culture), leaf 7DPS (days post sowing), leaf 15DPS, leaf 30DPS, node, internode, sheath, flag leaf, panicle, peduncle and root of 60DPS plant, and 30DPH (days post harvest) seeds. A total of 108 tissues samples comprising of 12 vegetative and reproductive stages from three genotypes in three biological replicates were harvested by immediate quick freezing in liquid nitrogen in 2 ml SealRite microcentrifuge tubes.

Hormone treatments

Seeds of 30DPH were soaked in 70% (v/v) ethanol for 1 min followed by washing in 2.5% (v/v) sodium hypochlorite solution containing 0.1% (v/v) Tween 20 for 15 min and rinsed thoroughly with sterile distilled water. Surface sterilized seeds were grown in PhytoCon culture vessels (Phytotechnology Laboratories, Overland Park, KS, USA) containing half strength Murashige and Skoog (MS) medium for 14 days. Seedlings were kept in sucrose free liquid half strength MS medium for 24 h. Seedlings of 15DPG (days post germination) were transferred to PhytoCon culture

vessels (Phytotechnology Laboratories) containing liquid half strength MS supplemented with 100 μM abscisic acid (ABA, Sigma, St. Louis, MO, USA), 50 μM brassinolide (Bra, Sigma), 50 μM gibberellic acid (GA, Sigma), 50 μM indole-3-acetic acid (IAA, Sigma), 100 μM methyl jasmonate (MeJa, Sigma), 100 μM salicylic acid (SA, Sigma), 100 μM Zeatin (Zea, Sigma) and incubated for 6 h. Leaves from a total of 72 samples from seven treatments in three biological replicates including one untreated control of three genotypes were harvested and immediately frozen as mentioned in the earlier section.

Abiotic stress conditions

In the dehydration stress treatments, seedlings of 15DPG (same as hormone treatments) were kept in 400 μM mannitol solution for 6 h. For drought and salinity stresses, water supply was withheld and 300 mM sodium chloride (NaCl) solution was provided for 5 days to 30DPS plants, respectively. Cold and heat stresses were carried out by maintaining 30DPS plants at 4±1°C and 42±1°C, respectively for 6 h for 3 consecutive days. Stress symptoms were monitored visually by the appearance of leaf rolling and yellowing, as well as by measuring stomatal conductance and photosynthesis rates of plants using a LI-COR 6400-40 with an integrated fluorescence chamber head (LI-COR, Lincoln, Nebraska, USA) after the stress treatments.

Candidate reference genes selection and primer design

Locus identifiers (IDs) of Arabidopsis and rice potential candidate reference genes were obtained from previously published work (Table 1). Orthologous locus IDs from foxtail millet (*Setaria italica*) were identified using locus search from Phytozome. GenBank accession numbers were obtained from National Center for Biotechnology Information (NCBI) using BLASTN.

A total of eighteen genes were chosen for primer design using Primer3Plus software (http://www.bioinformatics.nl/cgi-bin/primer3plus/primer3plus.cgi) [34] considering the parameters specific for qRT-PCR. The sequences with detailed parameters for each primer pair are given in Table S1.

RNA isolation and cDNA synthesis

A total of 100 mg of frozen plant material was ground to fine powder in a 2 ml SealRite microcentrifuge tube using 3.2 mm stainless steel beads and an automated shaker SO-10M (Fluid Management, Wheeling, IL, USA). Total RNA was isolated from plant samples using the RNeasy plant mini kit (Qiagen, Valencia, CA, USA) according to the manufacturer's procedure. A first set of on-column DNAse I (Qiagen) digestion was carried out during the RNA extraction steps. The integrity of RNA samples were checked by 1% (w/v) agarose gel. The quantity and quality of RNA samples were also checked using a NanoDrop ND-1000 (NanoDrop Technologies, Wilmington, DE, USA). RNA samples with 260/280 ratio between 1.9 to 2.2 and 260/230 ratio between 2.0–2.5 were used for cDNA synthesis. To completely eliminate DNA contamination, 1 μg of total RNA was subjected to gDNA wipeout reaction using the QuantiTect reverse transcription kit (Qiagen) followed by first strand cDNA synthesis in a 20 μl reaction mixture using an optimized blend of oligo-dT and random primers according to manufacturer's instructions and stored at −20°C.

PCR and qRT-PCR

Specific amplification from cDNA was checked by PCR following the protocol described earlier [35] using 1 μl of cDNA, 10 mM dNTPs, 1 μM each of forward and reverse primers and

Table 1. Expression level of the selected candidate reference genes tested in pearl millet.

Genes	Description	Arabidopsis	Rice	Foxtail millet/Pearl millet[a]	Ct[b]±SD	CV±SD
ACT	Actin	At1g22620	LOC_Os05g36290	Si022372m.g/HM243500	28.2±2.7	1.8±0.5
CYC	Cyclophilin, peptidyl-prolyl cis-trans isomerase	At5g35100	LOC_Os08g19610	Si014078m.g	31.5±3.0	3.3±1.0
eEF1α	Eukaryotic elongation factor 1 alpha	At5g60390	LOC_Os03g08050	Si022039m.g/EF694165	24.1±3.2	1.8±0.5
FBX	F-box domain containing protein	At5g15710	LOC_Os04g57290	Si022138m.g	25.3±2.8	2.6±0.7
GAPDH	Glyceraldehyde-3-phosphate dehydrogenase	At3g04120	LOC_Os08g03290	Si014034m.g/GQ398107	22.8±3.1	2.6±0.6
eIF4a2	Eukaryotic initiation factor 4a2	At1g54270	N	Si006546m.g/EU856535	23.2±2.1	2.9±0.7
PEPKR	Phosphoenolpyruvate carboxylase kinase related	At1g12580	LOC_Os06g03682	Si006273m.g/FR872788	25.6±1.5	1.3±0.4
PP2A	Protein phosphatase 2A	At1g10430	LOC_Os02g12580	Si017892m.g	25.6±2.5	1.3±0.3
RCA	Rubisco activase	At2g39730	LOC_Os11g47970	Si026414m.g	24.9±2.9	2.4±0.6
SAMDc	S-adenosyl methionine decarboxylase	At3g25570	LOC_Os04g42090	Si010282m.g	26.2±5.2	4.8±1.2
TUA	Tubulin alpha	At1g04820	LOC_Os03g51600	Si035654m.g	23.1±3.4	2.1±0.5
TIP41	Tonoplast intrinsic protein	At4g34270	LOC_Os03g55270	Si036884m.g	28.5±1.5	1.1±0.3
UBC2	Ubiquitin-conjugating enzyme 2	At5g25760	LOC_Os02g42314	Si018564m.g	29.8±3.1	1.7±0.5
UBC18	Ubiquitin-conjugating enzyme 18	At5g42990	LOC_Os12g44000	Si023498m.g	26.3±2.2	2.7±0.6
UBQ5	Ubiquitin 5	At2g47110	LOC_Os01g22490	Si003209m.g	23.5±2.1	1.3±0.4
UNK	Transmembrane protein 56	At1g31300	LOC_Os01g56230	Si002525m.g	27.9±1.7	2.6±0.6
18S rRNA	18S ribosomal RNA	N	N	KC201690	24.0±4.9	5.7±1.3
25S rRNA	25S ribosomal RNA	N	N	AB197128	9.1±1.8	3.6±0.3

[a] Locus identifiers of selected candidate reference genes for foxtail millet and/or GenBank accession numbers for pearl millet with orthologous from Arabidopsis and rice are listed.
[b] The expression levels of the candidate genes obtained during qRT-PCR experiments of total samples (n = 234) are presented as mean threshold cycle (Ct) values. SD, standard deviation; CV, coefficient of variance; N, no corresponding locus identifier or accession number.

one unit Taq polymerase in a 10 μl total reaction mixture. The amplification program was as follows: 5 min at 95°C; followed by 30 cycles of 95°C for 30 sec, 58°C for 15 sec, 72°C for 30 sec; and a final extension of 72°C for 10 min followed by electrophoresis on 3% (w/v) agarose gel.

For qRT-PCR, cDNAs were diluted to 20 times into a final volume of 400 μl and the reactions were performed as described previously [36] in an optical 96 well plate (Applied Biosystems, Foster City, CA, USA) containing 1 μl of diluted cDNA, 200 nM of each gene specific primer and 2.5 μl of 2X Fast SYBR Green PCR master mix in a 5 μl total volume using a StepOnePlus[TM] real time PCR system (Applied Biosystems) equipment. The qRT-PCR reactions were conducted following the fast thermal cycles: 50°C for 2 min, 95°C for 20 sec, followed by 40 cycles of 95°C for 3 sec and 60°C for 30 sec. After 40 cycles, the specificity of the amplifications was tested by heating from 60°C to 95°C with a ramp speed of 1.9°C/min, resulting in melting curves. The threshold cycle (Ct) value was automatically determined for each reaction by the real time PCR system with default parameters. Raw data (not baseline corrected) of fluorescence levels and the specificity of the amplicons were checked by qRT-PCR dissociation curve analysis using StepOne Software (v2.3). The baseline correction and linear regression analysis on each amplification curve including the efficiencies (E) of the polymerase chain reactions were calculated based on the slope of the line ($E = 10^{slope}$), considering an ideal value range ($1.8 \leq E \geq 2$) and correlation ($R^2 \geq 0.9$) using the LinRegPCR software [37].The final Ct values were the mean of three biological replicates and the coefficient of variance (CV) was calculated to evaluate the variation of Ct values for each gene. Each qRT-PCR reaction set included water as a negative no-template control (NTC) instead of cDNA.

Analysis for expression stability of reference genes

Five different types of computer-based programs, Stability Index [28], delta (Δ)Ct [29], BestKeeper [30], geNorm [31] and NormFinder [32] methods were used to rank and compare the stability of candidate reference genes across all the experimental sets. For Stability Index, ΔCt, BestKeeper programs, the Ct value for each candidate reference gene was used to determine its relative expression stability. For NormFinder and geNorm, relative expression values were calculated from $2^{-\Delta\Delta Ct}$ using the formula applied before [31]. Overall recommended comprehensive geomean ranking values of the best reference genes were obtained using the ranking results of four algorithms, except Stability Index, in RefFinder [33]. The pairwise variation (Vn/Vn+1) between two sequential normalization factors (NFn and NFn+1) were estimated using geNorm software provided in qBasePlus (v2.4) [38] package for best and minimal number of reference genes needed to calculate an optimal normalization.

Validation of reference genes

Six genes were chosen to determine their differential expression after accurate normalization across five experimental sets using single and/or best combinations of reference genes (Table S2). Primer design and qRT-PCR reactions were followed as mentioned before. The average Ct value was calculated from three biological replicates and used for relative expression analyses. Normalization of the gene of interest in developmental tissue samples was calculated using the ΔCt values as previously described [12], while relative expression of genes of interest in hormone treatments and abiotic stress conditions was measured as suggested before [39]. The expression fold change value was represented as relative expression ($2^{-\Delta\Delta Ct}$). Statistical significant differences in gene expression patterns were evaluated using Tukey's range test in JMP (v7.0.2).

Transformation of pearl millet

Particle bombardment-mediated transformation of immature zygotic embryo derived calli was carried out using PDS-1000 He biolistic device (Bio-Rad, Hercules, CA) following the protocol described earlier [6]. Zygotic embryos were isolated from surface sterilized seeds and cultured on MS medium supplemented with 2,4-D (2.5 mg/l), maltose (30 g/l), pH 5.8 for callus formation. Particle bombardments were conducted using pCAMBIA1201 and pCAMBIA1302 vectors plasmid DNA (250 ng/shot) precipitated onto 0.6 μm gold particles (Bio-Rad) at a helium pressure of 1,100 psi following the protocol describe previously [6]. Expression of *β-glucuronidase* (*gus*) reporter gene was performed as mentioned earlier [40], while *green fluorescent protein* (*gfp*) expression was monitored using a fluorescence stereomicroscope (Leica MZ FLIII) coupled with a SPOT Insight CCD camera.

Results

Identification of candidate reference genes

We found locus identifiers and/or GenBank accession numbers of selected Arabidopsis and rice candidate reference genes from previous published work (Table 1). We identified orthologous locus IDs and/or GenBank accession numbers of these potential candidate reference genes from foxtail millet, a close relative of pearl millet, using orthologous group search in Phytozome and/or BLASTN search in NCBI GenBank. We selected a total of 18 genes for accurate transcripts normalization during gene expression study using qRT-PCR in pearl millet. These genes included both traditional housekeeping as well as several new reference genes namely, *actin* (*ACT*), *cyclophilin* (*CYC*), *eukaryotic elongation factor 1 alpha* (*eEF1α*), *F-box domain containing protein* (*FBX*), *glyceraldehyde-3-phosphate dehydrogenase* (*GAPDH*), *eukaryotic initiation factor 4a2* (*eIF4a2*), *phosphoenolpyruvate carboxylase-related kinase* (*PEPKR*), *protein phosphatase 2A* (*PP2A*), *rubisco activase* (*RCA*), *S-adenosyl methionine decarboxylase* (*SAMDc*), *alpha tubulin* (*TUA*), *tonoplast intrinsic protein* (*TIP41*), *ubiquitin-conjugating enzyme 2* (*UBC2*), *ubiquitin-conjugating enzyme 18* (*UBC18*), *ubiquitin 5* (*UBQ5*), *transmembrane protein 56* (*UNK*), *18S ribosomal RNA* (*18S rRNA*) and *25S ribosomal RNA* (*25S rRNA*) (Table 1).

Due to insufficient availability of sequence information in the NCBI GenBank, in addition to the sequences of pearl millet obtained from Genbank, we used full length transcript sequences from the foxtail millet to design the gene specific primers for qRT-PCR. Primer pairs were designed to anneal near the 3′ end or at the 3′ UTR of each gene using the Primer3Plus software following the parameters: length: 20±3 mer; product size range: 50–200 base pair; melting temperature: 60°C±3°, guanine-cytosine (GC) content: ~50% including absence for hairpin structures, self-dimers and weak or no self-complementarities at the 3′ end (Table S1).

Sample size, RNA quality and qRT-PCR conditions

We tested the expression of these potential candidate reference genes and quantified the Ct values using qRT-PCR in a total experimental set of 234 samples (Table 1). These included developmental tissues, hormone treatments and abiotic stress conditions from three pearl millet genotypes ICMR01004, IPCI1466 and IP300088 (Table 2). The developmental tissues experimental set included 108 samples from 12 vegetative and reproductive stages [callus 30DPC, seed 30DPH, leaf 7DPS, leaf

15DPS, leaf 30DPS, and node, internode, sheath, flag leaf, panicle, peduncle and root from 60DPS plants], whereas hormone treatments and abiotic stress conditions included 72 and 54 samples from 8 [control (without treatment), ABA, Bra, GA, IAA, MeJa, SA and Zea] and 6 [control (without stress), dehydration (mannitol), drought (no water), cold, heat and salinity] sets of samples (Tables 2, S3 to S5), respectively. The fifth experimental set comprised of Ct values from 78 tissue samples from each of the three pearl millet genotypes (Table 2). We isolated high quantity [368.7±63.3 ng/μl (mean±standard deviation, SD where n = 234)] and quality (average 260/280 ratio of 2.0±0.1 and 260/230 ratio of 2.1±1.6) of total RNA using the guanidinium thiocyanate-based RNeasy plant mini kit. The complete absence of DNA contamination was confirmed by qRT-PCR after two steps of DNAse treatments (first on-column and second gDNA wipeout reaction) for each sample. The reverse transcriptase reactions were primed using an optimized blend of oligo-dT and random primers provided in the kit in order to amplify transcripts from both highly and weakly expressed genes.

Accuracy and efficiency of amplification

To determine the accuracy of primers designed in this study to specifically amplify potential target candidate reference genes we performed PCR and qRT-PCR using either crude and/or diluted cDNAs. We obtained a single amplified product of the expected size in agarose gel electrophoresis (Figure S1) and the presence of one dominant peak of the specific amplicon in melt curve analysis (Figure S2), respectively. Further, a two-step qRT-PCR protocol for cDNA synthesis and cDNA amplification in successive steps reduced the undesired primer dimer formation using SYBR Green. No detectable amplifications in the no-template controls (NTCs) confirmed the absence of primer dimers or non-specific products (Figure S2). We determined the PCR efficiency (*E*) of each primer pair from the amplification plots of all amplification profiles using LinRegPCR software. The mean *E* values with SD for all primer pairs across all experimental samples from three biological replicates are given in Table S1. Primer pairs of most of the genes exhibited no significant differences in E values and displayed PCR efficiencies of more than 1.90, while primer pair for *CYC* and *25S rRNA* showed PCR efficiencies of 1.87±0.03 and 1.85±0.04 (Table S1). We further calculated correlation coefficients (R^2) of PCR efficiency values to evaluate the amplification curves. Except *TAU* ($R^2 = 0.89$), the rest of the primer pairs revealed $R^2 > 0.90$ from all reactions (Table S1).

Expression levels of candidate reference genes

Expression levels of all the candidate reference genes were measured by monitoring the Ct values in the qRT-PCR reactions. We analyzed all the Ct values under five groups which included total [first experimental set (n = 234), Table 1], developmental tissues [second experimental set (n = 108), Tables 2 and S3], hormone treatments [third experimental set (n = 72), Tables 2 and S4], abiotic stress conditions [fourth experimental set (n = 54), Tables 2 and S5] and genotypes [fifth experimental set (n = 78), Table 2]. In the first total experimental set the mean Ct values of the 18 candidate reference genes revealed a minimum of 9.1±1.8 and a maximum of 31.5±3.0 for highest and lowest expression levels for *25S rRNA* and *CYC* genes, respectively, while most of the values were distributed between 22.8±3.1 to 29.8±3.1 (Table 1). The mean Ct values of *SAMDc* (26.2±5.2) with highest SD indicated less stability as compared to *TIP41* (28.5±1.5) with lowest SD showing relatively stable expression in the total experimental set (Table 1). Similarly, we noticed large SDs of Ct values for *SAMDc* and *18S rRNA* indicating a more variable

Table 2. Expression levels of candidate reference genes across four experimental sets of pearl millet.

Genes	Developmental tissues Ct±SD*	Hormone treatments Ct±SD	Abiotic stresses Ct±SD	Genotypes Ct±SD
ACT	28.1±2.8	27.6±1.7	29.3±3.5	28.3±0.7
CYC	29.8±3.2	33.2±1.1	32.3±3.0	31.7±1.4
eEF1α	23.0±3.6	25.3±1.9	24.3±3.1	24.2±0.9
FBX	25.1±3.7	24.5±1.3	26.8±2.2	25.4±0.9
GAPDH	23.1±3.4	21.6±1.2	24.0±3.6	22.9±1.0
eIF4a2	23.0±2.1	22.5±1.1	24.7±2.4	23.3±1.0
PEPKR	25.3±1.3	25.4±0.7	26.3±2.3	25.6±0.5
PP2A	25.8±2.9	24.8±1.2	26.5±2.6	25.7±0.7
RCA	25.8±3.5	24.1±0.9	24.5±3.0	24.8±0.7
SAMDc	27.0±5.5	24.9±4.0	26.6±5.8	26.2±0.9
TUA	22.8±3.9	21.4±1.0	26.0±2.8	23.4±1.9
TIP41	28.3±1.7	28.1±0.7	29.3±1.7	28.6±0.5
UBC2	29.9±3.3	28.8±2.0	30.8±3.5	29.8±0.8
UBC18	26.3±2.7	25.8±0.7	26.9±2.4	26.3±0.4
UBQ5	23.7±2.5	22.3±0.8	24.7±2.0	23.5±1.0
UNK	27.7±2.0	27.6±1.1	28.7±1.7	28.0±0.5
18S rRNA	22.5±5.4	26.4±3.0	23.4±5.0	24.1±1.7
25S rRNA	9.0±2.3	8.7±0.6	9.8±1.9	9.1±0.5

*, Data are represented as mean threshold cycle (Ct) values from all analyzed samples in each individual experimental set with standard deviation (SD).

expression in the other four experimental sets, while little variation of Ct values was detected for rest of the genes (Table 2). In the second experimental set, high Ct values suggested low expression of all the candidate reference genes in the seeds compared to other developmental tissues (Table S3). The third and fourth experimental sets revealed elevated Ct values of these genes under SA treatment (Table S4), and heat and salinity stress conditions (Table S5) as compared to their respective controls. Furthermore, we evaluated the expression levels of candidate reference genes by calculating the CV of the Ct values. Among the four experimental sets, TIP41 showed the lowest CV value (1.1±0.3), while SAMDc and 18S rRNA revealed a greater variation in expression levels due to their high CV values (4.8±1.2 and 5.7±1.3, Table 1).

Stability ranking of the candidate reference genes

We used Stability Index (SI), BestKeeper, ΔCt, NormFinder, geNorm and RefFinder programs to identify the best reference genes for qRT-PCR data normalization in pearl millet. These programs allowed us to establish a stability ranking of each candidate reference gene using Ct values across the experimental sets and condition-specific levels (Tables 3–7).

The SI was calculated from the multiplication of the slope (a value of the regression analysis of geomeans and overall means) with CV considering the fact that gene with lowest SI from low slope and low CV provided the best reference gene. In the first total experimental set, PEPKR, PP2A and TIP41 with SI values of 0.05, 0.06 and 0.09 were the top three candidates, respectively, whereas 18S rRNA with highest SI value of 1.97 was the least preferred choice of reference gene (Table 3). Based on SI values, PEPKR (SI of 0.06) and TIP41 (SI of 0.16) were the two best candidates, while SAMDc (SI of 2. 40) was the worst candidate for normalization of gene expression in developmental tissue samples (Table 4). Analysis of SI values of reference genes in the third and fourth experimental sets revealed TIP41 as the superior candidate

with the smallest SI of 0.02 and 0.18, respectively for transcript normalization under hormone treatments and abiotic stress conditions (Tables 5 and 6). Among the three genotypes of pearl millet, TIP41 (SI of 0.16) was the top ranked reference gene for normalization of gene expression (Table 7).

The BestKeeper program determines the stability ranking of the reference genes based on the percentage of crossing point (%CP) to the BestKeeper Index and the SD from the geometric mean of the candidate reference genes Ct values, where the genes with lowest CP and SD values are identified as the best reference genes for normalization. In this study, BestKeeper analyses of the total experimental samples identified PEPKR (2.72±0.69), TIP41 (3.16±0.90) and PP2A (3.24±0.90) with lowest CP±SD values (Table 3), where genes with SD<1 are considered as stable. In the developmental tissues, hormone treated, abiotic stressed and genotype experimental sets many genes showed SD<1, while the most stable reference genes were PEPKR (2.78±0.70), TUA (0.70±0.23), TIP41 (1.60±0.47) and ACT (1.09±0.29), respectively (Tables 4–7).

The ΔCt method compared the relative expression of a reference gene with other candidate reference genes within each sample, thereby ranked the genes based on the average of STDEV or SD. Analyses using this program exhibited PP2A, UBC2, TIP41, UBQ5 and TIP41 with average STDEV values of 1.32, 1.33, 0.56, 0.97 and 0.64 as the most suitable reference genes for normalization in total, developmental, hormone treated, abiotic stress and genotypes experimental sets of pearl millet, respectively (Tables 3–7).

NormFinder ranks all candidate reference genes based on intra- and inter-group variations of expression stabilities by measuring the stability value (SV) for each reference gene. In our study, the NormFinder identified PP2A, TIP41 and PEPKR with SV of 0.43, 0.57 and 0.61, as the top three optimal reference genes for transcript normalization in the total tissue samples (Table 3). The

Table 3. Stability ranking of the reference genes in all tissue and experimental sets studied in pearl millet.

Rank	Stability Index		BestKeeper		ΔCT		NormFinder		geNorm		RefFinder	
	Genes	Index Value (SI)	Genes	CP(%)±SD	Genes	Ave of STDEV	Genes	Stability Value (SV)	Genes	Normalization Value (MV)	Genes	Geomean of ranking values
1	PEPKR	0.05	PEPKR	2.72±0.69	PP2A	1.32	PP2A	0.43	PP2A | TIP41	0.46	PP2A	1.68
2	PP2A	0.06	TIP41	3.16±0.90	TIP41	1.39	TIP41	0.57			TIP41	2.63
3	TIP41	0.09	PP2A	3.24±0.90	UBC2	1.40	PEPKR	0.61	UBC2	0.49	UBC2	3.41
4	UBC2	0.21	UBC2	4.24±1.26	TUA	1.41	UBC2	0.61	PEPKR	0.73	PEPKR	4.12
5	TUA	0.21	ACT	4.26±1.20	eEF1α	1.42	UBC18	0.68	UBQ5	0.83	UBQ5	4.16
6	eIF4a2	0.23	TUA	4.48±1.17	UBQ5	1.48	eEF1α	0.82	eEF1α	0.88	eEF1α	5.63
7	ACT	0.29	eIF4a2	4.60±1.07	ACT	1.51	UBQ5	0.86	ACT	0.93	UBC18	6.40
8	UBQ5	0.31	eEF1α	4.90±1.25	eIF4a2	1.55	eIF4a2	0.93	UBC18	0.96	TUA	6.45
9	UBC18	0.33	UNK	5.57±1.46	PEPKR	1.56	ACT	0.94	TUA	0.99	ACT	7.33
10	eEF1α	0.39	UBQ5	5.57±1.31	GAPDH	1.69	TUA	1.16	GAPDH	1.04	eIF4a2	7.94
11	SAMDc	0.67	GAPDH	6.76±1.54	SAMDc	1.72	SAMDc	1.17	eIF4a2	1.09	GAPDH	10.47
12	GAPDH	0.68	UBC18	7.14±1.81	UNK	1.73	UNK	1.18	UNK	1.15	UNK	11.00
13	UNK	0.83	RCA	7.26±1.80	FBX	1.87	FBX	1.40	FBX	1.22	FBX	13.47
14	FBX	0.86	SAMDc	7.46±1.79	25S rRNA	1.96	25S rRNA	1.45	25S rRNA	1.31	RCA	13.74
15	RCA	0.87	CYC	7.51±2.36	UBC18	2.22	GAPDH	1.86	SAMDc	1.40	SAMDc	15.24
16	25S rRNA	1.02	FBX	10.18±2.35	RCA	2.53	RCA	2.22	RCA	1.53	25S rRNA	15.47
17	CYC	1.83	25S rRNA	10.24±0.93	CYC	2.72	CYC	2.42	CYC	1.67	CYC	17.00
18	18S rRNA	1.97	18S rRNA	10.95±2.62	18S rRNA	2.84	18S rRNA	2.59	18S rRNA	1.80	18S rRNA	18.00

CP, crossing point; STDEV and SD, standard deviation.

Table 4. Stability ranking of the reference genes in developmental tissues of pearl millet.

Rank	Stability Index		BestKeeper		ΔCT		Normfinder		geNorm		RefFinder	
	Genes	Index value (SI)	Genes	CP(%)±SD	Genes	Ave of STDEV	Genes	Stability Value (SV)	Genes	Normalization Value (MV)	Genes	Geomean of ranking values
1	PEPKR	0.06	PEPKR	2.78±0.70	UBC2	1.33	TIP41	0.31	PP2A | TIP41	0.32	PP2A	2.28
2	TIP41	0.16	UBC2	4.02±1.14	PP2A	1.34	PP2A	0.38	UBC2	0.36	TIP41	2.51
3	PP2A	0.17	eEF1α	4.27±1.18	TUA	1.41	UBC2	0.52	PEPKR	0.46	UBC2	3.72
4	eEF1α	0.23	PP2A	4.66±1.07	UBQ5	1.43	PEPKR	0.66	eEF1α	0.64	TUA	4.46
5	ACT	0.35	ACT	5.02±1.41	eEF1α	1.44	eEF1α	0.69	ACT	0.79	eEF1α	5.42
6	UBC2	0.36	TIP41	5.10±1.52	TIP41	1.48	TUA	0.76	UBQ5	0.86	ACT	5.44
7	TUA	0.57	UBQ5	6.21±1.43	UBC18	1.48	ACT	0.81	TUA	0.94	PEPKR	6.45
8	UNK	0.63	elF4a2	6.26±1.61	PEPKR	1.50	UBQ5	0.81	UNK	0.98	UBQ5	6.97
9	UBQ5	0.63	TUA	6.29±1.69	ACT	1.53	elF4a2	0.82	GAPDH	1.02	UBC18	7.48
10	elF4a2	0.68	UNK	6.32±1.49	GAPDH	1.68	UNK	1.11	25S rRNA	1.09	UNK	7.64
11	UBC18	0.73	UBC18	6.60±1.73	25S rRNA	1.70	GAPDH	1.12	elF4a2	1.15	elF4a2	7.67
12	CYC	1.04	GAPDH	7.70±1.77	elF4a2	1.76	25S rRNA	1.27	UBC18	1.22	25S rRNA	8.82
13	GAPDH	1.08	CYC	8.03±2.39	18S rRNA	1.95	UBC18	1.55	18S rRNA	1.30	GAPDH	10.94
14	18S rRNA	1.70	18S rRNA	8.52±1.91	UNK	1.99	18S rRNA	1.59	FBX	1.40	18S rRNA	13.74
15	FBX	1.83	FBX	9.62±2.42	SAMDc	2.23	SAMDc	1.88	SAMDc	1.48	SAMDc	15.24
16	25S rRNA	2.09	SAMDc	10.45±2.38	FBX	2.24	FBX	1.94	CYC	1.63	FBX	15.98
17	RCA	2.31	RCA	11.56±2.94	CYC	2.75	CYC	2.46	RCA	1.80	CYC	16.74
18	SAMDc	2.40	25S rRNA	13.08±1.17	RCA	3.22	RCA	3.02			RCA	18.00

CP, crossing point; STDEV and SD, standard deviation.

Table 5. Stability ranking of the reference genes in hormone treatments of pearl millet.

Rank	Stability Index		BestKeeper		ΔCT		Normfinder		geNorm		RefFinder	
	Genes	Index value (SI)	Genes	CP(%)±SD	Genes	Ave of STDEV	Genes	Stability Value (SV)	Genes	Normalization Value (MV)	Genes	Geomean of ranking values
1	TIP41	0.02	TUA	0.70±0.23	TIP41	0.56	TIP41	0.13	TIP41 | UBQ5	0.16	TIP41	1.73
2	ACT	0.02	TIP41	0.88±0.25	PP2A	0.57	PEPKR	0.16	PP2A	0.17	UBQ5	2.91
3	PEPKR	0.02	UBQ5	0.91±0.26	UBC2	0.57	UBQ5	0.18	ACT	0.17	PP2A	4.43
4	TUA	0.05	ACT	1.02±0.23	TUA	0.59	PP2A	0.18	eEF1α	0.21	PEPKR	4.61
5	UBQ5	0.06	UBC2	1.06±0.27	PEPKR	0.60	eEF1α	0.20	UBC2	0.24	ACT	4.68
6	UBC2	0.06	eEF1α	1.07±0.29	eEF1α	0.62	ACT	0.22	PEPKR	0.26	eEF1α	5.13
7	eEF1α	0.06	PP2A	1.14±0.28	ACT	0.63	eIF4a2	0.26	TUA	0.27	TUA	6.70
8	PP2A	0.07	UBC18	1.24±0.32	UBQ5	0.64	UBC2	0.33	GAPDH	0.29	UBC2	7.00
9	UNK	0.07	eIF4a2	1.27±0.31	eIF4a2	0.65	TUA	0.34	UBC18	0.31	eIF4a2	7.43
10	25S rRNA	0.10	GAPDH	1.37±0.30	GAPDH	0.66	UBC18	0.41	eIF4a2	0.32	CYC	7.65
11	eIF4a2	0.14	PEPKR	1.65±0.35	UNK	0.67	GAPDH	0.46	25S rRNA	0.34	UBC18	7.88
12	FBX	0.16	CYC	1.76±0.39	25S rRNA	0.67	CYC	0.48	CYC	0.38	GAPDH	9.12
13	GAPDH	0.20	SAMDc	2.02±0.50	CYC	0.74	25S rRNA	0.70	FBX	0.43	25S rRNA	11.06
14	UBC18	0.20	RCA	2.14±0.59	FBX	0.97	SAMDc	0.83	UNK	0.49	SAMDc	14.73
15	RCA	0.46	GAPC2	2.79±0.68	SAMDc	0.98	GAPC2	0.95	SAMDc	0.55	FBX	14.73
16	CYC	0.47	25S rRNA	3.28±0.28	UBC18	1.06	RCA	1.37	RCA	0.66	UNK	15.49
17	18S rRNA	1.57	UNK	3.60±0.91	RCA	1.49	UNK		18S rRNA	0.82	RCA	17.00
18	SAMDc	4.93	18S rRNA	5.04±1.33	18S rRNA	2.07	18S rRNA	2.05			18S rRNA	18.00

CP, crossing point; STDEV and SD, standard deviation.

Table 6. Stability ranking of the reference genes in abiotic stress conditions of pearl millet.

Rank	Stability Index		BestKeeper		ΔCT		Normfinder		geNorm		RefFinder	
	Genes	Index value (SI)	Genes	CP(%)±SD	Genes	Ave of STDEV	Genes	Stability Value (SV)	Genes	Normalization Value (MV)	Genes	Geomean of ranking values
1	TIP41	0.18	TIP41	1.60±0.47	UBQ5	0.97	TIP41	0.28	PP2A | TIP41	0.39	PP2A	1.50
2	TUA	0.25	eEF1α	2.42±0.60	TIP41	1.01	PP2A	0.35	UBQ5	0.42	TIP41	2.94
3	UBQ5	0.50	TUA	2.65±0.76	PP2A	1.03	UBQ5	0.35	ACT	0.47	UBQ5	3.36
4	eEF1α	0.52	UBQ5	2.76±0.68	ACT	1.03	TUA	0.47	TUA	0.56	ACT	4.23
5	PP2A	0.56	PP2A	3.48±0.85	eEF1α	1.10	ACT	0.64	UBC2	0.63	TUA	4.36
6	UBC2	0.59	PEPKR	3.49±0.92	UBC2	1.14	eEF1α	0.73	eEF1α	0.69	eEF1α	4.41
7	UBC18	0.84	UBC2	3.75±1.00	PEPKR	1.17	UBC2	0.74	PEPKR	0.74	UBC2	5.24
8	PEPKR	0.86	UNK	3.81±0.89	TUA	1.19	elF4a2	0.79	UBC18	0.81	PEPKR	6.96
9	ACT	1.07	elF4a2	4.01±1.30	GAPDH	1.25	PEPKR	0.88	18S rRNA	0.88	UBC18	8.32
10	UNK	1.15	FBX	4.17±1.12	elF4a2	1.31	UNK	0.94	25S rRNA	0.93	GAPDH	9.46
11	RCA	1.67	CYC	4.44±1.36	UBC18	1.32	25S rRNA	0.97	UNK	0.98	18S rRNA	12.26
12	elF4a2	1.79	UBC18	4.61±1.24	25S rRNA	1.39	18S rRNA	1.09	GAPDH	1.02	25S rRNA	12.74
13	FBX	2.00	ACT	4.96±1.45	UNK	1.42	RCA	1.10	RCA	1.06	UNK	12.85
14	CYC	2.09	SAMDc	5.52±1.47	RCA	1.42	SAMDc	1.13	elF4a2	1.12	elF4a2	13.45
15	25S rRNA	2.17	18S rRNA	5.78±1.50	18S rRNA	1.57	GAPDH	1.33	SAMDc	1.18	RCA	13.93
16	GAPDH	3.29	RCA	5.99±1.47	SAMDc	1.63	UBC18	1.41	FBX	1.24	SAMDc	14.74
17	18S rRNA	10.59	GAPDH	8.04±1.93	FBX	1.75	FBX	1.54	CYC	1.31	CYC	16.26
18	SAMDc	12.24	25S rRNA	9.14±0.89	CYC	1.83	CYC	1.62			FBX	17.24

CP, crossing point; STDEV and SD, standard deviation.

Table 7. Stability ranking of the reference genes among three genotypes of pearl millet.

Rank	Stability Index		BestKeeper		ΔCT		Normfinder		geNorm		RefFinder	
	Genes	Index value (SI)	Genes	CP(%)±SD	Genes	Ave of STDEV	Genes	Stability Value (SV)	Genes	Normalization Value (MV)	Genes	Geomean of ranking values
1	TIP41	0.16	ACT	1.09±0.29	TIP41	0.64	TIP41	0.03	TIP41	ACT	0.05	TIP41	1.50
2	UBQ5	0.16	TIP41	1.31±0.37	ACT	0.65	PP2A	0.03			ACT	2.38
3	TUA	0.24	UBQ5	1.33±0.38	eEF1α	0.65	ACT	0.03	PP2A	0.08	PP2A	2.91
4	ACT	0.42	PEPKR	1.34±0.34	TUA	0.66	eEF1α	0.04	TUA	0.11	TUA	3.31
5	PEPKR	0.59	UBC2	1.74±0.52	PP2A	0.67	TUA	0.05	PEPKR	0.15	eEF1α	3.98
6	eEF1α	0.62	eEF1α	1.76±0.50	UBQ5	0.67	PEPKR	0.06	eEF1α	0.18	PEPKR	5.18
7	PP2A	0.78	PP2A	1.79±0.46	UBC2	0.69	UBQ5	0.27	UBQ5	0.22	UBQ5	6.19
8	UBC2	1.00	RCA	2.12±0.53	PEPKR	0.71	UBC2	0.35	UBC2	0.25	UBC2	7.74
9	UBC18	1.19	TUA	2.40±0.63	FBX	0.74	FBX	0.39	FBX	0.29	FBX	10.05
10	FBX	1.19	UBC18	2.55±0.62	25S rRNA	0.75	25S rRNA	0.39	UBC18	0.31	elF4a2	11.07
11	RCA	1.26	FBX	2.61±0.66	UBC18	0.78	UNK	0.54	UNK	0.33	UBC18	11.24
12	CYC	1.32	UNK	2.72±0.64	GAPDH	0.79	GAPDH	0.56	GAPDH	0.35	GAPDH	12.24
13	25S rRNA	1.68	GAPDH	2.82±0.64	UNK	1.04	UBC18	0.87	elF4a2	0.41	UNK	12.47
14	UNK	1.85	elF4a2	2.90±0.68	RCA	1.19	RCA	1.00	25S rRNA	0.48	RCA	12.54
15	elF4a2	2.01	CYC	3.25±1.03	elF4a2	1.34	elF4a2	1.13	RCA	0.60	25S rRNA	13.77
16	GAPDH	2.42	25S rRNA	3.57±0.33	SAMDc	1.57	SAMDc	1.47	SAMDc	0.70	SAMDc	16.21
17	SAMDc	9.55	18S rRNA	4.80±1.16	CYC	1.66	CYC	1.53	CYC	0.82	CYC	16.74
18	18S rRNA	9.86	SAMDc	5.73±1.34	18S rRNA	2.10	18S rRNA	2.05	18S rRNA	0.96	18S rRNA	17.74

CP, crossing point; STDEV and SD, standard deviation.

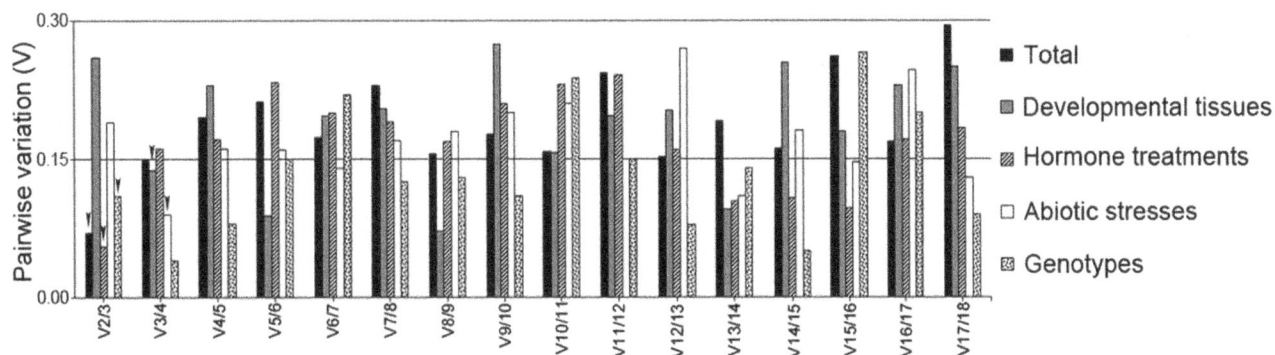

Figure 1. Estimation of pairwise variation to determine the optimal number of control reference genes required for accurate normalization using geNorm. Pairwise variation (V, Vn/Vn+1) was calculated between successively ranked normalization factors NFn and NFn+1. Arrowheads on the bar graph indicate the minimum number of genes required at the cut-off value 0.15 [31]. The V between the normalization factors of the two first-ranked and the three first-ranked is represented by V2/3 and so on, respectively.

NormFinder analyses in the developmental tissues (SV of 0.31), hormone treatments (SV of 0.13), abiotic stress conditions (SV of 0.28) and genotypes (SV of 0.03) experimental sets of pearl millet recognized *TIP41* as the most suitable reference gene (Tables 4–7).

In addition, we also examined the stability ranking of candidate reference genes using geNorm program (Tables 3–7). The geNorm statistical algorithm determines the normalization value (MV) based on the geometric mean of multiple reference genes and mean pair-wise variation of a gene from all other reference genes in each set of samples. In both first and second experimental sets, the two best reference genes were *PP2A | TIP41* with the lowest MV of 0.46 and 0.32, whereas *UBC2* with MV of 0.49 and 0.36 remained the third most suitable gene for transcript normalization in total and developmental tissues, respectively, as determined by the geNorm (Tables 3–4). The most preferred genes for normalization in hormone treatments and abiotic stress conditions were *TIP41 | UBQ5* (MV of 0.16) and *PP2A | TIP41* (MV of 0.39), respectively (Tables 5–6), while *TIP41 | ACT* had the lowest MV of 0.05 in the genotypes of pearl millet (Table 7). In addition, geNorm analyses revealed significantly high stability of several reference genes with MV of less than the cut-off range of 1.5 (Tables 3–7).

We further compared all the data generated by SI, BestKeeper, ΔCt, NormFinder and geNorm programs using recommended comprehensive ranking method in RefFinder software to confirm the stability ranking of reference genes for accurate transcript normalization across the experimental sets (Tables 3–7). The overall ranking of the best reference genes in total and categorized experimental sets according to RefFinder are given in Tables 3–8.

We next applied the geNorm software to calculate the Vn/Vn+1 between NFn and NFn+1 to determine the best combination of reference genes required for precise transcript quantification across different sets of experiments. Figure 1 summarizes the V values from the combination of reference genes and shows that a number of genes are required for reliable normalization of gene expression data among different experimental sets (Table 8).

Accurate normalization of gene and transgene expression using optimal combination of reference genes

In order to validate the selection of the best reference genes for accurate normalization of gene expression, we chose *PEPC* (phosphoenolpyruvate carboxylase), *ERF* (ethylene response

factor) and *DREB* (dehydration responsive element binding) genes to determine the relative transcript levels using qRT-PCR (Table S2). We monitored the expression of *PEPC*, an essential gene for C4 photosynthesis, in developmental tissue samples, whereas the expression pattern of two transcription factors, *ERF* and *DREB*, known to be regulated during abiotic and biotic stresses, were examined in hormone treated and abiotic stressed samples. Relative transcript levels of these genes were calculated after normalizing with the best ranked candidate reference genes as determined by geNorm and recommended by RefFinder (Table 8). Transcript abundance of *PEPC* when normalized using single top ranked reference genes, *PP2A*, *TIP41* and *UBC2*, revealed bias effect on the relative expression patterns (Figure 2). Furthermore, transcript normalization using a combination of two (*PP2A+TIP41*) and three (*PP2A+TIP41+UBC2*) reference genes showed much stable and constant expression profiles across tissues (Figure 2). Similarly, relative expression patterns of *ERF* and *DREB* in hormone treatments and abiotic stress conditions were affected by the selection of the reference gene or combination of genes, respectively (Figures 3–4). As predicted, a strong bias in the relative expression pattern of *PEPC*, *ERF* and *DREB* was obtained when the least stable gene was used for normalization. Incorporation of *TIP41* and *UBC2* or *UBQ5* during expression analyses neutralized the unwanted changes of transcript abundance to allow accurate normalization of *PEPC*, *ERF* and *DREB*. Overall expression of *PEPC* was significantly high in flag leaf and sheath as compared to nodal tissues of pearl millet genotypes (Figure 2). In the hormone treatments experimental set, Zea enhanced 2-fold expression of *ERF* in pearl millet genotype ICMT01004 and IPCI1466 compared to other hormones tested (Figure 3). The expression of *DREB* was up-regulated during drought followed by heat stresses in all the three genotypes (Figure 4). Genotypes showed differential expression patterns of these genes as well (Figures 2–4).

We also monitored the transcript abundance pattern of *β-glucuronidase* (*gus*), *green fluorescent protein* (*gfp*) and *hygromycin phosphotransferase* (*hpt*) expressing transgenes in transgenic pearl millet calli. Calli of three pearl millet genotypes were bombarded with *CaMV35S::gus* (pCAMBIA1201) and *CaMV35S::gfp* (pCAMBIA1302) constructs and transient expression of both *gus* and *gfp* reporter genes were visualized after 5 days (Figure S3). Expressions of *gus*, *gfp* and *hpt* genes were examined in transformed calli selected on hygromycin (30 mg/l) after 30

Table 8. Summary of the best combination of reference genes for accurate normalization across five experimental sets of pearl millet using geNorm and RefFinder programs.

Experimental sets	Total	Development tissues	Hormone treatments	Abiotic stresses	Genotypes
Best combination	V2/3	V3/4	V2/3	V3/4	V2/3
Pairwise variation (V)[a]	0.070	0.138	0.056	0.090	0.110
Reference control genes	*PP2A*	*PP2A*	*TIP41*	*PP2A*	*TIP41*
	TIP41	*TIP41*	*UBQ5*	*TIP41*	*ACT*
		UBC2		*UBQ5*	

[a] Pairwise variation (V) represents the optimal combination of reference control genes required to pass the suggested cut-off value 0.15 [31]. A single common reference control gene for expression study across experimental sets is highlighted in gray.

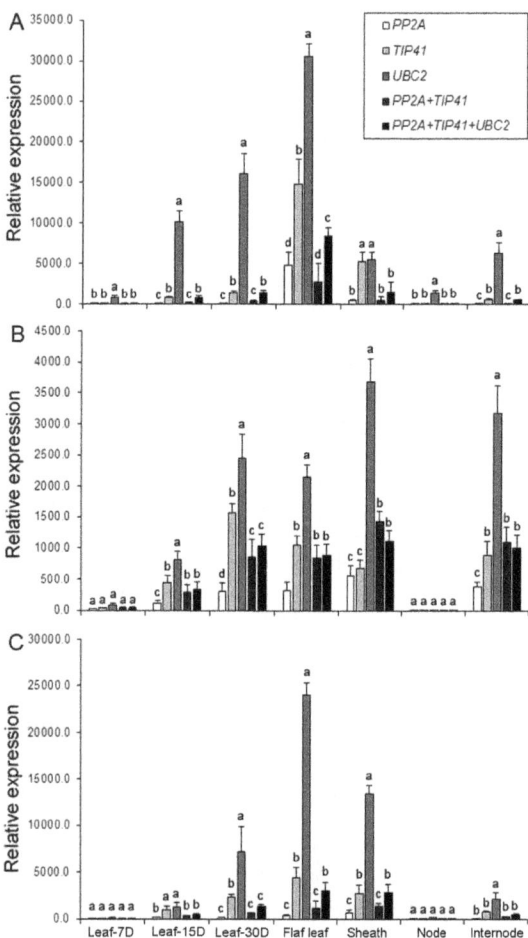

Figure 2. Validation of *PEPC* gene expression after normalization using optimal number of control reference genes in developmental tissue samples from genotypes (A) ICMR01004, (B) IPCI1466 and (C) IP300088. Results are presented as mean relative expression with SD from three biological replicates after normalization using the best combination of reference genes recommended by geNorm and RefFinder (see Table 8) for developmental tissue samples. Leaf- 7D, 15D and 30D represent 7DPS, 15DPS and 30DPS leaf samples while flag leaf, sheath, node and internode are from 60DPS plants. Different letters on the bars indicate significant differences at the $P \le 0.05$ level as tested by Tukey's Range Test.

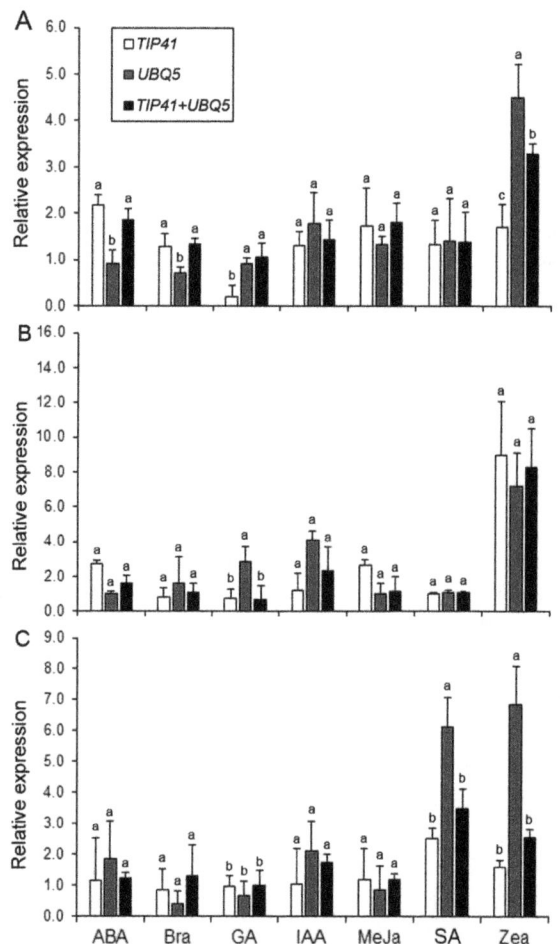

Figure 3. Validation of *ERF* gene expression after normalization using optimal number of control reference genes in hormone treated samples from genotypes (A) ICMR01004, (B) IPCI1466 and (C) IP300088. Data are presented as mean relative expression with SD from three biological replicates after normalization using the best combination of reference genes recommended by geNorm and RefFinder (see Table 8) for hormone treatments. ABA (abscisic acid), Bra (brassinolide), GA (gibberellic acid), IAA (indole-3-acetic acid), MeJa (methyl jasmonate), SA (salicylic acid) and Zea (zeatin) treatments of 15DPG plants. Different letters on the bars indicate significant differences at the $P \le 0.05$ level as tested by Tukey's Range Test.

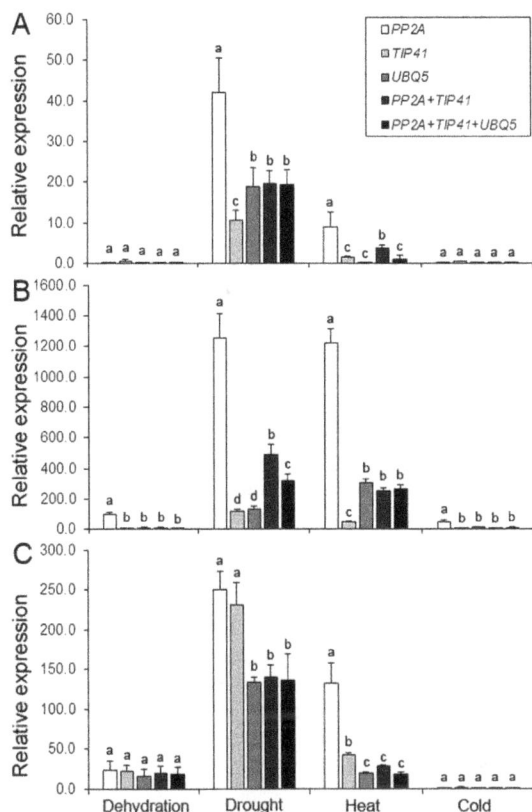

Figure 4. Validation of *DREB* gene expression after normalization using optimal number of control reference genes in genotypes (A) ICMR01004, (B) IPCI1466 and (C) IP300088 subjected to abiotic stress conditions. Results are presented as mean relative expression with SD from three biological replicates after normalization using the best combination of reference genes recommended by geNorm and RefFinder (see Table 8) for abiotic stress conditions. Dehydration (mannitol), drought (no water), heat (42°C) and cold (4°C) stresses are presented. Different letters on the bars indicate significant differences at the $P \leq 0.05$ level as tested by Tukey's Range Test.

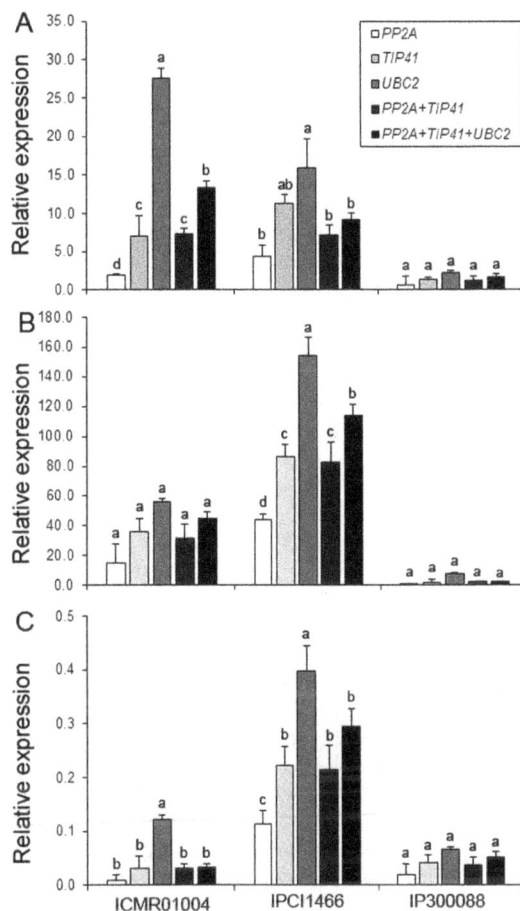

Figure 5. Validation of expression of *gus, gfp* and *hpt* transgenes using optimal number of control reference genes in hygromycin resistant calli from genotypes (A) ICMR01004, (B) IPCI1466 and (C) IP300088 after 30 days post particle bombardment-mediated transformation using pCAMBIA1201 and pCAMBIA1302, respectively. Results are presented as mean relative expression with SD from three biological replicates after normalization using the best combination of reference genes recommended by geNorm and RefFinder (see Table 8) for developmental tissue samples. Different letters on the bars indicate significant differences at the $P \leq 0.05$ level as tested by Tukey's Range Test.

days post bombardment using qRT-PCR. Normalization with the recommended reference genes (*PP2A, TIP41* and *UBC2*) showed similar effects on the relative expression patterns of *gus, gfp* and *hpt* transgenes in the calli of all three genotypes (Figure 5) as observed for *PEPC* in leaves (Figure 2), whereas the combination of the two (*PP2A+TIP41*) and the three (*PP2A+TIP41+UBC2*) reference genes exhibited more reliable transcript quantification. In general, expression analyses revealed that relative quantification of all three transgenes were higher in pearl millet genotype ICMT01004 and IPCI1466 compared to IP300088 (Figure 5).

Discussion

Transcriptome changes occurring during developmental processes and/or adverse environmental conditions are experiencing a growing research interest to understand the gene regulatory networks that control agronomically and economically important traits e.g. enhanced crop yield and biomass production under high atmospheric CO_2 or abiotic stress in the Panicoideae grasses including pearl millet. Transcriptomics data from microarray and next generation sequencing analyses should be validated using qRT-PCR [41]. QRT-PCR provides a useful tool to study

transcriptome changes in pearl millet because no genome sequence or microarray chip is available. Moreover, reliable transcript measurements using qRT-PCR analysis require accurate normalization against an appropriate internal control reference gene [9,28]. Normalization is important to adjust the variation introduced by various steps involved in the qRT-PCR such as quantity and quality of RNA samples, cDNAs, fluorescent fluctuations, sample-to-sample and/or well-to-well volume variations [10]. Therefore, pearl millet requires an assessment of appropriate reference genes for accurate transcript normalization in gene expression studies using qRT-PCR.

In this study, we demonstrated a comprehensive analysis of 18 potential candidate reference genes which included both traditional housekeeping genes like *ACT, eEF1α, GAPDH, TUA, UBC* and *UBQ5* and new candidate reference genes e.g. *PEPKR, PP2A, TIP41* on 234 samples from developmental tissues, hormone treatments and abiotic stress conditions of three pearl millet genotypes. We carried out simple total RNA extraction protocols using the guanidinium thiocyanate-based kit [42], which yielded acceptable RNA quality and quantity from all samples including roots and seeds of three pearl millet genotypes as mentioned by the golden rules of qRT-PCR [11]. Previously published protocols using same guanidinium thiocyanate-based kit demonstrated satisfactory amount of high quality RNA from rice [43]. Since DNA contamination can result in inaccurate quantification of RNA abundance [44], we conducted a second gDNA wipeout reaction on the isolated RNAs after on-column DNase treatment following manufacturer recommended protocol to completely eliminate the detectable genomic DNA contamination as verified by qRT-PCR for absence of any non-specific amplification. We primed cDNA synthesis using an optimized blend of oligo-dT and random primers to preferentially amplify the lowly abundant transcripts such as *CYC* in this study. In support of our finding, a weak expression of *CYC* was observed in rice [43]. However, the abundance of *25S rRNA* in different tissue, physiological conditions and pearl millet genotypes suggested the use of random hexamers to prime the reverse transcription reaction in our study.

We performed the two-step qRT-PCR method to reduce the unwanted primer dimer formation using SYBR Green detection dye [8]. This method was also followed for the large scale expression profiling of transcription factors in rice [43]. Specific amplification with expected amplicon size of each primer pair from the RT-PCR was confirmed by agarose gel electrophoresis (Figure S1). In addition, the single peak melting curves in the qRT-PCR with no amplicon peak in the NTCs proved the absence of primer dimers or non-specific products (Figure S2). The PCR efficiency of each primer pair was calculated from the raw amplification curves (absolute fluorescence data) captured during the exponential phase of amplification of each qRT-PCR reaction using LinRegPCR [37]. Except for *CYC* and *25S rRNA*, which showed an average efficiency of 1.87 ± 0.03 and 1.86 ± 0.04, all the candidate reference genes exhibited mean efficiency values greater than 1.90 (Table S1) suggesting specific transcripts being amplified at least at 90% efficiency per cycle in the qRT-PCR reactions [45]. An identical range of PCR efficiencies were reported for many orthologous of selected candidate reference genes from Arabidopsis [12], rice [43] and common bean [46]. In this study the average Ct (Tables 1–2 and S3–S5) values of candidate reference genes varied within the recommended range of 22.8 ± 3.1 to 31.5 ± 3.0 by qRT-PCR [47], except for *25S rRNA* which showed Ct of 9.1 ± 1.8 (Table 1). In support of our results, a low Ct (average Ct value of 8) of *25S rRNA* gene was also observed in rice [13]. The ΔCt was calculated using the previously published method [39] and the precision of the assay was assessed using the CV. In general, our candidate reference genes showed CV<5% of Ct values, suggesting higher stability in expression levels under all experimental conditions. Therefore, our data demonstrated that the selected reference genes in this study are potential candidates for accurate normalization of gene expression by qRT-PCR after proper validation. In conjunction of our study, low CV<5% of Cq values of reference genes under abiotic stress conditions in common bean was also reported [46].

It has been suggested that the selection of optimal number of reference genes must be experimentally determined [48]. However, no single reference gene was found to have a stable expression under different experimental conditions [10,33] and nor a single method is enough to test for the stability of the candidate reference genes [31,33]. We used the algorithms executed by six different programs for proper stability ranking of the candidate reference genes. The SI [28] and ΔCt [29] methods calculate the variation of Ct and ΔCt values in pairwise genes, whereas BestKeeper estimates the variation in Ct values and reference genes showing SD<1 are considered the most stable [30]. However, the NormFinder [32] and geNorm [31] statistical algorithms allowed us to determine the stability ranking by calculating the SV and MV of each reference gene, respectively (Tables 3–7). In our study geNorm analyses revealed MV<1.5 for most of the genes under different experimental conditions (Tables 3–7), suggesting the potential stability of reference genes [31]. However, in the total experimental set *PEPKR* was the first ranked candidate gene by SI and BestKeeper, but ranked third by geNorm (Table 3); this could be due to the sensitivity of geNorm to the co-regulation of genes with similar expression patterns. In addition, geNorm is less affected by expression intensity of the reference genes [49] and allowed us to determine the optimal number of genes required to accurately normalize qRT-PCR data based on the V values [31]. We applied RefFinder [33] for recommended comprehensive ranking by combining all five above programs. Earlier reports on bamboo [17], strawberry [49] and leafy spurge [50] showed that these computational programs did not place the top ranked genes in identical order. According to our analysis, the six statistical programs ranked the candidate reference genes in various orders from best to worst, which could be due to different algorithm used by each program. Overall, new reference genes ranked better than the traditional housekeeping genes by most of the programs (Tables 3–8). Normalization using multiple reference genes is critical not only to obtain reliable gene expression results since normalization using single gene can be erroneous [9], but it also evaluates the expression stability of the selected reference genes during qRT-PCR. The geNorm analyses allowed us to identify optimal number of reference genes (Table 8) required for accurate normalization by calculating the V values at the suggested cut-off range of 0.15 [31].

In this study all the six computational methods suggested that *PP2A*, *TIP41*, *UBC2*, *UBQ5* and *ACT* are the top 5 superior reference genes for accurate transcript normalization in pearl millet under different experimental conditions (Table 8). None of the traditional housekeeping genes qualified as the best reference gene for transcript normalization in total tissue across all the five experimental sets of pearl millet. Moreover, only *UBQ5* and *ACT* were found to be suitable for hormone treated, stress conditions and genotypes of pearl millet (Table 8), respectively. This is because expression stability of many housekeeping genes vary considerably owning to their involvement in the cellular metabolism and functions [26]. In accordance to our study, *ACT* was one of the best reference genes in foxtail millet [20]. In addition, *ACT* was shown be a good candidate reference gene for normalization of transcript data in rice [43] and strawberry [49]. Moreover, *UBQ* was found to be a suitable reference gene in mustard [21], poplar [28] and rice [13]. In the current study, *18S rRNA*, *25S rRNA* and *SAMDc* were consistently categorized as unsuitable, perhaps due to their inconsistency in gene expression by all the six programs (Tables 3–7), thereby rendering them inappropriate to use as reference gene. Similarly, poor stability of *18S rRNA* under abiotic stress conditions was reported in foxtail millet [20]. In conjunction with rice the high expression of *25S rRNA* in this study makes it inappropriate for normalization of weakly expressed genes [13]. We observed significant variation of *SAMDc* expression pattern, which has been shown recently to be a poor reference gene in switchgrass [25]. The *CYC*, *eEF1α* and *eIF4a* were listed as variable genes in many studies [24,43], thereby limiting their use as reference genes in pearl millet as well. We found *GAPDH* as an inappropriate reference gene, which was also ranked unsuitable for bamboo [17], brachypodium [14] and rice [13]. In our study, another traditional housekeeping gene *UBC2* ranked the third best reference genes after two novel candidate reference genes, *PP2A* and *TIP41* for normalization in developmental tissue samples. The UBC encodes an ubiquitin-conjugating enzyme E2 involved in protein degradation through ubiquitination reactions and performed best among the three traditional housekeeping genes in leafy spurge [50]. However, in the current study two novel candidate genes, *PP2A* and *TIP41* resulted as superior reference genes compared to traditional housekeeping genes tested under different experimental conditions. This finding is in agreement with previous reports where *PP2A* and *TIP41* combination was most suitable for abiotic stress conditions in caragana [27]. Recent reports demonstrated that *PP2A* and *TIP41* were the most recommended stable reference genes for transcript normalization in tissue samples of numerous plant species [17,19,27].

The suitability of these reference genes to conduct transcriptomics studies was assessed by monitoring the expression profiles of three endogenous genes and transgenes in both untransformed and genetically transformed pearl millet tissues. The *PEPC* encodes a ubiquitous cytosolic enzyme in higher plants which catalyzes the irreversible carboxylation of phosphoenolpyruvate (PEP) to oxaloacetate (OAA), a four carbon compound, in the initial fixation of atmospheric CO_2 during C4 photosynthesis [51]. We noticed that transcript levels of *PEPC* were high in the flag leaf compared to nodal tissue in all the pearl millet genotypes studied (Figure 2). The ERF and DREB are AP2 binding transcription factors which regulate plant responses to several environmental stress conditions [52] and up-regulated under abiotic stresses [52] and hormone signaling [53], respectively. Transcript abundance of *ERF* illustrated differential expression pattern after accurate quantification using *TIP41* and *UBQ5* under different hormone stimuli conditions (Figure 3). Currently, several reports have validated the optimum relative expression of *DREB* using appropriate reference genes under abiotic stress conditions [21,27]. In agreement with previous reports, we found *DREB* expression was up-regulated many fold in drought and heat stress conditions after accurate normalization using combination of reference genes (Figure 4). In addition, we provided evidence that these set of reference genes are also useful for transcript quantification in transformed pearl millet tissues, while incorporation of multiple reference genes provides the most reliable expression pattern after precise normalization.

Conclusions

To the best of our knowledge this is the first comprehensive assessment of appropriate reference genes for accurate transcript normalization using qRT-PCR analyses in pearl millet. Stability ranking using computer based Stability Index, ΔCt, BestKeeper, NormFinder, geNorm and RefFinder programs recommended *TIP41*, *PP2A*, *UBC2*, *UBQ5* and *ACT* as the best reference genes out of 18 potential candidate genes tested on different developmental and experimental conditions. This work will facilitate the developmental gene expression studies on C4 photosynthesis and hormone cross-talk during abiotic stress conditions in pearl millet, a crop with limited genomic and transcriptomics information, and also benefit the scientific community for conducting experiments on related bioenergy crop species.

References

1. Ejeta G, Hassen MM, Mertz ET (1987) In vitro digestibility and amino acid composition of pearl millet (Pennisetum typhoides) and other cereals. Proc Natl Acad Sci U S A 84: 6016–6019.

Supporting Information

Figure S1 Reverse transcription (RT)-PCR conformation of individual candidate reference gene showing specific amplification of the expected amplicon size from each primer pair in 3% (w/v) agarose gel. cDNAs prepared from RNA samples isolated from leaves of 30D old plants from three biological replicates were pooled together and PCR reactions were conducted using primer pair specific for each candidate reference gene. Lane name corresponds to each reference gene used for RT-PCR. M1 and M2 are 50 base pair (bp) and 100 bp DNA ladder, respectively.

Figure S2 Dissociation curve analyses for conformation of specific real-time PCR amplification with single peak for each primer pair. cDNAs were prepared from RNA samples isolated from flag leaves in three biological replicates and melt curves generated after qRT-PCR using primer pair specific for each gene with no template controls (NTC) are presented.

Figure S3 Expression of reporter genes in particle bombarded pearl millet genotype ICMR01004 calli. (A) *gus* reporter gene expression in calli bombarded with pCAMBIA1201 plasmid, (B) *gfp* reporter gene expression in calli after bombardment with pCAMBIA1302 plasmid. Both the reporter genes were driven by CaMV35S promoter and the expression was monitored after 5 days post bombardment.

Table S1 Primer sequences of candidate reference genes used for qRT-PCR.

Table S2 Information of selected endogenous genes and transgenes with primer sequences for validation of accurate normalization using suitable reference genes.

Table S3 Distribution of the Ct values of each candidate reference genes across the developmental tissue samples of pearl millet.

Table S4 Distribution of Ct values of each candidate reference genes in pearl millet samples subjected to hormone treatments.

Table S5 Distribution of Ct values of each candidate reference genes in pearl millet samples subjected to abiotic stress conditions.

Acknowledgments

Authors are thankful to Dr. Ellen Tumimbang, Mrs Elham Abed and Yrian Hong for technical support.

Author Contributions

Conceived and designed the experiments: PS EB. Performed the experiments: PS. Analyzed the data: PS EB. Contributed reagents/materials/analysis tools: EB. Contributed to the writing of the manuscript: PS EB.

2. Hill GM, Hanna WW (1990) Nutritive characteristics of pearl millet grain in beef cattle diets. J Anim Sci 68: 2061–2066.

3. Li P, Brutnell TP (2011) Setaria viridis and Setaria italica, model genetic systems for the Panicoid grasses. J Exp Bot 62: 3031–3037.

4. Upadhyaya HD, Reddy KN, Gowda CLL (2007) Pearl millet germplasm at ICRISAT genebank-status and impact. SAT eJournal 3: 1–5.

5. Yadav RS, Sehgal D, Vadez V (2011) Using genetic mapping and genomics approaches in understanding and improving drought tolerance in pearl millet. J Exp Bot 62: 397–408.

6. O'Kennedy MM, Stark HC, Dube N (2011) Biolistic-mediated transformation protocols for maize and pearl millet using pre-cultured immature zygotic embryos and embryogenic tissue. Methods Mol Biol 710: 343–354.

7. Mishra RN, Reddy PS, Nair S, Markandeya G, Reddy AR, et al. (2007) Isolation and characterization of expressed sequence tags (ESTs) from subtracted cDNA libraries of Pennisetum glaucum seedlings. Plant Mol Biol 64: 713–732.

8. Wong ML, Medrano JF (2005) Real-time PCR for mRNA quantitation. Biotechniques 39: 75–85.

9. Guenin S, Mauriat M, Pelloux J, Van Wuytswinkel O, Bellini C, et al. (2009) Normalization of qRT-PCR data: the necessity of adopting a systematic, experimental conditions-specific, validation of references. J Exp Bot 60: 487–493.

10. Huggett J, Dheda K, Bustin S, Zumla A (2005) Real-time RT-PCR normalisation; strategies and considerations. Genes Immun 6: 279–284.

11. Udvardi MK, Czechowski T, Scheible W-R (2008) Eleven Golden Rules of Quantitative RT-PCR. The Plant Cell Online 20: 1736–1737.

12. Czechowski T, Stitt M, Altmann T, Udvardi MK, Scheible WR (2005) Genome-wide identification and testing of superior reference genes for transcript normalization in Arabidopsis. Plant Physiol 139: 5–17.

13. Jain M, Nijhawan A, Tyagi AK, Khurana JP (2006) Validation of housekeeping genes as internal control for studying gene expression in rice by quantitative real-time PCR. Biochem Biophys Res Commun 345: 646–651.

14. Hong S-Y, Seo P, Yang M-S, Xiang F, Park C-M (2008) Exploring valid reference genes for gene expression studies in Brachypodium distachyon by real-time PCR. BMC Plant Biology 8: 112.

15. Petit C, Pernin F, Heydel JM, Delye C (2012) Validation of a set of reference genes to study response to herbicide stress in grasses. BMC Res Notes 5: 18.

16. Gutierrez L, Mauriat M, Guenin S, Pelloux J, Lefebvre JF, et al. (2008) The lack of a systematic validation of reference genes: a serious pitfall undervalued in reverse transcription-polymerase chain reaction (RT-PCR) analysis in plants. Plant Biotechnol J 6: 609–618.

17. Fan C, Ma J, Guo Q, Li X, Wang H, et al. (2013) Selection of Reference Genes for Quantitative Real-Time PCR in Bamboo (Phyllostachys edulis). PLoS ONE 8: e56573.

18. Ovesná J, Kučera L, Vaculová K, Štrymplová K, Svobodová I, et al. (2012) Validation of the β-amy1 Transcription Profiling Assay and Selection of Reference Genes Suited for a RT-qPCR Assay in Developing Barley Caryopsis. PLoS ONE 7: e41886.

19. Artico S, Nardeli SM, Brilhante O, Grossi-de-Sa MF, Alves-Ferreira M (2010) Identification and evaluation of new reference genes in Gossypium hirsutum for accurate normalization of real-time quantitative RT-PCR data. BMC Plant Biol 10: 49.

20. Kumar K, Muthamilarasan M, Prasad M (2013) Reference genes for quantitative real-time PCR analysis in the model plant foxtail millet (Setaria italica L.) subjected to abiotic stress conditions. Plant Cell, Tissue and Organ Culture (PCTOC) 115: 13–22.

21. Chandna R, Augustine R, Bisht NC (2012) Evaluation of Candidate Reference Genes for Gene Expression Normalization in Brassica juncea Using Real Time Quantitative RT-PCR. PLoS ONE 7: e36918.

22. Chi X, Hu R, Yang Q, Zhang X, Pan L, et al. (2012) Validation of reference genes for gene expression studies in peanut by quantitative real-time RT-PCR. Mol Genet Genomics 287: 167–176.

23. Paolacci A, Tanzarella O, Porceddu E, Ciaffi M (2009) Identification and validation of reference genes for quantitative RT-PCR normalization in wheat. BMC Molecular Biology 10: 11.

24. Tenea G, Peres Bota A, Cordeiro Raposo F, Maquet A (2011) Reference genes for gene expression studies in wheat flag leaves grown under different farming conditions. BMC Research Notes 4: 373.

25. Gimeno J, Eattock N, Van Deynze A, Blumwald E (2014) Selection and Validation of Reference Genes for Gene Expression Analysis in Switchgrass (Panicum virgatum) Using Quantitative Real-Time RT-PCR. PLoS ONE 9: e91474.

26. Thellin O, Zorzi W, Lakaye B, De Borman B, Coumans B, et al. (1999) Housekeeping genes as internal standards: use and limits. J Biotechnol 75: 291–295.

27. Zhu J, Zhang L, Li W, Han S, Yang W, et al. (2013) Reference gene selection for quantitative real-time PCR normalization in Caragana intermedia under different abiotic stress conditions. PLoS One 8: e53196.

28. Brunner AM, Yakovlev IA, Strauss SH (2004) Validating internal controls for quantitative plant gene expression studies. BMC Plant Biol 4: 14.

29. Silver N, Best S, Jiang J, Thein SL (2006) Selection of housekeeping genes for gene expression studies in human reticulocytes using real-time PCR. BMC Mol Biol 7: 33.

30. Pfaffl MW, Tichopad A, Prgomet C, Neuvians TP (2004) Determination of stable housekeeping genes, differentially regulated target genes and sample integrity: BestKeeper–Excel-based tool using pair-wise correlations. Biotechnol Lett 26: 509–515.

31. Vandesompele J, De Preter K, Pattyn F, Poppe B, Van Roy N, et al. (2002) Accurate normalization of real-time quantitative RT-PCR data by geometric averaging of multiple internal control genes. Genome Biol 3: Research0034.

32. Andersen CL, Jensen JL, Orntoft TF (2004) Normalization of real-time quantitative reverse transcription-PCR data: a model-based variance estimation approach to identify genes suited for normalization, applied to bladder and colon cancer data sets. Cancer Res 64: 5245–5250.

33. Chen D, Pan X, Xiao P, Farwell MA, Zhang B (2011) Evaluation and identification of reliable reference genes for pharmacogenomics, toxicogenomics, and small RNA expression analysis. J Cell Physiol 226: 2469–2477.

34. Untergasser A, Nijveen H, Rao X, Bisseling T, Geurts R, et al. (2007) Primer3Plus, an enhanced web interface to Primer3. Nucleic Acids Res 35: W71–74.

35. Saha P, Majumder P, Dutta I, Ray T, Roy SC, et al. (2006) Transgenic rice expressing Allium sativum leaf lectin with enhanced resistance against sap-sucking insect pests. Planta 223: 1329–1343.

36. Saha P, Ray T, Tang Y, Dutta I, Evangelous NR, et al. (2013) Self-rescue of an EXTENSIN mutant reveals alternative gene expression programs and candidate proteins for new cell wall assembly in Arabidopsis. The Plant Journal 75: 104–116.

37. Ramakers C, Ruijter JM, Deprez RH, Moorman AF (2003) Assumption-free analysis of quantitative real-time polymerase chain reaction (PCR) data. Neurosci Lett 339: 62–66.

38. Hellemans J, Mortier G, De Paepe A, Speleman F, Vandesompele J (2007) qBase relative quantification framework and software for management and automated analysis of real-time quantitative PCR data. Genome Biology 8: R19.

39. Pfaffl MW, Horgan GW, Dempfle L (2002) Relative expression software tool (REST) for group-wise comparison and statistical analysis of relative expression results in real-time PCR. Nucleic Acids Res 30: e36.

40. Jefferson RA, Kavanagh TA, Bevan MW (1987) GUS fusions: beta-glucuronidase as a sensitive and versatile gene fusion marker in higher plants. Embo j 6: 3901–3907.

41. Wang Y, Barbacioru C, Hyland F, Xiao W, Hunkapiller KL, et al. (2006) Large scale real-time PCR validation on gene expression measurements from two commercial long-oligonucleotide microarrays. BMC Genomics 7: 59.

42. Chomczynski P, Sacchi N (1987) Single-step method of RNA isolation by acid guanidinium thiocyanate-phenol-chloroform extraction. Anal Biochem 162: 156–159.

43. Caldana C, Scheible W-R, Mueller-Roeber B, Ruzicic S (2007) A quantitative RT-PCR platform for high-throughput expression profiling of 2500 rice transcription factors. Plant Methods 3: 7.

44. Bustin SA (2002) Quantification of mRNA using real-time reverse transcription PCR (RT-PCR): trends and problems. J Mol Endocrinol 29: 23–39.

45. Ruijter JM, Ramakers C, Hoogaars WMH, Karlen Y, Bakker O, et al. (2009) Amplification efficiency: linking baseline and bias in the analysis of quantitative PCR data. Nucleic Acids Research 37: e45.

46. Borges A, Tsai SM, Caldas DG (2012) Validation of reference genes for RT-qPCR normalization in common bean during biotic and abiotic stresses. Plant Cell Rep 31: 827–838.

47. Karlen Y, McNair A, Perseguers S, Mazza C, Mermod N (2007) Statistical significance of quantitative PCR. BMC Bioinformatics 8: 131.

48. Gimenez MJ, Piston F, Atienza SG (2011) Identification of suitable reference genes for normalization of qPCR data in comparative transcriptomics analyses in the Triticeae. Planta 233: 163–173.

49. Amil-Ruiz F, Garrido-Gala J, Blanco-Portales R, Folta KM, Munoz-Blanco J, et al. (2013) Identification and Validation of Reference Genes for Transcript Normalization in Strawberry (Fragaria x ananassa) Defense Responses. PLoS One 8: e70603.

50. Chao WS, Doğramaci M, Foley ME, Horvath DP, Anderson JV (2012) Selection and Validation of Endogenous Reference Genes for qRT-PCR Analysis in Leafy Spurge (Euphorbia esula). PLoS ONE 7: e42839.

51. Chollet R, Vidal J, O'Leary MH (1996) PHOSPHOENOLPYRUVATE CARBOXYLASE: A Ubiquitous, Highly Regulated Enzyme in Plants. Annu Rev Plant Physiol Plant Mol Biol 47: 273–298.

52. Agarwal PK, Agarwal P, Reddy MK, Sopory SK (2006) Role of DREB transcription factors in abiotic and biotic stress tolerance in plants. Plant Cell Rep 25: 1263–1274.

53. Cheng MC, Liao PM, Kuo WW, Lin TP (2013) The Arabidopsis ETHYLENE RESPONSE FACTOR1 regulates abiotic stress-responsive gene expression by binding to different cis-acting elements in response to different stress signals. Plant Physiol 162: 1566–1582.

3

Associations between Rice, Noodle, and Bread Intake and Sleep Quality in Japanese Men and Women

Satoko Yoneyama[1]*, **Masaru Sakurai**[1], **Koshi Nakamura**[1], **Yuko Morikawa**[1], **Katsuyuki Miura**[2], **Motoko Nakashima**[3], **Katsushi Yoshita**[4], **Masao Ishizaki**[5], **Teruhiko Kido**[6], **Yuchi Naruse**[7], **Kazuhiro Nogawa**[8], **Yasushi Suwazono**[8], **Satoshi Sasaki**[9], **Hideaki Nakagawa**[1]

1 Department of Epidemiology and Public Health, Kanazawa Medical University, Ishikawa, Japan, 2 Department of Health Science, Shiga University of Medical Science, Otsu, Japan, 3 Department of Community Health Nursing, School of Nursing, Kanazawa Medical University, Ishikawa, Japan, 4 Department of Food and Human Health Science Osaka City University, Graduate School of Human Life Science, Osaka, Japan, 5 Department of Social and Environmental Medicine, Kanazawa Medical University, Ishikawa, Japan, 6 School of Health Science, College of Medical, Pharmaceutical and Health Science, Kanazawa University, Kanazawa, Japan, 7 Department of Human Science and Fundamental Nursing, Toyama University School of Nursing, Toyama, Japan, 8 Department of Occupational and Environmental Medicine, Graduate School of Medicine, Chiba University, Chiba, Japan, 9 Department of Social and Preventive Epidemiology, the University of Tokyo, Tokyo, Japan

Abstract

Background: Previous studies have shown that a diet with a high-glycemic index is associated with good sleep quality. Therefore, we investigated the association of sleep quality with the intake of 3 common starchy foods with different glycemic indexes–rice, bread, and noodles–as well as the dietary glycemic index in a Japanese population.

Methods: The participants were 1,848 men and women between 20 and 60 years of age. Rice, bread, and noodle consumption was evaluated using a self-administered diet history questionnaire. Sleep quality was evaluated by using the Japanese version of the Pittsburgh Sleep Quality Index, and a global score >5.5 was considered to indicate poor sleep.

Results: Multivariate-adjusted odds ratios (95% confidence intervals) for poor sleep across the quintiles of rice consumption were 1.00 (reference), 0.68 (0.49–0.93), 0.61 (0.43–0.85), 0.59 (0.42–0.85), and 0.54 (0.37–0.81) (*p* for trend = 0.015); those for the quintiles of noodle consumption were 1.00 (reference), 1.25 (0.90–1.74), 1.05 (0.75–1.47), 1.31 (0.94–1.82), and 1.82 (1.31–2.51) (*p* for trend = 0.002). Bread intake was not associated with sleep quality. A higher dietary glycemic index was significantly associated with a lower risk of poor sleep (*p* for trend = 0.020).

Conclusion: A high dietary glycemic index and high rice consumption are significantly associated with good sleep in Japanese men and women, whereas bread intake is not associated with sleep quality and noodle consumption is associated with poor sleep. The different associations of these starchy foods with sleep quality might be attributable to the different glycemic index of each food.

Editor: Guy Brock, University of Louisville, United States of America

Funding: This research was supported by a Grant-in-Aid from the Ministry of Health, Labour, and Welfare, Health and Labor Sciences research grants, Japan (H18-Junkankitou[Seishuu]- Ippan-012, H19-Junkankitou [Seishuu]-Ippan-012, H19-Junkankitou [Seishuu]-Ippan- 021, H20-Junkankitou [Seishuu]-Ippan-013, H22-Junkankitou [Seishuu]-Ippan-005, H23-Junkankitou [Seishuu]-Ippan-005) and the Japan Arteriosclerosis Prevention Fund. The funders had no role in study design, data collection and analysis.

Competing Interests: The authers have declared that no competing interests exist.

* Email: yoneyama@kanazawa-med.ac.jp

Introduction

Sleep quality is known to be a function of sleep duration and latency [1]. Epidemiological studies have shown that short sleep duration is associated with increased mortality [2,3], poor mood [2,4], chronic health conditions (e.g., obesity and metabolic syndrome) [2,5], cardiovascular disease and diabetes [2,6], hypertension [2,7], and poor self-related health and quality of life [2,8]. Moreover, previous studies have shown that dietary factors affect sleep quality. A cross-sectional study of children younger than 2 years of age showed that the consumption of a meal with a high-glycemic index (GI) in the evening was associated with longer sleep duration [9]. In addition a clinical trial showed that sleep onset latency is reduced by approximately 10 minutes

after the consumption of a carbohydrate-rich evening meal with a high-GI compared with an evening meal with a low-GI [10]. These data suggest that sleep quality is influenced by the carbohydrate-based GI of the meals.

The dietary GI of the general Japanese population is approximately 70 [11–13], this is considerably higher than the dietary GI in predominantly Western populations (i.e., European, Australian, and North American), which ranges from 48 to 60 [14–16]. This difference may be due to disparities in the average intakes of different foods that contribute to the GI. Rice is a common starchy food in the Japanese diet, as approximately 70% of the cereals consumed are rice [17], and rice accounts for 59% of the dietary GI [11–13]. However, no studies have evaluated the associations of rice or other common staple foods such as bread and noodles with sleep quality in a Japanese population.

This study investigated the associations between sleep quality and intake of carbohydrate-based staple foods (i.e., rice, bread, and noodles) as well as the dietary GI and glycemic load (GL) in a Japanese population.

Methods

Participants

The study population comprised 7,306 employees of a factory that produces zippers and aluminum sashes in Toyama Prefecture, Japan. The Industrial Health and Safety Law in Japan requires that employers offer annual health examinations to all of their employees. The present study included data on 2,255 white-collar daytime workers between 20 and 60 years of age. White-collar workers were studied because many blue-collar workers are involved in shift work, making it difficult to evaluate sleep quality.

A questionnaire about diet was completed by 1,977 (88%) of the white-collar daytime workers in 2003, and a questionnaire about sleep was completed by 2003 (94%) of the worker in 2004. In total, 1,858 (82%) of the workers provided complete data on both questionnaires. Ten participants with extremely low or high energy intake (i.e., <500 or >4,000 kcal/day) were excluded from the study. Thus, data on 1,848 white-collar daytime workers (1,164 men and 684 women) were included in the analysis.

Data collection

The annual health examination included a medical history, physical examination, and anthropometric measurements. Body mass index was calculated as weight divided by height squared (kg/m^2). A questionnaire was used to identify health-related behaviors including smoking status (i.e., current, previous, or never), and habitual exercise; habitual exercise was assessed as hours per week spent on leisure time physical activities and was expressed as metabolic equivalent hours per week (MET-h/week).

Dietary assessment and calculation of the dietary GI and GL

A self-administered diet history questionnaire (DHQ) was used to assess dietary habits during the preceding month. [18,19]. The DHQ was developed to estimate the respondents' dietary intake of macronutrients and micronutrients for use in epidemiological studies in Japan. Detailed descriptions of the methods used to calculate dietary intakes and the validity of the DHQ have been reported previously [18,19]. Estimates of dietary intake for 147 food and beverage items, energy, protein, fat, total carbohydrate, and alcohol in 2003 were calculated using an ad hoc computer algorithm developed for the DHQ that was based on the Standard Tables of Food Composition in Japan [20]. The DHQ evaluates the consumption of 19 staple foods (i.e., rice, noodles, and other wheat foods). The intake frequency of each staple food for breakfast, lunch, dinner, and snack/midnight snack in 1 week was evaluated. For rice, the type of rice (i.e., white rice, white rice mixed with barley, white rice with germ, 50% polished rice, 70% polished rice, or brown rice) and serving size (i.e., number and size; cups for children, women, and men, and small and large bowls were defined as 110, 140, 170, 220, and 250 g, respectively) were evaluated. Similarly, the types of bread and noodles were evaluated. Bread was classified as white bread, buttered bread, cake bread, bread containing cream and sweet bean paste, pizza, *okonomiyaki* (Japanese "pizza," which contains shredded cabbage and dough cooked in a frying pan), or Japanese-style pancakes (small pancakes containing flour, sugar, and egg, cooked in a

frying pan). Noodles were classified as Japanese noodles (i.e., buckwheat and Japanese white noodles), instant noodles, Chinese noodles, or pasta [12]. The DHQ also includes the frequencies of skipping breakfast, lunch, dinner per week. Of the 147 food and beverage items included in the DHQ, 6 (4.1%) are alcoholic beverages, 8 (5.4%) contain no available carbohydrates, and 63 (42.9%) contain less than 3.5 g available carbohydrate per serving. Therefore, the dietary GI and GL were calculated on the basis of the remaining 70 items [11,21]. The GI databases used were an international table of GI [22], a report on the GI values of Japanese foods [23], a report on GI values published after the publication of the international GI tables [24], and an online database provided by the Sydney University Glycemic Index Research Service [25]. The GIs of all foods in the DHQ have been published elsewhere [11]. Although there are concerns regarding the utility of the GI for mixed meals (i.e., overall diet) [26,27], many studies have shown that the GI of mixed meals can be predicted on the basis of the GI value of each of the component foods [28]. We calculated the dietary GI as the sum of the percentage contribution of each food multiplied by their respective GI values. Dietary GL was calculated by multiplying the dietary GI by the total daily carbohydrate intake and dividing by 100. We used energy-adjusted values calculated using the density method (per 1,000 kcal) for GL [21]. The reproducibility and relative validity of the dietary GI and GL assessed using the DHQ have been reported elsewhere [21].

Sleep assessment

Sleep quality in the previous month was assessed by using the Japanese version of the Pittsburgh Sleep Quality Index questionnaire (PSQI-J) [29], which was developed from the original questionnaire, the PSQI [1]. In brief, the PSQI-J is a standardized self-administered questionnaire for assessing sleep quality that includes the following 7 components: subjective sleep quality, sleep latency, sleep duration, habitual sleep efficiency, sleep disturbances, use of sleep medication, and daytime dysfunction. Each component is weighted equally on a scale of 0 to 3, and the scores for each component are then summed to yield a PSQI-J global score ranging from 0 to 21, higher scores indicate poorer sleep quality. Participants completed the PSQI-J at home in 5–10 minutes, and the responses were then reviewed by a well-trained nurse.

Statistical analysis

When the PSQI-J component scores of the male and female participants were compared, sleep duration was significantly shorter in men than in women ($p = 0.011$); however, there were no significant differences in the other component scores including the PSQI-J global score between sexes. Therefore, all analyses were performed in the whole population, (i.e., not stratified by sex). In this study, we used energy density values for macronutrients and alcohol (% of energy [% energy]) and for food intakes (weight per 1000 kcal [g/1000 kcal]).

"Poor sleep," was defined as a PSQI-J global score >5.5 [29]. In a previous study, using a cut-off of 5.5 for the PSQI-J global score provided estimates with a sensitivity and specificity of 85.7% and 86.6% for primary insomnia, 80.0% and 86.6% for major depression, and 83.3% and 86.6% for schizophrenia, respectively [29]. Rice, bread, and noodle intake as well as the GI and GL were categorized into quintiles. Analysis of covariance was used to evaluate the means of each PSQI-J component score adjusted for age, sex and total energy intake. The prevalence of poor sleep in each quintile was compared using the χ^2 test. Odds ratios (ORs) with 95% confidence intervals (95% CIs) for poor sleep were

Table 1. Characteristics of study participants by quintiles of dietary rice, bread, and noodle intake as well as dietary glycemic index and glycemic load (N = 1,848).

	Q1 (lowest)	Q2	Q3 (middle)	Q4	Q5 (highest)	p value[a]
Rice (range, g/1,000 kcal)	<132.7	132.7–168.4	168.5–202.2	202.3–249.1	≥249.2	
Male (%)	48.0	54.5	61.7	74.1	76.7	<0.001
Age (years)	37.6 ± 10.1	39.2 ± 9.5	39.9 ± 9.8	39.9 ± 9.9	41.7 ± 10.2	0.001
Body mass index (kg/m²)	22.4 ± 3.1	22.5 ± 3.0	23.1 ± 3.2	22.9 ± 3.2	23.0 ± 3.0	0.060
Habitual exercise level (METs/week)	1.0 ± 12.1	1.1 ± 8.7	0.6 ± 4.0	0.8 ± 9.5	1.2 ± 8.5	0.823
Current smokers (%)	28.0	26.6	30.4	32.8	37.5	0.026
Skip breakfast (%)	64.5	42.1	35.9	18.3	11.7	<0.001
Total energy intake (kcal/day)	2012 ± 582	2034 ± 491	1942 ± 452	1870 ± 410	1731 ± 414	<0.001
Protein intake (%kcal)	17.9 ± 2.9	17.8 ± 2.6	17.2 ± 2.5	16.5 ± 2.4	15.5 ± 2.4	<0.001
Fat intake (%kcal)	31.8 ± 6.4	29.7 ± 5.4	26.9 ± 5.4	23.8 ± 5.0	19.8 ± 4.9	<0.001
Carbohydrate intake (%kcal)	52.0 ± 7.1	54.5 ± 5.6	57.3 ± 5.4	60.3 ± 5.6	65.3 ± 5.9	<0.001
Alcohol consumption (%kcal)	5.0 ± 7.5	4.4 ± 6.4	4.5 ± 6.6	4.7 ± 5.8	4.1 ± 6.1	0.603
Bread intake (g/1,000 kcal)	44.9 ± 32.2	35.4 ± 24.5	30.2 ± 24.0	20.9 ± 21.8	12.9 ± 18.9	<0.001
Noodle intake (g/1,000 kcal)	47.9 ± 38.1	37.7 ± 30.4	34.4 ± 29.6	37.3 ± 35.7	27.0 ± 29.0	<0.001
Dietary GI	63.8 ± 4.1	66.6 ± 2.7	68.1 ± 2.9	69.7 ± 2.5	71.6 ± 3.0	<0.001
Dietary GL (/1,000 kcal)	141 ± 46	157 ± 40	163 ± 42	171 ± 39	179 ± 44	<0.001
Bread (range, g/1,000 kcal)	0	0.1–14.8	14.9–30.1	30.2–47.2	≥47.3	
Male (%)	73.9	66.1	58.0	58.0	58.5	<0.001
Age (years)	41.1 ± 10.0	39.9 ± 9.8	39.7 ± 9.9	39.5 ± 10.2	38.9 ± 9.8	0.045
Body mass index (kg/m²)	23.1 ± 3.0	23.0 ± 3.3	22.8 ± 3.2	22.6 ± 3.2	22.5 ± 2.9	0.082
Habitual exercise level (METs/week)	1.3 ± 10.7	0.5 ± 2.4	1.3 ± 10.5	0.7 ± 5.1	0.9 ± 10.3	0.610
Current smokers (%)	40.7	29.5	29.2	26.0	29.2	0.284
Skip breakfast (%)	17.3	10.7	20.6	50.9	72.4	<0.001
Total energy intake (kcal/day)	1811 ± 445	2031 ± 508	1995 ± 508	1927 ± 405	1803 ± 472	<0.001
Protein intake (%kcal)	16.3 ± 3.0	17.0 ± 2.7	17.2 ± 2.7	17.3 ± 2.4	16.8 ± 2.3	<0.001
Fat intake (%kcal)	22.5 ± 7.1	25.7 ± 6.5	27.3 ± 6.5	27.8 ± 6.2	26.6 ± 6.2	<0.001
Carbohydrate intake (%kcal)	59.0 ± 8.6	57.8 ± 7.5	58.2 ± 7.0	57.4 ± 6.5	59.2 ± 6.9	0.003
Alcohol consumption (%kcal)	7.43 ± 8.36	5.15 ± 6.80	3.12 ± 4.61	3.40 ± 4.86	3.28 ± 5.38	<0.001
Rice intake (g/1,000 kcal)	239.1 ± 71.4	206.3 ± 62.9	185.6 ± 61.3	165.1 ± 52.7	154.8 ± 59.5	<0.001
Noodle intake (g/1,000 kcal)	35.3 ± 39.6	35.6 ± 27.5	37.9 ± 29.8	36.1 ± 32.1	34.7 ± 33.0	0.733
Dietary GI	70.1 ± 4.2	68.9 ± 3.6	67.9 ± 3.6	67.3 ± 3.6	67.2 ± 3.6	<0.001
Dietary GL (/1,000 kcal)	160 ± 44	172 ± 45	170 ± 46	161 ± 39	157 ± 42	<0.001
Noodles (range, g/1,000 kcal)	<7.4	7.4–23.3	23.4–36.5	36.6–57.4	≥57.5	
Male (%)	66.0	56.5	62.8	63.3	66.4	0.330
Age (years)	40.9 ± 10.2	39.9 ± 10.0	40.2 ± 9.5	38.6 ± 10.0	39.7 ± 9.9	0.037

Table 1. Cont.

	Q1 (lowest)	Q2	Q3 (middle)	Q4	Q5 (highest)	p value[a]
Body mass index (kg/m²)	22.9 ± 3.1	22.9 ± 3.1	22.6 ± 2.9	22.7 ± 3.1	23.0 ± 3.4	0.498
Habitual exercise level (METs/week)	0.6 ± 7.2	1.2 ± 10.9	0.8 ± 5.4	1.2 ± 8.7	0.8 ± 9.8	0.864
Current smokers (%)	26.2	27.3	32.2	33.0	27.9	0.801
Skip breakfast (%)	39.5	37.8	28.8	33.4	32.8	0.584
Total energy intake (kcal/day)	1833 ± 455	1945 ± 470	1997 ± 421	1917 ± 509	1861 ± 509	<0.001
Protein intake (%kcal)	16.6 ± 2.9	17.3 ± 2.6	17.0 ± 2.6	16.9 ± 2.8	16.8 ± 2.4	0.006
Fat intake (%kcal)	25.5 ± 6.9	27.4 ± 6.5	26.9 ± 6.6	26.0 ± 6.7	23.9 ± 6.5	<0.001
Carbohydrate intake (%kcal)	58.8 ± 7.6	57.8 ± 6.9	57.1 ± 6.8	58.3 ± 7.3	59.8 ± 7.9	<0.001
Alcohol consumption (%kcal)	4.72 ± 6.90	3.48 ± 5.39	4.86 ± 6.77	4.52 ± 6.38	4.95 ± 6.40	0.012
Rice intake (g/1,000 kcal)	220.1 ± 72.8	188.8 ± 66.3	185.1 ± 64.0	184.3 ± 64.5	174.7 ± 69.0	<0.001
Bread intake (g/1,000 kcal)	25.9 ± 28.2	31.9 ± 27.9	25.6 ± 22.1	27.0 ± 25.0	27.4 ± 26.7	0.008
Dietary GI	70.5 ± 4.0	68.9 ± 3.5	68.5 ± 3.4	67.6 ± 3.4	66.0 ± 3.7	<0.001
Dietary GL (/1,000 kcal)	163 ± 41	166 ± 43	168 ± 41	163 ± 44	159 ± 50	0.094
Dietary GI (range)	<65.3	65.3–67.7	67.8–69.5	69.6–71.6	≥71.7	
Male (%)	52.4	53.5	62.3	69.2	77.5	0.011
Age (years)	38.7 ± 10.3	40.2 ± 9.6	39.1 ± 9.6	39.6 ± 10.0	41.6 ± 9.9	<0.001
Body mass index (kg/m²)	22.6 ± 3.3	22.7 ± 3.2	22.7 ± 2.9	23.0 ± 3.3	23.1 ± 2.8	0.172
Habitual exercise level (METs/week)	1.3 ± 12.7	0.4 ± 1.9	1.1 ± 10.6	0.6 ± 3.8	1.2 ± 8.9	0.466
Current smokers (%)	27.5	24.9	33.7	31.5	38.3	0.605
Skip breakfast (%)	50.8	39.2	35.5	28.4	18.4	<0.001
Total energy intake (kcal/day)	2017 ± 560	1977 ± 469	1896 ± 449	1858 ± 426	1804 ± 439	<0.001
Protein intake (%kcal)	17.7 ± 2.7	17.9 ± 2.5	16.8 ± 2.3	16.5 ± 2.5	15.7 ± 2.6	<0.001
Fat intake (%kcal)	28.7 ± 6.6	27.9 ± 6.2	26.4 ± 6.4	24.5 ± 6.2	22.3 ± 6.4	<0.001
Carbohydrate intake (%kcal)	56.0 ± 7.4	56.9 ± 6.9	58.8 ± 7.1	59.0 ± 6.9	61.1 ± 7.4	<0.001
Alcohol consumption (%kcal)	4.1 ± 6.3	3.6 ± 5.3	3.6 ± 5.2	5.5 ± 7.5	5.7 ± 7.1	<0.001
Rice intake (g/1,000 kcal)	128.9 ± 55.5	160.4 ± 44.0	187.3 ± 50.2	214.9 ± 46.8	261.7 ± 61.5	<0.001
Bread intake (g/1,000 kcal)	35.5 ± 29.2	32.6 ± 24.8	30.6 ± 26.0	24.5 ± 24.1	14.6 ± 20.6	<0.001
Noodle intake (g/1,000 kcal)	53.9 ± 40.2	46.5 ± 34.0	34.5 ± 28.8	27.3 ± 23.7	17.4 ± 19.3	<0.001
Dietary GL (range,/1,000 kcal)	<128	128–149	150–169	170–194	≥195	
Male (%)	47.8	48.1	64.8	70.3	84.0	<0.001
Age (years)	37.7 ± 9.7	39.4 ± 9.7	40.4 ± 9.6	41.3 ± 10.1	40.4 ± 10.2	<0.001
Body mass index (kg/m²)	22.0 ± 3.1	22.5 ± 3.1	23.1 ± 3.3	23.1 ± 3.0	23.4 ± 2.9	<0.001
Habitual exercise level (METs/week)	0.4 ± 2.6	0.8 ± 7.7	0.3 ± 1.1	1.3 ± 12.4	2.0 ± 12.3	<0.001
Current smokers (%)	33.7	27.4	27.7	33.9	33.2	0.301
Skip breakfast (%)	58.9	36.8	31.7	24.1	20.9	<0.001

Table 1. Cont.

	Q1 (lowest)	Q2	Q3 (middle)	Q4	Q5 (highest)	p value[a]
Total energy intake (kcal/day)	1472 ± 305	1695 ± 272	1870 ± 302.9	2052 ± 331	2465 ± 450	<0.001
Protein intake (%kcal)	17.5 ± 3.0	17.2 ± 2.6	17.1 ± 2.6	16.8 ± 2.5	15.9 ± 2.4	<0.001
Fat intake (%kcal)	28.2 ± 7.1	26.6 ± 6.4	25.9 ± 6.6	25.1 ± 6.6	24.0 ± 6.4	<0.001
Carbohydrate intake (%kcal)	53.8 ± 7.4	57.6 ± 6.5	58.5 ± 6.8	59.8 ± 6.8	62.1 ± 6.7	<0.001
Alcohol consumption (%kcal)	6.5 ± 8.8	4.3 ± 5.7	4.2 ± 6.1	4.0 ± 5.2	3.5 ± 5.0	<0.001
Rice intake (g/1,000 kcal)	154.3 ± 64.4	189.4 ± 64.3	197.1 ± 66.5	203.4 ± 65.0	208.9 ± 71.6	<0.001
Bread intake (g/1,000 kcal)	30.7 ± 28.5	27.1 ± 24.7	29.8 ± 28.1	24.4 ± 23.5	25.9 ± 25.2	0.004
Noodle intake (g/1,000 kcal)	41.8 ± 35.7	36.3 ± 30.9	31.1 ± 30.9	34.3 ± 31.1	36.1 ± 34.4	<0.001

GI: glycemic index, GL: glycemic load, METs/week: metabolic equivalent hours per week; Q: quintile.
[a]The χ^2 test was used to analyze categorical variables, and linear regression analysis was used to calculate p-values for trends for continuous variables.

calculated using multiple logistic regression analyses. ORs were first adjusted for age (years; continuous) and sex (age- and sex-adjusted model) and then for body mass index (kg/m^2; continuous), smoking status (i.e., current, previous, or never; dummy variable), habitual exercise (MET-h/week; continuous), alcohol consumption (percentage of energy; continuous), frequency of breakfast consumption per week (i.e., 0–3, 4–6, or 7 days/week; dummy variable), rice intake (for the multivariate analyses of bread and noodle intake; continuous), bread intake (for the multivariate analyses of rice and noodle intake; continuous), and noodle intake (for the multivariate analyses of rice and bread intake; continuous). The p-values for linear trends were calculated by using the median value of each quintile. Furthermore, each food, dietary GI, and dietary GL was included in the logistic regression analyses as continuous variables, and the ORs for an increment of 1 standard deviation in these variables were calculated. Statistical analyses were performed with Statistical Analysis System version 9.3 (SAS Institute Inc., Cary, NC, USA). The level of significance was set at p<0.05.

Ethical considerations

Written informed consent was not obtained from the participants. The design of the present study was approved by the occupational safety and health committee of the subject company, which consisted of employee representatives. Employees were informed of the study design and of the right to refuse to participate in the study in the study documents. Participants who answered the questionnaire were regarded as having consented to the survey. Linkable anonymized data were provided by the company to ensure that individuals would not be identifiable by the researchers. This study was approved by the Institutional Review Committee of Kanazawa Medical University for Ethical Issues.

Results

Table 1 shows the characteristics of the 1,848 study participants stratified by rice, bread, and noodle consumption quintiles as well as dietary GI and GL quintiles. Higher rice intake was significantly associated with older age ($p = 0.001$), higher carbohydrate intake ($p<0.001$), higher dietary GI ($p<0.001$), higher GL ($p<0.001$), lower bread intake ($p<0.001$), lower noodle intake ($p<0.001$), lower frequency of breakfast consumption ($p<0.001$), and a lower probability of being a current smoker ($p = 0.026$). Body mass index, alcohol consumption, and habitual exercise were not significantly associated with rice intake. Higher bread intake was significantly associated with a higher frequency of breakfast consumption ($p<0.001$), female sex ($p<0.001$), younger age ($p = 0.045$), lower alcohol consumption ($p<0.001$), lower rice intake ($p<0.001$), lower dietary GI ($p<0.001$), and lower dietary GL ($p<0.001$). Habitual exercise and smoking status were not significantly associated with bread intake. Higher noodle intake was significantly associated with higher carbohydrate intake ($p<0.001$), higher bread intake ($p = 0.008$), higher alcohol consumption ($p = 0.012$), lower rice intake ($p<0.001$), and lower dietary GI ($p<0.001$). Sex, smoking status, and frequency of breakfast consumption were not significantly associated with noodle intake. Increasing GI quintiles were significantly associated with older age ($p<0.001$), alcohol consumption ($p<0.001$), higher carbohydrate intake ($p<0.001$), higher rice intake ($p<0.001$), lower bread intake ($p<0.001$), lower noodle intake ($p<0.001$), and lower frequency of breakfast consumption ($p<0.001$). Habitual exercise and smoking status were not significantly associated with the dietary GI.

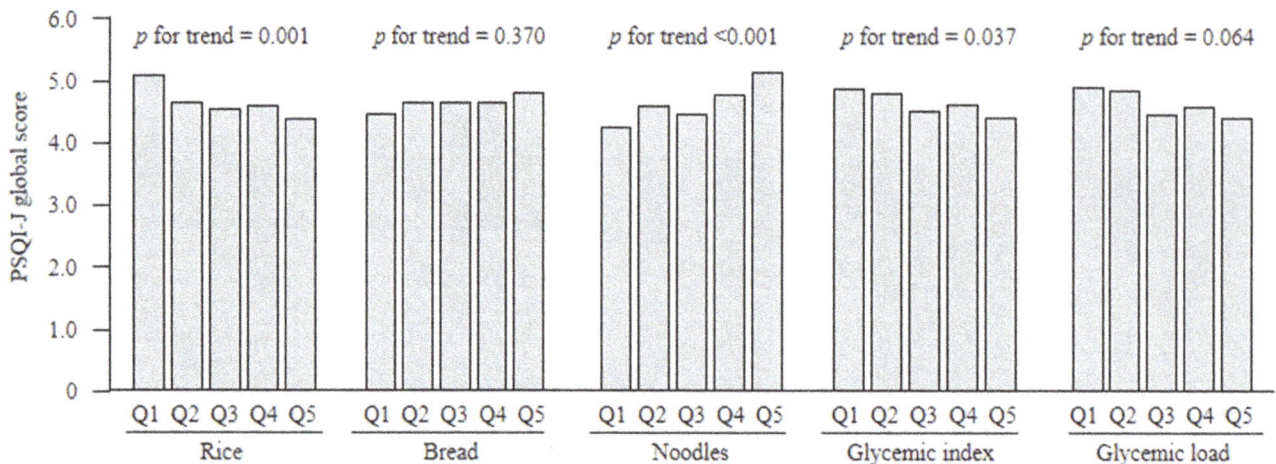

Figure 1. PSQI-J global scores for quintiles of starchy food intake, dietary glycemic index, and glycemic load. Mean PSQI-J global scores were adjusted for age, sex, and total energy intake (kcal/day, continuous), by using the analysis of covariance model. PSQI-J: Japanese version of the Pittsburgh Quality Index; Q: quintile.

Figure 1 shows the mean PSQI-J global score according to rice, bread, and noodle consumption quintile as well as dietary GI and GL quintiles adjusted for age, sex and total energy intake. Higher rice intake and higher GI were significantly associated with a lower PSQI-J global score ($p = 0.001$ and $p = 0.037$, respectively), and higher noodle intake was significantly associated with a higher PSQI-J global score ($p<0.001$). Table 2 shows sleep duration and PSQI-J component scores according to rice, bread, and noodle consumption quintile as well as dietary GI and GL quintile. Higher rice intake was significantly associated with lower scores for poor sleep duration ($p = 0.003$) but was not associated with any other component of sleep quality. Higher noodle intake was significantly associated with a higher frequency of sleep disturbance ($p<0.001$), higher levels of daytime dysfunction ($p = 0.005$), increased use of sleep medication ($p = 0.008$), poorer subjective sleep quality ($p = 0.021$), and longer sleep latency ($p = 0.049$). A higher dietary GI was significantly associated with lower scores for poor sleep duration ($p = 0.013$) but not with other PSQI-J components. These associations remained even after adjusting for age, sex, and total energy intake (data not tabulated).

The multivariate-adjusted ORs (quintile, 95%CI) for the prevalence of poor sleep across the quintiles of rice intake were 1.00 (reference), 0.68 (0.49–0.93), 0.61 (0.43–0.85), 0.59 (0.42–0.85), and 0.54 (0.37–0.81), respectively, indicating that rice intake had a significant positive association with better sleep (p for linear trend = 0.015) (Table 3). In contrast, the multivariate-adjusted ORs across the quintiles of noodle intake were 1.00 (reference), 1.25 (0.90–1.74), 1.05 (0.75–1.47), 1.31 (0.94–1.82), and 1.82 (1.31–2.51), respectively, indicating that noodle consumption was significantly associated with poor sleep (p for linear trend = 0.002) (Table 3). These associations did not change even after further adjusting for vegetable, meat, and fish intake (data not shown). Dietary GI was also associated with good sleep (p for trend = 0.020), whereas dietary GL was not (p for trend = 0.092).

Discussion

This study evaluated the association between sleep quality and the intake of common starchy foods (i.e., rice, bread, and noodles) as well as the dietary GI and GL in a Japanese population. Rice consumption was positively associated with sleep quality. In contrast, noodle consumption had a significant inverse association

with sleep quality. Furthermore, a significant positive relationship between dietary GI and sleep quality was observed. Because rice is a major contributor to the dietary GI among Japanese people, differences between the GI values of rice, bread, and noodles may influence their sleep quality.

The present results show that higher consumption of rice, which is the main contributor to the dietary GI in Japanese foods, and the dietary GI itself are closely associated with a lower PSQI-J global score, i.e., good sleep. In a previous cross-sectional study of children younger than 2 years of age, the consumption of an evening meal with a high-GI was associated with longer sleep duration compared with the consumption of an evening meal with a low-GI [9]. The present results are consistent with these previous findings, because rice intake and the dietary GI were associated with sleep duration but not sleep latency in the present study. In contrast, in a clinical trial of 12 healthy young men, a carbohydrate-based meal with a high-GI was significantly associated with a shortening of sleep onset latency compared with a meal with a low-GI and was most effective when consumed 4 hours before going to sleep [10]. We assessed daily rice consumption and the dietary GI, but not rice consumption or the GI of evening meals, which may have affected the present results.

A high dietary GI may affect sleep quality via the effects of tryptophan (TRP) and melatonin [30–36]. A previous study showed that both carbohydrate intake and a meal with a high-GI increase the ratio of TRP to other large neutral amino acids (TRP/LNAA) after the meal, compared with a meal with a low-GI [37]. LNAA and TRP are competitively transported across the blood–brain barrier, and a higher TRP/LNAA ratio would result in more TRP being transported into the brain. In the brain, TRP is converted into serotonin and then to melatonin, which induces sleep [30–36].

Similar to the dietary GI, rice consumption was significantly associated with a low PSQI-J global score in the present study. Rice, especially white rice, is a common starchy food eaten by Japanese people; it accounts for approximately 28% of the daily energy intake and 70% of cereal intake [17]. Furthermore, white rice accounts for 59% of the dietary GI for in the Japanese diet [11–13]. Therefore, rice intake would affect sleep quality via the effect of the GI. In addition, rice contains high levels of melatonin [38], which may also favor good sleep.

Table 2. PSQI-J component scores by quintiles of rice, bread, and noodle intake as well as dietary glycemic index and glycemic load.

	Q1 (lowest)	Q2	Q3 (middle)	Q4	Q5 (highest)	P value[a]
Rice						
Sleep duration (hours)	6.2 ± 0.9	6.3 ± 0.8	6.4 ± 0.8	6.4 ± 0.9	6.4 ± 0.8	0.008
PSQI-J						
Subjective sleep quality	1.26 ± 0.69	1.16 ± 0.64	1.18 ± 0.68	1.15 ± 0.64	1.08 ± 0.63	0.063
Sleep latency	0.72 ± 0.77	0.70 ± 0.74	0.70 ± 0.77	0.67 ± 0.74	0.67 ± 0.75	0.630
Sleep duration	1.04 ± 0.84	0.91 ± 0.74	0.81 ± 0.70	0.84 ± 0.75	0.79 ± 0.73	0.003
Habitual sleep efficiency	0.13 ± 0.50	0.06 ± 0.25	0.07 ± 0.31	0.08 ± 0.34	0.09 ± 0.40	0.787
Sleep disturbances	0.93 ± 0.51	0.94 ± 0.51	0.90 ± 0.50	0.89 ± 0.49	0.86 ± 0.51	0.290
Use of sleep medication	0.08 ± 0.40	0.04 ± 0.26	0.05 ± 0.33	0.06 ± 0.36	0.03 ± 0.29	0.773
Daytime dysfunction	0.99 ± 0.73	0.91 ± 0.67	0.87 ± 0.74	0.96 ± 0.72	0.86 ± 0.71	0.211
Bread						
Sleep duration (hours)	6.4 ± 0.8	6.4 ± 0.8	6.3 ± 0.8	6.3 ± 0.8	6.2 ± 0.8	0.007
PSQI-J						
Subjective sleep quality	1.13 ± 0.69	1.17 ± 0.60	1.14 ± 0.64	1.17 ± 0.65	1.22 ± 0.69	0.407
Sleep latency	0.71 ± 0.76	0.71 ± 0.75	0.64 ± 0.73	0.69 ± 0.73	0.72 ± 0.78	0.570
Sleep duration	0.77 ± 0.75	0.83 ± 0.74	0.90 ± 0.74	0.92 ± 0.78	0.97 ± 0.78	0.003
Habitual sleep efficiency	0.07 ± 0.30	0.10 ± 0.39	0.10 ± 0.37	0.09 ± 0.41	0.08 ± 0.37	0.706
Sleep disturbances	0.88 ± 0.50	0.94 ± 0.52	0.93 ± 0.50	0.89 ± 0.50	0.90 ± 0.51	0.383
Use of sleep medication	0.05 ± 0.33	0.05 ± 0.32	0.05 ± 0.33	0.04 ± 0.29	0.06 ± 0.38	0.905
Daytime dysfunction	0.89 ± 0.72	0.94 ± 0.71	0.97 ± 0.73	0.91 ± 0.71	0.89 ± 0.69	0.491
Noodles						
Sleep duration (hours)	6.38 ± 0.83	6.32 ± 0.81	6.37 ± 0.80	6.30 ± 0.86	6.27 ± 0.82	0.275
PSQI-J						
Subjective sleep quality	1.11 ± 0.66	1.14 ± 0.66	1.13 ± 0.63	1.22 ± 0.68	1.23 ± 0.65	0.021
Sleep latency	0.64 ± 0.72	0.65 ± 0.70	0.67 ± 0.75	0.72 ± 0.79	0.79 ± 0.78	0.049
Sleep duration	0.82 ± 0.75	0.91 ± 0.76	0.82 ± 0.73	0.90 ± 0.77	0.94 ± 0.78	0.114
Habitual sleep efficiency	0.05 ± 0.28	0.07 ± 0.37	0.10 ± 0.39	0.09 ± 0.39	0.12 ± 0.40	0.069
Sleep disturbances	0.81 ± 0.52	0.94 ± 0.51	0.87 ± 0.50	0.95 ± 0.49	0.96 ± 0.50	<0.001
Use of sleep medication	0.02 ± 0.23	0.03 ± 0.27	0.03 ± 0.28	0.06 ± 0.36	0.10 ± 0.46	0.008
Daytime dysfunction	0.83 ± 0.72	0.91 ± 0.70	0.92 ± 0.72	0.90 ± 0.71	1.03 ± 0.70	0.005
Dietary GI						
Sleep duration (hours)	6.3 ± 0.8	6.3 ± 0.8	6.3 ± 0.8	6.4 ± 0.8	6.4 ± 0.8	0.025
PSQI-J						
Subjective sleep quality	1.21 ± 0.68	1.17 ± 0.62	1.16 ± 0.65	1.17 ± 0.70	1.11 ± 0.63	0.279
Sleep latency	0.73 ± 0.76	0.74 ± 0.76	0.64 ± 0.76	0.66 ± 0.74	0.69 ± 0.72	0.382

Table 2. Cont.

	Q1 (lowest)	Q2	Q3 (middle)	Q4	Q5 (highest)	P value[a]
Sleep duration	0.98 ± 0.78	0.91 ± 0.78	0.88 ± 0.74	0.84 ± 0.76	0.79 ± 0.73	0.013
Habitual sleep efficiency	0.09 ± 0.38	0.10 ± 0.42	0.09 ± 0.36	0.08 ± 0.34	0.07 ± 0.36	0.923
Sleep disturbances	0.93 ± 0.51	0.94 ± 0.50	0.88 ± 0.51	0.91 ± 0.50	0.87 ± 0.50	0.204
Use of sleep medication	0.06 ± 0.32	0.09 ± 0.45	0.03 ± 0.29	0.05 ± 0.35	0.02 ± 0.21	0.065
Daytime dysfunction	0.96 ± 0.71	0.91 ± 0.71	0.88 ± 0.71	0.94 ± 0.72	0.89 ± 0.71	0.561
Dietary GL						
Sleep duration (hours)	6.3 ± 0.8	6.3 ± 0.8	6.3 ± 0.8	6.4 ± 0.8	6.4 ± 0.8	0.067
PSQI-J						
Subjective sleep quality	1.15 ± 0.66	1.19 ± 0.66	1.13 ± 0.62	1.15 ± 0.64	1.20 ± 0.71	0.515
Sleep latency	0.71 ± 0.75	0.76 ± 0.77	0.62 ± 0.74	0.67 ± 0.73	0.71 ± 0.77	0.159
Sleep duration	0.94 ± 0.81	0.91 ± 0.74	0.87 ± 0.72	0.82 ± 0.76	0.85 ± 0.77	0.241
Habitual sleep efficiency	0.12 ± 0.46	0.09 ± 0.39	0.07 ± 0.34	0.09 ± 0.35	0.07 ± 0.30	0.370
Sleep disturbances	0.87 ± 0.50	0.92 ± 0.48	0.86 ± 0.52	0.96 ± 0.48	0.91 ± 0.53	0.041
Use of sleep medication	0.04 ± 0.33	0.07 ± 0.37	0.05 ± 0.32	0.06 ± 0.40	0.02 ± 0.20	0.388
Daytime dysfunction	0.94 ± 0.72	0.88 ± 0.70	0.89 ± 0.71	0.92 ± 0.71	0.96 ± 0.73	0.537

PSQI-J, the Japanese version of the Pittsburgh Sleep Quality Index; GI, glycemic index; GL, glycemic load; Q: quintile.
[a]Linear regression analyses were used to assess the linear trends between sleep duration and each PSQI-J component score across the quintiles of starchy food intake, dietary GI, and dietary GL by using the median value of each quintile.

Table 3. ORs for the prevalence of poor sleep[a] in each quintile of dietary rice, bread, and noodle intake as well as dietary glycemic index and glycemic load.

	Q1 (lowest)	Q2	Q3 (middle)	Q4	Q5 (highest)	P[b] value	Continuous[c] (1 SD increment)	P value
Rice								
Prevalence of poor sleep (%)	39.0	30.7	28.0	28.0	26.0	<0.001		
Age-, sex-adjusted OR	1	0.69	0.60	0.59	0.54	0.001	0.85	0.010
(95%CI)	(reference)	(0.51, 0.94)	(0.44, 0.82)	(0.43, 0.81)	(0.39, 0.74)		(0.76, 0.96)	
Multivariate-adjusted OR[d]	1	0.68	0.61	0.59	0.54	0.015	0.87	0.045
(95%CI)	(reference)	(0.49, 0.93)	(0.43, 0.85)	(0.42, 0.85)	(0.37, 0.81)		(0.76, 1.00)	
Bread								
Prevalence of poor sleep (%)	27.7	30.1	29.5	31.3	33.3	0.284		
Age-, sex-adjusted OR	1	1.13	1.10	1.19	1.31	0.545	1.06	0.312
(95%CI)	(reference)	(0.82, 1.55)	(0.80, 1.51)	(0.87, 1.63)	(0.96, 1.78)		(0.95, 1.19)	
Multivariate-adjusted OR[d]	1	1.14	1.04	1.05	1.14	0.921	1.01	0.885
(95%CI)	(reference)	(0.81, 1.60)	(0.74, 1.47)	(0.74, 1.50)	(0.79, 1.63)		(0.89, 1.14)	
Noodle								
Prevalence of poor sleep (%)	24.6	30.5	26.4	31.3	39.0	<0.001		
Age-, sex-adjusted OR	1	1.35	1.10	1.38	1.95	<0.001	1.21	0.001
(95%CI)	(reference)	(0.97, 1.87)	(0.79, 1.53)	(1.00, 1.91)	(1.42, 2.67)		(1.09, 1.35)	
Multivariate-adjusted OR[d]	1	1.25	1.05	1.31	1.82	0.002	1.21	<0.001
(95%CI)	(reference)	(0.90, 1.74)	(0.75, 1.47)	(0.94, 1.82)	(1.31, 2.51)		(1.09, 1.35)	
Dietary GI								
Prevalence of poor sleep (%)	35.4	34.3	24.7	28.9	28.5	0.016		
Age-, sex-adjusted OR	1	0.96	0.59	0.73	0.72	0.006	0.82	<0.001
(95%CI)	(reference)	(0.71, 1.30)	(0.43, 0.81)	(0.54, 1.00)	(0.52, 0.99)		(0.74, 0.91)	
Multivariate-adjusted OR[d]	1	0.96	0.60	0.77	0.77	0.020	0.85	0.006
(95%CI)	(reference)	(0.71, 1.31)	(0.44, 0.84)	(0.56, 1.07)	(0.55, 1.07)		(0.76, 0.95)	
Dietary GL								
Prevalence of poor sleep (%)	31.1	30.0	25.8	31.1	33.9	0.379		
Age-, sex-adjusted OR	1	0.96	0.78	1.02	1.14	0.215	1.02	0.721
(95%CI)	(reference)	(0.70, 1.32)	(0.56, 1.08)	(0.74, 1.40)	(0.83, 1.57)		(0.91, 1.14)	
Multivariate-adjusted OR[d]	1	0.98	0.79	1.09	1.27	0.092	1.07	0.284
(95%CI)	(reference)	(0.71, 1.35)	(0.56, 1.11)	(0.77, 1.54)	(0.88, 1.82)		(0.95, 1.21)	

GI, glycemic index; GL, glycemic load, CI, confidence interval; SD, standard deviation; OR, odds ratio; PSQI-J, Japanese version of the Pittsburgh Sleep Quality Index; Q, quintile;

[a]A PSQI-J global score >5.5 indicate poor sleep.

[b]The χ^2 test was used to analyze the prevalence of poor sleep, and logistic regression analysis was used to assess the linear trends of ORs by using the median value of each quintile.

[c]Differences of SD for rice, bread, noodles, dietary GI, and dietary GL were 69.1 g/1,000 kcal, 26.2 g/1,000 kcal, 32.8 g/1,000 kcal, 3.9, and 43.8/1,000 kcal, respectively.

[d]Multivariate models included age (continuous), sex (continuous), Body mass index (kg/m², continuous), smoking status (i.e., current, previous, or never; dummy variable), alcohol consumption (percentage of energy; continuous), frequency of breakfast consumption (i.e., 0–3, 4–6, or 7 days/week; dummy variable), habitual exercise (MET-h/week; continuous), rice intake (for the multivariate analyses of bread and noodles; continuous), bread intake (for the multivariate analyses of rice and noodles; continuous), and noodle intake (for the multivariate analyses of rice and bread; continuous).

In contrast, bread intake was not significantly associated with sleep quality, whereas noodle intake was significantly associated with poor sleep. The GIs of the breads and noodles used in the analyses ranged from 51 to 74 and 46 to 47, respectively; these values are lower than 77, which is the GI of Japanese white rice [11]. In a previous study, the TRP/LNAA ratio increased by 17% after a mixed-macronutrient meal with a high-GI (GI = 70) and was higher than that after a mixed-macronutrient meal with a low-GI (GI of 50; 8% increase), even though the amount of carbohydrates was the same in both meals (66.5% energy) [37]. Although noodles are major starchy foods, the GI of noodles is too low to increase the postprandial TRP/LNAA ratio. Furthermore, noodle intake was inversely associated with rice intake. In the present study, higher noodle intake was associated with poor sleep even after adjusting for rice intake. However, it is possible that the adjustment using statistical models is insufficient; thus, the low sleep quality of subjects with higher noodle intake may be due to a lower rice intake. An interventional study is required to investigate the differences in the associations of starchy foods with sleep quality.

Shorter sleep duration is associated with a relative increase and decrease in calories derived from fat and carbohydrates, respectively [39]. In Japan, breakfast often consists of foods low in fat but high in carbohydrates and fiber. Japanese people who eat breakfast generally consume more rice [40]. People with good sleep quality tend to eat breakfast, which may affect the association between rice intake and sleep quality. However, in the present study, lower rice intake was significantly associated with poor sleep even after adjusting for the frequency of breakfast consumption.

In this study, diets with high rice intake and a high GI were significantly associated with good sleep; however, such diets are also reported to be associated with several health problems including obesity, diabetes mellitus, cardiovascular disease, and some cancers [11–16,41–43]. Furthermore, obesity induced by the long-term consumption of a diet with a high-GI may cause sleep apnea syndrome, which may also affect sleep quality. Accordingly, the association between the long-term consumption of meals with a high-GI and sleep quality should be analyzed in greater detail.

One of the strengths of the present study is that we examined rice consumption in a large Japanese population. Japanese people consume approximately 10 times more rice than European and North American people [44]. Thus, it is important to evaluate the association between rice and sleep in people with high rice consumption. Further, this is the first study to investigate the association between rice, bread, and noodle consumption and sleep quality. In addition, the GI and GL were calculated by using

responses to a validated questionnaire [21]. Nevertheless, the present study also has several limitations. First, we restricted the final study population to white-collar workers; white-collar work is reported to be strongly correlated with poor sleep quality [45]. Compared with the general Japanese population, the study participants had a similar mean PSQI-J global score but shorter mean sleep duration [46]. Therefore, the results of this study cannot necessarily be generalized to the overall Japanese population. Second, sleep quality is reported to be related to physiological actions and eating behaviors such as skipping meals, eating speed, and watching television during meals [47]; we did not have data on these variables. Third, the DHQ and PSQI-J were evaluated approximately 1 year apart. However, lifestyle factors such as dietary habits and sleep quality are unlikely to change much in 1 year among steadily employed middle-aged people. Fourth, women are reported to have difficulty sleeping at the beginning and end of their menstrual cycle [48,49]; in the present study, we did not obtain data on the menstrual cycles of the female participants. However, when we compared the PSQI-J component and global scores of men and women, nearly identical trends were observed (data not shown). Fifth, the dietary GI and rice consumption data used in this study were daily values, not evening values. As mentioned in the preceding text, the consumption of a meal with a high-GI within 4 hours of going to bed may be an effective way of facilitating sleep [10]. In the present study, the dietary GI and rice intake were significantly associated with sleep duration but not sleep latency. However, dietary intake at dinner may be more closely associated with sleep quality.

In conclusion, the present study indicate that high consumption of rice and a high dietary GI are associated with good sleep, especially good sleep duration. Meanwhile, higher noodle consumption is associated with poor sleep quality. The effects of starchy foods on sleep may differ according to their GI values. Diets with a high-GI, especially those with high rice intake, may contribute to good sleep. Nevertheless, further interventional studies are required to determine appropriate carbohydrate intake during the evening meal to facilitate good sleep.

Author Contributions

Conceived and designed the experiments: SY MS K. Nakamura YM KM MN HN. Performed the experiments: MS K. Nakamura YM KM MN KY MI TK YN HN. Analyzed the data: SY. Contributed reagents/materials/analysis tools: SY MS K. Nakamura YM KM MN KY MI TK YN K. Nogawa YS SS HN. Wrote the paper: SY MS.

References

1. Buysse DJ, Reynolds CF III, Monk TH, Berman SR, Kupfer DJ (1989) The Pittsburgh Sleep Quality Index: a new instrument for psychiatric practice and research. Psychiatry Res 28: 193–213.
2. Bixler E (2009) Sleep and society: an epidemiological perspective. Sleep Med 10: S3–S6.
3. Hublin C, Parlinen M, Koskenvuo M, Kaprio J (2007) Sleep and mortality: a population-based 22-year follow-up study. Sleep 30: 1245–1253.
4. Banks S, Dinges DF (2007) Behavioral and physiological consequences of sleep restriction. J Clin Sleep Med 3: 519–528.
5. Hall MH, Muldoon MF, Jennings JR, Buysse DJ, Flory JD, et al. (2008) Self-reported sleep duration is associated with the metabolic syndrome in midlife adults. Sleep 31: 635–643.
6. Gangwisch JE, Heymsfield SB, Boden-Albala B, Bujs RM, Kreier F, et al. (2007) Sleep duration as a risk factor for diabetes incidence in a large US sample. Sleep 30: 1667–1673.
7. Gangwisch JE, Heymsfield SB, Boden-Albala B, Buijs RM, Kreier F, et al. (2006) Short sleep duration as a risk factor for hypertension. Analyses of the first national health and nutrition examination survey. Hypertension 47: 833–839.
8. Magee CA, Caputi P, Iverson DC (2011) Relationships between self-rated health, quality of life and sleep duration in middle aged and elderly Australians. Sleep Med 12: 346–350.
9. Diethelm K, Remer T, Jilani H, Kunz C, Buyken AE (2011) Associations between the macronutrient composition of the evening meal and average daily sleep duration in early childhood. Clin Nutr 30: 640–646.
10. Afaghi A, O'Connor H, Chow CM (2007) High-glycemic-index carbohydrate meals shorten sleep onset. Am J Clin Nutr 85: 426–430.
11. Murakami K, Sasaki S, Takahashi Y, Okubo H, Hosoi Y, et al. (2006) Dietary glycemic index and load in relation to metabolic risk factors in Japanese female farmers with traditional dietary habits. Am J Clin Nutr 83: 1161–1169.
12. Nakashima M, Sakurai M, Nakamura K, Miura K, Yoshita K, et al. (2010) Dietary glycemic index, glycemic load and blood lipid levels in middle-aged Japanese men and women. J Atheroscler Thromb 17: 1082–1095.
13. Sakurai M, Nakamura K, Miura K, Takamura T, Yoshita K, et al. (2012) Dietary glycemic index and risk of type 2 diabetes mellitus in middle-aged Japanese men. Metabolism 61: 47–55.
14. Willett W, Manson J, Liu S (2002) Glycemic index, glycemic load, and risk of type 2 diabetes. Am J Clin Nutr 76: 274S–280S.

15. Barclay AW, Petocz P, McMillan-Price J, Flood VM, Prvan T, et al. (2008) Glycemic index, glycemic load, and chronic disease risk–a meta-analysis of observational studies. Am J Clin Nutr 87: 627–637.
16. Krishnan S, Rosenberg L, Singer M, Hu FB, Djousse L, et al. (2007) Glycemic index, glycemic load, and cereal fiber intake and risk of type 2 diabetes in US black women. Arch Intern Med 167: 2304–2309.
17. Office for Life-style Related Diseases Control, General Affairs Division, Health Service Bureau, Ministry of Health, Labour and Welfare (2013) The National Health and Nutrition Survey in Japan, 2011. (in Japanese). Ministry of Health, Labour and Welfare.
18. Sasaki S, Yanagibori R, Amano K (1988) Self-administered diet history questionnaire developed for health education: a relative validation of the test-version by comparison with 3-day diet record in women. J Epidemiol 8: 203–215.
19. Sasaki S, Ushio F, Amano K, Morihara M, Todoroki O, et al. (2000) Serum biomarker-based validation of a self-administered diet history questionnaire for Japanese subjects. J Nutr Sci Vitaminol (Tokyo) 46: 285–296.
20. Science and Technology Agency, Japan (2000) Standard Tables of Food Composition in Japan, 5th ed. Tokyo: Printing Bureau of the Ministry of Finance. [In Japanese.].
21. Murakami K, Sasaki S, Takahashi Y, Okubo H, Hirota N, et al. (2008) Reproducibility and relative validity of dietary glycemic index and load assessed with a self-administered diet-history questionnaire in Japanese adults. Br J Nutr 99: 639–648.
22. Foster-Powell K, Holt SH, Brand-Miller JC (2002) International table of glycemic index and glycemic load values. Am J Clin Nutr 76: 5–56.
23. Sugiyama M, Tang AC, Wakaki Y, Koyama W (2003) Glycemic index of single and mixed meal foods among common Japanese foods with white rice as a reference food. Eur J Clin Nutr 57: 743–752.
24. Fernandes G, Velangi A, Wolever TM (2005) Glycemic index of potatoes commonly consumed in North America. J Am Diet Assoc 105: 557–562.
25. Sydney University Glycemic Index Research Service (2007) The official website of glycemic index and GI database. Available: http://www.glycemicindex.com. Accessed February 1, 2007.
26. Henry CJ, Lightowler HJ, Strik CM, Renton H, Hails S (2005) Glycaemic index and glycaemic load values of commercially available products in the UK. Br J Nutr 94: 922–930.
27. Coulston AM, Hollenbeck CB, Swislocki AL, Reaven GM (1987) Effect of source of dietary carbohydrate on plasma glucose and insulin responses to mixed meals in subjects with NIDDM. Diabetes Care 10: 395–400.
28. Wolever TM, Jenkins DJ, Jenkins AL, Josse RG (1991) The glycemic index: methodology and clinical implications. Am J Clin Nutr 54: 846–854.
29. Doi Y, Minowa M, Uchiyama M, Okawa M, Kim K, et al. (2000) Psychometric assessment of subjective sleep quality using the Japanese version of the Pittsburgh Sleep Quality Index (PSQI-J) in psychiatric disordered and control subjects. Psychiatry Res 97: 165–172.
30. Fernstrom JD, Wurtman RJ (1971) Brain serotonin content: physiological dependence on plasma tryptophan levels. Science 173: 149–152.
31. Fernstrom JD, Wurtman RJ (1972) Brain serotonin content: physiological regulation by plasma neutral amino acids. Science 178: 414–416.
32. Madras BK, Cohen EL, Fernstrom JD, Larin F, Munro HN, et al. (1973) Letter: Dietary carbohydrate increases brain tryptophan and decreases free plasma tryptophan. Nature 244(5410): 34–35.
33. Hartmann E, Spinweber CL (1979) Sleep induced by L-tryptophan. Effect of dosages within the normal dietary intake. J Nerv Ment Dis 167: 497–499.
34. Wurtman RJ, Wurtman JJ, Regan MM, McDermott JM, Tsay RH, et al. (2003) Effects of normal meals rich in carbohydrates or proteins on plasma tryptophan and tyrosine ratios. Am J Clin Nutr 77: 128–132.
35. Berry EM, Growdon JH, Wurtman JJ, Caballero B, Wurtman RJ (1991) A balanced carbohydrate: protein diet in the management of Parkinson's disease. Neurology 41: 1295–1297.
36. Lyons PM, Truswell AS (1988) Serotonin precursor influenced by type of carbohydrate meal in healthy adults. Am J Clin Nutr 47: 433–439.
37. Herrera CP, Smith K, Atkinson F, Ruell P, Chow CM, et al. (2011) High-glycemic index and -glycaemic load meals increase the availability of tryptophan in healthy volunteers. Brit J Nutr 105: 1601–1606.
38. Badria FA (2002) Melatonin, serotonin, and tryptamine in some Egyptian food and medicinal plants. J Med Food 5: 153–157.
39. Weiss A, Xu F, Storfer-lsser A, Thomas A, Ievers-Landis CE, et al. (2010) The association of sleep duration with adolescents' fat and carbohydrate consumption. Sleep 33: 1201–1209.
40. Taniguchi A, Yamanaka-Okumura H, Nishida Y, Yamamoto H, Kaketani Y, et al. (2008) Natto and viscous vegetables in a Japanese style meal suppress postprandial glucose and insulin response. Asia Pac J Clin Nutr 17: 663–668.
41. Esfahani A, Wong JM, Mirrahimi A, Srichaikul K, Jenkins DJ, et al. (2009) The glycemic index: physiological significance. J Am Coll Nutr Suppl 28: 439S–445S.
42. Nanri A, Mizoue T, Noda M, Takahashi Y, Kato M, et al. (2010) Rice intake and type 2 diabetes in Japanese men and women: the Japan Public Health Center-based Prospective Study. Am J Clin Nutr 92: 1468–1477.
43. Eshak ES, Iso H, Date C, Yamagishi K, Kikuchi S, et al. (2011) Rice intake is associated with reduced risk of mortality from cardiovascular disease in Japanese men but not women. J Nutr 141: 595–602.
44. Food and Agriculture Organization of the United Nations (FAOSTAT) (2014) Available: http://faostat.fao.org/#. Accessed March 20, 2014.
45. Doi Y, Minowa M, Tango T (2003) Impact and correlates of poor sleep quality in Japanese white-collar employees. Sleep 26: 467–471.
46. Doi Y, Minowa M, Uchiyama M, Okawa M (2001) Subjective sleep quality and sleep problems in the general Japanese adult population. Pshychiatry Clin Neurosci 3: 213–215.
47. Sato-Mito N, Sasaki S, Murakami K, Okubo H, Takahashi Y, et al. (2011) The midpoint of sleep is associated with dietary intake and dietary behavior among young Japanese women. Sleep Med 12: 289–294.
48. Kravitz HM, Janssen I, Santoro N, Bromberger JT, Schocken M, et al. (2005) Relationship of day-to-day reproductive hormone levels to sleep in midlife women. Arch Intern Med 165: 2370–2376.
49. Baker FC, Driver HS (2004) Self-reported sleep across the menstrual cycle in young, healthy women. J Psychosom Res 56: 239–243.

4

Virus-Mediated Chemical Changes in Rice Plants Impact the Relationship between Non-Vector Planthopper *Nilaparvata lugens* Stål and Its Egg Parasitoid *Anagrus nilaparvatae* Pang et Wang

Xiaochan He[1,2][9], Hongxing Xu[1][9], Guanchun Gao[3], Xiaojun Zhou[2], Xusong Zheng[1], Yujian Sun[2], Yajun Yang[1], Junce Tian[1], Zhongxian Lu[1]*

1 State Key Laboratory Breeding Base for Zhejiang Sustainable Pest and Disease Control, Institute of Plant Protection and Microbiology, Zhejiang Academy of Agriculture Sciences, Hangzhou, China, 2 Jinhua Research Academy of Agricultural Sciences, Jinhua, China, 3 School of Medicine Science, Jiaxing University, Jiaxing, China

Abstract

In order to clarify the impacts of southern rice black-streaked dwarf virus (SRBSDV) infection on rice plants, rice planthoppers and natural enemies, differences in nutrients and volatile secondary metabolites between infected and healthy rice plants were examined. Furthermore, the impacts of virus-mediated changes in plants on the population growth of non-vector brown planthopper (BPH), *Nilaparvata lugens*, and the selectivity and parasitic capability of planthopper egg parasitoid *Anagrus nilaparvatae* were studied. The results showed that rice plants had no significant changes in amino acid and soluble sugar contents after SRBSDV infection, and SRBSDV-infected plants had no significant effect on population growth of non-vector BPH. *A. nilaparvatae* preferred BPH eggs both in infected and healthy rice plants, and tended to parasitize eggs on infected plants, but it had no significant preference for infected plants or healthy plants. GC-MS analysis showed that tridecylic aldehyde occurred only in rice plants infected with SRBSDV, whereas octanal, undecane, methyl salicylate and hexadecane occurred only in healthy rice plants. However, in tests of behavioral responses to these five volatile substances using a Y-tube olfactometer, *A. nilaparvatae* did not show obvious selectivity between single volatile substances at different concentrations and liquid paraffin in the control group. The parasitic capability of *A. nilaparvatae* did not differ between SRBSDV-infected plants and healthy plant seedlings. The results suggested that SRBSDV-infected plants have no significant impacts on the non-vector planthopper and its egg parasitoid, *A. nilaparvatae*.

Editor: Youjun Zhang, Institute of Vegetables and Flowers, Chinese Academy of Agricultural Science, China

Funding: This study was supported by the Agro-Industry R & D Special Fund of China (Grant No. 201003031), and the National Basic Research Program of China (973, Grant No. 2010CB126202). The funders had no role in study design, data collection and analysis, decision to publish, or preparation of the manuscript.

Competing Interests: The authors have declared that no competing interests exist.

* Email: luzxmh@gmail.com

[9] These authors contributed equally to this work.

Introduction

The multi-trophic relationship involving plants, herbivorous insects, and natural enemies is the most basic component of nearly all ecosystems. Any change in the factors affecting plant growth can change the interactive relationships among the three trophic levels through a variety of mechanisms [1]. Plant viruses transmitted by arthropods are an important biological factor in agro-ecosystems. Viruses can affect not only the yield and quality of host plants, but also the growth, physiological and biochemical changes as well as the ecological characteristics of arthropods serving as the vector. Furthermore, they can have direct or indirect effects on non-vector herbivorous arthropods and their natural enemies, potentially impacting entire agro-ecosystems [2–4].

Viral infection in plants can cause changes in nutrient components of host plants. After being infected with rice black-streaked dwarf virus (RBSDV), diseased rice plants had a 31.13% increase in free amino acid levels, and soluble sugar content was three times higher than that in healthy rice plants [5]. The total amino acid content in the phloem of wheat (*Triticum aestivum*) declined after infection with barley yellow dwarf virus (BYDV) [6]. Plant secondary compounds are key information factors linking trophic relationships among plants, pests and natural enemies [7,8]. After infection, the type and content of volatile substances in host plants changed, which in turn affected the behaviors of herbivorous arthropods and their natural enemies [9]. Mexican bean beetles (*Epilachna varivestis*) tended to choose those tissues infected with southern bean mosaic virus (SBMV) or bean pod mottle virus (BPMV), which were both caused by plant secondary compounds under viral infection [10]. Squash (*Cucurbita pepo*), infected with cucumber mosaic virus (CMV), exhibited a significant increase in the content of volatile secondary metabolites, and its attraction to aphis (*Myzus persicae*) and its parasitoid (*Aphid gossypii*) was strongly enhanced [11]. Many studies have focused on the interaction between viruses and host plants [6,12,13], and between viruses and vector insects [10,14–16].

However, only a small number of studies have included the three trophic levels, consisting of viruses, insects (especially non-vector insects) and natural enemies [11].

The southern rice black-streaked dwarf virus (SRBSDV) was first discovered in Guangdong China in 2001 as a new rice virus transmitted by rice whitebacked planthopper (WBPH), *Sogatella furcifera* (Horváth) [17,18]. In 2011 and 2012, its total distribution areas in China and Vietnam were 700,000 and 500,000 hectares, respectively [19,20]. Currently, the virus is also found in Japan [21]. It was reported that the viruliferous WBPH laid significantly fewer eggs than non-viruliferous hoppers. There were no significant differences in the hatchability of eggs laid by virulifierous and non viruliferous females [22]. This study using paired viruliferous and nonviruliferous WBPH showed that both infected females and males had significantly reduced fecundity and F1 egg hatchability. When paired with either a non viruliferous female or male, there were no significant effects in fecundity and egg hatchability [23]. In paddy fields, the brown planthopper (BPH), *Nilaparvata lugens*, normally coexists with WBPH and always shares its host rice plants with WBPH in east China [24,25]; however, BPH is not the vector of SRBSDV. To further understand the ecological impacts of the plant virus, this study investigated the effects of the changes in nutrients and volatile secondary metabolites of host plants after SRBSDV infection on the population growth of non-vector brown planthopper (BPH), *Nilaparvata lugens*, as well as the selectivity and parasitic capability of its egg parasitoid *Anagrus nilaparvatae*. We sought to clarify the wider ecological role of the plant virus in each trophic level of rice plants and to provide a solid basis for better prevention and sustainable management of rice planthoppers during SRBSDV epidemics.

Materials and Methods

Rice plants

The rice variety for rearing BPH and WBPH, susceptible TN1, was provided by the International Rice Research Institute (IRRI), the Philippines. After indoor seed germination, seedlings were planted in a cement sink in a netted room free of insects. At 3-leaf stage, seedlings were transplanted into pots (diameter 9 cm) in the netted room. 45–60 d old plants were fed to BPH and WBPH.

Rice variety Y-Liangyou 1 was used in experiment. It is a rice variety susceptible to SRBSDV and the dominant *indica* hybrid rice in Zhejiang province, China.

Rice planthoppers

Vector WBPH and non-vector BPH were collected from paddy fields of the China National Rice Research Institute (CNRRI), Hangzhou (119.95°E, 30.07°N). Dr. Fu Qiang (fuqiang@caas.cn) of CNRRI should be contacted for future permissions and there are no endangered or protected species involved. The planthoppers were maintained continuously on susceptible rice TN1 in an artificial climate chamber under the conditions of 26±1°C, 70–90% relative humidity, and L12:D12.

Parasitoid

The egg parasitoid of rice planthoppers, *A. nilaparvatae* was trapped in the rice fields and bred continuously indoors. Gravid BPH female adults were caged for oviposition on potted rice plants for 24 h. After removal of the cage and BPH, the potted plants with BPH eggs were transferred to paddy fields in the experimental farm of the Zhejiang Academy of Agricultural Sciences, Hangzhou (120.18° E, 30.27° N). After exposing them in thefield for 48 h, plants with parasitized eggs were collected and returned

to the artificial climate chamber under the conditions of 26±1°C, 70–90% relative humidity, and L12:D12. Each potted rice plant was covered with a polyethylene cage (height 60 cm, diameter 9 cm) and an outer black cloth. Its top had an opening of 1 cm in diameter and was connected to an inverted transparent glass tube (diameter 1 cm, length 6 cm). After the emergence of *A. nilaparvatae*, the glass tubes were replaced daily. In addition, after sexual identification, *A. nilaparvatae* in the tubes were placed into cages containing rice plants with fresh BPH eggs in several batches. The next generation of *A. nilaparvatae* was used for testing.

Infected and healthy rice plants

Rice variety Y-Liangyou 1 was used in this experiment. It is a rice variety susceptible to SRBSDV and the dominant *indica* hybrid rice in Wuyi county (119.81°E, 28.9°N), Zhejiang province, China. The area has been subject to frequent SRBSDV outbreaks in recent years [26]. To obtain SRBSDV-infected plants, 2nd instar nymphs of WBPH were placed in a beaker padded with wet filter paper to make them hungry. After 2 h without feeding, they were transferred to SRBSDV-infected plants (Y-Liangyou 1) collected from a paddy field in Wuyi county for 2–3 d. They were then transferred and bred on healthy TN1 seedlings at the tillering stage for a circulative period of one week. Lastly, they were inoculated to Y-Liangyou 1 plants at 3-leaf stage [27,28]. After the appearance of the typical SRBSDV-infected symptoms on the rice plants about 40 d old, the plants were marked and the leaves were individually sampled and molecularly identified by the methods of Li et al. (2012) [29]. 45–60 d old healthy and infected plants were used for testing.

Chemicals used in the experiment

Standard samples of volatiles, including octanal (purity≥99%), methyl salicylate (analytical standard), undecane (analytical standard), hexadecane (analytical standard), and tridecanal (purity≥95%) were purchased from the Sigma-Aldrich Corporation. Using liquid paraffin as a solvent, solutions diluted 10^2, 10^4 and 10^6 times were prepared for testing. Internal standards n-octane and nonyl acetate were purchased from Sigma-Aldrich; 400 ng of octane and 400 ng of nonyl acetate were weighed and mixed with 20 μL hexane as the internal standard.

Determination of amino acid content in rice plants

Leaf sheaths of SRBSDV-infected 60 d old rice plants and corresponding healthy rice plants were sampled after RT-PCR determination. After de-enzyming for 1 h at 110 °C, they were dried at 80 °C to constant weight. One gram dried leaf was ground into powder, and 0.1% HCl was added to the volume of 25 ml. The solution was filtered after complete dissolution; 2 ml of supernatant was mixed with 4 ml 0.1% TFA solution by rapid shaking. After purification with a SEP-PAK column, the solution was loaded onto an amino acid analyzer (Sykam S433D) for determination of amino acid content.

Determination of soluble sugar content in rice plants

Soluble sugar content in rice plants was determined following the methods of Wang et al. (2010). Leaf sheaths from SRBSDV-infected plants and from corresponding healthy rice plants were sampled and dried to constant weight and ground into powder. About 0.1 g dried powder was placed into a large test tube, into which 15 ml of distilled water was added. After 20 min of boiling in a water bath, the sample was cooled and filtered into a 100 ml volumetric flask. The residue was washed with distilled water several times to constant volume; 1.0 ml of sample extract was added to 5 ml of anthrone reagent. After rapid shaking and mixing of each tube, they were heated in a boiling water bath for 10 min, and cooled afterwards for measurement of OD_{620}. AR anhydrous glucose was plotted as the standard curve to calculate soluble sugar content of the sample.

Extraction and analysis of plant volatiles

Instruments and materials included a GCMS-QP2010 gas chromatograph - mass spectrometer (Shimadzu Corporation), a solid phase micro extraction (SPME) device (Supelco Inc. USA), an extraction head (Polyacrylate (PA) 85 μm, polydimethylsiloxane (PDMS) 100 μm, polydimethylsiloxane/divinybenzene (PDMS/DVB) 65 μm), and a capillary column Rtx-5MS (0.25 μm×30.0 m×0.25 mm).

Plants infected with SRBSDV and healthy plants of the same age were put into the adsorption device. The roots were wrapped with freshness-preserving film to prevent dirt and other air flow from mixing with the rice volatiles. Air from a blower was purified by passing it through distilled water and activated carbon. Afterwards, it entered a glass cylinder (diameter 10 cm, height 50 cm) from the top at 800 ml/min. After passing through the entire cylinder, it entered an adsorption column from the lower side at 600 ml/min. The inflowing air was greater than the outflowing air, which ensured that the air filled the entire glass cylinder. The adsorption column was connected to a pump, and the volatiles were continuously collected for 4 h (10:00–14:00).

After 4 h of collection, the adsorption column was removed and rinsed with 800 μL hexane. The eluent was put into a 1500 μL storage vial and added to 10 μL of internal standard (200 ng n-octane and 200 ng nonyl acetate were added to 20 μL n-hexane). The measurement was made after mixing evenly.

A micro-injector was used to load 1 μL of sample; a gas chromatograph - mass spectrometry (GC-MS) was used for analysis. Tests were conducted at 26±1°C and 60% relative humidity.

The components were analyzed qualitatively based on the degree of agreement between the mass spectrum of components in the sample and the spectrum generated by GC-MS ChemStation software as well as the degree of agreement between the retention time of the components in the sample and that of the standard compounds in GC-MS. Relative quantification was performed based on the area ratio of component peaks in the sample to the internal standard peak.

For chromatography, a capillary column (Rtx-5MS; 30 m×0.25 mm×0.25 μm) was used with carrier gas helium at 24 cm/sec (99.999%) and an initial column temperature of 40°C, which was maintained for 3 min, then increased to 230°C at 8°C/min, and maintained for 9.5 min. Splitless injection was performed.

For mass spectrometry an EI ion source was used with an ionization energy of 70 eV, an ion source temperature of 200°C, an inlet temperature of 250°C, and a mass scan range of m/z 45~500. A library search was conducted using NIST08.L and NIST08s.L.

Population growth of non-vector BPH on SRBSDV-infected and healthy plants

SRBSDV-infected plants and healthy rice plants were uprooted and separated into single plants. The outer sheathes and the inactive roots were removed. Afterwards, they were rinsed with tap water and placed in tubes (diameter 1.5 cm, height 15.0 cm) containing 1.5 cm deep Kimura B nutrient solution. Ten newly hatched nymphs within 24 h of BPH were inoculated into each tube, which were then sealed with degreased cotton. This procedure was repeated ten times for infected and healthy plants, respectively. Their growth and development was observed daily. The rice plants were replaced as needed until planthopper emergence. Within 12 h after adult emergence, a pair of BPH adults were inoculated into tubes containing either healthy or infected plants. They were then put into a biochemical incubator under conditions of 26±1 °C and 12 L: 12D, where they were allowed to mate and lay eggs. When the nymphs hatched, their numbers were counted each day and these nymphs were removed until no more nymphs hatched for 5 consecutive days. Afterwards, rice plants were dissected under a microscope. The number of unhatched eggs was recorded. Based on the number of hatched nymphs and the number of unhatched eggs, the hatching rate of eggs was calculated.

Host selectivity of A. nilaparvatae for BPH eggs in SRBSDV-infected and healthy plants

A single SRBSDV-infected plant and a healthy rice plant of the same age were transferred into the same clay pot covered with a cage. After mating, three gravid BPH female adults for each plant were introduced into the cages for 2 d. After the BPH were removed, two pairs of newly emerged A. nilaparvatae were inoculated, and the parasitoids were removed after 24 h. After 5 d, the numbers of parasitized eggs and healthy eggs were counted by dissecting plants under a microscope. This experiment was conducted inan artificial climate chamber under the conditions of 26±1 °C, 70–90% relative humidity, and L12:D12. Replications were conducted ten times for each treatment.

Preference of A. nilaparvatae for odor sources from SRBSDV-infected and healthy plants

The behavioral responses of A. nilaparvatae to different odor sources (as described in "Extraction and analysis of plant volatiles") from SRBSDV-infected and healthy plants containing BPH eggs were measured using a Y-tube olfactometer. Lengths of the two arms and the straight tube of the Y-tube olfactometer were 10.0 cm, the inner diameter was 1.0 cm, and the inclusion angle between the two arms was 75°. An open ampoule holding 1 ml of odor source solution was placed inside the odor source bottle, and another opening ampoule holding 1 ml of liquid paraffin was placed in the control bottle. The two arms of the Y tube were in turn connected to the odor source bottle (or the control bottle), humidification bottle, air filter (with activated carbon) and flow meter by a Teflon tube, and the base of the Y tube was connected to a pump for air injection. During the measurement of behavior, air was pumped from the base of the Y tube. The flow rate of air in the two arms was regulated at 150 ml/min. When the Y tube was filled with a volatile odor source, one newly emerged A. nilaparvatae was introduced into the Y tube through its base port. Each A. nilaparvatae that entered one arm of the Y tube and moved upwind more than 5 cm was counted; otherwise it was not counted if there was no response after more than 10 min. For each odor source at each concentration level, 60 A. nilaparvatae were measured. For every 10 insects measured, anhydrous ethanol was used to wash and dry the tube. Afterwards, the two arms were swapped, and the connecting positions at the odor source bottles as well as CK bottles were adjusted to eliminate the possible effects of slight geometric differences in the two arms on the behavior of A. nilaparvatae. After finishing the determination with each concentration of each odor source, the Y tube and odor source bottles were washed, and then dried in an oven at 120 °C. The activated carbon for air filtering was heated in an oven at 100 °C for later use. Three treatment groups were set up: 1) healthy plants and clean air as control, 2) SRBSDV-infected plants and clean air as a control, and 3) SRBSDV-infected plants and healthy plants.

Behavioral responses of A. nilaparvatae to single substances at different concentrations

Based on the results of the analysis of plant volatiles from infected and healthy rice plants, octanal, methyl salicylate, undecane and hexadecane were not detected from the SRBSDV-infected rice plants, and tridecanal was not found from healthy rice plant. In order to clarify the functions of those five chemical substances, behavioral responses of A. nilaparvatae to single substances at different concentrations were determined by using a Y-tube olfactometer. The method was described in detail above.

Parasitic capability of A. nilaparvatae on BPH eggs in SRBSDV-infected and healthy plants

SRBSDV-infected plants and healthy rice plants were uprooted and separated into single plants. The outer sheaths and the inactive roots were removed. Afterwards, they were rinsed with tap water and individually placed into tubes (diameter 1.5 cm, height 15.0 cm) containing 1.5 cm deep Kimura B nutrient solution, which were then sealed with degreased cotton. Three gravid female BPH were introduced into each tube, and 24 h later, one female parasitoid emerging within 4 h was then inoculated into the tube. The tubes were kept in the artificial climate chamber under the conditions of 26 ± 1 °C, 70–90% relative humidity, and L12:D12. Rice plants were dissected under a microscope on the 6th d after A. nilaparvatae died. Parasitic capability of A. nilaparvatae was measured by the number of parasitic eggs. Twenty-two replications were conducted for each treatment.

Statistics and analysis

Isolation and identification of volatile compounds released by the rice plants were performed using the NIST08 mass spectral library. BPH population growth rate was calculated as follows:

Population growth rate = nymph survival × ratio of female adult × number of eggs laid by each female × egg hatching rate.

SPSS18.0 was used for independent samples t tests. A binomial distribution was used to test preference for different odor sources. Descriptive statistics were expressed as the mean ± standard error, and the significance level was set at $\alpha = 0.05$.

Results

Changes in amino acid and soluble sugar contents in rice plants after SRBSDV infection

After SRBSDV infection, the content of various amino acids and the total amino acid content in the rice plants did not change significantly ($t = -0.144$, $df = 6$, $P = 0.899$) (Table 1). Soluble sugar content in the SRBSDV-infected plants was 6.69%, whereas the content in healthy plants was 6.00%. The difference was not significant ($t = -1.060$, $df = 6$, $P = 0.330$).

Plant volatiles from SRBSDV-infected and healthy rice plants

The results shown in Table 2 indicated that total 24 types of substances were identified as volatile components released by rice plants, including 23 types in healthy rice plants and 20 types in SRBSDV-infected plants. They were mainly composed of aldehydes, alcohols, alkanes, and lipids. Tridecylic aldehyde was collected only from SRBSDV-infected plants, whereas octanal,

undecane methyl salicylate and hexadecane were only from healthy rice plants. The relative contents of methyl benzene, (Z)-3-hexenal, (E)-2-hexenal, 3-hexanol, 1-hexanol, heptyl aldehyde, 2-ethyl hexanol and nonyl aldehyde were elevated after SRBSDV infection. The contents of methyl benzene, (Z)-3-hexenal, (E)-2-hexenal, and 3-hexanol were significantly different ($P < 0.05$).

Population growth of non-vector BPH on SRBSDV-infected and healthy plants

The population growth rate of non-vector BPH on SRBSDV-infected plants was 52.20 ± 9.75, which had no obvious difference than its population growth rate of 57.75 ± 2.28 on healthy plants ($t = 0.501$, $df = 15.69$, $P = 0.623$).

Host selectivity and parasitic capability of A. nilaparvatae to BPH eggs on infected and healthy rice plants

As shown in Figure 1, the parasitism rate of BPH eggs on SRBSDV-infected plants was 39.97%, whereas the parasitism rate on the corresponding healthy plants was 36.08%. The difference was not significant ($t = 0.454$, $df = 14$, $P = 0.657$).

Parasitic capability of A. nilaparvatae to BPH eggs on SRBSDV-infected plants was 21.32 per wasp, and that on healthy plants was 18.38. This difference was not significant ($t = 0.941$, $df = 36.33$, $P = 0.353$).

Preference of A. nilaparvatae for odor sources from SRBSDV-infected and healthy plants

The results of preference testing of A. nilaparvatae on infected and healthy plants showed that A. nilaparvatae preferred rice plants with BPH eggs ($t = 2.136$, $df = 4$, $P_{healthy} = 0.018$; $t = 2.771$, $df = 4$, $P_{infected} = 0.004$). However, there was no significant preference for infected plants or healthy plants ($P = 1.000$) (Fig. 2).

Behavioral responses of A. nilaparvatae to single volatile substances at different concentrations

The results in figure 3 indicated that A. nilaparvatae had no obvious selectivity between single volatile substances at different concentrations and liquid paraffin in the control group. When the test substance was diluted 10^2 times, A. nilaparvatae tended to choose hexadecane or methyl salicylate compared with the control group. When diluted 10^4 times, four of the substances had a certain attractiveness for A. nilaparvatae. Octanal was an exception. When it was diluted 10^6 times, the proportions of A. nilaparvatae between the two odor sources were close to equal.

When pure products were diluted 10^2–10^6 times, methyl salicylate, undecane and tridecanal had the strongest attraction to A. nilaparvatae. With decreasing dilution, the attraction of octanal to A. nilaparvatae also decreased; the opposite was true for hexadecane.

Discussion

Plant viruses can have positive [16,30,31] or negative [32,33] impacts on the growth, survival and reproductive capability of herbivorous insects. For persistent viruses, the vector has a long time to acquire viruses, allowing viruses to induce changes in plant physiology and biochemistry that will affect vector insects. This is conducive to the spread of viruses [15,34]. Our results showed that after SRBSDV infection, rice plants had no significant changes in amino acids and soluble sugar contents. Tu et al. (2013) introduced WBPH on SRBSDV-infected rice plants to feed for 48 h, and later transferred them to healthy plants. Their growth

Table 1. Amino acid and soluble sugar in infected and healthy rice plants.

Amino acids	Healthy plants (% dry weight)	Infected plants (% dry weight)	Change rate (%)	P
Asp	2.24±0.04	3.04±0.10	35.71	0.050
Thr	0.88±0.02	0.85±0.03	3.98	0.393
Ser	0.86±0.02	0.87±0.03	1.75	0.712
Glu	2.33±0.05	2.40±0.08	3.23	0.519
Pro	0.88±0.02	0.86±0.03	1.71	0.712
Gly	1.00±0.02	0.91±0.03	8.54	0.170
Ala	1.22±0.02	1.09±0.04	11.07	0.107
Val	1.02±0.02	0.95±0.02	6.86	0.132
Met	0.18±0.00	0.17±0.01	5.56	0.423
Ile	0.80±0.02	0.72±0.03	10.06	0.137
Leu	1.64±0.03	1.42±0.05	13.46	0.077
Tyr	0.47±0.01	0.46±0.02	3.19	0.504
Phe	1.02±0.02	0.89±0.03	12.75	0.084
His	0.66±0.01	0.65±0.03	2.27	0.657
Lys	1.12±0.02	0.99±0.03	11.61	0.084
Arg	0.94±0.02	0.90±0.03	4.26	0.397
Total amino acids	17.23±0.28	17.14±0.56	0.52	0.899
Soluble sugar	6.00±0.35	6.69±0.56	11.50	0.330

Table 2. Comparison of volatile compounds emitted from healthy and SRBSDV-infected plants.

Chemicals	Healthy plants (ng/g)	SRBSDV-infected plants (ng/g)	P
methyl benzene	8.40±1.08	17.38±2.86	0.013
(Z)-3-hexenal	5.65±1.42	10.22±0.90	0.013
(E)-2-hexenal	7.54±2.19	12.91±3.26	0.035
3-hexanol	3.68±0.89	6.27±0.62	0.027
ethylbenzene	5.43±1.50	6.56±1.33	0.576
1-hexanol	1.08±0.23	1.24±0.11	0.539
heptanal	2.64±0.67	3.28±0.59	0.483
octanal	4.33±1.09	/	
2-ethyl hexanol	13.58±3.50	24.05±6.80	0.203
undecane (H)	7.92±2.28	/	
nonanal	7.95±2.45	8.72±1.77	0.799
menthol	3.00±1.27	2.07±0.66	0.510
naphthalene	4.19±1.02	3.89±0.99	0.839
methyl salicylate	1.04±0.78	/	
dodecane	2.18±0.60	1.70±0.51	0.548
decanal	5.39±1.60	4.13±0.95	0.498
dodecanal	4.45±1.62	4.48±0.95	0.985
(+)-longifolene	12.82±3.72	7.50±2.21	0.240
α-cedrene	21.25±6.38	8.13±2.61	0.084
β-caryophyllen+β-cedrene	10.24±3.13	3.15±0.91	0.056
dimethyl phthalate	112.35±32.98	94.73±38.36	0.735
hexadecane	26.72±7.87	/	
tetradecanal	4.72±1.87	3.51±0.71	0.538
α-cedrol	9.73±2.80	4.72±1.26	0.109
tridecanal	/	0.63±0.28	

Note: "/" indicates not be detected.

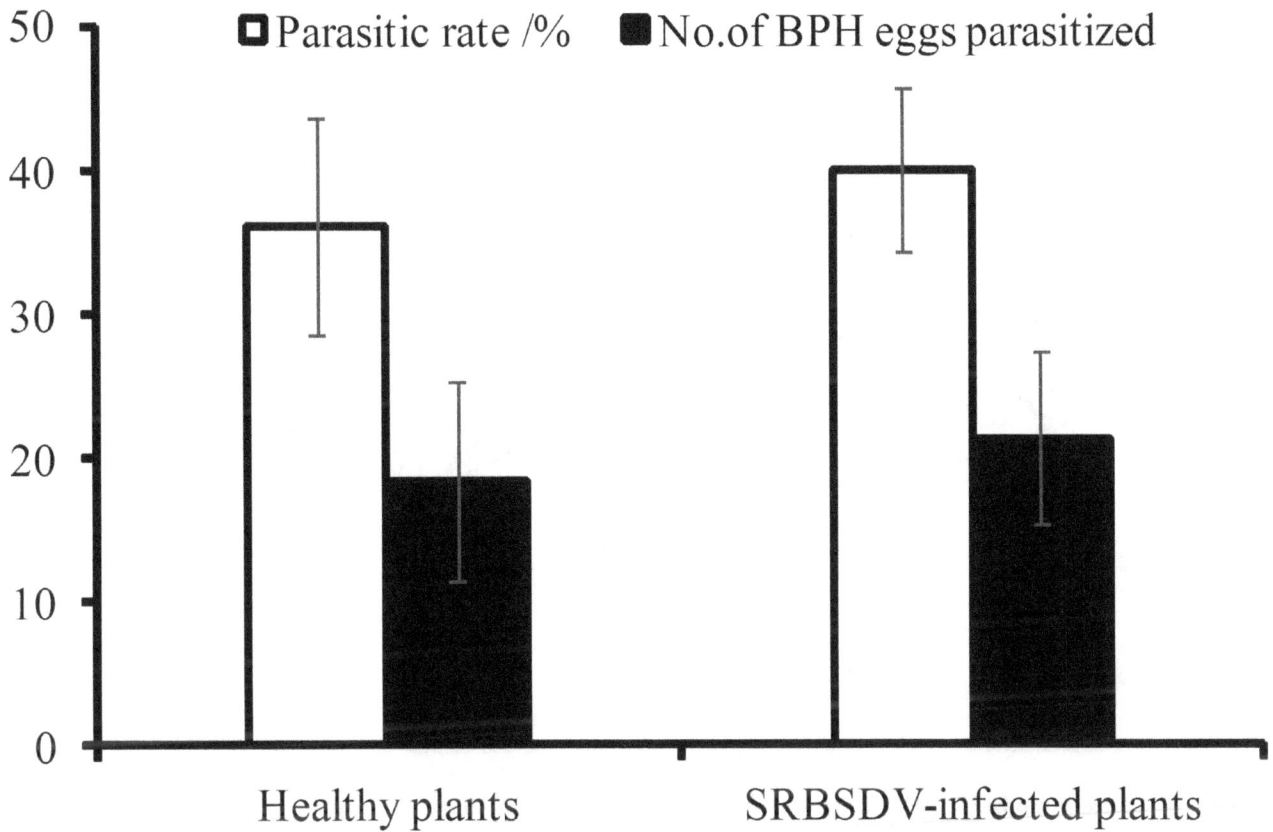

Figure 1. Effects of SRBSDV-infected rice plants on parasitic selectivity and parasitic capability of *Anagrus nilaparvatae.*

and development were then observed [22]. They found that, compared with WBPH feeding on healthy plants (16.8±0.4 d), the nymph duration of planthoppers feeding on SRBSDV-infected plants were prolonged (23.6±0.5 d) at 20°C, though there was no significant difference at 25 and 28°C. The duration of infection on SRBSDV-infected plants may have been responsible for these

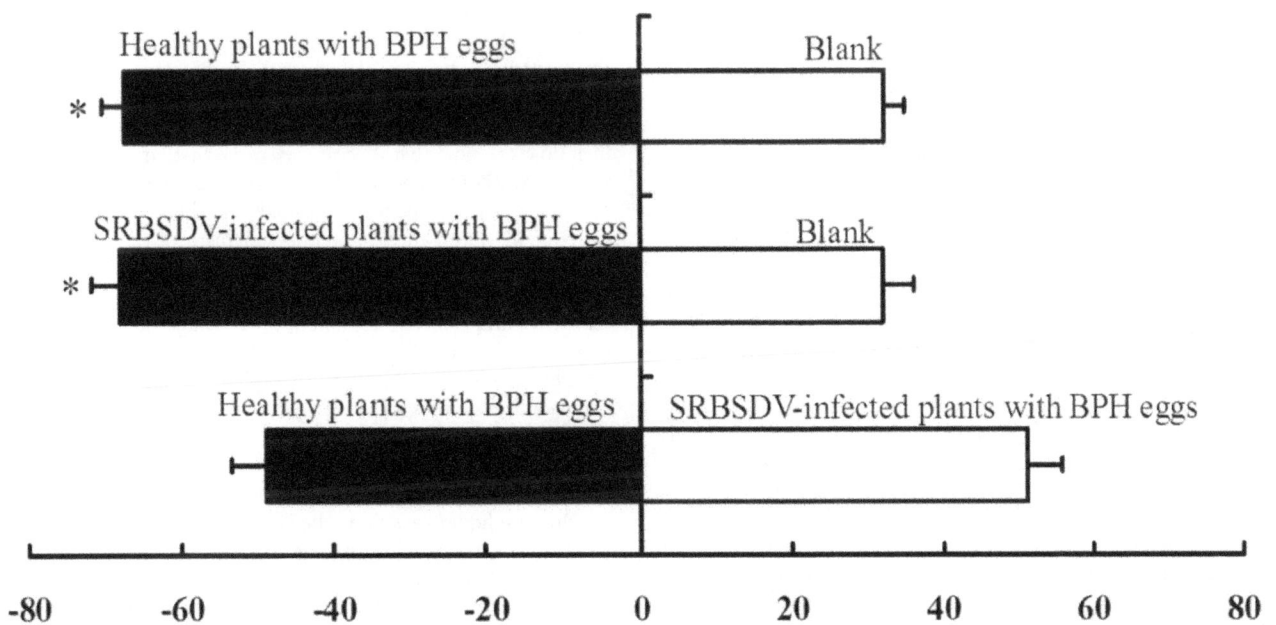

Figure 2. Preference of *Anagrus nilaparvatae* **for different odor sources.**

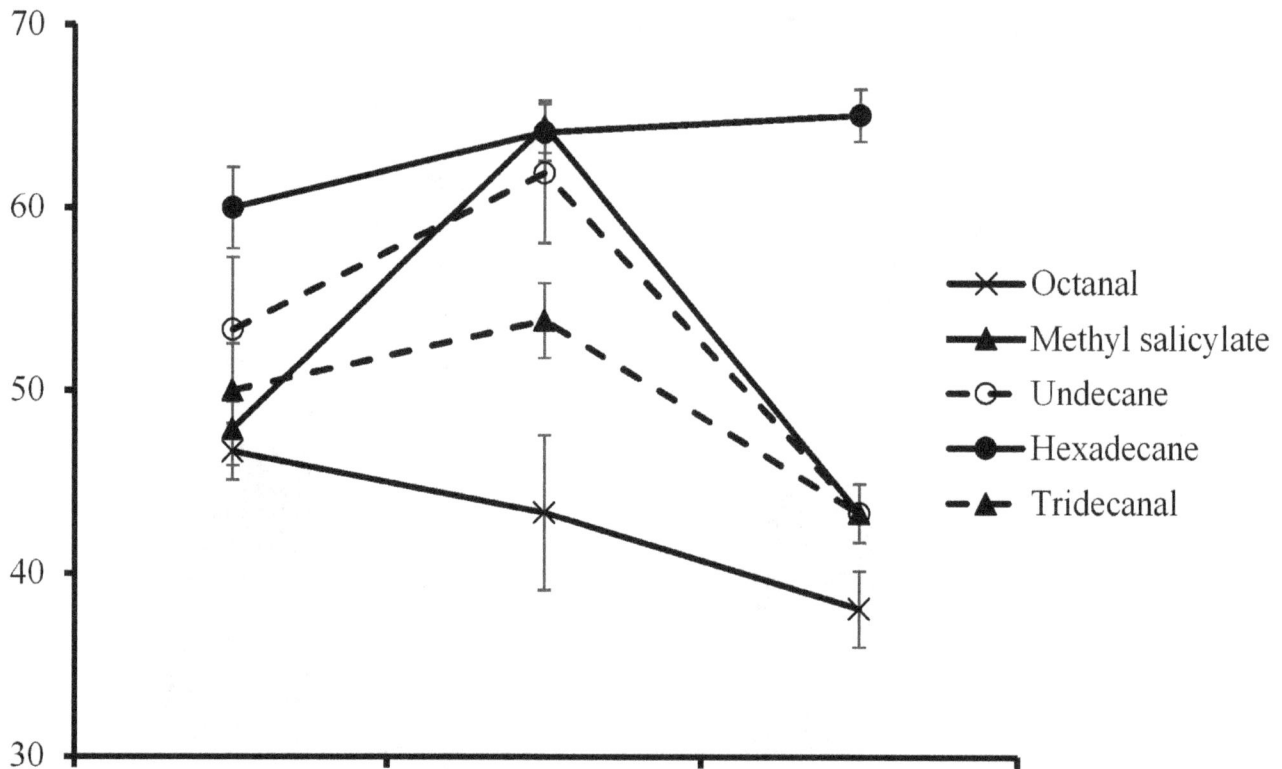

Figure 3. Behavioral responses of *Anagrus nilaparvatae* to single substance at different concentrations.

differences. Tu et al. (2013) showed the direct impact of a virus on vector insects, which was similar to our previous result that virus-carrying WBPH can significantly inhibit the reproductive ability of WBPH serving as the vector [22,23]. In addition, SRBSDV infection could affect the feeding behavior of vector WBPH. Viruliferous WBPH fed in phloem more frequently than non-viruliferous WBPH, which would increase the probability of virus inoculation [23]. However, SRBSDV-infected plants had no significant impacts on the non-vector planthopper and its egg parasitoid, *A. nilaparvatae*, indicating no change happened in biological control by egg parasitoid of vector WBPH, since WBPH shares the same natural enemies with BPH [35]. Meanwhile WBPH should get more attention than BPH in the paddy fields of epidemic of SRBSDV.

There are few reports on the impact of plant viruses on non-vector insects. In a study of non-vector Q-type *Bemisia tabaci* feeding on tobacco infected with tomato yellow leaf curl china virus (TYLCCNV), the adult longevity and fecundity of *B. tabaci* were higher than those feeding on healthy tobacco [36]. Non-vector BPH and WBPH feeding on RBSDV-infected rice plants had improved ecological fitness, and this promoted the expansion of non-vector insect populations under natural conditions [5,37]. However, Iris yellow spot virus (IYSV) had no significant impact on the survival of its non-vectors, *Frankliniella occidentalis*, *Frankliniella schultzei*, and *Frankliniella schultzei* [38]. Our present results showed that SRBSDV had no significant impact on the population growth of non-vector BPH. This finding can be explained by the fact that both amino acid and soluble sugar contents in rice plants did not change after SRBSDV infection. However, a previous study found that both amino acid and soluble sugar contents were significantly increased in the rice plants with RBSDV infection than in healthy plants, resulting in higher

nymphal survival rate, female adult weight and egg hatchability of non-vector BPH, as well as its higher activities of defense enzymes and detoxifying enzymes [5].

Plant viruses can have indirect effects on natural enemies through vector insects. They can also have direct effects on natural enemies' growth and reproduction. For example, larval development of *Aphidius ervi* in *Sitobion avenae* was significantly delayed when barley yellow dwarf virus (BYDV) acquisition took place before or shortly after the parasitoid had hatched, but not when the parasitoid was at the second larval stage during virus acquisition. Similarly, the presence of BYDV led to significantly higher aphid mortality when they acquired virus up to and including the time that *A. ervi* was at the first larval stage. Adult female parasitoids deposited fewer eggs in viruliferous aphids [39]. Host selection by female parasitoids consists of host location, recognition, acceptance, as well as judgment of the suitability of the host for its development [40]. Our study showed that SRBSDV-infected plants have no significant impact on host selectivity and parasitic capability of *A. nilaparvatae*. To clarify differences in volatile secondary metabolites between SRBSDV-infected plants and healthy rice plants using GC-MS, the behavioral responses of *A. nilaparvatae* to octanal, undecane, methyl salicylate, hexadecane and tridecanal at different concentrations were tested. *A. nilaparvatae* exhibited no significant selectivity between single volatile substances at different concentrations and liquid paraffin in the control group. Rice volatiles are the major carriers in compound communication between rice plants, planthoppers and *A. nilaparvatae* [41]. It has been found that SRBSDV-infected plants are attractive to vector WBPH [42], whereas we found in this study that the egg parasitoid *A. nilaparvatae* has no significant selectivity to odor sources from infected and healthy plants, as well as no differences in behavior responses to single substances. A possible reason is that *A. nilaparvatae* is a parasitoid, and SRBSDV does not spread throughout the eggs. In comparison, the number and quality of eggs may have a larger impact. Among the three virus-mediated trophic levels of rice plants, planthoppers and parasitoid, SRBSDV-infected plants do not have obvious impacts on the non-vector BPH and its parasitoid.

With regard to spread of the virus, the number of vector insects is only one factor affecting the efficiency of spread. Through predation or parasitism, predators can also affect the spread of viral diseases [43]. When *Aphidius ervi* was introduced to the system of *Vicia faba*, pea enation mosaic virus (PEMV), and *Acyrthosiphon pisum*, the number of pea aphids, their longevity were reduced, which in turn lowered damage to host plants and plant susceptibility. Furthermore, *Acyrthosiphon pisum* parasitized by *Aphidius ervi* are more active, accelerating the spread of viral disease [43]. Introduction of the predator *Coccinella septempunc-*

tata improves the infection rate of barley yellow dwarf virus (BYDV) spread by *Rhopalosiphum padi*, however, the parasitoid *Aphidius gifuensis* has no significant impact on the infection rate of healthy plants [44]. It is clear that spread of the virus is induced indirectly by predators in the system of predators, vector insects and host plants. Impacts of plant viruses on host plants, herbivorous insects, natural enemies and the entire ecosystem can be complex. Further study is needed to fully understand the interactions involved.

Author Contributions

Conceived and designed the experiments: ZL. Performed the experiments: XH HX GG. Analyzed the data: XH HX X. Zheng YY JT ZL. Contributed reagents/materials/analysis tools: ZL X. Zheng YS. Contributed to the writing of the manuscript: HX XH ZL GG YY YS X. Zhou X. Zheng.

References

1. Chen YG, Olson DM, Ruberson JR (2010) Effects of nitrogen fertilization on tritrophic interactions. Arthroped-Plant Interact 4: 81–94.
2. Hurd H (2003) Manipulation of medically important insect vectors by their parasites. Annu Rev Entomol 48:141–161.
3. Lefévre T, Koella JC, Renaud F, Hurd H, Brion DG, et al. (2006) New prospects for research on manipulation of insect vectors by pathogens. PloS Pathog 2(7): e72.
4. Lefévre T, Lebarbenchon C, Gauthier-Clerc M, Missé D, Poulin R, et al. (2009) The ecological significance of manipulative parasites. Trends Ecol Evol 24: 41–48.
5. Xu HX, He XC, Zheng XS, Yang YJ, Lu ZX (2013) Influence of rice black streaked dwarf virus on the ecological fitness of non-vector planthopper *Nilaparvata lugens* (Stål) (Hemiptera: Delphacidae). Insect Sci 00: 1–8, doi10.1111/1744-7917.12045.
6. Fiebig M, Poehling HM, Borgemeister C (2004) Barley yellow dwarf virus, wheat, and Sitobion avenae: a case of trilateral interations. Entomol Exp Appl 110(1): 11–21.
7. Yan SC, Zhang DD, Chi DF (2003) Advances of studies on the effects of plant volatiles on insect behavior. Chinese J Appl Ecol 14: 310–313.
8. Zhou Q, Xu T, Luo SM (2004) Effects of rice volatile infochemicals on insect. Chin J Appl Ecol 15: 345–348.
9. Obara N, Hasegawa M, Kodama O (2002) Induced volatiles in elicitor-treated and rice blast fungus-inoculated rice leaves. Biosci Biotech Biochem 66: 2549–2559.
10. Musser RO, Hum-Musser SM, Felton GW, Gergerich RC (2003) Increased larval growth and preference for virus-infected leaves by the Mexican bean beetle, Epilachna varivestis mulsant, a plant virus vector. J Insect Behav 16: 247–256.
11. Mauck KE, de Moraes CM, Mescher MC (2010) Deceptive chemical signals induced by a plant virus attract insect vectors to inferior hosts. PNAS 107: 3600–3605.
12. Jiménez-Martínez ES, Bosque-Perez NA, Berger PH, Zenmetra RS (2004) Life history of the bird cherry-oat aphid, Rhopalosiphum padi(Homoptera: Aphididae), on transgenic and untransformed wheat challenged with Barley yellow dwarf virus. J Econ Entomol 97(2): 203–212.
13. Colvin J, Omongo CA, Govindappa MR, Stevenson PC, Maruthi MN, et al. (2006) Host-plant viral infection effects on arthropod-vector population growth, development and behavior: management and epidemiological implication. Adv Virus Res 67: 419–452.
14. Alvarez AE, Garzo E, Verbeek M, Vosman B, Dicke M, et al. (2007) Infection of potato plants with potato leafroll virus changes attraction and feeding behavior of Myzus persicae. Entomol Exp Appl 125: 134–144.
15. Stafford CA, Walker GP, Ullman DE (2011) Infection with a plant virus modifies vector feeding behavior. PNAS 108: 9350–9355.
16. Belliure B, Sabelis MW, Janssen A (2010) Vector and virus induced plant responses that benefit a non-vector herbivore. Basic Appl Ecol 11: 162–169.
17. Zhang HM, Yang J, Chen JP, Adams MJ (2008) A black-streaked dwarf disease on rice in China is caused by a novel fijivirus. Arch Virol 153(10): 1893–198.
18. Zhou GH, Wen JJ, Cai DJ, Li P, Xu DL, et al. (2008) Southern rice black-streaked dwarf virus: A new proposed Fijivirus species in the family Reoviridae. Chin Sci Bull 53: 3677–3685.
19. Hoang AT, Zhang HM, Yang J, Chen JP, Hébrard E, et al. (2011) Identification, characterization, and distribution of Southern rice black streaked dwarf virus in Vietnam. Plant Dis 95(9): 1063–1069.
20. Zhou GH, Xu DL, Xu DG, Zhang MX (2013) Southern rice black-streaked dwarf virus: a white-backed planthopper transmitted fijivirus threatening rice production in Asia. Front Microbiol 4: 270.
21. Matsukura K, Towata T, Sakai J, Onuki M, Okuda M, et al. (2013) Dynamics of Southern rice black-streaked dwarf virus in rice and implication for virus acquisition. Phytopathol 103: 509–512.
22. Tu Z, Ling B, Xu DL, Zhang MX, Zhou GH (2013) Effects of southern rice black-streaked dwarf virus on the development and fecundity of its vector, Sogatella furcifera. Virol J 10: 145.
23. Cheng XN, Wu JC, Ma F (2003) Brown planthopper: occurrence and control. Beijing: China Agriculture Press.
24. Cheng JA (2009) Rice planthopper problems and relevant causes in China. In: Heong KL, Hardy B, editors. Planthoppers: new threats to the sustainability of intensive rice production systems in Asia. Los Baños (Philippines): International Rice Research Institute. 157–177 p.
25. Xu HX, He XC, Zheng XS, Yang YJ, Tian JC, et al. (2014) Southern rice black-streaked dwarf virus (SRBSDV) directly affects the feeding and reproduction behavior of its vector, Sogatella furcifera (Horváth) (Hemiptera: Delphacidae). Virol J 11: 55
26. Zhao Y, Wu CX, Zhu XD, Jiang XH, Zhang XX, et al. (2011) Tracking the source regions of southern rice black-streaked dwarf virus (SRBSDV) occurred in Wuyi county, Zhejiang province, China in 2009, transmitted by Sogatella furcifera (Horváth) (Homoptera: Delphacidae). Acta Entomol Sin 54(8): 949–959.
27. Zhang SX, Li L, Wang XF, Zhou GH (2007) Transmission of Rice stripe virus acquired from frozen infected leaves by the small brown small brown planthopper (Laodelphax striatellus Fallen). J Virol Meth 146: 359–362.
28. Cao Y, Pan F, Zhou Q, Li GH, Liu SQ, et al. (2011) Transmission characteristics of Sogatella furcifera: A vector of the southern rice black streaked dwarf virus. Chin J Appl Entomol 48(5): 1314–1320.
29. Li YZ, Cao Y, Zhou Q, Guo HM, Ou GC (2012) The efficiency of southern rice black-streaked dwarf virus transmission by the vector Sogatella furcifera to different host plant species. J Integr Agri 11(4): 621–627.
30. Belliure B, Janssen A, Maris PC, Peters D, Sabelis MW (2005) Herbivore arthropods benefit from vectoring plant virus. Ecol Lett 8: 70–79.
31. Eubanks MD, Carr DE, Murphy JF (2005) Variation in the response of *Mimulus guttatus* (Scrophulariaceae) to herbivore and virus attack. Evol Ecol 19: 15–27.
32. Röder G, Rahier M, Naisbit RE (2007) Coping with an antagonist: The impact of a phytopathogenic fungus on the development and behavior of two species of alpine leaf beetle. Oikos 116: 1514–1523.
33. Rayapuram C, Baldwin IT (2008) Host-plant-mediated effects of Nadefensin on herbivore and pathogen resistance in Nicotiana attenuate. BMC Plant Biol 8: 109.
34. Ogada PA, Maiss E, Poehling HM (2012) Influence of tomato spotted wilt virus on performance and behaviour of western flower thrips (Frankliniella occidentalis). J Appl Entomol 137: 488–498.
35. Barrion AT, Litsinger A (1994) Taxonomy of rice insect pests and their arthropod parasites and predators. In: Heinrich EA editor. Biology and management of rice insects. Wiley Eastern Limited, International Rice Research Institute Press, Los Baños (Philippines): 13–359 p
36. Liu J, Li M, Li JM, Huang CJ, Zhou XP, et al. (2010) Viral infection of tobacco plants improves the performance of Bemisia tabaci but more so for an invasive than for an indigenous biotype of the whitefly. J Zhejiang Univ Sci B 11: 30–40.
37. He XC, Xu HX, Zheng XS, Yang YJ, Gao GC, et al. (2011) Effects of rice streaked dwarf virus on ecological fitness of non-vector planthopper, Sogatella fucifera. Chin J Rice Sci 25: 654–658.
38. Birithia R, Subramanian S, Pappu HR, Muthomi J, Narla RD (2013) Analysis of Iris yellow spot virus replication in vector and non-vector thrips species. Plant Pathol 62: 1407–1414.
39. Christiansen-Weniger P, Powell G, Hardie J (1998) Plant virus and parasitoid interactions in a shared insect vector/host. Entomol Exp Appl 86: 205–213.

40. Vinson BS (1976) Host selection by insect parasitoids. Annu Rev Entomol 21: 109–133.

41. Lou YG, Cheng JA (2001) Host-recognition kairomone from *Sogatella furcifera* for parasitoid *Anagrus nilaparvatae*. Entomol Exper Appl 101: 59–68.

42. Wang H, Xu DL, Pu LL, Zhou GH (2014) Southern rice black-streaked dwarf virus alters insect vectors' host orientation preferences to enhance spread and increase rice ragged stunt virus co-infection. Phytopath 104(2): 196–201.

43. Hodge S, Powell G (2008) Complex interaction between a plant pathogen and insect parasitoid via the shared vector-host: Consequences for host plant infection. Oecologia 157: 387–397.

44. Smyrnioudis IN, Harrington R, Clark SJ, Katis SJ (2001) The effects of natural enemies on the spread of barley yellow dwarf virus (BYDV) by *Rhopalosiphum padi* (Hemiptera: Aphididae). Bull Entomol Res 91: 301–306.

5

Identification of Target Genes of the bZIP Transcription Factor OsTGAP1, Whose Overexpression Causes Elicitor-Induced Hyperaccumulation of Diterpenoid Phytoalexins in Rice Cells

Koji Miyamoto[1,2], Takashi Matsumoto[3], Atsushi Okada[2], Kohei Komiyama[2], Tetsuya Chujo[2¤],
Hirofumi Yoshikawa[3,4], Hideaki Nojiri[2], Hisakazu Yamane[1,2], Kazunori Okada[2*]

1 Department of Biosciences, Teikyo University, Utsunomiya, Tochigi, Japan, 2 Biotechnology Research Center, The University of Tokyo, Bunkyo-ku, Tokyo, Japan,
3 Genome Research Center, NODAI Research Institute, Tokyo University of Agriculture, Setagaya-ku, Tokyo, Japan, 4 Department of Bioscience, Tokyo University of
Agriculture, Setagaya-ku, Tokyo, Japan

Abstract

Phytoalexins are specialised antimicrobial metabolites that are produced by plants in response to pathogen attack. Momilactones and phytocassanes are the major diterpenoid phytoalexins in rice and are synthesised from geranylgeranyl diphosphate, which is derived from the methylerythritol phosphate (MEP) pathway. The hyperaccumulation of momilactones and phytocassanes due to the hyperinductive expression of the relevant biosynthetic genes and the MEP pathway gene OsDXS3 in OsTGAP1-overexpressing (OsTGAP1ox) rice cells has previously been shown to be stimulated by the chitin oligosaccharide elicitor. In this study, to clarify the mechanisms of the elicitor-stimulated coordinated hyperinduction of these phytoalexin biosynthetic genes in OsTGAP1ox cells, transcriptome analysis and chromatin immunoprecipitation with next-generation sequencing were performed, resulting in the identification of 122 OsTGAP1 target genes. Transcriptome analysis revealed that nearly all of the momilactone and phytocassane biosynthetic genes, which are clustered on chromosomes 4 and 2, respectively, and the MEP pathway genes were hyperinductively expressed in the elicitor-stimulated OsTGAP1ox cells. Unexpectedly, none of the clustered genes was included among the OsTGAP1 target genes, suggesting that OsTGAP1 did not directly regulate the expression of these biosynthetic genes through binding to each promoter region. Interestingly, however, several OsTGAP1-binding regions were found in the intergenic regions among and near the cluster regions. Concerning the MEP pathway genes, only OsDXS3, which encodes a key enzyme of the MEP pathway, possessed an OsTGAP1-binding region in its upstream region. A subsequent transactivation assay further confirmed the direct regulation of OsDXS3 expression by OsTGAP1, but other MEP pathway genes were not included among the OsTGAP1 target genes. Collectively, these results suggest that OsTGAP1 participates in the enhanced accumulation of diterpenoid phytoalexins, primarily through mechanisms other than the direct transcriptional regulation of the genes involved in the biosynthetic pathway of these phytoalexins.

Editor: Pankaj K. Singh, University of Nebraska Medical Center, United States of America

Funding: This work was supported by a JSPS Grant-in-Aid for Scientific Research to H.Y. (No. 22380066), by a Grant-in-Aid for JSPS Fellows to K.M. (No. 09J01084), and by the Program for Promotion of Basic Research Activities for Innovative Biosciences (PROBRAIN) to K.O. This work was funded in part by Ministry of Education, Culture, Sports, Science and Technology-Supported Program for the Strategic Research Foundation at Private Universities [2013–2017] to H.Y. [project number S131052A01]. The funders had no role in study design, data collection and analysis, decision to publish, or preparation of the manuscript.

Competing Interests: The authors have declared that no competing interests exist.

* Email: ukokada@mail.ecc.u-tokyo.ac.jp

¤ Current address: Plant-Microbe Interactions Research Unit, Division of Plant Sciences, National Institute of Agrobiological Sciences, Tsukuba, Ibaraki, Japan

Introduction

Phytoalexins are specialised antimicrobial metabolites that are produced by plants in response to pathogen attack [1]. In rice, momilactones and phytocassanes are recognised as the major diterpenoid phytoalexins [2–5].

In plants, isopentenyl diphosphate and dimethylallyl diphosphate, which are the basic C5 precursors for terpenoid biosynthesis, are produced by two distinct pathways: the mevalonate pathway and methylerythritol phosphate (MEP) pathway [6]. In suspension-cultured rice cells, the MEP pathway genes exhibit elicitor-induced expression, and the MEP pathway is required for the production of sufficient amounts of diterpenoid phytoalexins [7]. However, the transcriptional regulatory mechanisms of the MEP pathway genes remain unknown.

In the biosynthesis of diterpenoid phytoalexins, the common precursor geranylgeranyl diphosphate is sequentially cyclised by OsCPS2, OsCPS4, OsKSL7, and OsKSL4 into two distinct diterpene hydrocarbons: *ent*-cassa-12,15-diene and 9βH-pimara-7,15-diene [8–10]. For momilactone biosynthesis, two P450 monooxygenases (CYP99A2 and CYP99A3) and a dehydrogenase (OsMAS) are involved in the downstream oxidation of 9βH-pimara-7,15-diene [11,12]. For phytocassane biosynthesis, four P450 monooxygenases (CYP71Z7, CYP76M7, CYP76M8, and CYP701A8/OsKOL4) are involved in the oxidation of *ent*-cassa-12,15-diene [13–16]. The momilactone and phytocassane biosynthetic genes are localized in narrow regions of chromosomes 4 and 2, respectively, creating functional gene clusters [11,13]. These biosynthetic genes exhibit the temporally coordinated expression of mRNAs after treatment with a biotic elicitor in suspension-cultured rice cells [7,11].

The basic leucine zipper (bZIP) transcription factor OsTGAP1 has been shown to be involved in the regulation of the production of momilactones and phytocassanes. OsTGAP1-overexpressing (OsTGAP1ox) rice cells exhibit the hyperaccumulation of momilactones and phytocassanes as well as the enhanced expression of all momilactone biosynthetic genes, the phytocassane biosynthetic gene *OsKSL7*, and the MEP pathway gene *OsDXS3*, upon treatment with an elicitor [17]. However, the details of the regulation of these genes by OsTGAP1 remain unknown.

In *Arabidopsis thaliana*, 75 members of the bZIP transcription factor family have been identified and classified into ten groups (group A–I and S) based on sequence similarity among their basic regions and the presence of additional conserved motifs [18]. TGA factors (AtTGA1–7), which belong to the group D bZIP transcription factors, regulate pathogenesis-related genes such as *PR-1* through binding to the TGACG-motif (*as-1*-like element, TGACG[T/G]) on the promoter region and mediate salicylic acid–induced defence responses [19]. The transcriptional regulatory mechanism of *PR-1* has been well studied [20]. However, knowledge regarding the genome-wide binding regions of these TGA factors is limited; only AtTGA2 binding regions have been identified by chromatin immunoprecipitation (ChIP) with tiling arrays containing probes representing the 2-kbp upstream regions of *Arabidopsis* genes [21]. In rice, 89 members of the bZIP transcription factor family have been identified, among which 14 share conserved motifs with the *Arabidopsis* group D bZIP transcription factors [22]. However, the biological functions of only four TGA factors, including OsTGAP1, have been determined. Three TGA factors in rice, viz. rTGA2.1, rTGA2.2, and rTGA2.3, are involved in the regulation of defence responses against *Magnaporthe oryzae* and the rice bacterial blight *Xanthomonas oryzae* pv. *oryzae* [23–26]. However, the target genes of these three TGA factors remain unknown.

In the present study, transcriptome analysis and ChIP analysis with next-generation sequencing (ChIP-seq) using untreated and elicitor-treated OsTGAP1ox cells were performed, resulting in the identification of OsTGAP1 target genes. Interestingly, the clustered diterpenoid phytoalexin biosynthetic genes were not included among the OsTGAP1 target genes. However, it should be noted that several OsTGAP1-binding regions were found in the intergenic regions among and near the cluster regions. Concerning the MEP pathway genes, it was found that only *OsDXS3* possessed an OsTGAP1-binding region in its upstream region and that OsTGAP1 directly regulated *OsDXS3* expression, while other MEP pathway genes were not included among the OsTGAP1 target genes. Possible mechanisms by which OsTGAP1 may regulate the production of diterpenoid phytoalexins are also discussed.

Materials and Methods

Plants, chemical treatment, and rice transformation

Oryza sativa L. 'Nipponbare' was used as the wild-type strain. Suspension-cultured rice cells were maintained as described in a previous paper [8]. *N*-acetylchitooctaose was prepared and used to treat the rice cells as a chitin oligosaccharide elicitor, as described previously [17,27]. Rice transformation was performed as described previously [17].

Plasmid construction

The *Zea mays* polyubiquitin promoter was amplified by PCR using the primers UBQp attB4 F and UBQp attB1 R from pANDA [28] and cloned into pDONRP4-P1R (Invitrogen, CA, USA), resulting in pDONR-UBQp. The polyubiquitin promoter and OsTGAP1 open reading frame (ORF) were then cloned into R4pGWB501 [29] from pDONR-UBQp and pENTR-TGA [17] using LR Clonase II Plus Enzyme mix (Invitrogen). The resulting plasmid was designated as R4pGWB-UBQp-TGA and used for rice transformation.

The 2-kbp upstream region of *OsDXS3* was amplified by PCR using the primers DXS3p 2k F and DXS3p R and cloned into the *Kpn*I and *Hin*dIII sites of the pGL3-Basic vector (Promega, WI, USA), resulting in pGL3-DXS3p-2k. Next, three mutated constructs were generated: one of the two TGACGT sequences was mutated in m1 and m2, while both TGACGT sequences were mutated in m3. The TGACGT sequences in the *OsDXS3* promoter were mutated by PCR using the following primer pairs: DXS3p-m1-F and DXS3p-m1-R, DXS3p-m2-F and DXS3p-m2-R, and DXS3p-m3-F and DXS3p-m3-R. Each mutated *OsDXS3* promoter was then amplified by PCR using the primers DXS3p 2k F and DXS3p R. The amplified DNA fragments were cloned into the *Kpn*I and *Hin*dIII sites of the pGL3-Basic vector and then sequenced, resulting in pGL3-DXS3p-2k-m1, pGL3-DXS3p-2k-m2, and pGL3-DXS3p-2k-m3.

Fragments of the *OsDXS3* promoter (250 bp and 240 bp) were amplified by PCR using the primers DXS3p 250 F, DXS3p 240 F, and DXS3p R from pGL3-DXS3p-2k. A 250-bp fragment of the *OsDXS3* promoter with mutated TGACGT sequences was also amplified by PCR using the primers DXS3p 250 m1 F, DXS3p 250 m2 F, DXS3p 250 m3 F, and DXS3p R. These amplified DNA fragments were cloned into the *Bgl*II and *Hin*dIII sites of the pGL3-Basic vector and sequenced. The resulting plasmids were designated as pGL3-DXS3p-250, pGL3-DXS3p-240, pGL3-DXS3p-250-m1, pGL3-DXS3p-250-m2, and pGL3-DXS3p-250-m3.

The OsTGAP1 ORF was cloned into pUbi_RfA_Tnos [30] from pENTR-TGA using LR Clonase II Enzyme mix (Invitrogen). The resultant plasmid was designated pUbi_TGA_Tnos and used as the effector plasmid.

A summary of the plasmids used in this study is provided in Table S1, and the sequences of PCR primers used for plasmid construction are provided in Table S2.

Antibody generation and purification

An OsTGAP1-specific antibody was generated by immunizing rabbits with the keyhole limpet hemocyanin-coupled synthetic peptide MELYPGYLEDHFNIHK corresponding to the N-terminal peptide sequence (residues 1–16) of OsTGAP1. Cys residues were added to the N-terminus of the peptides to ensure efficient coupling to the keyhole limpet hemocyanin carrier protein. The OsTGAP1-specific antibody was further purified using an antigen affinity column.

Nuclear extraction

Rice cells (approximately 3 g) were ground and suspended in 30 ml of 70% glycerol buffer (20 mM Hepes/NaOH, pH 7.4, 5 mM $MgCl_2$, 5 mM KCl, 50 mM saccharose, 70% [v/v] glycerol, 1 mM DTT, and 200 mM phenylmethylsulfonyl fluoride) and incubated for 20 min at 4°C under mild shaking. After the filtration of the slurry through a quadrilayer of Miracloth (Merck, UK), the nuclei were collected by centrifugation (4,000×g, 60 min, 4°C) and washed in 10% glycerol buffer (20 mM Hepes/NaOH, pH 7.4, 5 mM $MgCl_2$, 5 mM KCl, 50 mM saccharose, 10% [v/v] glycerol, 1 mM DTT, and 200 mM phenylmethylsulfonyl fluoride) by gentle resuspension followed by centrifugation (4,000×g, 20 min, 4°C). The nuclei were resuspended in 500 µl of TE buffer (pH 8.0) and sonicated for 1 min with a Sonifier 250D (Branson, CT, USA) at output setting 1 with a pulse of 1 s and duty cycle of 50%. The sonicated nuclei were cleared by centrifugation (20,000×g, 15 min, 4°C) followed by ultracentrifugation (100,000×g, 60 min, 4°C). The supernatant was then subjected to immunoprecipitation (IP). For protein gel blot analysis, acetone precipitation (95% [v/v] acetone, −20°C, overnight) was performed. The precipitated protein was suspended in Laemmli SDS sample buffer and boiled.

Conjugation of IgG to Dynabeads Protein G

Conjugation of the anti-OsTGAP1 antibody or normal rabbit IgG to Dynabeads Protein G (Invitrogen) was performed using bis[sulfosuccinimidyl] suberate (Thermo Fisher, MA, USA) as the cross-linking reagent according to the manufacturer's protocol.

Immunoprecipitation

NaCl and NP-40 were added to the nuclear protein at final concentrations of 100 mM and 0.4% (v/v), respectively. Next, 50 µl of Dynabeads Protein G washed with IP buffer (10 mM Tris-HCl [pH 8.0], 1 mM EDTA, 100 mM NaCl, 0.4% [v/v] NP-40) was added to the nuclear protein. After incubation at 4°C for 60 min with rotation, the Dynabeads Protein G was removed. A small portion of precleared nuclear protein was collected as an 'Input' control. The 'Input' control was mixed with Laemmli SDS sample buffer and boiled. The precleared nuclear protein was divided into two aliquots. Dynabeads Protein G cross-linked normal IgG was added to one sample, and Dynabeads Protein G cross-linked anti-OsTGAP1 antibody was added to the other sample. After overnight incubation at 4°C with rotation, the Dynabeads Protein G was collected, and a flow-through sample was then harvested. The Dynabeads Protein G was washed five times in 1 ml of IP buffer and resuspended in 20 µl of 0.2 M glycine-HCl (pH 2.5). After incubation at room temperature for 5 min, the eluted protein was collected. This elution step was performed twice. The eluted protein was neutralized by adding 5 µl of 1 M Tris (pH 11). The resulting sample was then subjected to acetone precipitation. The precipitated protein was suspended in Laemmli SDS sample buffer and boiled.

Protein gel blot analysis

The boiled samples were subjected to SDS-PAGE on 8% (w/v) polyacrylamide gels and then transferred to Amersham Hybond ECL Nitrocellulose Membranes (GE Healthcare, UK). OsTGAP1 was detected using the anti-OsTGAP1 antibody (dilution 1:2,500 [v/v]) as the primary antibody and the ECL anti-rabbit IgG horseradish peroxidase–linked species-specific whole antibody (dilution 1:25,000 [v/v]) (GE Healthcare) as the secondary antibody. Chemiluminescent detection was conducted using the Immobilon Western Chemiluminescent HRP Substrate (Millipore, MA, USA).

Formaldehyde fixation

Formaldehyde was added to the liquid medium of cultured rice cells at a final concentration of 1% (v/v). After incubation at 25°C for 7 min, 1 M glycine was added to a final concentration of 100 mM to stop fixation. After incubation at 25°C for 5 min, the rice cells were collected by filtration using a Buechner funnel and then flash-frozen in liquid N_2.

Chromatin immunoprecipitation

Formaldehyde cross-linked nuclei were extracted from fixed OsTGAP1ox rice cells (6 g) as described in the 'Nuclear extraction' section above. After the nuclei were resuspended in 500 µl of TE buffer (pH 8.0), their chromatin was sheared by repetitive sonication with a Bioruptor UCW-201 (Tosho Denki, Japan; power 5, ON time 30 s, OFF time 60 s, 30 cycles). The sonicated chromatin was cleared by centrifugation (20,000×g, 15 min, 4°C) followed by ultracentrifugation (100,000×g, 60 min, 4°C). NaCl and NP-40 were then added at final concentrations of 100 mM and 0.4% (v/v), respectively, and IP buffer was added to the sonicated chromatin to a final volume of 2.5 ml. Then, 250 µl of Dynabeads Protein G washed with IP buffer was added to the sonicated chromatin. After incubation at 4°C for 60 min with rotation, the Dynabeads Protein G was removed, and 500 µl of precleared sample was collected as an 'Input' control. The precleared nuclear protein was then divided into two aliquots. Rabbit normal IgG (2.5 µg) was added to one sample, and 2.5 µg of the anti-OsTGAP1 antibody was added to the other. After overnight incubation at 4°C with rotation, 25 µl of Dynabeads Protein G washed with IP buffer was added. After incubation at 4°C for 60 min with rotation, the Dynabeads Protein G was washed seven times in 10 ml of IP buffer. After the wash, the Dynabeads Protein G was resuspended in 500 µl of TE buffer (pH 8.0), and 20 µl of 5 M NaCl were added. The Dynabeads Protein G was then incubated at 65°C overnight. The 'Input' control was also amended with 10 µl of 5 M NaCl and incubated at 65°C overnight. The samples were treated with RNase and Proteinase K, followed by phenol-chloroform extraction and isopropanol precipitation. The precipitated DNA was then dissolved in TE buffer (pH 8.0).

Next-generation sequencing

The construction of DNA libraries was performed using a Paired-End DNA Sample Prep Kit (Illumina, CA, USA) and a Multiplexing Sample Preparation Oligonucleotide Kit (Illumina) according to the manufacturer's instructions. In this step, ~350-bp fragments were collected from each sample using E-Gel SizeSelect 2% (Invitrogen). The constructed libraries were subsequently subjected to deep sequencing using a Genome Analyzer II (Illumina). These reads were mapped to the rice genome with the Burrows-Wheeler Aligner (BWA) software package [31] using the International Rice Genome Sequencing Project genome sequence (build 5) from the Rice Annotation Project Database (RAP-DB: http://rapdb.dna.affrc.go.jp) as the reference genome sequence. The OsTGAP1-binding regions in each sample were then detected using Partek Genomics Suite (ver. 6.5; http://www.partek.com/, Partek Software, MO, USA) according to the following thresholds: window size = 100, peak cut-off false discovery rate (FDR) <0.001, strand separation FDR<0.05, and significant enrichment in ChIP DNA compared to the 'Input' control (FDR<0.05). The sequence data were deposited in the

DDBJ Sequence Read Archive (http://trace.ddbj.nig.ac.jp/dra/index.html; ID: DRA001274).

DNA microarray analysis

Wild-type (WT) and OsTGAP1ox rice cells were treated with a chitin oligosaccharide elicitor. Total RNA was then isolated from a small portion of the rice cells using an RNeasy Plant Mini Kit (Qiagen, Germany) at 0, 6, and 24 h after the elicitor treatment and labelled with Cyanine 3 dye (Cy3) or Cyanine 5 dye (Cy5). Aliquots of Cy3-labelled cRNA and Cy5-labelled cRNA (825 ng each) were used for hybridization in a 60-mer rice oligo microarray with 44 k features (Agilent Technologies, CA, USA). Two series of microarray analyses were performed as follow. First, to investigate elicitor responsiveness, a time-course analysis of the elicitor treatment of WT cells was performed. The Cy3-labelled cRNA probe from the 0 h time point of the WT cells was used as a reference, and the Cy5-labelled cRNA probes from the 6 h and 24 h time point samples were compared against the 0 h reference. Second, comparison analysis was performed between the WT and OsTGAP1ox cells. The Cy3-labelled cRNA probes from the WT cells (0, 6, and 24 h after elicitor treatment) were used as references, and the Cy5-labelled cRNA probes from the OsTGAP1ox cells (0, 6, and 24 h after elicitor treatment) were compared against the references at each time point. The glass slides were scanned using a microarray scanner (G2565, Agilent), and the resulting output files were imported into Feature Extraction software (ver. 11; Agilent). These data were deposited into the Gene Expression Omnibus of the NCBI (http://www.ncbi.nlm.nih.gov/geo/; ID: GSE53414 and GSE53417). Statistical analyses were performed using Partek Genomics Suite. The expression levels of the OsTGAP1 target genes at each time point were calculated from fold changes relative to those of the WT cells 0 h after elicitor treatment, and the calculated data were imported into MultiExperiment Viewer (MeV v4.6, http://www.tm4.org/mev/) for cluster analysis. Hierarchical cluster analysis based on the average linkage and cosine correlation was used to cluster the genes on the y-axis using MeV.

ChIP-PCR

Quantitative PCR was performed using Power SYBR Green PCR Master Mix (Applied Biosystems, CA, USA) on an ABI PRISM 7300 Real-Time PCR System (Applied Biosystems). The DNA amount was determined by generating standard curves using a series of known concentrations of the target sequence. The ratio of DNA present in the ChIP DNA to that in the 'Input' control was calculated. The primers DXS3p TGACGT F and DXS3p TGACGT R were used for the OsDXS3 promoter region.

Gel mobility shift assay (GMSA)

The DNA probe was amplified by PCR from pGL3-DXS3p-2k using the primers DXS3p TGACGT F and DXS3p TGACGT R. Mutated probes in the TGACGT sequence (m1, m2, and m3) were also amplified by PCR from pGL3-DXS3p-2k-m1, pGL3-DXS3p-2k-m2, and pGL3-DXS3p-2k-m3, respectively, using the same primers. These amplified probes were end-labelled with ^{32}P by T4 polynucleotide kinase (Takara Bio, Japan). Recombinant N-terminal glutathione S-transferase (GST)-fused OsTGAP1 (GST-OsTGAP1) and GST protein were expressed in E. coli Rosetta 2 (DE3) harbouring pDEST15-TGA [17] or pGEX-6p2 (GE Healthcare) and purified using Glutathione Sepharose 4B (GE Healthcare). The reaction mixture comprised 20 mM Hepes (pH 7.6), 1 mM EDTA, 10 mM $(NH_4)_2SO_4$, 1 mM DTT, 0.2% (v/v) Tween 20, 30 mM KCl, 0.5 pmol of recombinant GST or GST-OsTGAP1, and 0.05 pmol of the probe in a final volume of 20 μl.

The above mixed samples were incubated for 20 min at room temperature, and then 5 μl of loading buffer (0.25×TBE buffer, 40% [w/v] glycerol, 0.2% [w/v] bromophenol blue) was added. The samples were separated on a 4% polyacrylamide gel in 0.5×TBE at room temperature, and the bands were visualized by autoradiography.

Transactivation assay

pUbi_GUS_Tnos [30] and pUbi_OsTGAP1_Tnos, in which β-glucuronidase (GUS) and OsTGAP1, respectively, are under the control of the polyubiquitin promoter, were used as the effector plasmids. pGL3-DXS3p-2k, pGL3-DXS3p-250, pGL3-DXS3p-240, pGL3-DXS3p-250-m1, pGL3-DXS3p-250-m2, and pGL3-DXS3p-250-m3, in which the OsDXS3 promoter was fused to the firefly luciferase (FLUC) gene, were used as the reporter plasmids. The plasmid pPTRL [32], which contains the Renilla luciferase (RLUC) gene under the control of the cauliflower mosaic virus (CaMV) 35S promoter, was used as an internal control. Particle bombardment of rice cells was conducted as previously described [17]. In the cotransfection assays, 0.1 μg or 1 μg of the effector plasmid, 1.0 μg of the reporter plasmid, and 0.5 μg of pPTRL were used for each bombardment. Luciferase (LUC) assays were performed as previously described [17]. The ratio of LUC activity (FLUC/RLUC) was calculated to normalize the values after each assay.

Results

Genome-wide identification of in vivo OsTGAP1-binding regions

To locate in vivo OsTGAP1-binding regions, ChIP-seq analysis was performed. An antipeptide antibody directed against the N-terminal peptide sequence (residues 1–16) of OsTGAP1 was generated, as this sequence has high immunogenicity and low similarity to other proteins in rice. To assess the specificity of the generated anti-OsTGAP1 antibody, protein gel blot analysis of the nuclear extracts from the WT and OsTGAP1ox cells was performed. A single band (~45 kD) was detected from the nuclear extract of the WT cells, and this band was strengthened in the nuclear extract of the OsTGAP1ox cells (Figure S1A). IP, using the anti-OsTGAP1 antibody, was also performed from the nuclear extracts of the WT cells. When rabbit normal IgG was used for IP, no bands were detected by protein gel blot analysis using the anti-OsTGAP1 antibody. In contrast, a single band thought to represent OsTGAP1 was detected after IP by the anti-OsTGAP1 antibody (Figure S1B). These results indicated that the anti-OsTGAP1 antibody generated in this study had sufficient specificity.

The OsTGAP1ox cells were treated with or without the chitin elicitor for 6 h and then collected for ChIP-seq analysis using the anti-OsTGAP1 antibody. Two biological replicates were performed for each sample. DNA libraries of ChIP DNA from the untreated and elicitor-treated samples were then generated. DNA libraries of the 'Input' controls were also generated from each sample. These libraries were then subjected to next-generation sequencing using a Genome Analyzer II (Illumina). A total of 4.5 to 10 million reads (100 bp per read) were obtained from each library. These reads were mapped to the rice genome by BWA [31]. OsTGAP1-binding regions were then detected in each sample. Two biologically independent ChIP-Seq experiments were compared, and the regions that showed enrichment in both datasets were identified as OsTGAP1-binding regions. The results identified 2,763 and 2,777 binding regions from the untreated and elicitor-treated samples, respectively (Table S3, Table S4). A comparison of the binding regions in the two conditions revealed that a common set of 2,003 regions (approximately 70%) were bound by OsTGAP1 both with and without elicitation.

Subsequent analysis revealed that among all of these OsTGAP1-binding regions, 32% (i.e. 880 and 901 regions) were assigned to the upstream regions of particular rice genes (within 2 kbp of the transcription start site) in the untreated and elicitor-treated conditions, respectively; moreover, 10% and 9% (i.e. 272 and 248 regions) were assigned to the gene regions of particular rice genes (from the transcription start site to the transcription termination site), and 18% and 16% (i.e. 491 and 452 regions) were assigned to the downstream regions of particular rice genes (within 2 kbp of the transcription termination site) in the untreated and elicitor-treated conditions, respectively. Intriguingly, the remaining 41% and 42% of the regions (i.e. 1,120 and 1,176) were assigned to the intergenic regions in the untreated and elicitor-treated conditions, respectively. This proportion is larger than those of the other plant transcription factors (HY5, PIL5, AGL15, and BZR1) whose genome-wide binding regions have previously been reported [33–36]. These results are summarized in Figure 1A. The locations of the OsTGAP1-binding regions assigned to the upstream or downstream regions of particular genes were significantly concentrated within 400-bp of the transcription start sites (Figure 1B). This distribution pattern implies that OsTGAP1-binding regions located within 400-bp of the transcription start sites has the significant role on the transcriptional regulation of downstream genes. Therefore, this study focussed on the genes that possessed an OsTGAP1-binding region within the 400-bp upstream region in either the untreated or elicitor-treated condition. A total of 693 genes were identified as potential OsTGAP1 target genes (Table S5).

Analysis of OsTGAP1-binding motifs

Overrepresented motifs in the OsTGAP1-binding regions were analysed using Partek Genomics Suite based on Gibbs motif sampler [37]. The TGACGT sequence was found in both the untreated and elicitor-treated condition (Figure 1C). These results are consistent with those of a previous report, which observed that OsTGAP1 binds to the TGACGT sequence *in vitro* [17]. We also analyzed whether TGACGT sequences are overrepresented in the OsTGAP1-binding regions compared to the rest of genome. As a result, TGACGT sequences are enriched approximately 8-fold in the OsTGAP1-binding regions compared to the rest of genome, although known binding motifs of other transcription factors were not enriched. (Figure S2). Furthermore, analysis of the distribution pattern of the TGACGT sequence around the OsTGAP1-binding regions (−1,000 to+1,000 bp of each predicted binding region) revealed that the occurrence of the TGACGT sequence peaks at the predicted OsTGAP1-binding regions in both the untreated and elicitor-treated condition (Figure 1D).

Identification of OsTGAP1 target genes

To identify the genes regulated by OsTGAP1, a genome-wide DNA microarray analysis was performed using the WT and OsTGAP1ox cells, and gene expression was compared between these cells at 0, 6, and 24 h after the chitin elicitor treatment. Four biological replicates were used for each time point. Statistical analysis was performed using ANOVA-FDR (q value≤0.05), and genes with changes in expression were identified as those that experienced a two-fold increase or decrease of expression level in the OsTGAP1ox cells at least one time point compared to the expression levels in the WT cells. Based on this criterion, 1,352 genes, 1,539 genes, and 1,267 genes were upregulated in the OsTGAP1ox cells at 0, 6, and 24 h after the elicitor treatment,

respectively. By combining these three groups, 2,268 genes were identified that were upregulated in the OsTGAP1ox cells at least one time points (Figure S3A, Table S6). Conversely, 1,278 genes, 1,387 genes, and 1,624 genes were downregulated in the OsTGAP1ox cells at 0, 6, and 24 h after the elicitor treatment, respectively. By combining these three groups, 2,276 genes were identified that were downregulated in the OsTGAP1ox cells at least one time points (Figure S3B, Table S7).

To determine the OsTGAP1 target genes, the transcriptome data and ChIP-seq data were compared. Among the 2,763 upregulated genes, 86 (3.1%) possessed OsTGAP1-binding regions in each 400-bp upstream region (Figure 2A, Table S8). Moreover, among the 2,777 downregulated genes, 36 (1.3%) possessed OsTGAP1-binding regions in each 400-bp upstream region (Figure 2B, Table S9). Therefore, these 122 genes were tentatively identified as OsTGAP1 target genes.

Next, the 86 upregulated genes and 36 downregulated genes were subjected to hierarchical clustering. Based on their expression patterns in the WT and OsTGAP1ox cells, the 86 upregulated genes were classified into two groups: group I, whose expression decreased in WT cells after elicitor treatment; and group II, whose expression increased in WT cells after elicitor treatment (Figure 2C). Most of the genes (approximately 88%) were included in group II. The 36 downregulated genes were classified into three groups: group I, whose expression increased in WT cells after elicitor treatment; group II, whose expression was not largely changed in WT cells after elicitor treatment; and group III, whose expression decreased in WT cells after elicitor treatment (Figure 2D). Group I included 14 genes (39%), group II included 17 genes (47%), and group III included 5 genes (14%).

Furthermore, functional classification of these OsTGAP1 target genes was performed. The genes were classified into the following categories according to function: biosynthetic enzyme, defence response, transcription factor, protein kinase, transporter, other function, and unknown function. The results of the functional classification are summarized in Figure 2E. The 86 genes that were upregulated in the OsTGAP1ox cells were classified as follows: biosynthetic enzyme, 24 genes (27.9%); defence response, 6 genes (7.0%); transcription factor, 7 genes (8.1%); protein kinase, 8 genes (9.3%); transporter, 6 genes (7.0%); other function, 13 genes (15.1%); unknown function, 22 genes (25.6%). The 36 genes that were downregulated in the OsTGAP1ox cells were functionally classified as follows: biosynthetic enzyme, 12 genes (33.3%); defence response, 3 genes (8.3%); transcription factor, 9 genes (25%); protein kinase, 3 genes (8.3%); transporter, 1 gene (2.8%); other function, 4 genes (11.1%); unknown function, 4 genes (11.1%).

OsTGAP1-binding regions in phytoalexin biosynthetic gene clusters

Although nearly all of the momilactone and phytocassane biosynthetic genes were confirmed to be hyperinductively expressed in the elicitor-stimulated OsTGAP1ox cells through transcriptome analysis (Table S3), none of the momilactone and phytocassane biosynthetic genes in the clusters were identified as OsTGAP1 target genes (Figure S4 and S5). These results suggest that OsTGAP1 does not directly regulate the expression of these genes by binding to their promoter regions. Intriguingly, several OsTGAP1-binding regions were found in intergenic regions among and near the cluster regions (Figure 3). This result raises the possibility that OsTGAP1 positively effects the expression of clustered genes by binding to these regions, although it may also indirectly regulate the expression of these genes via other transcription factors.

A

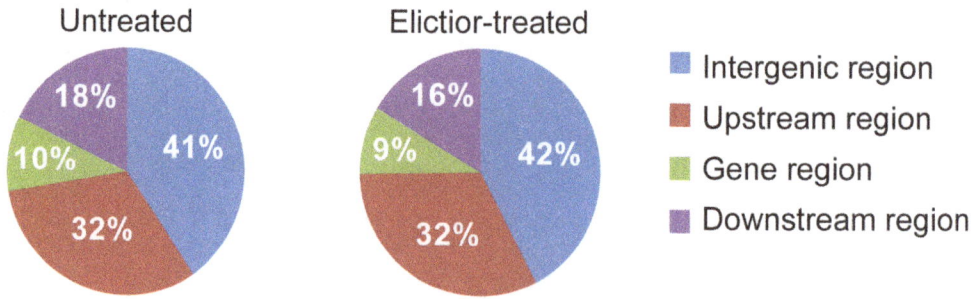

B

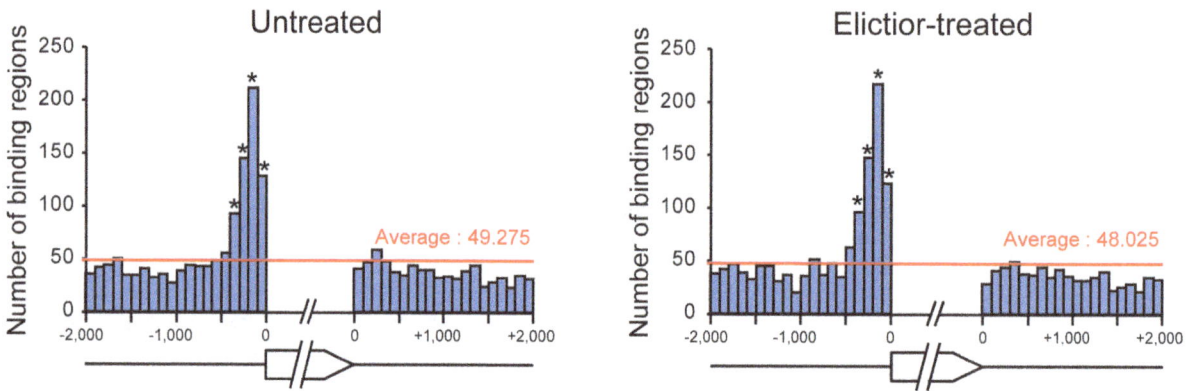

C

D

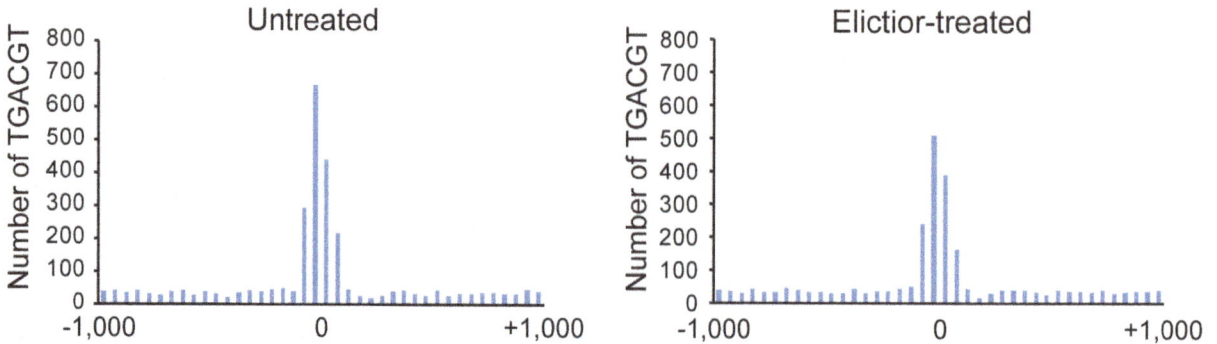

Figure 1. Overview of the results of ChIP-seq. (A) Distribution of OsTGAP1-binding regions in the rice genome. The upstream region includes the binding regions within 2 kbp of the transcription start site. The gene region includes the binding regions that are located between the transcription start site and the transcription termination site. The downstream region includes the binding regions within 2 kbp of the transcription termination site. The remaining binding regions were assigned to intergenic regions. (B) Distribution of OsTGAP1-binding regions in the upstream and downstream regions. Red lines indicate the average of number of OsTGAP1-binding regions. Statistical analysis was performed for each 100-bp region, and significantly enriched regions were indicated by asterisks (P<0.01 the binomial test and the Bonferroni correction). (C) Overrepresented motifs in the OsTGAP1-binding regions analysed in the untreated and elicitor-treated conditions using Partek Genomics Suite based on Gibbs motif sampler [37]. (D) Distribution of TGACGT sequences around the OsTGAP1-binding regions.

A

Genes that possess
OsTGAP1-binding region
within 500 bp upstream region
693

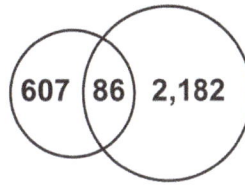

Up-regulated genes
in OsTGAP1ox
2,268

B

Genes that possess
OsTGAP1-binding region
within 500 bp upstream region
693

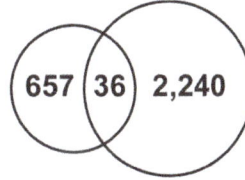

Down-regulated genes
in OsTGAP1ox
2,276

C

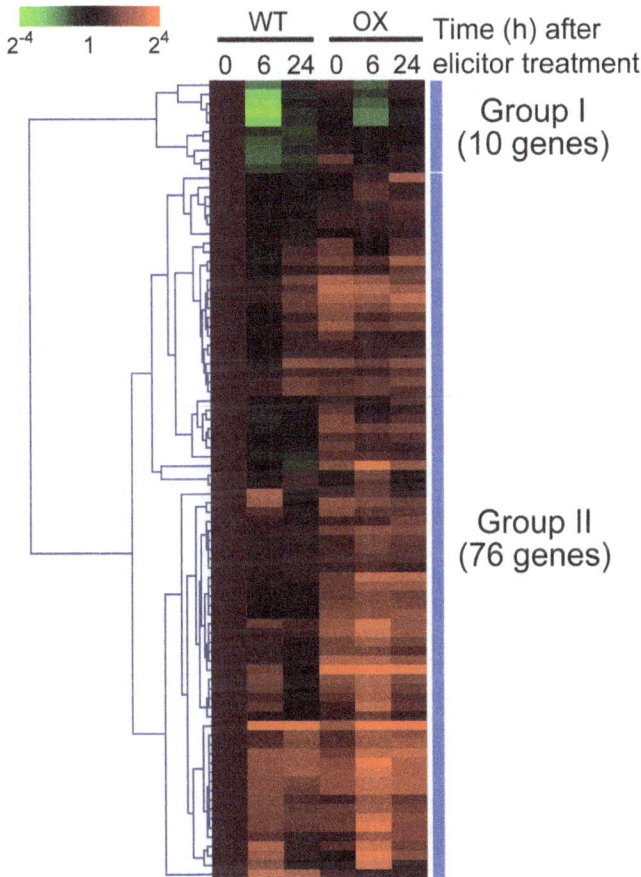

D

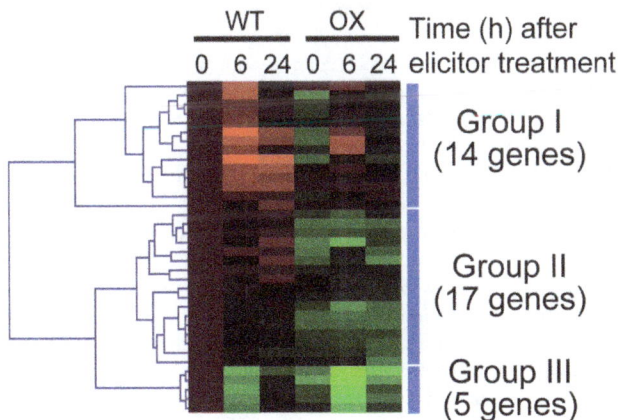

E

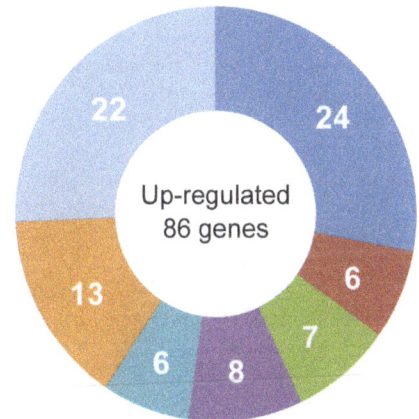

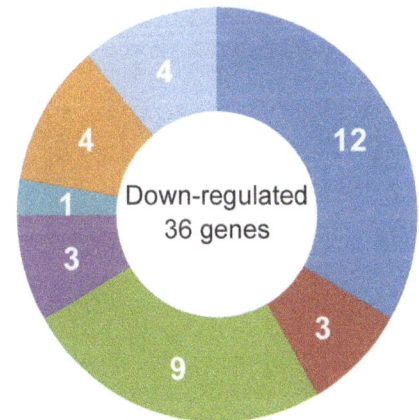

- Biosynthetic enzyme
- Defense response
- Transcription factor
- Protein kinase
- Transporter
- Other function
- Unknown function

Figure 2. Expression profiles and functional classification of the OsTGAP1 target genes. (A and B) Venn diagrams showing the overlap of genes that possess an OsTGAP1-binding regions within the 400-bp upstream region with the upregulated or downregulated genes in OsTGAP1-overexpressing rice cells. (C and D) Hierarchical clustering of the OsTGAP1 target genes using the cosine correlation and average linkage methods. WT and OX represent wild-type rice cells and OsTGAP1-overexpressing rice cells, respectively. Each column represents the mean of three biological replicates at each time point shown above the heat map. Colours represent induction (red) and repression (green), as indicated by the colour bar. The values of the heat maps are relative to those in the WT cells 0 h after the elicitor treatment. (E) Functional classification of the OsTGAP1 target genes. Categories are as follows: biosynthetic enzyme, defence response, transcription factor, protein kinase, transporter, other function, and unknown function.

OsTGAP1 binds to the *OsDXS3* promoter via the TGACGT sequence

OsDXS3, which encodes 1-deoxy-D-xylulose-5-phosphate synthase (DXS) (EC 2.2.1.7), was included among the OsTGAP1 target genes. The MEP pathway is involved in diterpenoid phytoalexin production in rice [7], and DXS catalyses a key step of this pathway [38]. Therefore, we dissected the transcriptional regulation of *OsDXS3* to understand the regulation of diterpenoid phytoalexin production by OsTGAP1.

OsTGAP1 bound to an approximately 100-bp upstream region of the transcription start site of *OsDXS3*, both with and without elicitation, in the ChIP-seq analysis (Figure 4A). This binding was confirmed by quantitative PCR using ChIP DNA from another biological replicate. The *OsDXS3* upstream region was enriched in the ChIP DNA immunoprecipitated by the anti-OsTGAP1 antibody compared to that immunoprecipitated by the normal rabbit IgG (Figure 4B), indicating that OsTGAP1 binds to the *OsDXS3* promoter. This OsTGAP1-binding region contained two TGACGT sequences (Figure 5A). GMSAs were performed using a DNA probe containing the *OsDXS3* promoter to investigate whether OsTGAP1 recognized these two TGACGT sequences. As shown in Figure 5B, the GST-fused OsTGAP1 (GST-OsTGAP1) recombinant protein could bind the DNA probe containing the *OsDXS3* promoter. When either TGACGT sequence in this region was mutated (m1 and m2), the binding of GST-OsTGAP1 to the DNA probe was weakened. When both TGACGT sequences were mutated (m3), GST-OsTGAP1 no longer bound to the DNA probe (Figure 5B). Taken together, these results indicate that OsTGAP1 binds to the *OsDXS3* promoter via two TGACGT sequences.

OsTGAP1 directly regulates *OsDXS3* expression

To further investigate whether the two TGACGT sequences on the *OsDXS3* promoter are involved in the transcriptional regulation of *OsDXS3*, a transactivation assay was performed. Fragment of 250 bp upstream from the ATG translation start site of *OsDXS3* was cloned and fused to the *FLUC* gene to produce a reporter plasmid (Figure 6A). Effector plasmids were constructed that contained a *GUS* or *OsTGAP1* gene under the control of the maize ubiquitin promoter. The reporter plasmid and either the *GUS* or *OsTGAP1* effector plasmid were delivered into cultured rice cells, along with an internal control plasmid that contained the *RLUC* gene under the control of the CaMV *35S* promoter, by particle bombardment. The LUC activities were quantified and calculated as described in the 'Materials and Methods'.

Rice cells with the reporter and *OsTGAP1* effector plasmids showed 7-fold higher LUC activity than those with the reporter and *GUS* effector plasmids (Figure 6B). Because *OsDXS3* mRNA level is increased by elicitor treatment, LUC activities after elicitor treatment were also measured. Although the enhancement of LUC activity by OsTGAP1 was detected, no change to LUC activity was observed following elicitor treatment (Figure 6B). These results suggest that a 250-bp fragment of the *OsDXS3* promoter contains elements contributing to the OsTGAP1-

dependent induction of *OsDXS3* expression but not to the gene's responsiveness to the elicitor. A transactivation assay was performed using a fragment of 2 kbp upstream from the ATG translation start site of *OsDXS3*. Rice cells with the reporter plasmid containing a 2-kbp fragment of the *OsDXS3* promoter and the *OsTGAP1* effector plasmid showed higher LUC activity than those with *GUS* effector plasmid. However, LUC activity showed no change under elicitor treatment, as was similar to the results observed for the 250-bp fragment of the *OsDXS3* promoter (Figure S6).

Next, a series of three mutants of the 250-bp fragment of the *OsDXS3* promoter fused to *FLUC* (m1, m2, and m3) were constructed. In m1 and m2, one of the two TGACGT sequences was mutated, while in m3, both TGACGT sequences were mutated. A *FLUC* reporter plasmid containing a 240-bp fragment of the *OsDXS3* promoter was also constructed, and this reporter plasmid did not contain any TGACGT sequences. This series of reporter plasmids was introduced to the cultured rice cells along with either the *GUS* or *OsTGAP1* effector plasmids, and the LUC activities were measured after incubation. The rice cells with m1 or m2 and the *OsTGAP1* effector plasmid exhibited almost the same level of LUC activity as those with the 250-bp promoter construct. However, rice cells with m3 or the 240-bp promoter construct showed significantly lower LUC activity compared to those with the 250-bp promoter construct (Figure 6C). These results indicate that the two TGACGT sequences on the *OsDXS3* promoter are involved in the OsTGAP1-dependent induction of *OsDXS3* expression. Taken together with the fact that OsTGAP1 binds to the *OsDXS3* promoter via these two TGACGT sequences, the results strongly suggest that OsTGAP1 directly regulates *OsDXS3* expression.

Discussion

Genome-wide analysis of OsTGAP1 target genes

In this study, ChIP-seq and transcriptome analyses were performed to identify the OsTGAP1 target genes. From the ChIP-seq analysis, the TGACGT sequence was found to be an overrepresented motif in the OsTGAP1-binding regions (Figure 1C, Figure S2). However, there also be the OsTGAP1-binding regions which did not contain any TGACGT sequences (Table S3, Table S4), suggesting that OsTGAP1 may bind to these regions through other motifs. OsTGAP1 may also interact with a DNA-binding protein that recognizes other motifs. However, known binding motifs of other transcription factors were not found in this study (Figure S2). To investigate how OsTGAP1 binds to these regions, a more detailed functional analysis of OsTGAP1 using a biochemical approach is required.

The locations of the OsTGAP1-binding regions assigned to the upstream region were found to be concentrated within 400 bp of the transcription start sites (Figure 1B), suggesting that OsTGAP1 binds near the transcription start sites and regulates the expression of downstream genes. Therefore, this study focussed on those genes that possessed an OsTGAP1-binding region within 400 bp of the upstream region and identified the OsTGAP1 target genes

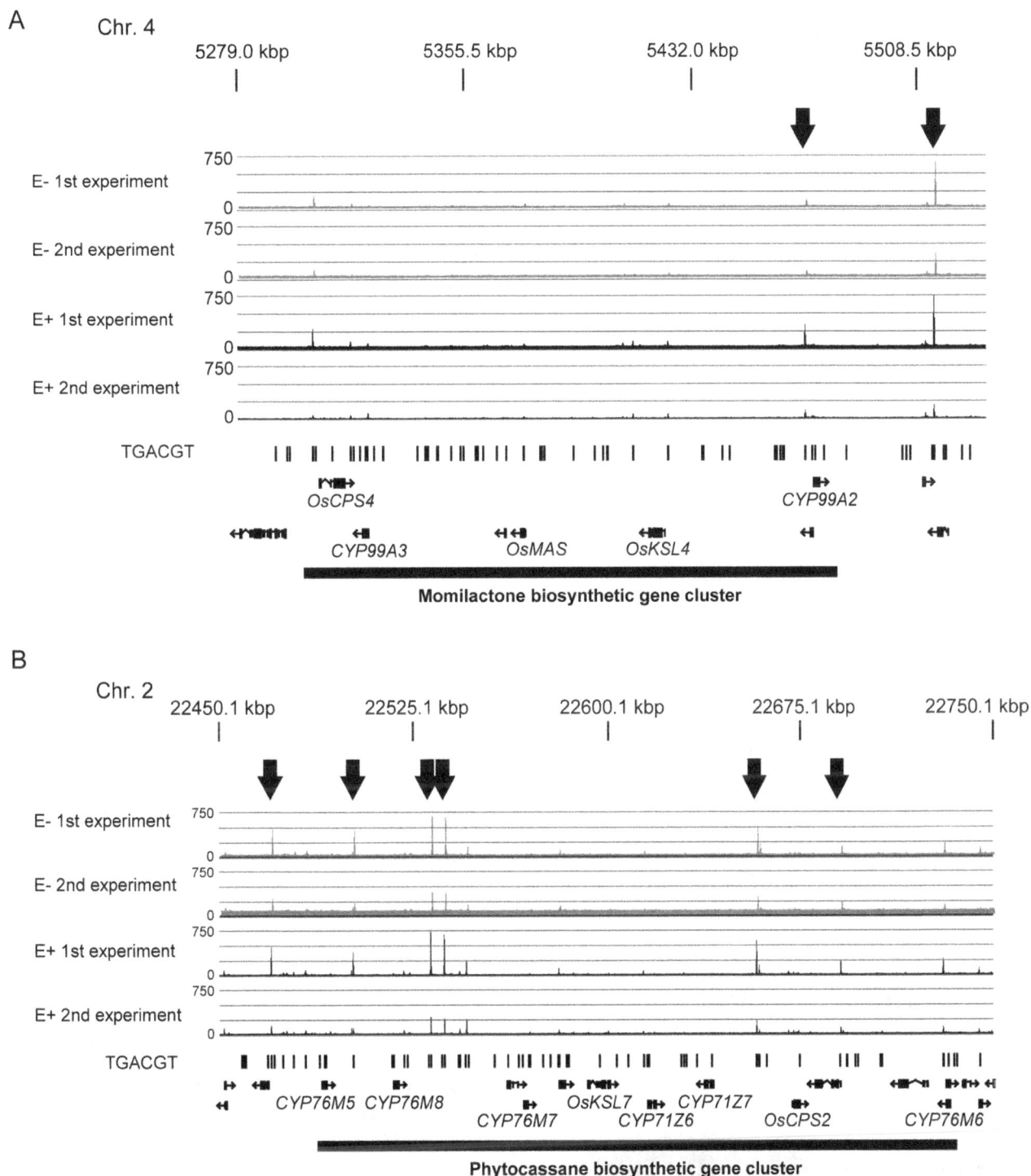

Figure 3. OsTGAP1-binding regions in diterpenoid phytoalexin biosynthetic gene clusters. Mapped ChIP-seq reads in the momilactone (A) and phytocassane biosynthetic gene clusters (B) in untreated (E−, grey) and elicitor-treated (E+, black) conditions, visualized using Partek Genomics Suite. Two biological replicates were performed. Black bars indicate the positions of TGACGT sequences. The genes in these regions are shown in the bottom row.

through comparisons to transcriptome data. A total of 86 upregulated genes, along with 36 downregulated genes, were identified in OsTGAP1ox cells (Table S8, Table S9). These target genes were classified based on their expression patterns. Most of the 86 upregulated genes (group II) exhibited an elicitor-induced

expression pattern in WT cells. Moreover, this elicitor-induced expression was enhanced in OsTGAP1ox cells (Figure 2C). In light of a previous report that OsTGAP1 shows transactivational activity [17], OsTGAP1 may function to enhance the expression of these genes, including *OsDXS3*. However, the regulatory

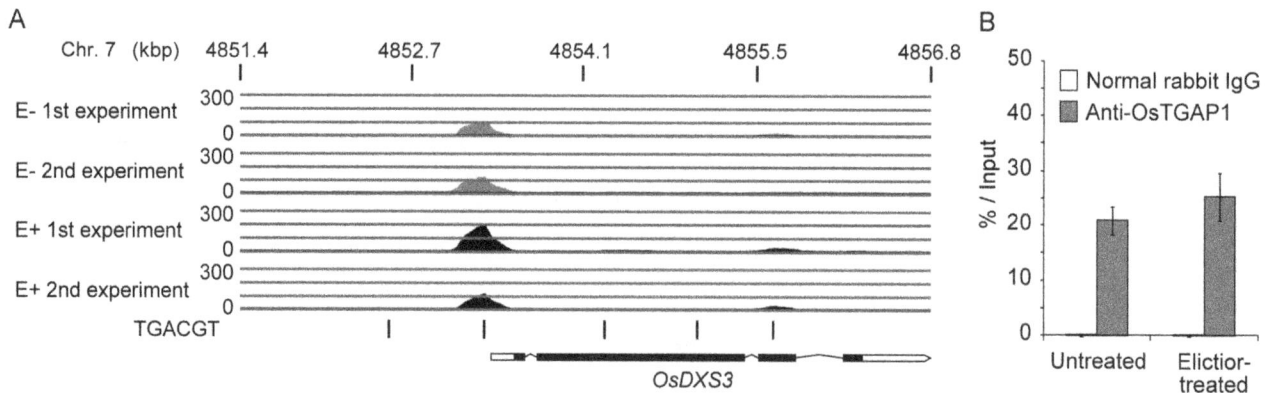

Figure 4. OsTGAP1-binding regions in the *OsDXS3* upstream region. (A) The mapped ChIP-seq reads in untreated (E−, grey) and elicitor-treated (E+, black) conditions were visualized using Partek Genomics Suite. Two biological replicates were performed. Black bars indicate the positions of TGACGT sequences. The gene structure of *OsDXS3* is shown in the bottom row. Open and closed squares indicate untranslated and coding regions, respectively. Lines indicate introns. (B) ChIP-PCR was performed using ChIP DNA immunoprecipitated by the anti-OsTGAP1 antibody and normal rabbit IgG. Values indicate the ratio of the amount of DNA in the ChIP DNA to the amount in the 'Input' control (n = 3); bars indicate the standard deviation of the mean.

mechanisms of the 36 genes that were downregulated in OsTGAP1ox cells remain unclear. As these 36 genes could be classified into three groups based on their expression patterns (Figure 2D), each group may be regulated by OsTGAP1 in a different manner. Recent studies have reported that several

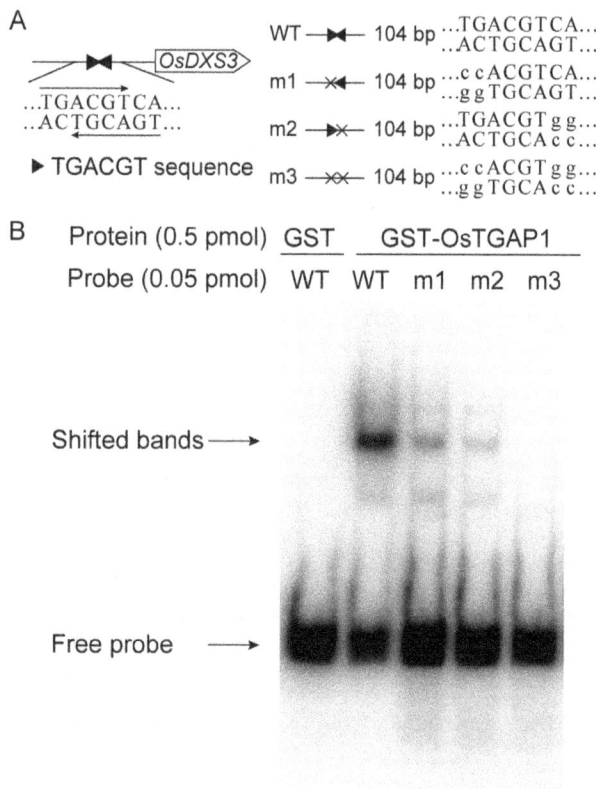

Figure 5. Gel mobility shift assay using the *OsDXS3* upstream region. (A) DNA probes used in GMSA. Closed triangles indicate TGACGT sequences. (B) GMSA was performed using purified recombinant GST-fused OsTGAP1 (GST-OsTGAP1) protein and ^{32}P-labelled DNA probes containing the TGACGT sequences in the *OsDXS3* promoter. WT: wild-type probe, m1–m3: mutated probes.

transcription factors in plants act as both activators and repressors [39,40]. These transcription factors are thought to interact with coactivators and corepressors, thereby altering these functions. OsTGAP1 may interact with as yet unknown corepressors to suppress the expression of these genes.

A substantial number of the OsTGAP1 target genes encode enzymes of biosynthetic pathways other than terpenoid biosynthesis, such as flavonoid biosynthesis and lipid metabolism (Figure 2E, Table S8, Table S9). This fact suggests that OsTGAP1 is involved not only in the regulation of diterpenoid phytoalexin production but also in the regulation of other biosynthetic processes. In addition, the 86 genes that were upregulated in OsTGAP1ox cells include several defence-related genes, such as glucanase (Os07g0168600) and cystatin (Os01g0803200 and Os01g0915200; Figure 2E, Table S8). Furthermore, *OsPLDbeta1* (Os10g0524400) and *OsWRKY76* (Os09g0417600) were included among the OsTGAP1 target genes that were downregulated in OsTGAP1ox cells (Table S9). As OsPLDbeta1 and OsWRKY76 negatively regulate the rice defence response against *M. oryzae* [41,42], OsTGAP1 may contribute to rice defence responses against pathogens by regulating the expression of these genes. In addition, the OsTGAP1 target genes upregulated in OsTGAP1ox cells included several genes related to plant hormone signalling: *DWARF AND LOW-TILLERING/OsGRAS32* (Os06g0127800), which relates to brassinosteroid signalling [43], and *OsBIF2* (Os12g0614600), which relates to auxin signalling [44] (Table S8). These results raise the possibility that OsTGAP1 impacts the signalling of these plant hormones.

Regulation of MEP pathway genes by OsTGAP1

OsTGAP1 binds to the *OsDXS3* promoter via two TGACGT sequences (Figure 5B), and these two TGACGT sequences are involved in the OsTGAP1-dependent induction of *OsDXS3* expression (Figure 6C). These results strongly suggest that OsTGAP1 directly regulates *OsDXS3* expression. However, a 250-bp fragment and a 2-kbp fragment of the *OsDXS3* promoter including the OsTGAP1-binding region exhibited no responsiveness to the elicitor (Figure 6B, Figure S6), despite the increase in *OsDXS3* mRNA level due to the elicitor treatment [7]. The regulatory mechanism of the elicitor-induced expression of *OsDXS3* requires further explanation.

A

Translation start site (+1)

-250

5′ ▶◀ OsDXS3 ▷ 3′ ▶ TGACGT sequence

(TATA box)

-250 **ATGACGTCA**TTATTCCCCTCTCGCGCGCC<u>TATAAAAT</u>CCG

⌐→ Transcription start site

-210 CCTCCATCTCTGCTTCTCCTCCCACTCCTCTTCCTCCTCC
-170 TCCTCCTCCTCCTGCGATCCCGGCGCAGCATCGATCGCCA
-130 TTGCCACTCGAGCTCCGAGCTCCTCGGAAAAGCAGTGAAC
-90 CCACCACTAAGCTCGCGACATCTCGCAACACAGCTTGCTT
-50 GCTGCGTCGCGGTATATAGAGCAGTAGCTAGCTTAGCTTG
-10 TCGATCTCCA

B

■ OsTGAP1 Elicitor +
▨ OsTGAP1 Elicitor -
▧ GUS Elicitor +
□ GUS Elicitor -

Relative LUC activity (firefly LUC / *Renilla* LUC)

C

■ OsTGAP1
□ GUS

Relative LUC activity (firefly LUC / *Renilla* LUC)

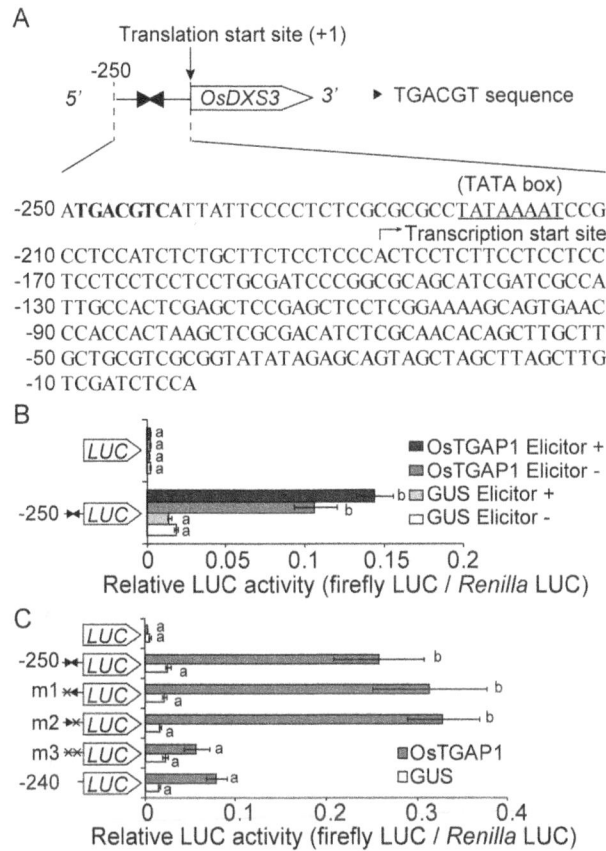

Figure 6. Transactivation assay using a 250-bp upstream region of *OsDXS3*. (A) Nucleotide sequence of the 250-bp upstream region of the ATG translation start site of *OsDXS3*. Closed triangles indicate TGACGT sequences. The TGACGT sequences in the nucleotide sequence are also indicated by bold characters. The putative TATA box is underlined. Transcription start sites on RAP-DB are indicated by black arrows. (B) Transactivation assay using a 250-bp fragment of the *OsDXS3* promoter with GUS or OsTGAP1 effector plasmid (0.1 µg per bombardment). Values indicate the relative luciferase (LUC) activities (firefly LUC/*Renilla* LUC) after 24 h incubation of rice cells with or without chitin elicitor treatment (n = 3); bars indicate the standard error of the mean. Statistically different data groups are indicated by different letters (P<0.01 by one-way ANOVA with a Tukey's *post hoc* test). (C) Transactivation assay using mutated or deleted *OsDXS3* promoters with GUS or OsTGAP1 effector plasmids (0.1 µg per bombardment). Values indicate the relative LUC activities (firefly LUC/*Renilla* LUC) after 24 h incubation of rice cells without chitin elicitor treatment (n = 12); bars indicate the standard error of the mean. Statistically different data groups are indicated by different letters (P<0.01 by one-way ANOVA with a Tukey's post *hoc* test).

The rice genome encodes three *DXS* genes: *OsDXS1*, *OsDXS2*, and *OsDXS3* [45]. Among these three *DXS* genes, *OsDXS3* is the only one upregulated by elicitor treatment [17]. This observation implies that *OsDXS3* is responsible for diterpenoid phytoalexin production and that *OsDXS1* and *OsDXS2* are involved in the biosynthesis of terpenoids for primary metabolism. The OsTGAP1 target genes identified in this study included *OsDXS3* but not *OsDXS1* and *OsDXS2*. The expression of *OsDXS1* and *OsDXS2* was also unchanged in OsTGAP1ox cells (Table S10). These results suggest that the transcriptional regulation of the MEP pathway by OsTGAP1 mainly contributes to diterpenoid phytoalexin production and that its contribution to the biosynthesis of terpenoids for primary metabolism is limited.

Of the MEP pathway genes, only *OsDXS3* was included among the OsTGAP1 target genes. Nevertheless, two MEP pathway genes (*OsHDS* and *OsHDR*) were found among the upregulated genes in OsTGAP1ox cells (Table S3). Other MEP pathway genes (*OsDXR*, *OsCMS*, *OsCMK*, and *OsMCS*) were also upregulated by approximately 1.5-fold 24 h after elicitor treatment in the OsTGAP1ox cells compared to their expression after treatment in WT cells, although these genes were eliminated by the threshold of the data analysis (Table S10). These results indicate that the expression of all seven MEP pathway genes is affected by OsTGAP1. However, *OsDXS3* was more strongly upregulated than were the other MEP pathway genes in OsTGAP1ox cells. Therefore, the enhanced expression of MEP pathway genes by OsTGAP1, except that of *OsDXS3*, may be indirect.

Possible mechanism for the regulation of diterpenoid phytoalexin biosynthetic gene clusters by OsTGAP1

OsTGAP1ox cells exhibit enhanced expression of all momilactone biosynthetic genes and the phytocassane biosynthetic gene *OsKSL7* [17]. In this study, other phytocassane biosynthetic genes in the cluster (*OsCPS2*, *CYP76M5-8*, and *CYP71Z6*) were found to be upregulated in the OsTGAP1ox cells (Table S3), suggesting that OsTGAP1 functions in the transcriptional regulation of the two diterpenoid phytoalexin biosynthetic gene clusters. However, a simple model in which OsTGAP1 directly regulates the expression of these biosynthetic genes through binding to each promoter region followed by transactivation was not supported by the ChIP-seq analysis (Figure 3, Figure S4 and Figure S5).

Several transcription factor genes were found among the candidate OsTGAP1 target genes, including *OsWRKY76* (Figure 2E, Table S8, Table S9). Recently, *OsWRKY76* has been reported to negatively regulate diterpenoid phytoalexin biosynthetic genes [42]. Therefore, the downregulation of *OsWRKY76* expression may be among the mechanisms explaining the enhanced expression of diterpenoid phytoalexin biosynthetic genes in OsTGAP1ox cells.

Another hypothesis is that the binding of OsTGAP1 to the intergenic regions of the gene clusters plays a particular role in the transcriptional regulation of diterpenoid phytoalexin biosynthetic genes. In this study, at least 40% of the OsTGAP1-binding regions were located within intergenic regions (Figure 1A). This proportion is larger than those observed for the other plant transcription factors (HY5, PIL5, AGL15, and BZR1) whose genome-wide binding regions have been previously reported [33–36]. However, in the case of FHY3, which is a component of phytochrome A signalling, approximately 40% of FHY3-binding regions are located in intergenic regions, and a large portion of these intergenic binding regions are localized in the centromeric regions [46]. FHY3 binds to the promoter regions of its target genes, thereby regulating phytochrome A signalling and the circadian clock. FHY3 also binds to the centromeric repeats, suggesting that FHY3 has a function beyond regulating the expression of target genes via binding to their promoter regions. Similar to that of FHY3, the binding of OsTGAP1 to intergenic regions may also have unknown but essential functions. The information presented in this study regarding OsTGAP1-binding regions near and in the diterpenoid phytoalexin biosynthetic gene cluster regions will contribute to future research investigating the regulation of these gene clusters by OsTGAP1.

Supporting Information

Figure S1 Specificity of anti-OsTGAP1 antibody.

Figure S2 The enrichment of each motif in OsTGAP1-binding regions.

Figure S3 Summary of genes whose expression was altered in OsTGAP1-overexpressing rice cells.

Figure S4 OsTGAP1-binding regions around momilactone biosynthetic genes.

Figure S5 OsTGAP1-binding regions around phytocassane biosynthetic genes.

Figure S6 Transactivation assay using the 2-kbp fragment of the *OsDXS3* promoter.

Table S1 Plasmids used in this study.

Table S2 Primers used in this study.

Table S3 List of OsTGAP1-binding regions in untreated condition.

Table S4 List of OsTGAP1-binding regions in elicitor-treated condition.

Table S5 List of genes that possess OsTGAP1-binding site within 400 bp upstream region.

Table S6 List of genes whose expression is significantly upregulated in OsTGAP1-overexpressinng rice cells.

Table S7 List of genes whose expression is significantly downregulated in OsTGAP1-overexpressinng rice cells.

Table S8 List of OsTGAP1 target genes whose expression is upregulated in OsTGAP1-overexpressinng rice cells.

Table S9 List of OsTGAP1 target genes whose expression is downregulated in OsTGAP1-overexpressinng rice cells.

Table S10 Expression profiles of MEP pathway genes from microarray analysis.

Acknowledgments

We thank Dr. N. Shibuya for the *N*-acetylchitooctaose; Dr. M. Kishi-Kaboshi and Dr. H. Takatsuji for the pUbi_RfA_Tnos plasmid; Dr. T. Nakagawa for the R4pGWB501 plasmid; Dr. K. Shimamoto for the pANDA plasmid; Dr. M. Takagi for the pPTRL plasmid; Mr N. Seki for his kind support with the data analysis using Partek Genomics Suite; and Dr. Y. Nagamura and Ms R. Motoyama for their technical support with the microarray analysis.

Author Contributions

Conceived and designed the experiments: KM HN H. Yamane KO. Performed the experiments: KM TM AO KK TC. Analyzed the data: KM TM. Contributed reagents/materials/analysis tools: H. Yoshikawa. Contributed to the writing of the manuscript: KM TC H. Yamane KO.

References

1. Ahuja I, Kissen R, Bones AM (2012) Phytoalexins in defense against pathogens. Trends Plant Sci 17: 73–90.
2. Kato T, Kabuto C, Sasaki N, Tsunagawa M, Aizawa H, et al. (1973) Momilactones, growth inhibitors from rice, oryza sativa L. Tetrahedron Lett 14: 3861–3864.
3. Cartwright DW, Langcake P, Pryce RJ, Leworthy DP, Ride JP (1981) Isolation and characterization of two phytoalexins from rice as momilactones A and B. Phytochemistry 20: 535–537.
4. Okada K (2011) The biosynthesis of isoprenoids and the mechanisms regulating it in plants. Biosci Biotechnol Biochem 75: 1219–1225.
5. Yamane H (2013) Biosynthesis of phytoalexins and regulatory mechanisms of it in rice. Biosci Biotechnol Biochem 77: 1141–1148.
6. Lichtenthaler HK, Schwender J, Disch A, Rohmer M (1997) Biosynthesis of isoprenoids in higher plant chloroplasts proceeds via a mevalonate-independent pathway. FEBS Lett 400: 271–274.
7. Okada A, Shimizu T, Okada K, Kuzuyama T, Koga J, et al. (2007) Elicitor induced activation of the methylerythritol phosphate pathway toward phyto-alexins biosynthesis in rice. Plant Mol Biol 65: 177–187.
8. Cho EM, Okada A, Kenmoku H, Otomo K, Toyomasu T, et al. (2004) Molecular cloning and characterization of a cDNA encoding ent-cassa-12,15-diene synthase, a putative diterpenoid phytoalexin biosynthetic enzyme, from suspension-cultured rice cells treated with a chitin elicitor. The Plant J 37: 1–8.
9. Otomo K, Kanno Y, Motegi A, Kenmoku H, Yamane H, et al. (2004) Diterpene cyclases responsible for the biosynthesis of phytoalexins, momilactones A, B, and oryzalexins A-F in rice. Biosci Biotechnol Biochem 68: 2001–2006.
10. Otomo K, Kenmoku H, Oikawa H, Konig WA, Toshima H, et al. (2004) Biological functions of ent- and syn-copalyl diphosphate synthases in rice: key enzymes for the branch point of gibberellin and phytoalexin biosynthesis. The Plant J 39: 886–893.
11. Shimura K, Okada A, Okada K, Jikumaru Y, Ko KW, et al. (2007) Identification of a biosynthetic gene cluster in rice for momilactones. J Biol Chem 282: 34013–34018.
12. Wang Q, Hillwig ML, Peters RJ (2011) CYP99A3: functional identification of a diterpene oxidase from the momilactone biosynthetic gene cluster in rice. The Plant J 65: 87–95.
13. Swaminathan S, Morrone D, Wang Q, Fulton DB, Peters RJ (2009) CYP76M7 is an ent-cassadiene C11alpha-hydroxylase defining a second multifunctional diterpenoid biosynthetic gene cluster in rice. Plant Cell 21: 3315–3325.
14. Wu Y, Hillwig ML, Wang Q, Peters RJ (2011) Parsing a multifunctional biosynthetic gene cluster from rice: Biochemical characterization of CYP71Z6 & 7. FEBS Lett 585: 3446–3451.
15. Wang Q, Hillwig ML, Okada K, Yamazaki K, Wu Y, et al. (2012) Characterization of CYP76M5–8 indicates metabolic plasticity within a plant biosynthetic gene cluster. J Biol Chem 287: 6159–6168.
16. Wang Q, Hillwig ML, Wu Y, Peters RJ (2012) CYP701A8: a rice ent-kaurene oxidase paralog diverted to more specialized diterpenoid metabolism. Plant Physiol 158: 1418–1425.
17. Okada A, Okada K, Miyamoto K, Koga J, Shibuya N, et al. (2009) OsTGAP1, a bZIP transcription factor, coordinately regulates the inductive production of diterpenoid phytoalexins in rice. J Biol Chem 284: 26510–26518.
18. Jakoby M, Weisshaar B, Droge-Laser W, Vicente-Carbajosa J, Tiedemann J, et al. (2002) bZIP transcription factors in Arabidopsis. Trends Plant Sci 7: 106–111.
19. Alves M, Dadalto S, Gonçalves A, De Souza G, Barros V, et al. (2013) Plant bZIP Transcription Factors Responsive to Pathogens: A Review. Int J Mol Sci 14: 7815–7828.
20. Moore JW, Loake GJ, Spoel SH (2011) Transcription dynamics in plant immunity. Plant Cell 23: 2809–2820.
21. Thibaud-Nissen F, Wu H, Richmond T, Redman JC, Johnson C, et al. (2006) Development of Arabidopsis whole-genome microarrays and their application to the discovery of binding sites for the TGA2 transcription factor in salicylic acid-treated plants. The Plant J 47: 152–162.
22. Nijhawan A, Jain M, Tyagi AK, Khurana JP (2008) Genomic Survey and Gene Expression Analysis of the Basic Leucine Zipper Transcription Factor Family in Rice. Plant Physiol 146: 333–350.
23. Chern MS, Fitzgerald HA, Yadav RC, Canlas PE, Dong X, et al. (2001) Evidence for a disease-resistance pathway in rice similar to the NPR1-mediated signaling pathway in Arabidopsis. The Plant J 27: 101–113.
24. Fitzgerald HA, Canlas PE, Chern MS, Ronald PC (2005) Alteration of TGA factor activity in rice results in enhanced tolerance to Xanthomonas oryzae pv. oryzae. The Plant J 43: 335–347.
25. Yuan Y, Zhong S, Li Q, Zhu Z, Lou Y, et al. (2007) Functional analysis of rice NPR1-like genes reveals that OsNPR1/NH1 is the rice orthologue conferring disease resistance with enhanced herbivore susceptibility. Plant Biotechnol J 5: 313–324.

26. Delteil A, Blein M, Faivre-Rampant O, Guellim A, Estevan J, et al. (2012) Building a mutant resource for the study of disease resistance in rice reveals the pivotal role of several genes involved in defence. Mol Plant Pathol 13: 72–82.

27. Ito Y, Kaku H, Shibuya N (1997) Identification of a high-affinity binding protein for N-acetylchitooligosaccharide elicitor in the plasma membrane of suspension-cultured rice cells by affinity labeling. The Plant J 12: 347–356.

28. Miki D, Shimamoto K (2004) Simple RNAi Vectors for Stable and Transient Suppression of Gene Function in Rice. Plant Cell Physiol 45: 490–495.

29. Nakagawa T, Nakamura S, Tanaka K, Kawamukai M, Suzuki T, et al. (2008) Development of R4 Gateway Binary Vectors (R4pGWB) Enabling High-Throughput Promoter Swapping for Plant Research. Biosci Biotechnol Biochem 72: 624–629.

30. Chujo T, Miyamoto K, Ogawa S, Masuda Y, Shimizu T, et al. (2014) Overexpression of Phosphomimic Mutated OsWRKY53 Leads to Enhanced Blast Resistance in Rice. PLoS ONE 9: e98737.

31. Li H, Durbin R (2009) Fast and accurate short read alignment with Burrows-Wheeler transform. Bioinformatics 25: 1754–1760.

32. Ohta M, Ohme-Takagi M, Shinshi H (2000) Three ethylene-responsive transcription factors in tobacco with distinct transactivation functions. The Plant J 22: 29–38.

33. Lee J, He K, Stolc V, Lee H, Figueroa P, et al. (2007) Analysis of Transcription Factor HY5 Genomic Binding Sites Revealed Its Hierarchical Role in Light Regulation of Development. Plant Cell 19: 731–749.

34. Oh E, Kang H, Yamaguchi S, Park J, Lee D, et al. (2009) Genome-Wide Analysis of Genes Targeted by PHYTOCHROME INTERACTING FACTOR 3-LIKE5 during Seed Germination in Arabidopsis. Plant Cell 21: 403–419.

35. Zheng Y, Ren N, Wang H, Stromberg AJ, Perry SE (2009) Global Identification of Targets of the Arabidopsis MADS Domain Protein AGAMOUS-Like15. Plant Cell 21: 2563–2577.

36. Sun Y, Fan X-Y, Cao D-M, Tang W, He K, et al. (2010) Integration of Brassinosteroid Signal Transduction with the Transcription Network for Plant Growth Regulation in Arabidopsis. Dev Cell 19: 765–777.

37. Neuwald AF, Liu JS, Lawrence CE (1995) Gibbs motif sampling: Detection of bacterial outer membrane protein repeats. Protein Sci 4: 1618–1632.

38. Lois LM, Rodríguez-Concepción M, Gallego F, Campos N, Boronat A (2000) Carotenoid biosynthesis during tomato fruit development: regulatory role of 1-deoxy-D-xylulose 5-phosphate synthase. The Plant J 22: 503–513.

39. Ikeda M, Mitsuda N, Ohme-Takagi M (2009) Arabidopsis WUSCHEL Is a Bifunctional Transcription Factor That Acts as a Repressor in Stem Cell Regulation and as an Activator in Floral Patterning. Plant Cell 21: 3493–3505.

40. Bonaccorso O, Lee J, Puah L, Scutt C, Golz J (2012) FILAMENTOUS FLOWER controls lateral organ development by acting as both an activator and a repressor. BMC Plant Biol 12: 176.

41. Yamaguchi T, Kuroda M, Yamakawa H, Ashizawa T, Hirayae K, et al. (2009) Suppression of a Phospholipase D Gene, OsPLDβ1, Activates Defense Responses and Increases Disease Resistance in Rice. Plant Physiol 150: 308–319.

42. Yokotani N, Sato Y, Tanabe S, Chujo T, Shimizu T, et al. (2013) WRKY76 is a rice transcriptional repressor playing opposite roles in blast disease resistance and cold stress tolerance. J Exp Bot 64: 5085–5097.

43. Tong H, Jin Y, Liu W, Li F, Fang J, et al. (2009) DWARF AND LOW-TILLERING, a new member of the GRAS family, plays positive roles in brassinosteroid signaling in rice. The Plant J 58: 803–816.

44. Morita Y, Kyozuka J (2007) Characterization of OsPID, the Rice Ortholog of PINOID, and its Possible Involvement in the Control of Polar Auxin Transport. Plant Cell Physiol 48: 540–549.

45. Kim B-R, Kim S-U, Chang Y-J (2005) Differential expression of three 1-deoxy-D-xylulose-5-phosphate synthase genes in rice. Biotechnol Lett 27: 997–1001.

46. Ouyang X, Li J, Li G, Li B, Chen B, et al. (2011) Genome-Wide Binding Site Analysis of FAR-RED ELONGATED HYPOCOTYL3 Reveals Its Novel Function in Arabidopsis Development. Plant Cell 23: 2514–2535.

6

Differential Activity of *Striga hermonthica* Seed Germination Stimulants and *Gigaspora rosea* Hyphal Branching Factors in Rice and Their Contribution to Underground Communication

Catarina Cardoso[1], Tatsiana Charnikhova[1], Muhammad Jamil[1¤a], Pierre-Marc Delaux[2,3¤b], Francel Verstappen[1,4], Maryam Amini[1¤c], Dominique Lauressergues[2,3], Carolien Ruyter-Spira[1,5], Harro Bouwmeester[1,4]*

1 Laboratory of Plant Physiology, Wageningen University, Wageningen, the Netherlands, 2 Laboratoire de Recherche en Sciences Végétales, Unité Mixte de Recherche (UMR) 5546, Université de Toulouse, Castanet-Tolosan, France, 3 Laboratoire de Recherche en Sciences Végétales, Unité Mixte de Recherche (UMR) 5546, Centre National de la Recherche Scientifique (CNRS), Castanet-Tolosan, France, 4 Centre for Biosystems Genomics, Wageningen, the Netherlands, 5 Bioscience, Plant Research International, Wageningen, the Netherlands

Abstract

Strigolactones (SLs) trigger germination of parasitic plant seeds and hyphal branching of symbiotic arbuscular mycorrhizal (AM) fungi. There is extensive structural variation in SLs and plants usually produce blends of different SLs. The structural variation among natural SLs has been shown to impact their biological activity as hyphal branching and parasitic plant seed germination stimulants. In this study, rice root exudates were fractioned by HPLC. The resulting fractions were analyzed by MRM-LC-MS to investigate the presence of SLs and tested using bioassays to assess their *Striga hermonthica* seed germination and *Gigaspora rosea* hyphal branching stimulatory activities. A substantial number of active fractions were revealed often with very different effect on seed germination and hyphal branching. Fractions containing $(-)-$orobanchol and *ent*-2'-*epi*-5-deoxystrigol contributed little to the induction of *S. hermonthica* seed germination but strongly stimulated AM fungal hyphal branching. Three SLs in one fraction, putative methoxy-5-deoxystrigol isomers, had moderate seed germination and hyphal branching inducing activity. Two fractions contained strong germination stimulants but displayed only modest hyphal branching activity. We provide evidence that these stimulants are likely SLs although no SL-representative masses could be detected using MRM-LC-MS. Our results show that seed germination and hyphal branching are induced to very different extents by the various SLs (or other stimulants) present in rice root exudates. We propose that the development of rice varieties with different SL composition is a promising strategy to reduce parasitic plant infestation while maintaining symbiosis with AM fungi.

Editor: Maarja Öpik, University of Tartu, Estonia

Funding: This work was financed by The Netherlands Organization for Scientific Research with VICI grant nr 865.06.002 and Equipment grant nr 834.08.001, attributed to HB, and co-financed by the Centre for Biosystems Genomics, a part of the Netherlands Genomics Initiative/Netherlands Organization for Scientific Research, attributed to HB. The funders had no role in study design, data collection and analysis, decision to publish, or preparation of the manuscript.

Competing Interests: The authors have declared that no competing interests exist.

* Email: harro.bouwmeester@wur.nl

¤a Current address: Department of Biosciences, COMSATS Institute of Information Technology, Islamabad, Pakistan
¤b Current address: Department of Agronomy, University of Wisconsin – Madison, Madison, Wisconsin, United States of America
¤c Current address: Syngenta Seeds B.V., Enkhuizen, the Netherlands

Introduction

Parasitic plants of the genus *Striga* are economically important species that parasitize the dicotyledonous cowpea, and cereal crops such as rice sorghum and maize [1]. In the most affected areas, parasitic plants constitute a major constraint to food production and efficient control methods are scant. *Striga* seeds will only germinate after exposure to host derived molecules, called germination stimulants that the parasite uses to detect host presence. The first phases of root parasitism occur underground

and the presence of the parasite is difficult to diagnose until the emergence of its shoots. However, crop yield is already compromised at that stage making timely control of this pest even more difficult [2,3]. It is therefore important to develop control strategies that act before infection is initiated, for example by avoiding or reducing germination of the parasites' seeds. Strigolactones (SLs) are the best described class of germination stimulants and a reduction in the production of these compounds indeed resulted in reduced *Striga* infection [4–6]. However, SLs are also signaling compounds for the establishment of symbiosis with arbuscular mycorrhizal (AM) fungi and are plant hormones that modulate plant architecture [7–12] and therefore, non-discriminate reduction of their production would likely have negative side effects. The symbiotic AM fungi perceive SLs and respond with extensive pre-symbiotic hyphal branching, thus increasing the efficiency of root colonization. In this symbiotic interaction, the fungus takes up nutrients (especially phosphate and nitrogen) and water from the soil and supplies them to the plant in exchange for carbon assimilates [13]. Plants under phosphate starvation increase production and release of SLs into the rhizosphere to promote the symbiosis [14–16]. In soils contaminated with seeds of the parasitic plants, low phosphate availability results in increased levels of infestation by parasitic plants [4].

Adaptation responses to low phosphate such as reduced shoot branching and root system expansion are mediated by SLs [11,17,18]. SL biosynthetic mutants suffer, to some extent, from reduced symbiosis with AM fungi and exhibit altered plant shoot and root architecture which may negatively affect crop yields [9–11,19,20].

SLs are derived from all-*trans*-β-carotene that is isomerized into 9-*cis*-β-carotene by β-carotene isomerase D27 (DWARF27) followed by two consecutive cleavage steps by CAROTENOID CLEAVAGE DIOXYGENASE 7 (CCD7; HIGH TILLERING DWARF1 - HTD1/DWARF17 – 17 in rice) and CAROTENOID CLEAVAGE DIOXYGENASE 8 (CCD8; DWARF10 – D10 in rice) resulting in the production of carlactone [21]. The biosynthetic steps that convert carlactone to SL are not yet elucidated. SLs are a reasonably large class of natural compounds consisting of over 15 structural variants, most of which differ only by having one instead of two methyl groups on the cyclohexenyl A-ring or by having various combinations of hydroxyl or acetoxyl substituents on the A- and B-rings [22]. SLs occur in two distinct stereochemical configurations and the stereochemistry of some SLs was recently revised [23]. SLs from the orobanchol-like family have an *ent* oriented C-ring (Figure 1, structures 1–5 and 8c). In the strigol-like family the C-ring has the opposite chirality of the orobanchol-like family (Figure 1, structures 6;7 and 8a) [23]. Plants produce a mixture of SLs that differs between and sometimes even within species [16,24,25]. So far, only orobanchol-like SLs have been identified in rice: (−)−orobanchol (1), *ent*-2'-*epi*-5-deoxystrigol (2), orobanchyl acetate (3), 7-oxoorobanchyl acetate (4) [10,23]. In addition, three putative methoxy-5-deoxystrigol isomers (5) have been reported with unknown structure and stereochemistry [4].

Parasitic plant seeds and AM fungi have different sensitivities to different SL variants [26,27]. Interestingly, it was reported that orobanchol-like SLs (of the same type as found in rice exudates) are considerably less active at inducing *Striga hermonthica* seed germination [27]. Here, we extensively survey the chemical composition (SL content) and biological activity of rice root exudates to understand the relevance of the different SLs, and possible other signalling molecules, in the establishment of mycorrhizal symbiosis with the AM fungus, *Gigaspora rosea*, and infection by the parasitic plant, *Striga hermonthica*.

Materials and Methods

Strigolactone standards

The synthetic SL GR24 (9a–d) and (±)-strigol (8a,d, $R^1 = CH_3$; $R^2 = OH$) were kindly provided by Prof. Binne Zwanenburg (Radboud University Nijmegen, Netherlands); (−)−orobanchol (1) and (+)-*ent*-2'-*epi*-orobanchol (6), solanacol, orobanchyl acetate (3), 7-oxoorobanchyl acetate (4), 7-oxoorobanchol, and sorgomol (7) were provided by Prof. Koichi Yoneyama

(Utsunomiya University, Japan); (±)-2'-*epi*-strigol (8b,c, $R^1 = CH_3$; $R^2 = OH$), (±)-2'-*epi*-5-deoxystrigol (8b,c, $R^1 = CH_3$; $R^2 = H$) and (±)-5-deoxystrigol (8a,d, $R^1 = CH_3$; $R^2 = H$) were a gift from Prof. Tadao Asami (University of Tokyo, Japan) (Structures 1 to 8a–d represented in Figure 1).

Plant growth and root exudate collection

The exudates were collected from rice seedlings of the variety Nipponbare and the SL biosynthetic mutant line *d10-2* with Nipponbare background, kindly provided by Prof. Junko Kyozuka (University of Tokyo, Japan) [10]. The seeds were sown in pots of 14 cm diameter filled with quartz sand. The experiment was conducted with three pots per treatment. One pot containing 25 plants represents one replicate. Plants were watered every three days during the first week and every two days during the remaining weeks to full substrate saturation with half-strength modified Hoagland nutrient solution containing NH_4NO_3 (5.6 mM), K_2HPO_4 (0.4 mM), $MgSO_4$ (0.8 mM), $FeSO_4$ (0.18 mM), $CaCl_2$ (1.6 mM), K_2SO_4 (0.8 mM), $MnCl_2$ (0.0045 mM), $CuSO_4$ (0.0003 mM), $ZnCl_2$ (0.0015 mM), Na_2MoO_4 (0.0001 mM). After 3 weeks, phosphate starvation and phosphate starvation in combination with 0.01 μM fluridone – an inhibitor of carotenoid and therefore SL biosynthesis – were applied. Control plants were watered with the half-strength modified Hoagland nutrient solution described above. For the phosphate starvation treatment, KNO_3 (0.8 mM) was substituted for K_2HPO_4 to maintain the same the K^+ concentration. Residual phosphate was removed from the pots by applying 1 L of the concentration nutrient solution and draining the pots. Six days after the start of the treatments the treatment was repeated. Root exudates were collected 24 hours later by applying 1 L of the corresponding nutrient solution and collecting the flow through.

Sample preparation

The root exudates were concentrated using an SPE cartridge (GracePure™ SPE C18 – Max 500 mg) and eluted in 4 mL of 100% acetone. For HPLC, 250 μL of water was added to 1 mL of this acetone eluent after which the acetone was evaporated under a flow of N_2. The remaining 250 μL sample was injected into the HPLC and 1 min fractions (corresponding to 1 mL) were collected. The fractions were evaporated to dryness and dissolved in 200 μL water for further analysis. For MRM-LC-MS analysis, 50 μL of the C18 acetone eluent was diluted 3-fold in water and HPLC fractions were diluted 2-fold in water. For seed germination bioassays, C18 acetone eluents (crude exudates) were diluted 32-fold in water and HPLC fractions 5-fold. For the AM fungal hyphal branching bioassay the HPLC fractions of each replicate were pooled and tested in the same concentration as in the seed germination bioassay.

Fractionation of root exudates

Root exudates were fractioned by HPLC. The samples were injected into a XBridge™ C18 column (4,6*150 mm from 5 μm, Waters) using a U6K injector (Waters). For the gradient model 510 pumps (Waters) were used. The mobile phase was water and the following gradient to acetonitrile used: 1 min 100% water, 2 min 27% acetonitrile, 15 min 45% acetonitrile, 24 min 80% acetonitrile and 24.2 min 100% acetonitrile which was maintained for 4 minutes to clean the column. The flow rate was 1 mL min^{-1} and the column temperature 25°C. Fractions of one minute were collected using a Biofrac fraction collector (Biorad).

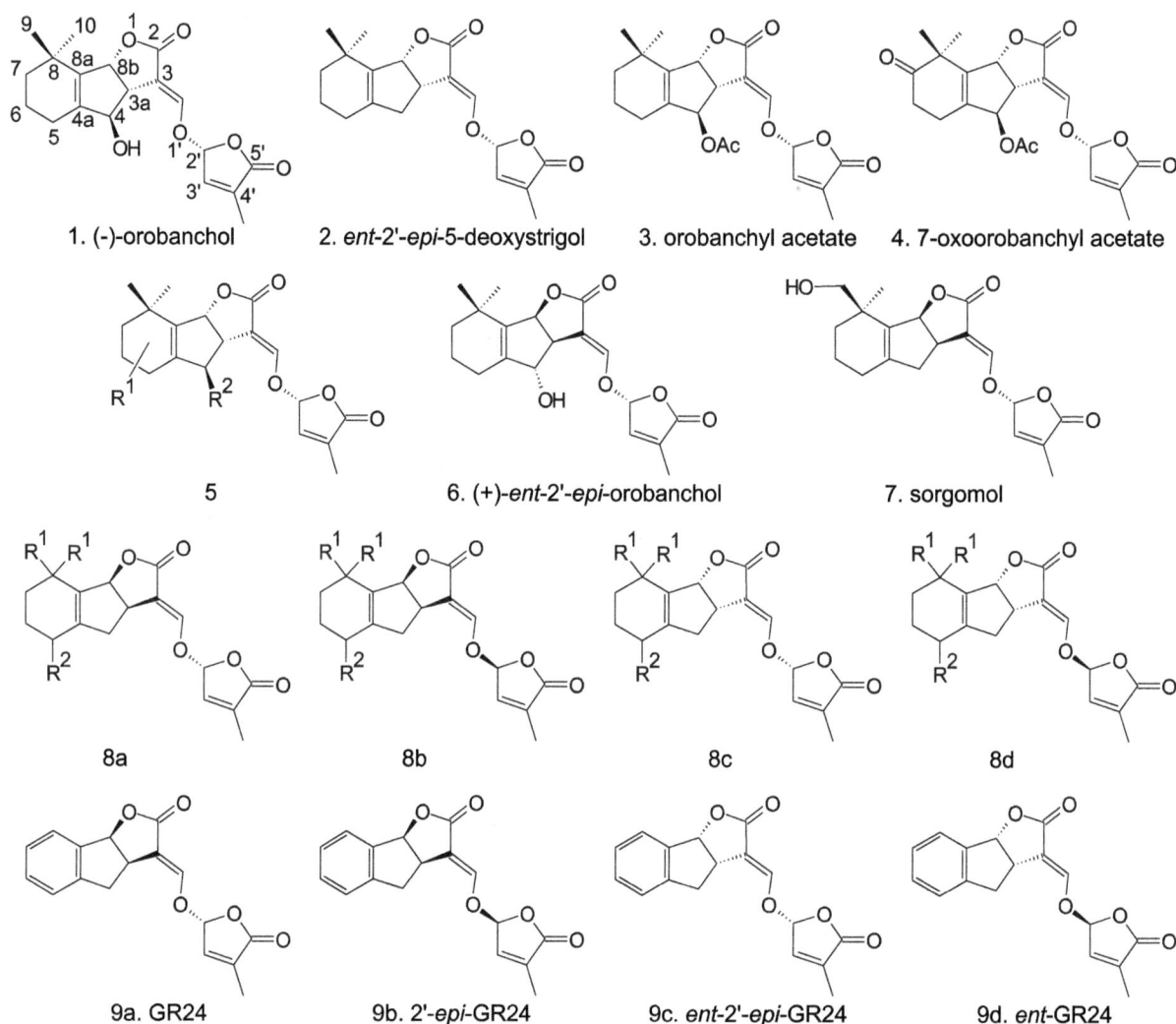

Figure 1. SL structures present in rice root exudates or tested in a seed germination or hyphal branching bioassay. (**1**) (−)−orobanchol; (**2**) ent-2'-epi-5-deoxystrigol; (**3**) orobanchyl acetate and (**4**) proposed structure of 7-oxoorobanchyl acetate [23]; (**5**) proposed structure of methoxy-5-deoxystrigol isomers (R^1 = OMe; R^2 = H) or the methyl ether of orobanchol (R^1 = H; R^2 = OMe); (**6**) (+)-ent-2'-epi-orobanchol; (**7**) sorgomol; (**8a–d**) stereoisomers of strigol (R^1 = CH$_3$; R^2 = OH); 5-deoxystrigol (R^1 = CH$_3$; R^2 = H), stereochemical configurations **8a** and **8c** are natural (strigol-type and orobanchol-type, respectively) while configurations **8b** and **8d** are not naturally occurring; (**9a–d**) stereoisomers of SL analogue GR24.

MRM-LC-MS analysis

For LC-MS analysis, samples were filtered through mini syringe filters (Minisart SRP4). The retention times, mass transitions and MS/MS spectra of available SL standards such as (+)-ent-2'-epi-orobanchol (**6**), (−)-orobanchol (**1**), (±)-5-deoxystrigol (**8a,d**, R^1 = CH$_3$; R^2 = H), (±)-2'-epi-5-deoxystrigol (**8b,c**, R^1 = CH$_3$; R^2 = H), (±)-sorgolactone, (±)-strigol (**8a,d**, R^1 = CH$_3$; R^2 = OH), solanacol, orobanchyl acetate (**3**), (±)-7-oxoorobanchol and (±)-7-oxoorobanchyl acetate (**4**) were compared with each sample to quantify SLs using ultra performance liquid chromatography coupled to tandem mass spectrometry (UPLC-MS/MS). Analyses were performed using a Waters Xevo tandem quadruple (TQ) mass spectrometer equipped with an ESI source. Chromatographic separation was achieved on an Acquity UPLC BEH C18 column (150 × 2.1 mm, 1.7 μm) (Waters) by applying a water/acetonitrile gradient to the column, starting from 5% (v/v) acetonitrile for 2.0 min and rising to 50% (v/v) acetonitrile at 8.0 min, followed by a 1.0 min gradient to 90% (v/v) acetonitrile, which was maintained for 0.1 min before going back to 5% (v/v)

acetonitrile using a 0.2 min gradient, prior to the next run. Finally, the column was equilibrated for 2.8 min, using this solvent composition. Operation temperature and flow-rate of the column were 50°C and 0.4 mL min^{-1}, respectively. Sample injection volume was 15 μL. The mass spectrometer was operated in positive electrospray ionization (ESI) mode. Cone and desolvation gas flows were set to 50 and 1000 L h^{-1}, respectively. The capillary voltage was set at 3.0 kV, the source temperature at 150°C and the desolvation temperature at 650°C. The cone voltage was optimized for each SL standard using the IntelliStart MS Console. Argon was used for fragmentation by MS/MS spectra in the collision cell.

The identification of SLs in rice root exudates and extracts was done using Multiple Reaction Monitoring (MRM) and by comparing retention times and MRM mass transitions with those of the available SL standards mentioned above. MRM transitions were optimized for each standard using the IntelliStart MS Console. The MRM transitions for putative 4-methoxy-5-deoxystrigol isomers were initially set based on the theoretically predicted fragmentation (see Results and Discussion section). MRM-transitions for the predicted putative SLs were incorporated in the MRM-method. The structures of all detected SLs were confirmed by MS/MS fragmentation spectra. Data acquisition and analysis were performed using MassLynx 4.1 software (Waters). Full mass scan and precursor ion scan for $m/z = 97$ were performed to search for unknown SLs in biologically active HPLC-fractions 15 to 19. The LC-MS results of the measurements of $(-)$-orobanchol (**1**) and *ent*-2'-*epi*-5-deoxystrigol (**2**) (of 3 biological replicates) were compared using ANOVA followed by pair wise comparisons with t-test (LSD values) in Genstat (Genstat for Windows 12th Edition; VSN International).

S. hermonthica seed germination bioassay

Seeds of *Striga hermonthica* used for the bioassay were kindly provided by Bob Vasey and originating from a sorghum field in Sudan, collected in 1995. The bioassays were performed as described [28]. The samples were tested in three technical replicates (three discs) and 3 biological replicates were tested, one independent bioassay per biological replicate. Given the binomial distribution nature of the measurements the mean values of the seed germination scores (3 replicates per treatment per fraction) were compared using a Chi-square test in Genstat (Genstat for Windows 12th Edition; VSN International).

AM fungal hyphal branching bioassay

For the AM branching bioassay spores of *Gigaspora rosea* (DAOM 194757) were used. The spores were routinely produced in pots containing leek and collected by wet sieving. They were washed in water/0.05% Tween 20 (v/v), soaked with 2% (w/v) Chloramine T (Sigma) for 10 min, washed again three times in sterile water for 30 s per wash, and stored in an antibiotic solution containing 100 mg L^{-1} gentamycin and 200 mg L^{-1} streptomycin. After 2 days at 4 °C, a second treatment with Chloramine T was carried out under the same conditions. They were then stored in the antibiotic solution at 4°C before use. Branching bioassays were carried out according to Buee *et al.* [29]. Four spores of *Gi. rosea* were germinated (in 2% CO_2 at 30 °C in dark) on M medium (Becard & Fortin, 1988) supplemented with 10 μM quercetin (Sigma) and solidified with 0.6% Phytagel (Sigma). Seven days after inoculation, each spore produced a single germ tube growing upwards. Two small wells on each half of the Petri dish, near the hyphal tip, were made in the gel with a Pasteur pipette tip and 5 μL of the test solution (SL analogue GR24 or purified fraction) or 10% acetonitrile (control) was injected in each well. After 24 h, hyphal branching was recorded by counting newly formed hyphal tips. Twenty to thirty spores were used for each treatment. Values of each tested fraction were compared with the corresponding control using the Student's t-test in Genstat (Genstat for Windows 12th Edition; VSN International).

Results

Rice root exudates were profiled to find compounds responsible for *S. hermonthica* seed germination and AM fungi hyphal branching. Plants were submitted to phosphate starvation with and without the application of 0.01 μM fluridone, an inhibitor of carotenoid and hence SL biosynthesis [28]. A reduction in seed germination stimulatory activity of root exudate fractions by fluridone treatment would suggest that the compound(s) responsible for the biological activity in that fraction is a (are) SL(s). The root exudates were fractioned by HPLC and the biological activity of the resulting fractions as well as the crude exudates (not fractioned) was assessed. Each of the three biological replicates was

tested in independent seed germination bioassays of which one is shown in Figure 2 A, and the remaining are shown in Figure S1. Preliminary results showed that fractions eluting before 14 min (fractions 1 to 14) do not exhibit seed germination stimulatory activity (data not shown). Fractioned and crude exudates of control plants with sufficient phosphate showed almost no activity across all fractions. The seed germination stimulatory activity was significantly increased by phosphate starvation in crude exudates and in fractions 15–21, 23–25 and 28 ($P<0.05$ using X^2 test). This was observed consistently in all replicates (Figure 2 A and Figure S1). Fluridone treatment significantly reduced the activity of crude exudates and of fractions 15 to 19 and 24) with 99.9% confidence ($P<0.001$ using X^2 test) in all biological replicates. In fractions 20, 21, and 25 fluridone treatment also reduced the activity but the effect was not consistently significant; in two replicates these fractions showed significance at $P<0.001$ using the X^2 test, but not in the third replicate, likely because the germination stimulatory activity in these fractions was sometimes low. Also in fraction 28 that induces low germination activity the fluridone treatment did not significantly reduce seed germination. The effect of crude and fractioned root exudates on AM fungal hyphal branching was also assessed using *Gi. rosea* spores (Figure 2 B and C). Just as for *S. hermonthica* seed germination, phosphate starvation significantly increased branching stimulatory activity of crude exudates and most fractions (16–20, 24–25 and 27–28, at $P<0.005$, using Student t-test). Crude and fractioned exudates of control, non-phosphate starved plants did not show a significant difference in branching activity compared with the negative control, 10% acetonitrile. All the active fractions (of phosphate starvation treated rice) induced hyphal branching to a similar level. Interestingly, however, fraction 18 - the most active in the *S. hermonthica* seed germination bioassay - induced less AM fungal hyphal branching than the other active fractions (Figure 2 C).

To gain insight into the identity of the compounds responsible for the biological activity, MRM-LC-MS analysis was performed. The MRM chromatograms of crude exudates revealed an intense peak in the channels m/z 347>233, 347>205 and 347>97 at retention time 8.05 min, which matches with an authentic standard of $(-)$-orobanchol (**1**) (Figure 3 A and B). In the channels m/z 331>234 and 331>97 there was a peak at 12.51 min, which matches with *ent*-2'-*epi*-5-deoxystrigol (**2**) (Figure 3 C and B). MS/MS fragmentation spectra, and the addition of authentic orobanchol and $(\pm)$-2'-*epi*-5-deoxystrigol (**8b,c**, $R^1 = CH_3$; $R^2 = H$) to the samples, confirmed that the compounds detected were indeed orobanchol and *ent*-2'-*epi*-5-deoxystrigol. The same two SLs were barely detectable in exudates of control plants supplied with full nutrient solution, were most abundant in the phosphate starvation treatment and were both significantly reduced by fluridone treatment ($P<0.001$) (Figure 4 A). In addition to $(-)$−orobanchol (**1**) and *ent*-2'-*epi*−5−deoxystrigol (**2**) three unknown peaks were detected using the channels for 7-oxoorobanchol (m/z 361>247 and 361>97) at the retention times 9.9, 10.3 and 10.9 min (Figure 3 E and F), which is substantially later than 7-oxoorobanchol, which elutes at 3.7 min. The three unknown compounds (from here on referred to as methoxy-5-deoxystrigol isomers; see Discussion) were most abundant in the phosphate starvation treatment and were reduced by fluridone treatment (Figure 4 B - D) suggesting they are SL-like compounds.

MRM-LC-MS analysis was also performed on the HPLC fractions to try to correlate the presence of SLs with the seed germination and hyphal branching activity. $(-)$−orobanchol (**1**) was detected in fractions 19 to 21 with highest abundance in fraction 20 and *ent*-2'-*epi*-5-deoxystrigol (**2**) in fractions 27 and 28 with highest signal in fraction 28 (data not shown). The three

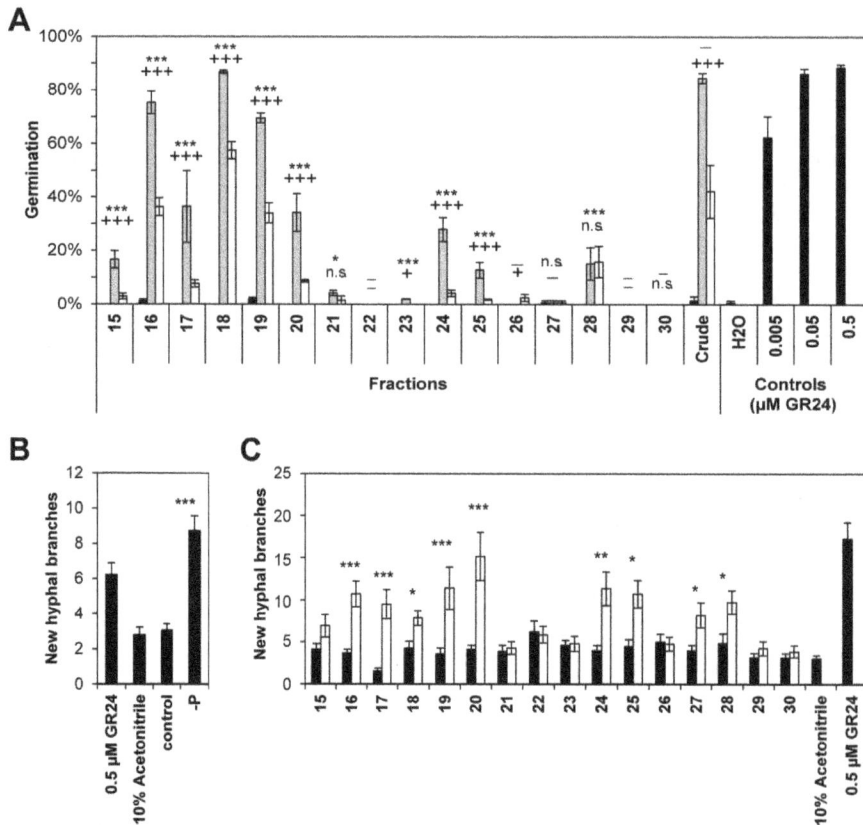

Figure 2. Activity profiles of rice root exudates. Germination of *S. hermonthica* (one biological replicate; the other two are shown in Figure S1) obtained with crude exudates and exudate fractions from rice plants (**A**) treated with full nutrition (black bars); phosphate starvation (grey bars) and phosphate starvation plus 0.01 μM fluridone (white bars). Water and SL analogue GR24 (0.005, 0.05 and 0.5 μM) were used as controls. The error bars represent the standard error of 3 technical replicates. Significance levels between treatments as determined using a X^2 test are indicated: */+ = $P< 0.05$; **/++ = $P<0.01$; ***/+++ = $P<0.001$; n.s. = $P>0.05$; * = control vs. phosphate starvation treatment; + = phosphate starvation vs. phosphate starvation plus fluridone treatment. When germination values are close to zero the statistical test cannot be performed, which is indicated with "−". AM fungal hyphal branching induced by crude exudates (**B**) and exudate fractions (**C**) of rice treated with full nutrition (black bars) and phosphate starvation (white bars) in germinating *Gi. rosea* spores. The assay was performed with pooled samples of three biological replicates. GR24 (0.005, 0.05 and 0.5 μM) and 10% acetonitrile in water were used as controls. The bars represent the mean of the total number of new branches, the error bar the standard error of the mean (n = 20). Significance values comparing means between control treatments and phosphate starvation treatment are indicated above the bars. (* = $P<0.05$, ** = $P<0.01$, *** = $P<0.001$).

methoxy-5-deoxystrigol isomers detected in the channel for 7-oxoorobanchol (m/z 361> 247 and 361>97) at the retention times 9.95, 10.3 and 10.9 min eluted in fractions 23-25 with highest abundance in fraction 24 and are likely responsible for the seed germination stimulant/hyphal branching activity peak in fractions 24-25. The activity of the three compounds could not be evaluated individually as they did not separate on HPLC due to their highly similar retention time.

The fractions with highest seed germination inducing activity (16 and 18) were also analyzed using known MRM transitions typical for SLs as well as full mass scan and precursor ion scan for $m/z = 97$. However, we could not detect any masses that could be indicative for SLs and displayed an expected abundance pattern across the treatments similar as the known SLs: low in control, high in P starvation, low upon fluridone treatment.

To further investigate the nature of the active compounds in fractions 16, 18, 24 and 25, exudates of the SL biosynthetic mutant *d10-2* were studied [10]. HPLC fractions 16–25 collected from *d10-2* and its background Nipponbare were tested using the seed germination bioassay. All active fractions in the wild type had reduced activity in the mutant (Figure 5 A) supporting the SL

(CCD8-dependent) nature of the compounds responsible for the biological activity of these fractions.

The MRM-LC-MS spectra of *d10-2* mutant root exudates confirmed that (−)-orobanchol (**1**) and *ent-2'-epi*-5-deoxystrigol (**2**) as well as the three methoxy-5-deoxystrigol isomers detected at 9.9, 10.35 and 10.95 were strongly decreased in *d10-2* mutant root exudates (Figure 5 B,C), further indicating that the latter three are SLs/require CCD8. Fractions 16 and 18 of *d10-2* exudates were also analyzed by LC-MS and compared with those from wild type plants using full-scan mass spectrometry, but no differential masses were found that could explain the seed germination activity in the wild type and give a hint on the identity of the seed germination stimulant(s) in these fractions.

The activity profiles obtained with the exudate fractions, when tested with the seed germination and hyphal branching are different. Some fractions that stimulate high seed germination percentages induce low fungal response and the contrary is also observed. To further investigate the differences in activity observed in our bioassays, we performed a seed germination bioassay using pure or racemic mixtures of SLs (Table 1). We observed that sorgomol (**7**) is the most active of the tested SLs

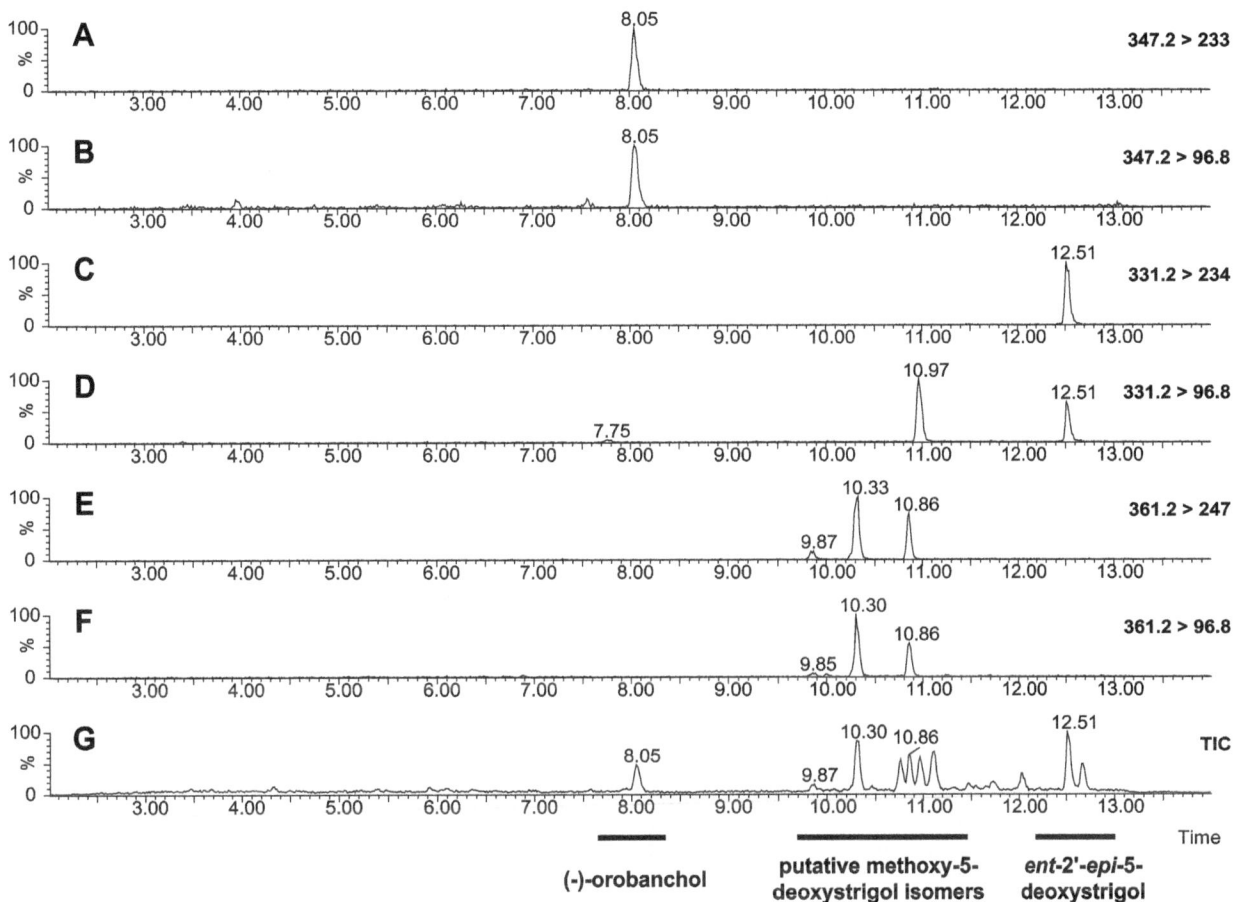

Figure 3. SL analysis of rice root exudates. Root exudates from rice plants grown under phosphate starvation were analyzed with liquid chromatography coupled with tandem mass spectrometry (LC-MS/MS) using multiple reaction monitoring (MRM). Chromatograms of (**A**) transitions 347.2 > 233 and (**B**) 347.2>96.8 for orobanchol; (**C**) transitions 331.2> 234 and (**D**) 331.2>96.8 for 2'-epi-5-deoxystrigol; (**E**) transitions 361.2>247 and (**F**) 361.2>96.8 for three putative methoxy-5-deoxystrigol isomers; (**G**) the total ion count (TIC) showing of all measured transitions and where orobanchol (8.05 min), ent-2'-epi-5-deoxystrigol (12.51 min) and the three putative methoxy-5-deoxystrigol isomers (9.87; 10.33; 10.86 min) are visible.

inducing 36% seed germination at 200 nM and 26% seed germination at 20 nM followed by (+)-*ent*-2'-epi-orobanchol (**6**, 34% seed germination at 200 nM and 8.7% at 20 nM). The racemates of (±)-strigol (**8a,d**, $R^1 = CH_3$; $R^2 = OH$); (±)-5-deoxystrigol (**8a,d**, $R^1 = CH_3$; $R^2 = H$) and the racemic mixture of all 4 stereoisomers of GR24 (**9a–d**) have intermediate activity inducing 9.3%, 6.0% and 2.0% seed germination at 20 nM and inducing 19%, 26% and 25% seed germination at 200 nM, respectively. The racemate of (±)-2'-epi-5-deoxystrigol (**8b,c**, $R^1 = CH_3$; $R^2 = H$) induced less seed germination (11 % at 200 nM) and was not active at 20 nM. The least active SLs were (−)-orobanchol (**1**) and the racemate of (±)-2'-epi-strigol (**8b,c**, $R^1 = CH_3$; $R^2 = OH$) that induced less than 1% seed germination in both concentrations. Table 1 also summarises data from a study by Akiyama et al. that analysed the *Gigaspora margarita* hyphal branching activity of a range of different SLs [26]. In contrast to what is observed with *S. hermonthica*, both orobanchol (**1**) and *ent*-2'-*epi*-5-deoxystrigol (**2**) are highly active at inducing hyphal branching and their activity is similar to their natural stereoisomers. Strigol (**8a**, $R^1 = CH_3$; $R^2 = OH$), sorgomol (**7**), GR24 (**9a**) and (±)-2'-epi-strigol (**8b,c**, $R^1 = CH_3$; $R^2 = OH$) were considerable less active (100 fold) than orobanchol (**1**), and the remaining

GR24 stereoisomers (**9b–d**) were 10000- to 1000-fold less active than GR24 (**9a**).

Discussion

Rice root exudates were fractioned to evaluate the contribution of SLs and potentially other signalling molecules to the *S. hermonthica* seed germination stimulant and AMF hyphal branching activity of rice root exudate. MRM–LC–MS analysis of these HPLC fractioned rice root exudates showed the presence of (−)−orobanchol (**1**) in fractions 19, 20 and 21 and *ent*-2'-*epi*-5-deoxystrigol (**2**) in fraction 28 suggesting that these SLs are responsible for the seed germination and hyphal branching stimulatory activities of these fractions. These results confirm the presence of SLs found previously in root exudates of the rice variety Nipponbare except for orobanchyl acetate (**3**) that was not detected in the present study but is reported by others [23]. A fourth SL -7-oxoorobanchyl acetate (**4**) – was also reported to be produced in Nipponbare between days 10 to 17 after germination [23]. In the present study the exudates were collected at a later stage and this SL was not detected.

The relative abundance of (−)−orobanchol (**1**) and *ent*-2'-*epi*-5-deoxystrigol (**2**), measured by MRM-LC-MS across the different

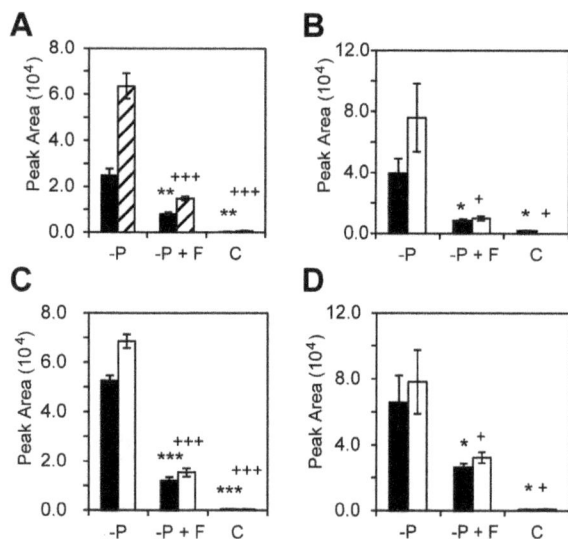

Figure 4. Abundance of (−)−orobanchol, ent-2'-epi-5-deoxystrigol and putative SL-like compounds in phosphate starvation and fluridone treatments. Peak areas obtained with liquid chromatography coupled with tandem mass spectrometry (LC-MS/MS) analysis using multiple reaction monitoring (MRM) of root exudates of rice. (A) (−)−orobanchol (MRM transition 347.2 >96.8; black bars) and ent-2'-epi-5-deoxystrigol (MRM transition 331.2>234; hatched bars); (B–D) three putative SL-like compounds measured in crude exudates with the retention times: rt = 9.87 (B); rt = 10.3 and (C) rt = 10.9 (D) and the MRM transitions 361>96.8 (black bars) and 361>247 (white bars). All measurements taken from crude exudates of plants grown in different treatments: phosphate starvation (−P); phosphate starvation combined with fluridone (−P+F) and control treatment with full nutrient supply (C). The error bars represent the standard error of 3 biological replicates. Significance values are indicated with * (for ent-2'-epi-orobanchol and for 361>96.8 transition) and + (for 2'-epi-5-deoxystrigol and for 331.2> 234 transition) and compare phosphate starvation (-P) treatment vs. phosphate starvation with fluridone (-P+F) and -P vs. full nutrition (C) (*/+ = P<0.05, **/++ = P<0.01, ***/+++ = P<0.001).

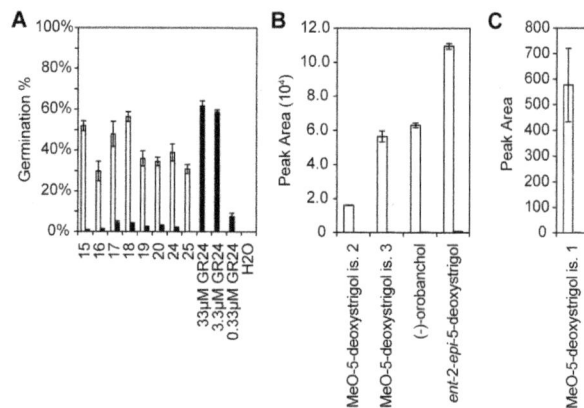

Figure 5. Characterization of rice mutant d10 exudate. Germination assay with S. hermonthica seeds on exudate fractions of full nutrition (black bars) and phosphate starvation (white bars) treated plants (A). Water and GR24 (0.33, 3.3 and 33 μM) were used as controls. Peak areas of ent-2'-epi-5-deoxystrigol, (−)−orobanchol and putative methoxy-5-deoxystrigol isomers 2 and 3 (B) and putative methoxy-5-deoxystrigol isomer 1 (C) obtained with liquid chromatography coupled with tandem mass spectrometry (LC-MS/MS) analysis using multiple reaction monitoring (MRM) of root exudates of d10-2 (black bars) and WT (white bars) under phosphate starvation treatment.

treatments matches the seed germination stimulatory activity of the fractions where these SLs elute (19–20 and 28 respectively). Phosphate starvation induced the highest production of (−)−orobanchol (1) and ent-2'-epi-5-deoxystrigol (2) which resulted in the highest germination of Striga seeds. Fluridone application inhibited the biosynthesis of these SLs which resulted in a lower biological activity of the fractions and crude exudates, confirming the inhibitory effect of fluridone on SL production that was previously described [28]. MRM-LC-MS analysis of fractions 24–25 revealed the presence of three compounds with the same mass m/z 361 showing up in the 361.2>247 and 361.2>96.8 MRM channels. These metabolites were most abundant in exudates of phosphate starved plants and were reduced by fluridone application. The seed germination activity obtained with fractions 24 and 25 correlates with the abundance of the detected masses. MS/MS analysis of the compounds eluting in fractions 24 and 25 shows fragmentation patterns typical for SLs [30]: loss of the D-ring and H_2O yields fragment ions $[M+H − D\text{-ring} - H_2O]^+$ with m/z = 247 and the fragment ion of the D-ring itself $C_5H_5O_2$ with m/z = 97 (Figure 6). The loss of methanol $[M+H −MeOH]^+$, yielding the fragment ion m/z = 329, is not typical for the fragmentation of known SLs and could indicate the presence of a methoxy-group in the molecule (Figure 6). This feature could explain the late retention time of these putative SLs compared with orobanchol (Figure 3 G) and other known SLs given that methyl ethers are less

polar than alcohols (Figure 3 E and G). The MS/MS fragmentation spectra of all three compounds are very similar (Figure 6). Based on these data we suggest that the compounds eluting at 9.5, 10.3 and 10.9 are methoxy-5-deoxystrigol isomers (5). Isolation followed by NMR or chemical synthesis should give the final proof of the structure of these three isomers. As we do not have this proof as yet, we will refer to these new compounds under the combined name methoxy-5-deoxystrigol isomers [4]. The absence of the putative methoxy-5-deoxystrigol isomers in d10-2 exudate further supports that these compounds are produced from the SL pathway (Figure 5 B and C).

Fractions 16 and 18 induced the highest level of S. hermonthica germination (Figure 2 and Figure S1) and do not contain any of the SLs discussed above. As mentioned above, two other SLs, orobanchyl acetate (3) and 7-oxoorobanchyl acetate (4), were recently reported in rice [23]. Orobanchyl acetate (3) elutes after (−)-orobanchol (1) and is unlikely to be responsible for the activity in fractions 16 and 18. 7-Oxoorobanchyl acetate (4) elutes before (−)−orobanchol (1) and could be present in fraction 16 to 18. This SL was previously detected in exudates collected 10 to 17 days after germination [23]. In the present study, exudates were collected in later stages and this SL was not detected in crude exudates nor in any of the fractions. We could also not detect any other known SLs in fractions 16 to 18; however the seed germination bioassays showed that the activity of fractions 16 and 18 followed the same trend across the treatments as the activity of SL containing fractions. They were increased by phosphate starvation and reduced by fluridone application (Figure 2 A). The seed germination stimulatory activity of these fractions in d10-2 root exudate was also clearly reduced (Figure 5 A). All this strongly suggests that the activity in these fractions is caused by compounds derived from the SL pathway after carlactone (as CCD8 is required for their production). Considering their high activity in the induction of S. hermonthica germination it is of great interest to identify these compounds.

Overall, the activity profiles for S. hermonthica seed germination and AM fungal hyphal branching are similar but not the same (Figure 2). All active fractions in the seed germination bioassay

Table 1. *Striga hermonthica* germination and *Gigaspora margarita* hyphal branching in the presence of SL standards.

	S. hermonthica germination (%)		*Gi. margarita* hyphal branching[1]
	200 nM	**20 nM**	**MEC[2] in pg per disc**
(±)-2'-*epi*-strigol	0.00 ± 0.00	0.00 ± 0.00	100
(−)−orobanchol[3]	0.67 ± 0.67	0.00 ± 0.00	1
ent-2'-epi-5-deoxystrigol	–	–	3
2'-epi-5-deoxystrigol	–	–	30
(±)-2'-epi-5-deoxystrigol	11.33 ± 1.76	0.00 ± 0.00	–
GR24	–	–	100
ent-GR24	–	–	10000
2'-epi-GR24	–	–	1000
ent-2'-epi-GR24	–	–	1000
GR24 (4 stereoisomers)	25.33 ± 1.76	2.00 ± 1.15	
ent-5-deoxystrigol	–	–	30
5-deoxystrigol	–	–	3
(±)-5-deoxystrigol	26.00 ± 9.87	6.00 ± 3.05	–
(±)-strigol	19.33 ± 8.82	9.33 ± 4.67	100
Ent-2'-epi-orobanchol[4]	34.00 ± 1.15	8.67 ± 2.40	1
sorgomol	36.00 ± 3.05	26.00 ± 5.03	100

[1]Results extracted from Akiyama et al. [26]; [2] MEC = minimum effective concentration; [3] in Akiyama et al. [26] these compounds are named (+)−orobanchol and (+)-2'-*epi*-orobanchol respectively, before revision of stereochemical structure whereas the present table indicates the revised stereochemistry [23].

exhibited hyphal branching stimulatory activity albeit to a different extent. Fraction 20 [(−)−orobanchol (**1**)] induced high AM fungal hyphal branching but stimulated little germination of *S. hermonthica* seeds compared with other fractions. Also, *ent*-2'-*epi*-5-deoxystrigol (**2**, fraction 28; low seed germination), the methoxy-5-deoxystrigol isomers (fractions 24–25; low seed germination) and fraction 16 and 18 (high seed germination) have very different activity with regard to the induction of seed germination whereas being quite similar in the induction of hyphal branching. Fraction 27 did not display seed germination stimulatory activity, but it did induce hyphal branching. This is probably due to the presence of *ent*-2'-*epi*-5-deoxystrigol (**2**) that is still detected in this fraction but at lower concentration than in fraction 28. Hence the concentration of *ent*-2'-*epi*-5-deoxystrigol (**2**) in fraction 27 may not be sufficient to induce *S. hermonthica* seed germination but is apparently high enough to induce AM fungal hyphal branching.

The separation on HPLC is not good enough to separate all the active compounds. This results in tailing peaks for example for (−)−orobanchol (**1**), present in fractions 19 to 21 and with highest abundance in fraction 20. Fraction 19 induces response of both *S. hermonthica* seeds and AM fungi to an extent that is intermediate to fractions 18 and 20. The activity in fraction 19 is probably a result from the cumulative effect of (−)−orobanchol (**1**) and the tail of the unknown active compound eluting mostly in fraction 18. Similarly, the activity observed in fraction 17 might also be due to fronting of fraction 18 and tailing of fraction 16. However, we cannot exclude the presence of other active compounds in fractions 17 and 19.

The two activity profiles show that some of the most active mycorrhizal hyphal branching stimulants present in rice root exudates play only a minor role in the induction of *S. hermonthica* germination. The seed germination stimulatory activity of known concentrations of SLs was assessed and compared with results of a study relating structural differences in SLs to AM fungal hyphal branching stimulatory activity [26] (Table 1). The structure of (−

)−orobanchol (**1**) and (+)-*ent*-2'-*epi*-orobanchol (**6**) have been revised after the study by Akiyama et al. [26] hence, these compounds were originally labeled (+)−orobanchol and (+)-2'-*epi*-orobanchol respectively [23,26,31]. As previously shown by Nomura et al. [27] sensitivity of *S. hermonthica* seeds is highly dependent on the orientation of the C-ring, and is more sensitive to the strigol–type configuration. In our bioassays this preference is confirmed, the highest seed germination was obtained with sorgomol (**7**) and (+)-*ent*-2'-epi-orobanchol (**6**) while (−)−orobanchol (**1**) hardly induced any seed germination (Table 1).

AM fungi also have different sensitivity to different SL structures [26]. The strigol–type configuration is sometimes more active, as was observed for GR24 (**9a**) but not always [26]. For instance, all strigol stereoisomers (**8a–d**, $R^1 = CH_3$; $R^2 = OH$) have equal activity just as (−)−orobanchol (**1**) and (+)-*ent*-2'-epi-orobanchol (**6**) (Table 1) [26]. Also the two natural stereoisomers of 5-deoxystrigol (**8a** and **8c**, $R^1 = CH_3$; $R^2 = H$) have each the same activity at inducing hyphal branching (Table 1) [26]. Indeed, the activity of SLs to stimulate hyphal branching seems to be more influenced by modifications in rings A and B than by stereochemical variation [26,32]. In our hyphal branching assay with a different AM species, *Gi. margarita*, we obtained a similar response to the different SLs as reported for *Gi. rosea* [26]. Fraction 20, where (−)−orobanchol (**1**) elutes, displays high activity in the branching bioassay whereas there is no clear activity peak in the seed germination bioassay (Figure 2). Similarly, *ent*-2'-*epi*-5-deoxystrigol (**2**) detected in fractions 27 and 28 induced hyphal branching and only fraction 28 with highest amounts of this SL induced low seed germination (Figure 2). SL activity is also affected by different chemical and structural properties that influence diffusion and stability [26,33]. However compared to other SLs, (−)−orobanchol (**1**) and *ent*-2'-*epi*-5-deoxystrigol (**2**) are highly active at stimulating hyphal branching. Therefore, the low activity of (−)−orobanchol (**1**) and *ent*-2'-*epi*-5-deoxystrigol (**2**) at inducing seed germination does not seem to be a result of

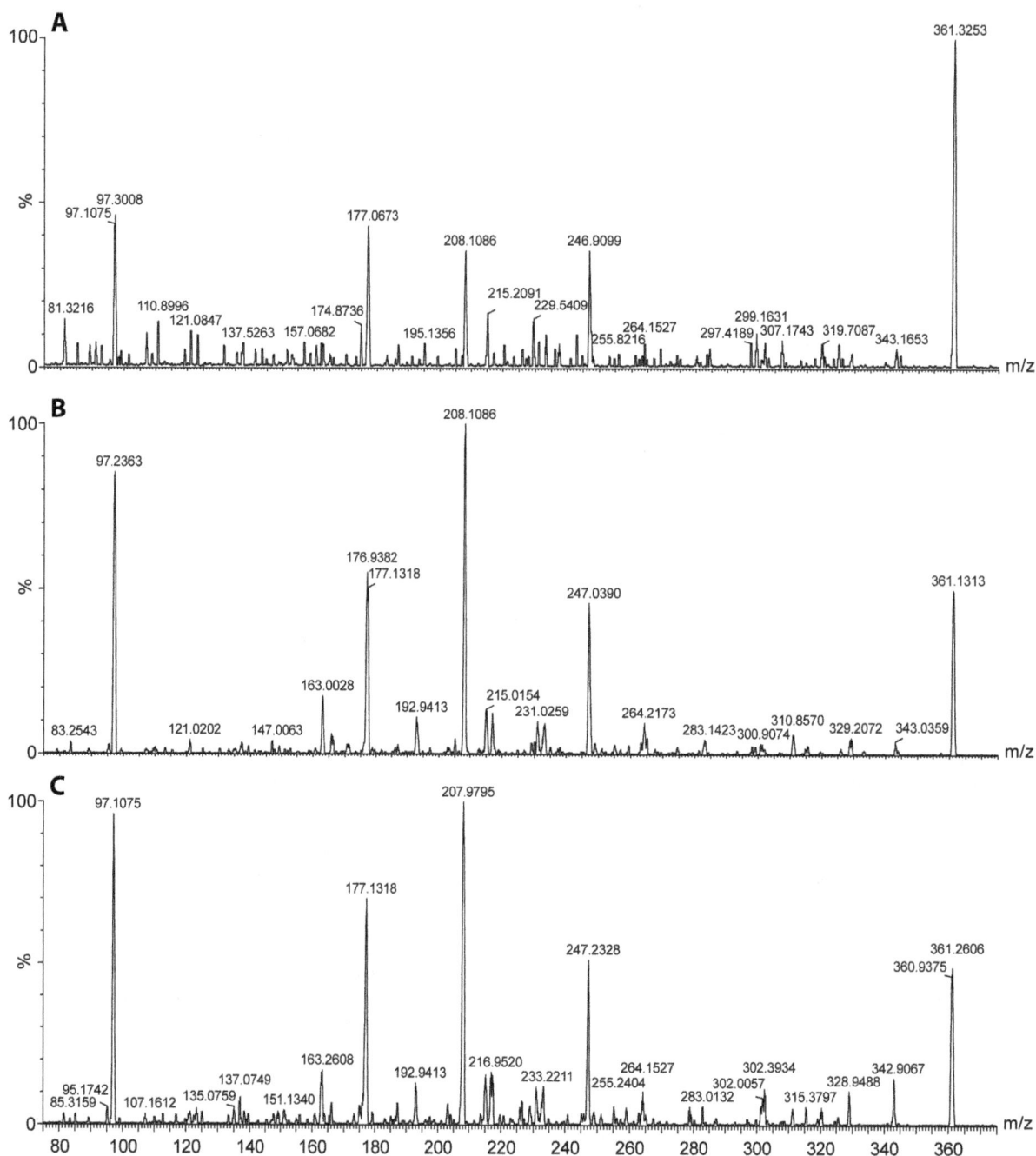

Figure 6. MS/MS spectra of putative SL-like compounds. The spectra were measured at the retention time of each isomer: 9.87 min (**A**), 10.35 min (**B**) and 10.95 min (**C**).

instability or poor diffusion of these two SLs but rather a result of lower sensitivity of the seeds to these compounds.

Our bioassays suggest that strong hyphal branching stimulators make little contribution to the overall stimulation of parasitic seed germination. Moreover, the fractions showing the largest effect on seed germination (fractions 16 and 18) contain stimulants of unknown structure. The reduction of activity in these fractions by fluridone and by mutation in *D10 (CCD8)* suggests that they are SL-like. The strong differences in activity across the exudate fractions suggest that *S. hermonthica* infection and potentially also the infection by other parasitic plant species could be reduced by

altering the qualitative composition of SLs rather than just quantitatively reducing their production. New varieties with such altered SL composition could maintain their ability to establish symbiosis with AM fungi while at the same time they induce less *Striga* seed germination. In a recent study, 20 rice cultivars were screened for the abundance of SLs in their root exudates [6]. The authors observed that the relative amounts of (−)−orobanchol (**1**) and ent-2'-*epi*-5-deoxystrigol (**2**) differ across cultivars, suggesting that selection for different SL composition is possible.

ent-2'-*epi*-5-Deoxystrigol (**2**) and 5-deoxystrigol (**8a**, R^1 = CH$_3$; R^2 = H) have the most simple structure of all SLs so far identified in plants. They have been suggested to be produced from carlactone through the action of MAX1 [and possibly additional enzyme(s)] [21,34] which would imply that the orientation of the C-ring, a structural feature of major importance for the induction of seed germination as well as hyphal branching, is determined by MAX1. We propose that ent-2'-epi-5-deoxystrigol (**2**) is the precursor for the remaining rice SLs. Further supporting this hypothesis, it has been recently shown that a sorghum enzyme(s) - likely a cytochrome P450 - converts ent-2'-epi-5-deoxystrigol (**1**) and 5-deoxystrigol (**8a**, $R^1 = CH_3$; $R^2 = H$) into ent-2'-epi-sorgomol and sorgomol respectively [35]. Breeding for a different SL composition would be greatly aided by the characterization of these later steps in SL biosynthesis, that is the decoration of the SLs' core-structure.

As a word of caution, the sensitivity to specific SLs may vary between *Striga* species and/or races [36]. Therefore, assessment of the seed germination and hyphal branching requirements for the host/parasite and host/AM combination present in a certain region would be necessary in order to direct the development of new, locally adapted, cultivars that are less affected by *Striga* parasitism but still efficient in AM symbiosis establishment. A study performed in sorghum has shown that host plants are especially vulnerable to plant parasitism in the early stages of their life cycle [37]. Since SL composition in rice exudates changes according to the age of the plants [23] efforts to produce new rice varieties should take into consideration SL variation throughout the life cycle.

Finally, we can only speculate about the driving forces for the diversification in SL structures that we see in rice. Unlike for the gibberellins, it seems that there are not just one or two active molecules accompanied by inactive precursors and degradation products. In their role as rhizosphere signaling molecules the different SLs all display activity, albeit admittedly with different efficiency. With regard to their endogenous function, as plant hormones regulating a suit of developmental processes, we only just begin to understand the structure-activity relationships [38]. As a result of all these different functions the consequences of evolutionary and human (breeding) selection pressures are complex – which is reflected in the large structural diversification - and the resulting structural diversification so far difficult to explain.

Supporting Information

Figure S1   Activity profiles of rice root exudates tested with *S. hermonthica* seed germination assay. Two biological replicates are shown here and the third replicate is shown in Figure 2 A. Crude exudates and exudate fractions from rice plants treated with full nutrition (black bars); phosphate starvation (grey bars) and phosphate starvation plus 0.01 µM fluridone (white bars). Water and SL analogue GR24 (0.005, 0.05 and 0.5 µM) were used as controls. The error bars represent the standard error of 3 technical replicates. Significance levels between treatments as determined using a X^2 test are indicated: */+ = $P<$ 0.05; **/++ = $P<0.01$; ***/+++ = $P<0.001$; n.s. = $P>0.05$; * = control vs. phosphate starvation treatment; + = phosphate starvation vs. phosphate starvation plus fluridone treatment. When germination values are close to zero the statistical test cannot be performed, which is indicated with "–".

Acknowledgments

We would like to thank, Bob Vasey for providing *S. hermonthica* seeds, Prof. Binne Zwanenburg, Prof. Koichi Yoneyama, Prof. Tadao Asami and Prof. Guillaume Bécard for providing SL standards. We also thank Jacques Withagen for helping with the statistical analysis.

Author Contributions

Conceived and designed the experiments: CC TC PMD CRS HB. Performed the experiments: CC TC MJ FV MA PMD DL. Analyzed the data: CC TC MJ PMD CRS HB. Contributed reagents/materials/analysis tools: CC TC FV PMD DL. Contributed to the writing of the manuscript: CC TC CRS HB.

References

1. Parker C (2009) Observations on the current status of *Orobanche* and *Striga* problems worldwide. Pest Manag Sci 65: 453–459. doi:10.1002/ps.1713

2. Cardoso C, Ruyter-Spira C, Bouwmeester HJ (2011) Strigolactones and root infestation by plant-parasitic *Striga*, *Orobanche* and *Phelipanche* spp. Plant Science 180: 414–420. doi:10.1016/j.plantsci.2010.11.007

3. Scholes JD, Press MC (2008) Striga infestation of cereal crops – an unsolved problem in resource limited agriculture. Current Opinion in Plant Biology 11: 180–186. doi:10.1016/j.pbi.2008.02.004

4. Jamil M, Charnikhova T, Cardoso C, Jamil T, Ueno K, et al. (2011) Quantification of the relationship between strigolactones and *Striga hermonthica* infection in rice under varying levels of nitrogen and phosphorus. Weed Research 51: 373–385. doi:10.1111/j.1365-3180.2011.00847.x

5. Jamil M, Rodenburg J, Charnikhova T, Bouwmeester HJ (2011) Pre-attachment *Striga hermonthica* resistance of new rice for Africa (NERICA) cultivars based on low strigolactone production. New Phytol 192: 964–975. doi:10.1111/j.1469-8137.2011.03850.x

6. Jamil M, Charnikhova T, Houshyani B, Ast A, Bouwmeester HJ (2012) Genetic variation in strigolactone production and tillering in rice and its effect on *Striga hermonthica* infection. Planta 235: 473–484. doi:10.1007/s00425-011-1520-y

7. Akiyama K, Matsuzaki K, Hayashi H (2005) Plant sesquiterpenes induce hyphal branching in arbuscular mycorrhizal fungi. Nature 435: 824–827. doi:10.1038/nature03608

8. Besserer A, Puech-Pagès V, Kiefer P, Gomez-Roldan V, Jauneau A, et al. (2006) Strigolactones stimulate arbuscular mycorrhizal fungi by activating mitochondria. PLoS Biology 4: e226. doi:10.1371/journal.pbio.0040226

9. Gomez-Roldan V, Fermas S, Brewer PB, Puech-Pagès V, Dun EA, et al. (2008) Strigolactone inhibition of shoot branching. Nature 455: 189–194. doi:10.1038/nature07271

10. Umehara M, Hanada A, Yoshida S, Akiyama K, Arite T, et al. (2008) Inhibition of shoot branching by new terpenoid plant hormones. Nature 455: 195–200. doi:10.1038/nature07272

11. Ruyter-Spira C, Kohlen W, Charnikhova T, van Zeijl A, van Bezouwen L, et al. (2011) Physiological effects of the synthetic strigolactone analog GR24 on root system architecture in *Arabidopsis*: another belowground role for strigolactones? Plant Physiology 155721734 . doi:10.1104/pp.110.166645

12. Kapulnik Y, Delaux P-M, Resnick N, Mayzlish-Gati E, Wininger S, et al. (2011) Strigolactones affect lateral root formation and root-hair elongation in *Arabidopsis*. Planta 233: 209–216. doi:10.1007/s00425-010-1310-y

13. Nadal M, Paszkowski U (2013) Polyphony in the rhizosphere: presymbiotic communication in arbuscular mycorrhizal symbiosis. Current Opinion in Plant Biology 16: 473–479. doi:10.1016/j.pbi.2013.06.005

14. Yoneyama K, Xie X, Kusumoto D, Sekimoto H, Sugimoto Y, et al. (2007) Nitrogen deficiency as well as phosphorus deficiency in sorghum promotes the production and exudation of 5-deoxystrigol, the host recognition signal for arbuscular mycorrhizal fungi and root parasites. Planta 227: 125–132. doi:10.1007/s00425-007-0600-5

15. Yoneyama K, Yoneyama K, Takeuchi Y, Sekimoto H (2007) Phosphorus deficiency in red clover promotes exudation of orobanchol, the signal for mycorrhizal symbionts and germination stimulant for root parasites. Planta 225: 1031–1038. doi:10.1007/s00425-006-0410-1

16. López-Ráez JA, Charnikhova T, Gómez-Roldán V, Matusova R, Kohlen W, et al. (2008) Tomato strigolactones are derived from carotenoids and their biosynthesis is promoted by phosphate starvation. New Phytol 178: 863–874. doi:10.1111/j.1469-8137.2008.02406.x

17. Umehara M, Hanada A, Magome H, Takeda-Kamiya N, Yamaguchi S (2010) Contribution of strigolactones to the inhibition of tiller bud outgrowth under phosphate deficiency in rice. Plant and Cell Physiology 51: 1118–1126. doi:10.1093/pcp/pcq084

18. Kohlen W, Charnikhova T, Liu Q, Bours R, Domagalska MA, et al. (2011) Strigolactones are transported through the xylem and play a key role in shoot architectural response to phosphate deficiency in nonarbuscular mycorrhizal host *Arabidopsis*. Plant Physiology 155: 974–987. doi:10.1104/pp.110.164640

19. Kohlen W, Charnikhova T, Lammers M, Pollina T, Tóth P, et al. (2012) The tomato CAROTENOID CLEAVAGE DIOXYGENASE8 (SlCCD8) regulates rhizosphere signaling, plant architecture and affects reproductive development through strigolactone biosynthesis. New Phytologist 196: 535–547. doi:10.1111/j.1469-8137.2012.04265.x

20. Koltai H (2011) Strigolactones are regulators of root development. New Phytol 190: 545–549. doi:10.1111/j.1469-8137.2011.03678.x

21. Alder A, Jamil M, Marzorati M, Bruno M, Vermathen M, et al. (2012) The path from β-carotene to carlactone, a strigolactone-like plant hormone. Science 335: 1348–1351. doi:10.1126/science.1218094

22. Yoneyama K, Xie X, Yoneyama K, Takeuchi Y (2009) Strigolactones: structures and biological activities. Pest Management Science 65: 467–470. doi:10.1002/ps.1726

23. Xie X, Yoneyama K, Kisugi T, Uchida K, Ito S, et al. (2013) Confirming stereochemical structures of strigolactones produced by rice and tobacco. Mol Plant 6: 153–163. doi:10.1093/mp/sss139

24. Awad AA, Sato D, Kusumoto D, Kamioka H, Takeuchi Y, et al. (2006) Characterization of strigolactones, germination stimulants for the root parasitic plants *Striga* and *Orobanche*, produced by maize, millet and sorghum. Plant Growth Regul 48: 221–227. doi:10.1007/s10725-006-0009-3

25. Xie X, Kusumoto D, Takeuchi Y, Yoneyama K, Yamada Y, et al. (2007) 2′-*epi*-Orobanchol and solanacol, two unique strigolactones, germination stimulants for root parasitic weeds, produced by tobacco. J Agric Food Chem 55: 8067–8072. doi:10.1021/jf0715121

26. Akiyama K, Ogasawara S, Ito S, Hayashi H (2010) Structural requirements of strigolactones for hyphal branching in AM fungi. Plant Cell Physiol 51: 1104–1117. doi:10.1093/pcp/pcq058

27. Nomura S, Nakashima H, Mizutani M, Takikawa H, Sugimoto Y (2013) Structural requirements of strigolactones for germination induction and inhibition of *Striga gesnerioides* seeds. Plant Cell Rep 32: 829–838. doi:10.1007/s00299-013-1429-y

28. Matusova R, Rani K, Verstappen FWA, Franssen MCR, Beale MH, et al. (2005) The strigolactone germination stimulants of the plant-parasitic *Striga* and *Orobanche* spp. are derived from the carotenoid pathway. Plant Physiology 139: 920–934. doi:10.1104/pp.105.061382

29. Buee M, Rossignol M, Jauneau A, Ranjeva R, Bécard G (2000) The pre-symbiotic growth of arbuscular mycorrhizal fungi is induced by a branching factor partially purified from plant root exudates. Mol Plant Microbe Interact 13: 693–698. doi:10.1094/MPMI.2000.13.6.693

30. Sato D, Awad AA, Chae SH, Yokota T, Sugimoto Y, et al. (2003) Analysis of strigolactones, germination stimulants for *Striga* and *Orobanche*, by high-performance liquid chromatography/tandem mass spectrometry. J Agric Food Chem 51: 1162–1168. doi:10.1021/jf025997z.

31. Ueno K, Nomura S, Muranaka S, Mizutani M, Takikawa H, et al. (2011) *Ent-2′-epi*-orobanchol and its acetate, as germination stimulants for *Striga gesnerioides* seeds isolated from cowpea and red clover. J Agric Food Chem 59: 10485–10490. doi:10.1021/jf2024193

32. Zwanenburg B, Pospíšil T (2013) Structure and activity of strigolactones: new plant hormones with a rich future. Mol Plant 6: 38–62. doi:10.1093/mp/sss141

33. Zwanenburg B, Mwakaboko AS, Reizelman A, Anilkumar G, Sethumadhavan D (2009) Structure and function of natural and synthetic signalling molecules in parasitic weed germination. Pest Manag Sci 65: 478–491. doi:10.1002/ps.1706

34. Seto Y, Sado A, Asami K, Hanada A, Umehara M, et al. (2014) Carlactone is an endogenous biosynthetic precursor for strigolactones. PNAS: 201314805. doi:10.1073/pnas.1314805111

35. Motonami N, Ueno K, Nakashima H, Nomura S, Mizutani M, et al. (2013) The bioconversion of 5-deoxystrigol to sorgomol by the sorghum, *Sorghum bicolor* (L.) Moench. Phytochemistry 93: 41–48. doi:10.1016/j.phytochem.2013.02.017

36. Matusova R, Bouwmeester HJ (2006) The effect of host-root-derived chemical signals on the germination of parasitic plants. Frontis 16: 39–54.

37. Gurney AL, Press MC, Scholes JD (1999) Infection time and density influence the response of sorghum to the parasitic angiosperm *Striga hermonthica*. New Phytologist 143: 573–580. doi:10.1046/j.1469-8137.1999.00467.x

38. Boyer F-D, Germain A de S, Pillot J-P, Pouvreau J-B, Chen VX, et al. (2012) Structure-activity relationship studies of strigolactone-related molecules for branching inhibition in garden pea: molecule design for shoot branching. Plant Physiol 159: 1524–1544. doi:10.1104/pp.112.195826.

7

Genetic Differentiation Revealed by Selective Loci of Drought-Responding EST-SSRs between Upland and Lowland Rice in China

Hui Xia[1], Xiaoguo Zheng[1], Liang Chen[1], Huan Gao[1], Hua Yang[1], Ping Long[1], Jun Rong[2], Baorong Lu[3], Jiajia Li[1], Lijun Luo[1]*

1 Shanghai Agrobiological Gene Center, Shanghai, China, 2 Center for Watershed Ecology, Institute of Life Science and Key Laboratory of Poyang Lake Environment and Resource Utilization, Ministry of Education, Nanchang University, Nanchang, China, 3 Ministry of Education Key Laboratory for Biodiversity and Ecological Engineering, Fudan University, Shanghai, China

Abstract

Upland and lowland rice (*Oryza sativa* L.) represent two of the most important rice ecotypes adapted to ago-ecosystems with contrasting soil-water conditions. Upland rice, domesticated in the water-limited environment, contains valuable drought-resistant characters that can be used in water-saving breeding. Knowledge about the divergence between upland and lowland rice will provide valuable cues for the evolution of drought-resistance in rice. Genetic differentiation between upland and lowland rice was explored by 47 Simple Sequence Repeats (SSRs) located in drought responding expressed sequence tags (ESTs) among 377 rice landraces. The morphological traits of drought-resistance were evaluated in the field experiments. Different outlier loci were detected in the *japonica* and *indica* subspecies, respectively. Considerable genetic differentiation between upland and lowland rice on these outlier loci was estimated in *japonica* ($Fst = 0.258$) and *indica* ($Fst = 0.127$). Furthermore, populations of the upland and lowland ecotypes were clustered separately on these outlier loci. A significant correlation between genetic distance matrices and the dissimilarity matrices of drought-resistant traits was determined, indicating a certain relationship between the upland-lowland rice differentiation and the drought-resistance. Divergent selections occur between upland and lowland rice on the drought-resistance as the $Qsts$ of some drought-resistant traits are significantly higher than the neutral Fst. In addition, the upland- and lowland-preferable alleles responded differently among ecotypes or allelic types under osmotic stress. This shows the evolutionary signature of drought resistance at the gene expression level. The findings of this study can strengthen our understanding of the evolution of drought-resistance in rice with significant implications in the improvement of rice drought-resistance.

Editor: Chengdao Li, Department of Agriculture and Food Western Australia, Australia

Funding: The work was supported by the following: Natural Science Foundation of China (Grant no. 31200279); Postdoctoral Science Foundation, Shanghai (No. 12R21421300); National Basic Research Program (Grant no. 2012CB114305); Shanghai "Phosphor" Science Foundation (Grant no. 11QA1405900); the National High-Tech Research and Development Program of China (863 Plan)(Grant No. 2012AA101102; 2014AA10A603); Shanghai Seed projects (2014.01); and Talented Person Project (2010CI120). The funders had no role in study design, data collection and analysis, decision to publish, or preparation of the manuscript.

Competing Interests: The authors have declared that no competing interests exist.

* Email: lijun@sagc.org.cn

Introduction

How plants have developed resistance to biotic and abiotic stresses is a fundamental question in plant biology. To address this question, plant ecotypes grown in contrasting environments are studied as they are likely under divergent selections, resulting in adaptive divergence [1]. Although ecotypes of wild species adapted to different ecosystems are served as ideal systems for studying the evolution of resistance to biotic and abiotic stresses [2–5], literatures on different ecotypes of a crop to a specific stress are still rare.

The Asian cultivated rice (*Oryza sativa* L.) is one of the most important cereal crops in the world by providing stable food for > 50% of the total global population (ref). However, rice production has encountered severe challenges due to frequent droughts and water shortages [6]. The utilization of natural variations of water-saving and drought-resistant characters in upland rice is an effective solution to improve the drought-resistance for rice [7–8].

Upland rice has adapted to the water-limited and rain-fed rice ecosystems, facing the higher risk of drought [9]. On the contrary, lowland rice is commonly planted in fields with irrigation facilities with relatively lower risks of drought. Upland rice may accumulate more drought-resistant genetic variances than lowland rice, leading to potential adaptive divergence of this rice ecotype on drought-resistance [7]. Therefore, studying differentiation between upland and lowland rice provides ideal systems for understanding the evolution of drought resistance and exploring drought-resistant genetic resources in crops. Previous attempts using neutral markers to explore differentiation between upland and lowland rice were not very successful [10–11], while another study using the functional-based markers has detected some ecotype-specific genetic features between upland and lowland rice [12].

The simple sequence repeats (SSRs) located in the ESTs (expressed sequence tags) are developed and widely applied in population genetics and breeding since the last decade [13]. The expression of drought-responding EST-SSRs is up- or down- regulated by drought stress. Although these genic SSRs are not always selective among ecotypes, they have a higher probability of being associated with the drought-resistance. In this study, 47 drought-responding EST-SSR loci were selected to explore the genetic differentiation between upland and lowland rice ecotypes from China. The questions to be addressed are: 1) Does considerable differentiation occur on drought-responding EST-SSR loci between upland and lowland rice ecotypes? 2) Is the differentiation between upland and lowland rice related to drought-resistance? 3) How do the upland- and lowland-differential alleles of the selective loci respond to drought stress? Answering these questions will facilitate our understanding of the evolution of drought-resistance in rice and give cues in rice breeding for drought-resistance.

Materials and Methods

Plant materials

A total of 377 rice landraces from Yunnan, Guizhou, Guangxi, Jiangsu, and Hebei province in China were used to study the differentiation between upland and lowland rice ecotypes (Table 1). Based on the information provided by the National Core Germplasm Database (http://crop.agridata.cn/A010110. asp), the landraces of upland rice from the five provinces were accounted for ~80% of the total upland rice germplasm in China. The predefined subspecies and ecotypes of rice landraces were provided by the institutes that collected them. Thus, crop materials in this experiment can be divided into four groups: *jap*-upland, *jap*-lowland, *ind*-upland, and *ind*-lowland. Additionally, 22 common wild rice strains were also used as reference while studying the differentiation among these two rice ecotypes. Twenty-two drought-resistant traits of 56 *japonica* and 49 *indica* materials were further evaluated in the field experiments rice landraces.

Rice DNA extraction and SSR genotyping

Rice genomic DNA was extracted from the 10-day-old seedlings following the common cetyltrimethyl ammonium bromide (CTAB) protocol. Three individual seedlings of a landrace were mixed together to include the genetic variations within a material. To focus on the drought-related genetic features, 47 polymorphic drought-responding EST-SSRs (SSRs locating in drought-responding ESTs) across 12 rice chromosomes were selected for genotyping (Table S1). The three primers system was applied in the polymerase chain reaction (PCR), containing a pair of EST-SSR primers and a rice-universal M13 sequence. The M13 sequence (5′ to 3′: CACGACGTTGTAAAACGAC) was labeled with fluorescent dyes at the 5′ end. The forward primer of the EST-SSR was also joint with the rice common M13 sequence at the 5′ end. The 20 ul reaction system contains 1×buffer (Mg2+), 2 mM each of dNTP, 50 ng of genomic DNA, 1 U of Taq polymerase, 16 mM of fluorescent dye labeled M13 sequence, 4 mM of the SSR primer liked with M13, and 20 mM of the other SSR primer. The PCR program started with a denaturation period of 4 min at 94°C, followed by 34 cycles of 40 s at 94°C, 30 s at their respective Tm (Table S1), and 40 s at 72°C, and ended after 10 min final extension at 72°C. The PCR products were then analyzed on ABI 3130XL (Applied Biosystems, USA) using ROX500 as the internal standard. The resulting chromatograms were analyzed and scored by Peakscanner ver. 1.0.

Field drought-resistance evaluation

The field evaluation of the drought-resistant traits was conducted at the Zhuanghang experimental station of Shanghai Academy of Agriculture Science in 2012. Two treatments were designed as control (CK) and drought (DT). Plants in CK were grown in a paddy field following the normal irrigated rice cultivation, while plants in DT were grown in a similar field nearby. The surface level of the DT field was approximately 2 meters higher than the CK field to simulate the upland condition and drain away the water more quickly. The DT field had a drought-resistance screen facility, which equipped a moveable platform capable of stretching out to keep the rainfall away when necessary. The germinated seeds were directly seeded both in the CK and DT on May 30th. However, water in the DT field was drained away on July 15th and it was never irrigated until August 25th. Individuals of each rice landrace were planted in a plot with 3×4 grids of 20 cm intervals between two seedlings. Consequently, there were 105 plots in one experimental block. Three replicates were included and arranged following the randomized complete block design. The number of stomas per leaf area, the excised leaf water (EWR) loss rate in 4 hours, and the root-shoot ratio were measured from leaves of main tillers in CK 30 days after the beginning of drought treatment. Relative water content (RWC) of leaf samples was measured 16 days after drought both in CK and DT. The content of malonaldehyde (MDA) was measured from fresh leaf tissues 40 days after drought following the protocol of the test kit (Nanjing Jiancheng Bioengineer Institute). The flag leaf length and width were measured at the heading stage both in CK and DT. All morphological traits were measured from three individuals from each replicate. The yield related traits were measured from four individuals of each replicate after harvest. The general drought resistance of a rice landrace was quantified by the drought-resistant index (DRI) [14]. It was calculated as $P_d/P_C*(P_d/P_{ad})$, where P_d is the yield production in drought treatment, P_C is the

Table 1. Summary of rice landraces used in the genotyping experiment and their inferred subspecies, ecotypes, and sources.

Groups Regions	Japonica		Indica	
	Upland	Lowland	Upland	Lowland
Guangxi	15	13	66	33
Yunan	15	6	15	24
Guizhou	42	17	12	9
Jiangsu	14	24	20	11
Hebei	22	19	0	0
Total	108	79	113	77

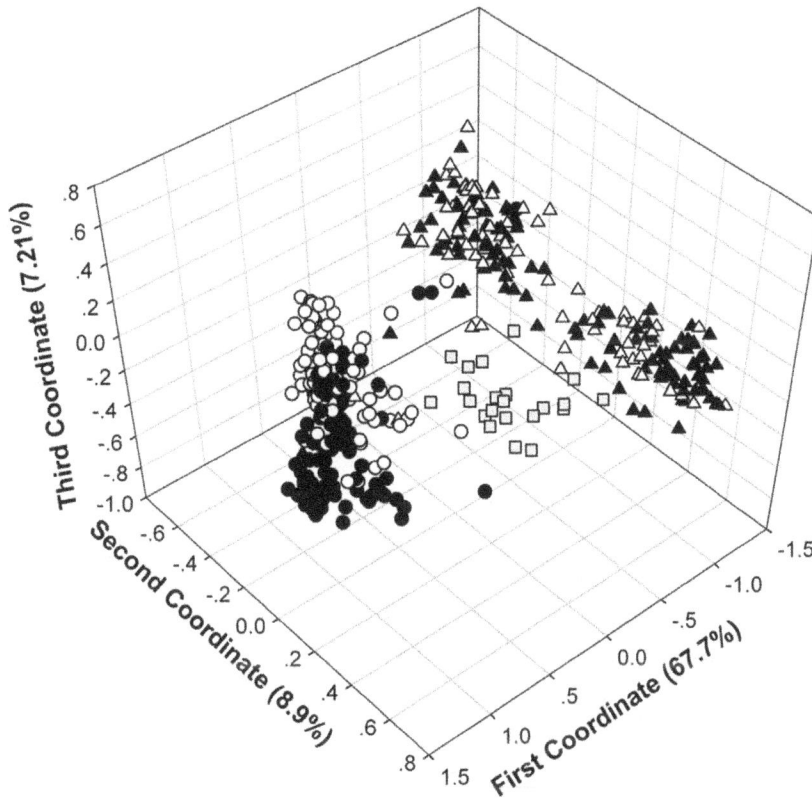

Figure 1. Population structure and differentiation investigated by Principal Coordinate Analysis. White color indicated lowland rice; dark color indicated upland rice; grey color and square indicated wild rice. Circles and triangles indicated *japonica* and *indica* subspecies, respectively.

yield production in control treatment, and P_{ad} is the average yield production of total materials in drought treatment.

Missing data and null alleles in data scoring

For genotyping, any null alleles were genotyped twice to avoid the manipulation errors. 337 putative null alleles were detected out of the total 18753 individual-loci combinations (1.8%), which was much lower than that expected by chance at the 5% level. The putative null alleles were treated as missing data in the further analyses.

Population structure and genetic diversity analysis

The model-based program STRUCTURE ver. 2.3.3 was used to infer population structure using the admixture model with a burn-in length of 50,000 and a run length of 50,000 for Markov Chain Monte Carlo iterations. Ten simulations were run for each K from K = 1 to K = 8 and the results were generated together using software CLUMPP_Windows.1.1.2. The Evanno's K was applied to determine the inferred K value [15]. The Principal Coordinate Analysis (PCoA) implemented in GenAlex ver. 6.43 was conducted to investigate the genetic distances between rice landraces and wild rice using the total 47 SSR loci. Genetic diversity was quantified by four estimators as gene diversity, number of alleles, heterozygosity, and polymorphism information

Table 2. Four genetic diversity estimators (mean ± standard deviations) analyzed in five groups using total loci set by PowerMarker ver. 3.25.

Group	Sample size	Allele No.	Gene Diversity	Heterozygosity	PIC
Jap-upland	108	6.88±0.59 b	0.435±0.030 a	0.036±0.008 b	0.399±0.032 a
Jap-lowland	79	6.49±0.46 a	0.428±0.040 a	0.024±0.004 a	0.402±0.033 a
Ind-upland	113	6.94±0.43 b	0.489±0.033 b	0.033±0.006 b	0.440±0.028 b
Ind-lowland	77	6.46±0.38 a	0.511±0.028 c	0.033±0.006 b	0.466±0.026 c
O. rufipogon	22	7.32±0.46 c	0.676±0.028d	0.336±0.023 c	0.636±0.025d
Overall	399	10.12±0.78	0.642±0.022	0.051±0.006	0.586±0.024

The different letters behind the values indicated significant differences (p<0.05) among groups by one-way ANOVA.

Table 3. Outlier loci detected under selection among upland and lowland ecotypes at 95% confidence by three outlier tests.

Method/Software	Code of locus	
	Japonica group	*Indica* group
Lositan	*E359, E647, E1238, E1899, E3735*	*E1177, E4208*
BayeScan	E214, E385, E674, E986, E1161, E1177, E1188, *E1238*, E1615	*E647*, E1188
Detsel	*E359, E647, E1899*, E1941, *E3735*,	E399, *E647, E1177*, E1350, E1719, *E4208*, E4632

Loci in **bold** and *italic* indicated they were detected at least twice.

content (PIC) using the software PowerMarker ver. 3.25 based on the total 47 EST-SSR loci. The standard deviations were calculated from 100 times bootstrap.

Neutrality tests for EST-SSR loci

Most of the genic markers used in this study were neutral among ecotypes as very low *Fst*s were recorded among rice ecotypes, similar with that using putatively neutral genomic SSRs [11]. Three *Fst*-based outlier tests were included to detect signs of selection among these genic SSRs respectively in the *indica* and *japonica* subspecies as indicated by the population structure. First, LOSITAN was used to detect loci under selection at the 95% confidence level [16]. An initial run with 100,000 simulations was conducted and followed by computing the distribution of neutral *Fst* using the putatively neutral loci derived from the initial run with 500,000 simulations. Second, the hierarchical-Bayesian method developed by Foll and Gaggiotti (2008) was applied to detect outlier loci using the software BAYESCAN (http://cmpg. unibe.ch/software/bayescan/) on the 95% posterior probabilities [17]. The parameters were set to 10 pilot runs of 5,000 iterations and additional burn-in of 50,000 iterations. Third, the software DETSEL 1.0 was applied to do pairwise comparisons to identify the outliers. For DETSEL, the coalescent simulation was conducted with the following parameters: mutation rate (infinite allele model) 0.005, 0.001, and 0.0001; ancestral population size $Ne = 500$, 1,000, and 10,000; population size before the split $N_0 = 100$; time since an assumed bottleneck $T_0 = 50$, 100, and 1,000 generations; and time since divergence $t = 100$ generations. Loci falling outside the specified "probability region (95% levels)" were considered as outliers [18]. Based on these outlier tests, two levels of loci sets could be defined from the total loci set as: (1) the neutral loci, which were not detected in any of the three tests, and (2) the decisive selective loci, which were detected at least twice by the three outlier tests.

Genetic differentiation among upland and lowland rice

The general level of genetic differentiation among different groups was quantified by pairwise *Fst* calculated in Arlequin ver. 3.1 using total loci set with 1,000 permutations. To estimate the level of differentiation on selective loci, *Fst* was calculated from the decisive selective loci set among upland and lowland ecotypes in *japonica* and *indica*, respectively.

Japonica landraces were from 5 provinces and *indica* landrace were from 4 provinces. Thus, there were 10 *japonica* and 8 *indica* populations included in this study. To explore the influence of the genetic differentiation on population structures, genetic distance (GD) matrices were calculated by GenAlex ver. 6.43 with 999 permutations using the corresponding neutral or decisive selective loci. The genetic distance matrix obtained form the calculation was used for the cluster analysis according to the un-weighted pair-group method with arithmetic averages (UPGMA) *via* the software NTSYS ver. 2.10e.

Testing whether the upland-lowland rice differentiation was related to drought-resistance

According to previous studies, the comparison of quantitative genetic divergence (*Qst*) and neutral genetic divergence (*Fst*) can be used to detect adaptive evolution. If the *Qst* is significantly higher than the *Fst* calculated by the neutral loci, it indicates that the directional selection drives phenotypic divergence and results in ecological adaptation [19]. In this study, the *Qst* of each trait was calculated as: $Qst = V_{pop}/(V_{pop}+2V_{ind})$, in which V_{pop} was the variance among-population and V_{ind} was the variance within-population. Although the SSRs located in the ESTs, the *Fst* generated among upland and lowland rice was as low as genomic SSRs [11]. This suggests most of these markers were neutral among the two ecotypes. The 95% confidence interval of *Qst* was calculated using R program with 1000 bootstraps. The *Fst* derived from neutral loci and its standard errors were calculated by Arlequin ver. 3.1 with 1000 permutations. Any significant differences between the *Qst* and *Fst* at $p = 0.05$ level was determined when $|Qst-Fst| > 2SQRT(SE_{Qst}^2+SE_{Fst}^2)$.

Table 4. Pairewise *Fst* among five experimental groups using total loci set.

Group	*Jap*-upland	*Jap*-lowland	*Ind*-upland	*Ind*-lowland	*O. rufipogon*
Jap-upland	0.0000	+	+	+	+
Jap-lowland	0.0714	0.0000	+	+	+
Ind-upland	0.4426	0.4649	0.0000	+	+
Ind-lowland	0.4255	0.4475	0.0302	0.0000	+
O. rufipogon	0.2683	0.2798	0.2405	0.2207	0.0000

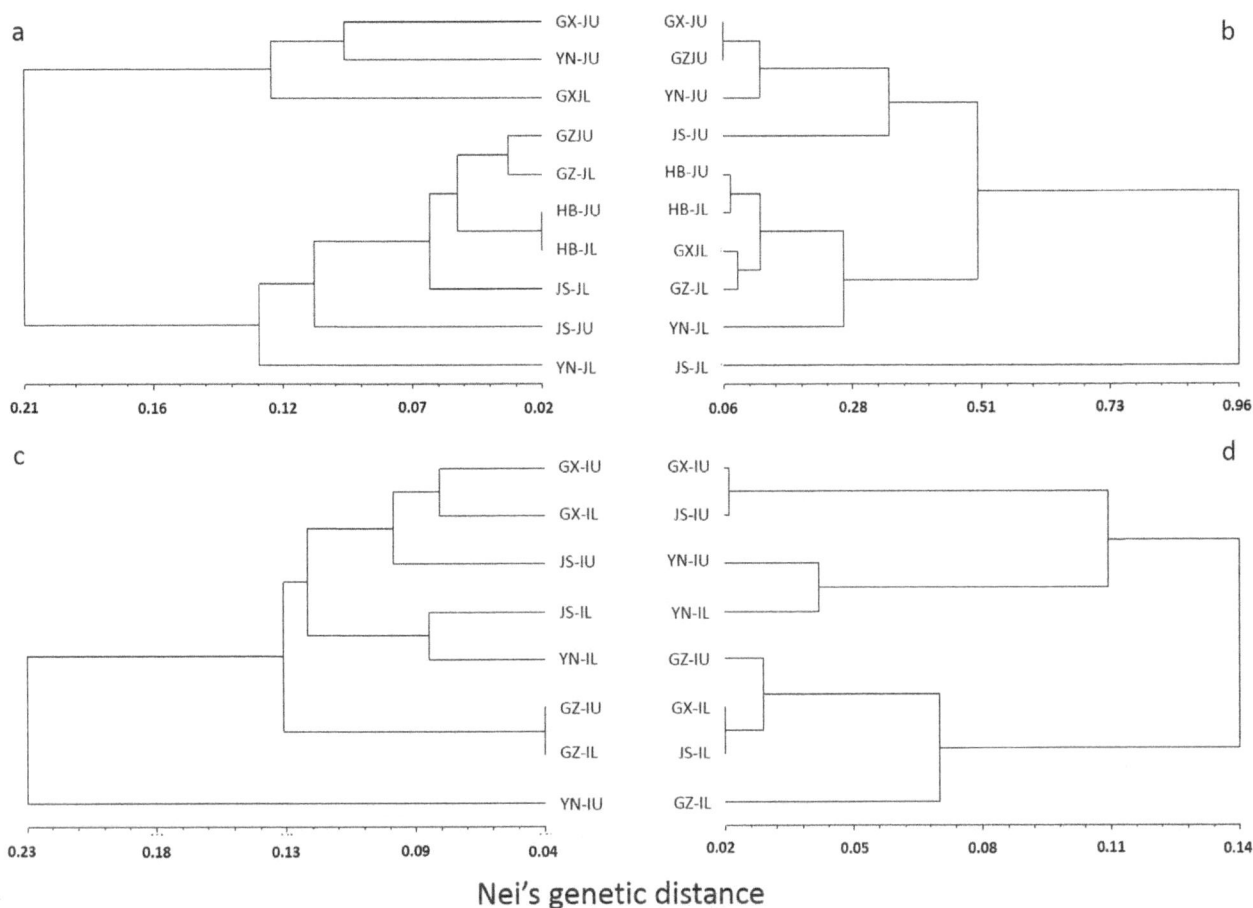

Figure 2. Regional populations cluster analysis using UPGMA method. a) *japonica* materials using neutral loci, b) *japonica* materials using decisive selective loci, c) *indica* materials using neutral loci, d) *indica* materials using decisive selective loci. GX, Guangxi; GZ, Guizhou; HB, Hebei; JS, Jiangsu; YN, Yunnan. U = upland; L = lowland.

To test whether the genetic differentiation between upland and lowland rice was resulting from divergent selections on drought-resistance, Mantel tests were conducted between the individual based GD matrix using the neutral or decisive selective loci and matrix constructed from differentiated morphological traits. The differentiated traits were defined as the traits having significant differences between upland and lowland rice by independent *t*-test using SPSS ver. 15.0. The individual based GD matrix was constructed by GenAlex ver. 6.43. The dissimilarity matrices of differentiated traits were constructed by Hierarchical Cluster Analysis on SPSS ver. 15.0.

Test the expression of ecotype-specific alleles of the selective loci

The decisive selective loci may play a role during upland rice adapting to water-limited upland conditions. To test this hypothesis, the expression levels of upland- and lowland-preferable alleles of the decisive selective loci (E647 and E1899 in *japonica*, E647 and E1177 in *indica*) were measured. Here, upland- or lowland-preferable allele was defined as a predominant allele in upland or lowland rice whose frequency was 20% higher than that in lowland or upland ecotype. The allele frequency was calculated as the number of counts of an allele/the total number of counts of all alleles in a given group *via* the software GenAlex ver. 6.43. Four landraces of each ecotype conferring the ecotype-preferable

alleles were included. Three-week-old seedlings in the growth chamber were treated with 20% polyethylene glycol (PEG) 6000 to cause osmotic stress (DT). Some materials were kept in nutrition solutions as controls (CK). When most of the seedlings showed signs of leaf rolling 5 hours after treatment, leaves of three seedlings in DT and CK were collected and kept in liquid nitrogen until RNA extraction. The RNA extraction was followed by the kits of TRNzol A$^+$ (TianGen Biotech Co. Ltd.). The level of gene expressions were quantified by RealTime-PCR using SYBR^R premix Ex Taq (Takara Bio. Inc.). *Actin* was used for reference. The expression change value was calculated as: the expression level in DT over that in CK. In addition, to test whether the ecotype-preferable alleles of the selective loci had any impacts on rice drought-resistance, the drought-resistant traits were further compared among upland and lowland rice containing differently expressed ecotype-preferable alleles (E647 and E1899 in *japonica*, and E1177 in *indica*).

Results

Population structure and general genetic diversity

A total of 477 alleles were detected in the 47 drought responding EST-SSR loci with number of alleles per locus ranging from 4–25. When running the STRUCTURE simulation using the total materials, there was a sharp peak in Evanno's ΔK at

Table 5. 22 drought-resistant traits measured in control treatment (CK) and drought treatment (DT) and their Qst calculated in *Jap*-upland/lowland and *Ind*-upland/lowland pairs (mean ± standard error).

Traits	Jap-upland	Jap-lowland	J-Qst	Ind-upland	Ind-lowland	I-Qst
RWC (CK)	**0.936±0.005***	**0.910±0.010***	0.052±0.0012	0.898±0.010	0.919±0.006	0.034±0.0007
RWC (DT)	0.871±0.006	0.853±0.011	0.032±0.0010	0.830±0.009	0.843±0.009	0.021±0.0007
No. of stomas per area	**13.9±0.3***	**15.1±0.5***	0.048±0.0012	14.8±0.3	14.7±0.4	0.011±0.0005
EWR	0.199±0.011	0.194±0.009	0.008±0.0004	0.219±0.011	0.219±0.011	0.010±0.0004
Root-shoot ratio	**0.215±0.006***	**0.259±0.014***	0.091±0.0016	0.225±0.009	0.239±0.014	0.017±0.0007
MDA (CK)	0.97±0.02	0.91±0.02	0.016±0.0006	**0.90±0.02***	**0.97±0.02***	**0.072±0.0016***
MDA (DT)	1.28±0.02	1.30±0.02	0.043±0.0010	1.27±0.02	1.26±0.02	0.010±0.0004
Flag leaf length (CK)	35.9±1.1	33.2±1.1	0.033±0.0010	36.0±1.5	38.0±1.6	0.020±0.0008
Flag leaf length (DT)	**32.1±1.2***	**24.5±1.1***	**0.171±0.0019***	32.1±1.6	28.5±2.2	0.032±0.0010
Flag leaf width (CK)	**1.62±0.05***	**1.33±0.05***	**0.131±0.0021***	1.45±0.05	1.51±0.05	0.016±0.0006
Flag leaf width (DT)	**1.70±0.06***	**1.31±0.07***	**0.151±0.0015***	1.33±0.07	1.35±0.06	0.009±0.0004
No. of panicles (CK)	**6.65±0.32***	**9.70±0.73***	**0.141±0.0021***	10.54±0.53	10.65±0.53	0.009±0.0004
No. of panicles (DT)	6.01±0.35	7.05±0.40	0.043±0.0011	11.12±0.67	10.24±0.85	0.018±0.0007
100-grain weight (CK)	2.36±0.06	2.50±0.07	0.030±0.0010	2.39±0.04	2.30±0.06	0.026±0.0009
100-grain weight (DT)	2.43±0.06	2.54±0.07	0.022±0.0008	**2.25±0.04***	**2.07±0.07***	**0.066±0.0015***
No. of seeds (CK)	**551.6±32.5***	**728.2±48.2***	**0.096±0.0017***	903.8±39.6	1079.7±70.2	**0.051±0.0013***
No. of seeds (DT)	480.0±29.3	445.7±29.5	0.014±0.0006	755.8±45.0	840.8±54.8	0.026±0.0010
Seed-set rate (CK)	**0.791±0.016***	**0.851±0.012***	0.086±0.0015	0.827±0.009	0.839±0.015	0.020±0.0008
Seed-set rate (DT)	0.822±0.009	0.835±0.020	0.018±0.0009	0.838±0.009	0.821±0.016	0.019±0.0006
Yield (CK)	**13.13±0.88***	**18.17±1.26***	**0.108±0.0018***	21.51±0.87	24.56±1.44	0.036±0.0010
Yield (DT)	11.25±0.53	11.21±0.80	0.011±0.0005	16.84±1.02	17.35±1.23	0.012±0.0006
DI	**0.89±0.13***	**0.57±0.07***	0.042±0.0009	1.02±0.11	0.93±0.12	0.017±0.0007

RWC: leaf relative water content, EWR: excised leaf water loss rate, MDA: malonaldehyde, DI: drought index.
The values in **bold** and with "*" indicated significant differences between upland and lowland ecotypes or significant difference between Qsts and the neutral Fst.

K = 2 (ΔK = 3718.5) which divided all materials into *japonica* and *indica* groups (Figure S1a). In *japonica*, the sharp peak occurs when K = 2 (ΔK = 7.6), which divided landraces into two groups as upland and lowland rice occupied the majority in each (Figure S1b). In *indica*, the sharp peak was also occurs when K = 2 (ΔK = 2760.1), which separated rice landraces of Guangxi from other regions (Figure S1c). However, there was a lower peak when K = 4 (ΔK = 78.0), in which upland rice (white) was distinguishable from lowland rice (thin gray) in Guangxi (Figure S1d). Similar to the results of STRUCTURE, PCoA indicated that *japonica* and *indica* landraces were separated along the first coordinate (x-axis) while common wild rice located at the center (Figure 1), suggesting that upland-lowland rice differentiation should be separately analyzed in *japonica* and *indica* subspecies. Differentiation between upland and lowland ecotypes was more apparent in *japonica* than that in *indica* (Figure 1). The level of genetic diversity was similar between upland and lowland rice ecotypes, while subspecies (*japonica* vs. *indica*) and species (*O. sativa* vs. *O. rufipogon*) showed considerable differences (Table 2).

Selective loci detected by neutrality tests

Among the 47 drought-induced EST-SSR loci, 5, 9, and 5 loci were detected to be selective by Lositan, BayeScan, and Detsel, respectively, in *japonica* subspecies. Five loci were detected to be the decisive selective loci (Table 3). In the *indica* subspecies, 2, 2 and 7 loci were detected to be selective by Lositan, BayeScan, and

Detsel, respectively. Three loci were considered as the decisive selective loci in *indica* (Table 3).

Genetic differentiation among upland and lowland rice

The values of Fst were as low as 0.0714 in *japonica* or 0.0302 in *indica* types of upland and lowland rice. The value was much lower than that between *japonica* and *indica* varieties (Table 4). This result was similar to a previous study in which Fst among different ecotype was recorded as 0.068 in *japonica* using putative neutral SSR markers [11]. However, the values of Fst between upland and lowland rice became much higher in *japonica* (0.285) and *indica* (0.127) when using their respective decisive selective loci. In the UPGMA clusters, upland and lowland rice populations were grouped separately when using the decisive selective loci, while populations of landraces from the same regions were clustered together when the neutral loci were used (Figure 2). These results suggest differentiation occurred between upland and lowland rice generally on the selective loci.

Divergent selection on drought-resistant traits

Drought resistant traits were evaluated in the field experiment. Significant differences were detected on 11 traits between upland and lowland ecotypes in *japonica*, while significant differences were detected on only 2 traits in *indica* (Table 5). The comparison between Qst with neutral Fst was used to test any divergent selection on the drought-resistant traits. The values of the Qst ranged from 0.009~0.171 in *japonica* and from 0.009~0.072 in

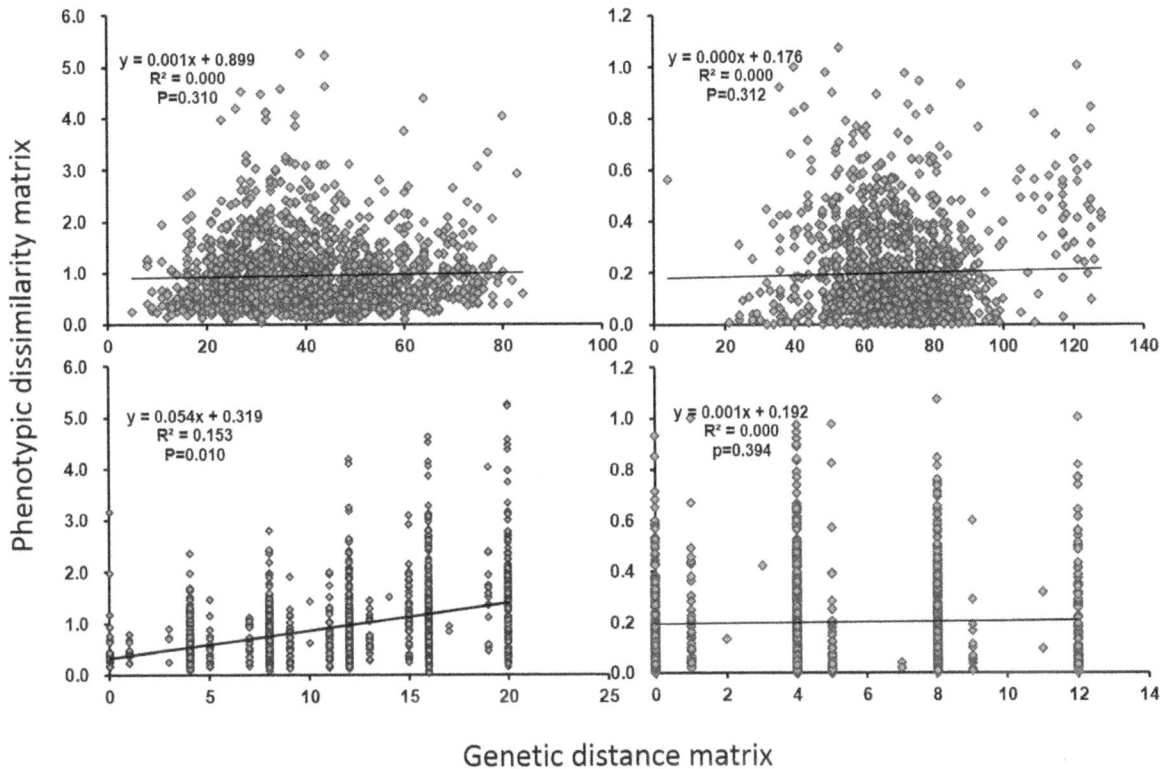

Figure 3. Correlations between matrices of genetic distance from different loci sets and the dissimilarity matrices from selective morphological traits. a) neutral loci in *japonica*; b) decisive selective loci in *japonica*; c) neutral loci in *indica*; d) decisive selective loci in in *indica*.

indica. The Q_{st}s of the most differentiated traits were significantly higher than the neutral F_{st}s in *japonica* (0.0898±0.0024) and *indica* (0.0363±0.0017) materials (Table 5). This result suggested that some of the drought-resistant traits were under divergent selection during domestication.

The genetic differentiation associated with morphological differentiation on drought-resistant traits between upland and lowland rice

Given that some drought-resistant traits were differentially selected in upland and lowland rice, their genetic differentiation may be associated with the evolution of drought-resistance. To test this, Mantel test was conducted between the individual-based genetic distance (DG) matrix and dissimilarity matrix constructed from morphological traits. In *japonica*, the dissimilarity matrix constructed from the differentiated traits was significantly correlated with the GD matrix using decisive selective loci (Figure 3b), while it was not correlated with the neutral GD matrix (Figure 3a). These results suggested a strong association between the differentiations on the drought-resistance and on the selective loci in *japonica*. However, the morphological dissimilarity matrix was neither significantly correlated with GD matrix from the selective loci nor with that from the neutral loci in *indica* (Figure 3c, 3d).

Expression of ecotype-preferable alleles and their impacts on drought-resistant traits

In *japonica* subspecies, the allele-6 of locus E647 was upland-preferable (taking account for 87.9% in upland rice but only 50.6% in lowland rice), while allele-2 was lowland-preferable (taking account for 40.4% in lowland rice but 0 in upland rice).

These two alleles expressed similarly in normal conditions (CK). However, the expression of allele-6 down-regulated largely (> 50%) while the expression of allele-2 remained at the same level in DT. Thus, the expression change values of the allele-2 (1.146±0.232) were marginally higher than the allele-6 (0.625±0.158, p<0.10) (Figure 4a). The allele-4 of E1899 was upland-preferable (taking account for 71.0% in upland rice but only 16.7% in lowland rice) and the allele-3 was lowland-preferable (taking account for 79.5% in lowland rice but 23.8% in upland rice). This EST was up-regulated in lowland rice ecotype while its expression was down-regulated in upland rice under DT. Thus, the expression change value of the allele-3 in lowland rice (1.832±0.583) was marginally higher than the allele-4 in upland rice (0.413±0.111, p<0.10) (Figure 4b).

In *indica* subspecies, the allele-6 of E647 was upland-preferable (taking account for 71.4% in upland rice but 50.7% in lowland rice), while the allele-7 was lowland-preferable (taking account for 49.3% in lowland rice but only 23.7% in upland rice). The allele-8 of E1177 was upland-preferable (taking account for 44.1% in upland rice but only 17.3% in lowland), while the allele-9 lowland-preferable (accounted for 74.7% in lowland rice but only 44.5% in upland). The expression change values of E647 were similar among ecotypes or allelic types (Figure 4c), while that of E1177 was much higher in lowland ecotype or lowland-preferable alleles than in upland rice or upland-preferable (Figure 4d).

As these ESTs differently expressed between allelic types or ecotypes, the drought-resistant traits were then compared among rice ecotypes conferring their ecotype-preferable alleles. As expected, many significant differences not previously detected among total upland and lowland rice were now detected (Table S2). For example, the upland rice conferring the allele-8 and

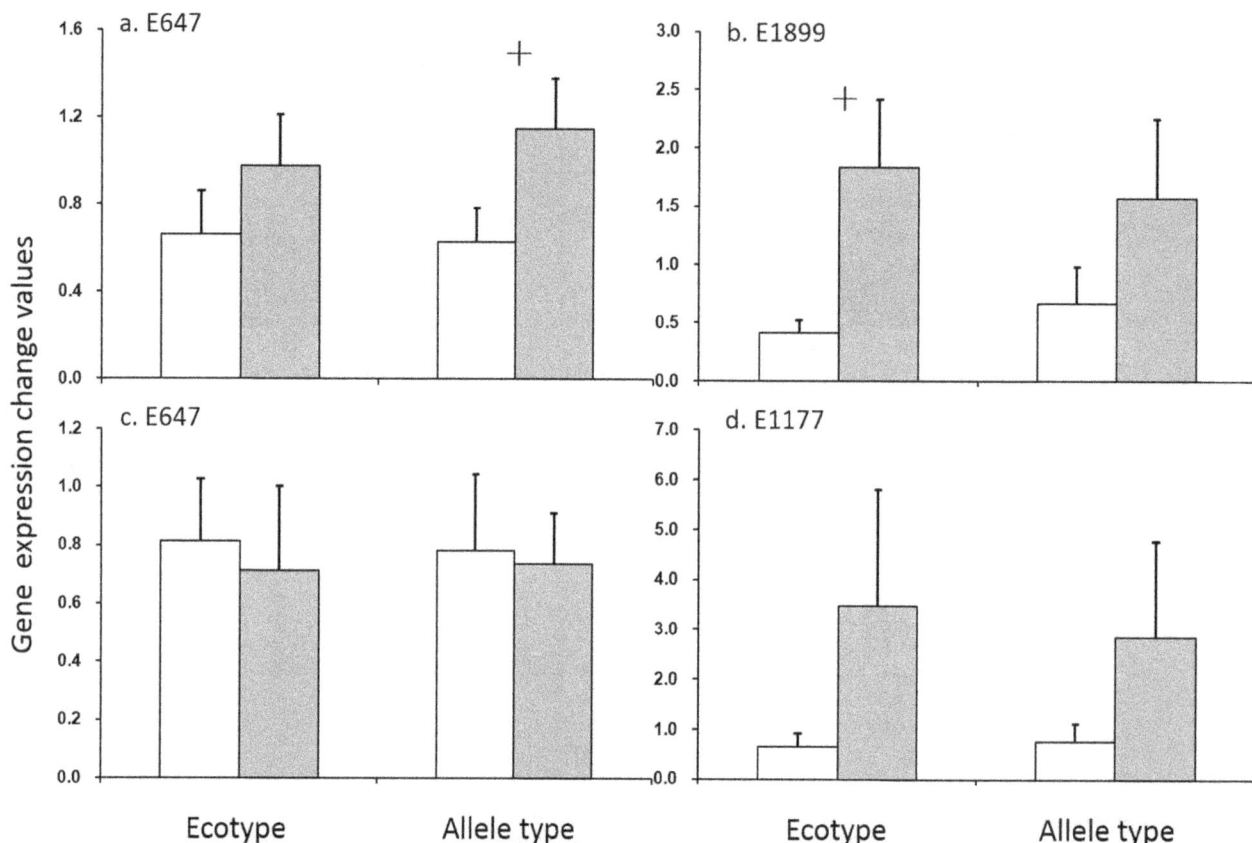

Figure 4. The expression change values (mean ± SE) between different ecotypes or alleles. The expression change values of E647 (a) and E1899 (b) between ecotypes or alleles in *japonica* and these of E647 (c) and E1177 (d) between ecotypes or alleles in *indica*. The expression change values were calculated as: gene expression in PEG/that in water. The white bar indicates upland rice or upland-specific allele, while the black bar indicates lowland rice or lowland-specific allele. "+" indicates differences at the significant level of p<0.10.

lowland rice conferring allele-9 of E1177 exhibited significant difference on the root-shoot ratio in *indica*. This was accordant to the annotated function of gene *Os06g0633300* (Table 6). Other genes containing the selective EST-SSRs were also considered to be related with resistance to abiotic stress given their annotations (Table 6), suggesting that these selective loci might play roles in rice drought-resistance.

Discussion

Considerable level of upland-lowland rice differentiation on selective loci

As in previously studies, obvious differentiation was reported on drought-resistant morphological traits between upland and lowland rice [9]. However, the general level of genetic differentiation between upland and lowland rice was very low in our study,

Table 6. Gene symbol, gene ID, and the annotated functions of the decisive selective loci detected in this study.

Locus	Gene symbol	Gene ID	Names	Predicted function
E647	Os01g0607400	4324222	hypothetical protein	Similar to STYLOSA protein
E359	Os06g0702600	4341978	hypothetical protein	Similar to Auxin response factor 7a
E1899	Os12g0563600	4352535	hypothetical protein	Protein of unknown function, DUF538 family protein
E3735	Os07g0260000	4342870	hypothetical protein	Protein prenyltransferase domain containing protein
E1238	Os10g0554200	4349339	hypothetical protein	TGF-beta receptor, type I/II extracellular region family protein
E1177	Os06g0633300	4341588	hypothetical protein	Phytosulfokines 1 precursor [Contains: Phyto sulfokine-alpha (PSK- alpha) (Phytosulfokine-a); Phytosulfokine-beta (PSK-beta) (Phytosulfokine-b)]
E4208	Os07g0546500	4343527	hypothetical protein	Conserved hypothetical protein

similar to the results in previous studies [10–11]. However, *Fst* calculated based on the outlier loci was considerably high between upland and lowland rice ecotypes, matching the level of *Fst* between *O. sativa* and *O. rufipogon* (0.252) in our study. These results suggested that adaptive divergence among upland and lowland rice occurred on these selective loci. Strong selections on the adaptive loci always affected population structures [20–21]. For example, wheat grown in different ecosystems was clustered separately based on the drought-responding gene TaSnRK2.7 [21]. Similar results were found in this study as revealed by the UPGMA cluster, providing evidences that these selective EST-SSRs received uniform selections among regions. It is noteworthy that morphological differentiation between upland and lowland ecotypes was also greater in *japonica* than in *indica*, consistent with the genetic data in these two subspecies.

Upland-lowland rice differentiation was driven by divergent selection on drought-resistance

Theoretically, drought-resistance alleles are more strongly selected in upland rice than that in lowland rice, due to its adaption to the water-limited environment. In this study, the general drought-resistance gained from field experiments was much higher in upland rice than that in lowland rice. The higher *Qst* than neutral *Fst* suggests some drought-resistant traits are under the divergent selections, leading genetic differentiation on the outlier loci between upland and lowland rice. The genetic differentiation between upland and lowland rice is likely associated with drought-resistance based on the Mantel tests between the genetic distance and morphological dissimilarity, especially in *japonica*. The formation of upland- or lowland-preferable alleles always result from divergent selections [22]. Different expression of the ecotype-preferable alleles under their favorable conditions could be considered as the signature of natural selection [23]. In this study, we found different expression patterns among upland- and lowland-preferable alleles encountering osmotic stress, adding further evidence of the adaptive divergence between upland and lowland rice. Based on these results, we conclude that genetic differentiation between upland and lowland rice is likely driven by divergent selection on the drought-resistance.

Candidate genes for drought-resistance gene from the selective loci

Recently studies disclosed that it was possible to indentify potential functional genes *via* studying adaptive divergence in natural populations [24], crops [5], and even in rice [25]. In this study, we found potential candidate genes that may play roles in rice drought-resistance. For example, *Os06g0633300* (E1177) encodes the rice phytosulfokine 1 precursor (*OsPSK1*). Its over-expression can promote rice cell division [26] and root development [27]. Interestingly, upland and lowland rice conferring different ecotype-referable alleles of E1177 exhibited significant differences on root-shoot ratio, a key character of drought-avoidance. Besides, genes with similar annotations as *Os01g0607400* (E647) and *Os12g0563600* (E1899) were also reported to be associated with plant stress responses by previous studies in other plant species [28–29]. These results provide strong evidences that these candidate genes should have played some

roles in rice drought-resistance. However, their functions in rice need further investigation. In a word, we can explore more drought-resistant genes by studying the genetic divergence among the upland and lowland rice.

Conclusions

Crops adapted to different agro-ecosystems always promote the variation of agricultural important genes. In this study, several outlier loci were detected between upland and low rice ecotypes. A considerable degree of genetic differentiation between upland and lowland rice was detected both at the DNA and gene expression level on these outlier loci. Results from this study reveals that the genetic differentiation among the two rice ecotypes is most likely driven by divergent selection on drought-resistant traits. The findings of this study not only help us to understand the underlying molecular basis of adaptive divergence, but also provide valuable implications for rice domestication and breeding, especially on the drought-resistance in rice.

Supporting Information

Figure S1 Population structures inferred by STRUCTURE. a) *Japonica* and *indica* subspecies were separated when K = 2. b) Inferred population structures in *japonica* subspecies when K = 2. c) Inferred population structures in *indica* subspecies when K = 2. d) Inferred population structures in *indica* subspecies when K = 4, in which some upland rice and lowland rice were separated.

Table S1 Basic information of the used 47 drought-induced EST-SSR loci.

Table S2 22 drought-resistant traits measured in upland and lowland rice materials containing their ecotype-preferable alleles (indicated in parentheses) in control treatment (CK) and drought treatment (DT) (mean ± standard error). The values in bold and with "*" indicated significant differences between upland and lowland ecotypes.

Table S3 Original data for SSR scoring and field evaluated traits (xls).

Acknowledgments

Rice materials were kindly provided by Rice Research Institute of Guangxi Academy of Agriculture Science, Hebei Academy of Agriculture Science, Yunnan Academy of Agriculture Science, and Jiangsu Academy of Agriculture Science.

Author Contributions

Conceived and designed the experiments: HX LC BRL LJL. Performed the experiments: HX XGZ HG HY PL JR JJL. Analyzed the data: HX XGZ LC PL JR. Contributed reagents/materials/analysis tools: HY JR LJL. Wrote the paper: HX LC JR BRL LJL.

References

1. Nosil P, Funk DJ, Ortiz-Barrientos D (2009) Divergent selection and heterogeneous genomic divergence. Mol Ecol 18: 375–402.

2. Shikano T, Ramadevi J, Merilä J (2010) Identification of local- and habitat-dependent selection: scanning functionally important genes in nine-spined sticklebacks (*Pungitius pungitius*). Mol Biol Evol 27: 2775–2789.

3. Hübner S, Höffken M, Oren E, Haseneyer G, Stein N, et al. (2009) Strong correlation of wild barley (*Hordeum spontaneum*) population structure with temperature and precipitation variation. Mol Ecol 18: 1523–1536.

4. Whitehead A, Triant DA, Champlin D, Nacci D (2010) Comparative transcriptomics implicates mechanisms of evolved pollution tolerance in a killifish population. Mol Ecol 19: 5186–5203.

5. Xia H, Camus-Kulandaivelu L, Stephan W, Tellier A, Zhang ZW (2010) Nucleotide diversity patterns of local adaptation at drought-related candidate genes in wild tomatoes. Mol Ecol 19: 4144–4154.

6. Bernier J, Atlin GN, Serraj R, Kumar A, Spaner D (2008) Breeding upland rice for drought resistance. J Sci Food Agr 88: 927–939.

7. Farooq M, Wahid A, Lee DJ, Ito O, Siddique KH (2009) Advances in drought resistance of rice. Crit Rev Plant Sci 28: 199–217.

8. Luo LJ (2010) Breeding for water-saving and drought-resistance rice (WDR) in China. J Exp Bot 61: 3509–3517.

9. IRRI (International Rice Research Institute) (1986) Upland rice:A global perspective. Los Baiios, Philippines.

10. Yu LX, Nguyen HT (1994) Genetic variation detected with RAPD markers among upland and lowland rice cultivars (*Oryza sativa* L.) Theor Appl Genet 87: 668–672.

11. Zhang DL, Zhang HL, Wang MX, Sun JL, et al. (2009) Genetic structure and differentiation of *Oryza sativa* L. in China revealed by microsatellites. Theor Appl Genet 119: 1105–1117.

12. Ishikawa R, Maeda K, Harada T, Niizeki M, Saito K (1992) Genotypic variation for 17 isozyme genes among Japanese upland varieties in rice. Japan J Breeding 42: 737–746.

13. Varshney RK, Graner A, Sorrells ME (2005) Genic microsatellite markers in plants: features and applications. TRENDS Biotech 23: 48–55.

14. Lan JS, Hu FS, Zhang JR (1990) The concept and statistical method of drought resistance index in crops. Acta Agriculturae Boreali-Sinica 5: 20–25 (in Chinese with English abstract)

15. Evanno G, Regnaut S, Goudet J (2005) Detecting the number of clusters of individuals using the software STRUCTURE: a simulation study. Mol Ecol 14: 2611–2620.

16. Antao T, Lopes A, Lopes RJ, Beja-Pereira A, Luikart G (2008) LOSITAN: a workbench to detect molecular adaptation based on a F_{ST}-outlier method. BMC Bioinformatics 9: 323.

17. Foll M, Gaggiotti O (2008) A genome-scan method to identify selected loci appropriate for both dominant and codominant markers: a Bayesian perspective. Genetics 180: 977–993.

18. Vitalis R, Dawson K, Boursot P, Belkhir K (2003) DetSel 1.0: a computer program to detect markers responding to selection. J Heredity 94: 429–431.

19. Miller JR, Wood BP, Hamilton MB (2008) *FST* and *QST* Under Neutrality. Genetics, 180, 1023–1037.

20. Martin MA, Mattioni C, Cherubini M, Taurchini D, Villani F (2010) Genetic diversity in European chestnut populations by means of genomic and genic microsatellite markers. Tree Genetics & Genomes 6: 735–744.

21. Zhang H, Mao X, Wu X, Wang C, Jing R (2011) An abiotic stress response gene *TaSnRK2.7-B* in wheat accessions: genetic diversity analysis and gene mapping based on SNPs. Gene 478: 28–34.

22. Steele KA, Edwards G, Zhu J, Witcombe JR (2004) Marker-evaluated selection in rice: shifts in allele frequency among bulks selected in contrasting agricultural environments identify genomic regions of importance to rice adaptation and breeding. Theor Appl Genet 109: 1247–1260.

23. Knight CA, Vogel H, Kroymann J, Shumate A, Witsenboer H, Mitchell-Olds T (2006) Expression profiling and local adaptation of *Boechera holboellii* populations for water use efficiency across a naturally occurring water stress gradient. Mol Ecol 15: 1229–1237.

24. Ross-Ibarra J, Morrell PL, Gaut BS (2007) Plant domestication, a unique opportunity to identify the genetic basis of adaptation. Proc Natl Acad Sci, USA 104: 8641–8648.

25. Lyu J, Zhang SL, Dong Y, et al. (2013) Analysis of elite variety tag SNPs reveals an important allele in upland rice. Nature Communications, 4: 2138. doi: 10.1038/ncomms3138.

26. Yang HP, Matsubayashi Y, Nakamura K, Sakagami Y (1999) *Oryza sativa PSK* gene encodes a precursor pf phytosulfokine-α, a sulfated peptide growth factor found in plants. Proc Natl Acad Sci, USA 96: 13560–13565.

27. Huang JY, Wang YF, Yang JS (2010) Over-expression of OsPSK3 increases chlorophyll content of leaves in rice. Hereditas 32: 1281–1289.

28. Ichitani K, Namigoshi K, Sato K, Taura S, Aoki M, et al. (2007) Fine mapping and allelic dosage effect of *Hwc1*, a complementary hybrid weakness gene in rice. Theor Appl Genet 114: 1407–1415.

29. Gholizadeh A (2011) Heterologous expression of stress-responsive duf538 domain containing protein and its morpho-biochemical consequences. Prot J 30: 351–358.

8

The Effects of Fluctuations in the Nutrient Supply on the Expression of Five Members of the *AGL17* Clade of MADS-Box Genes in Rice

Chunyan Yu, Sha Su, Yichun Xu, Yongqin Zhao, An Yan, Linli Huang, Imran Ali, Yinbo Gan*

Zhejiang Key Lab of Crop Germplasm, Department of Agronomy, College of Agriculture and Biotechnology, Zhejiang University, Hangzhou, China

Abstract

The *ANR1* MADS-box gene in *Arabidopsis* is a key gene involved in regulating lateral root development in response to the external nitrate supply. There are five *ANR1*-like genes in *Oryza sativa*, *OsMADS23*, *OsMADS25*, *OsMADS27*, *OsMADS57* and *OsMADS61*, all of which belong to the *AGL17* clade. Here we have investigated the responsiveness of these genes to fluctuations in nitrogen (N), phosphorus (P) and sulfur (S) mineral nutrient supply. The MADS-box genes have been shown to have a range of responses to the nutrient supply. The expression of *OsMADS61* was transiently induced by N deprivation but was not affected by re-supply with various N sources. The expression of *OsMADS25* and *OsMADS27* was induced by re-supplying with NO_3^- and NH_4NO_3, but downregulated by NH_4^+. The expression of *OsMADS57* was significantly downregulated by N starvation and upregulated by 3 h NO_3^- re-supply. *OsMADS23* was the only gene that showed no response to either N starvation nor NO_3^- re-supply. *OsMADS57* was the only gene not regulated by P fluctuation whereas the expression of *OsMADS23*, *OsMADS25* and *OsMADS27* was downregulated by P starvation and P re-supply. In contrast, all five *ANR1*-related genes were significantly upregulated by S starvation. Our results also indicated that there were interactions among nitrate, sulphate and phosphate transporters in rice.

Editor: John Schiefelbein, University of Michigan, United States of America

Funding: The research was supported by the International Scientific and Technological Cooperation Project of the Ministry of Science and Technology of China (grant number 2010DFA34430), International Scientific and Technological Cooperation Project of Science and Technology Department of Zhejiang Province (grant number 2013C34G2010017), National Natural Science Foundation of China (Grant No. 31370215; 31228002), and Zhejiang Provincial Natural Science Foundation of China (Grant No. Z3110004). The funders had no role in study design, data collection and analysis, decision to publish, or preparation of the manuscript.

Competing Interests: The authors have declared that no competing interests exist.

* Email: ygan@zju.edu.cn

Introduction

Nitrogen, phosphorus and sulfur are three major macronutrients essential for plant growth and development [1]. Nitrogen (N) deficiency is a vital factor limiting agricultural quality and productivity. In aerobic soil conditions, nitrate is the major source of nitrogen for many higher plant species [2–4]. Nitrate acts not only as a nutrient but also as a signal to regulate gene expression, energy transfer, protein activation, metabolic and physiological activities, and plant growth and development [5–8]. Phosphorus (P) is a component of many key biomolecules and participates in various enzymatic reactions and metabolic pathways [9]. Due to low availability of phosphate (Pi) in soil, it acts as the second most important limiting macronutrients for plant growth and development [10–12]. Plants can modify their root architecture to improve phosphate acquisition in phosphate deficient soil. Many studies have demonstrated that P deficiency affects root development, including root hair elongation, reduced primary root length and increased number and length of lateral roots in *Arabidopsis thaliana* [13–19]. Sulfur (S) is also an essential macronutrients, and its deficiency adversely affects plant growth and development as well as the quality of crops [20,21].

Plants can modify their root architecture to forage for sources of N that are distributed unevenly in soil [6]. One important kind of foraging response involves increased proliferation of lateral roots within soil patches enriched in certain nutrients, such as NH_4^+ and NO_3^- [22,23]. There are signaling mechanisms in roots to measure the levels of intrinsic and extrinsic nutritional factors so that they can modify their growth and development [24,25]. In *Arabidopsis*, localized nitrate treatment stimulated a localized increase in lateral root (LR) numbers and elongation. The stimulation of LR elongation was found to be the result of a signaling effect of external NO_3^- ion itself rather than downstream metabolites [26,27]. An important breakthrough in understanding NO_3^- in stimulating LR growth was the identification of the *ANR1* gene, which is a vital component in regulatory signaling pathway [26]. *ANR1* is a member of the plant MADS-box family of transcription factors, which covers more than 100 members in *Arabidopsis* [28–30]. In addition to their roles in regulation of reproductive development, the MADS-box transcription factors are also widely expressed in vegetative tissues [28,31]. Recent research has demonstrated that at least 50 MADS-box genes are expressed in roots of *Arabidopsis*, with *ANR1* as the only member so far to have a known function in lateral root development in *Arabidopsis* [32,33]. Previous reports

Table 1. List of primers used for Real-time PCR.

Gene	Forward primer (5′–3′)	Reverse primer (5′–3′)
OsMADS23	TCTTCTCCAGCACCAGCCGTCT	TGCTGCCTCCTGTTGCCAAAGC
OsMADS25	CCAGCTCAAGCATGAAATCAA	AAAGTTGCCTGTTGTTGTGGTGT
OsMADS27	GAAGCGGAGGAACGGGATCTTCAA	TGCCATACCGATCTATAACTGACT
OsMADS57	ACGAGCAGGCAGGTGACGTT	ACTCATAGAGCCTGCCGGTGCT
OsMADS61	GGGAGGGGCAAGATAGTGAT	TGGTGCTGGCATACTCGTAG
OsActin	CTTCATAGGAATGGAAGCTGCGGGT	CGACCACCTTGATCTTCATGCTGCT

revealed that *ANR1* is a positive regulator of lateral root growth and is not present in the primary root tip [34]. Initial studies using *Arabidopsis* root cultures had demonstrated that *ANR1* gene was NO_3^- inducible [26]. Subsequent results obtained from hydroponically culture experiments illustrated that the expression of *ANR1* was induced by N deprivation and rapidly downregulated when NO_3^- or other N source was re-supplied [29]. *OsMADS23, OsMADS25, OsMADS27, OsMADS57* and *OsMADS61* are five *ANR1*-like homologs in rice and their functions have not been well understood [35,36]. Recent work from Meng et al. (2013) has suggested that *OsMADS27* could play a key role in the response to cold and salt stress [37]. To gain further insight into the possible regulatory functions of *ANR1*-like genes in roots, we have investigated their root expression patterns in response to N, P, and S fluctuation using quantitative real-time PCR (qPCR).

Materials and Methods

Plant Materials

Rice seeds (*Oryza sativa* L. cv. *Nipponbare*) were used for all experiments.

Hydroponic culture

Rice seeds were surface-sterilized by treatment with 70% ethanol for 1 min and 10% sodium hypochlorite for 20 min, followed by five rinses with sterile distilled water. Seeds were germinated in the dark by placing into the incubator at 28°C for 2 days (d). Uniform seedlings were selected and transferred to black plastic buckets containing 4 L nutrient solution, where the growth conditions were 30/28°C day/night temperature with a 14/10 h light/dark at a relative humidity of 65–70%. The complete

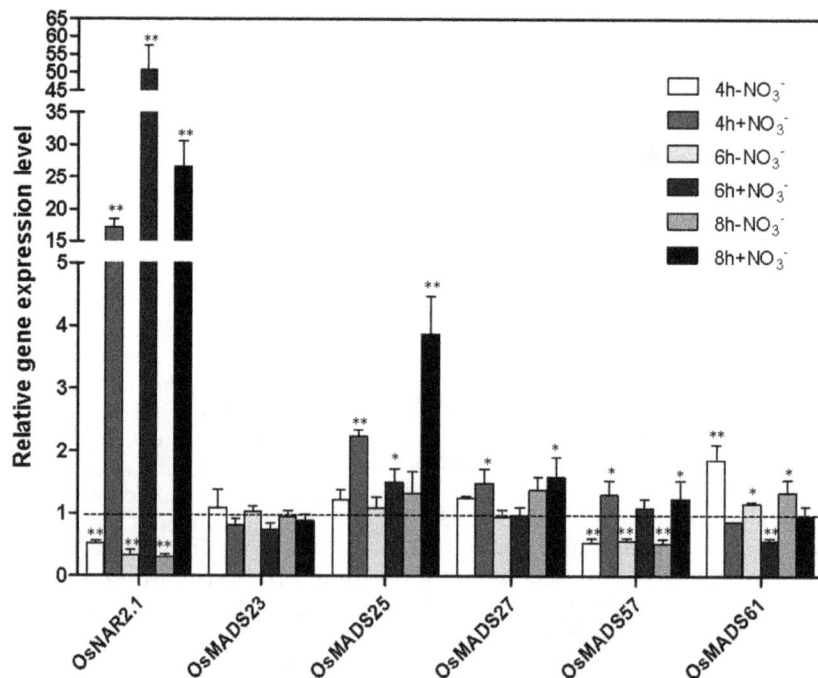

Figure 1. Effect of N deprivation and nitrate resupply on the expression of five *ANR1* related genes in rice roots. Ten-day old rice seedlings grown in complete nutrient solutions were transferred to modified nutrient solutions during which 2.88 mM KNO_3 was the sole nitrogen source for 4 days. Transcript abundance was assayed by qPCR and was expressed relative to the abundance in roots of plants of the same age grown under continuous N (CK). The *OsNAR2.1* gene, a known nitrate-regulated gene, was included for comparison. Treatments: CK: continuous KNO_3; − N4h: starved of N for 3 d and resupplied with KCl for 4 h; +N4h: resupplied with KNO_3 for 4 h; −N6h: starved of N for 3 d and resupplied with KCl for 6 h; +N6h: resupplied with KNO_3 for 6 h. The mRNA of *OsActin* was used as the reference. A Student's t-test was calculated at the probability of either 5% (*, p<0.05) or (**, P<0.01).

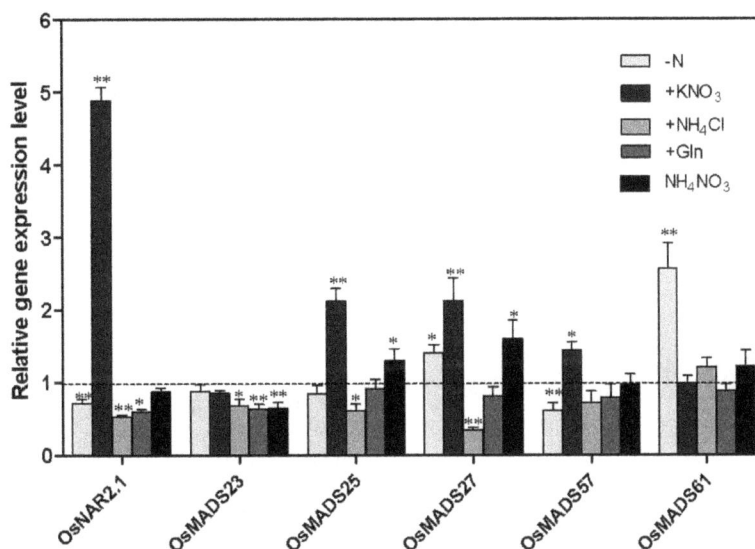

Figure 2. Effect of different N sources on the expression of five *ANR1*-related genes in rice roots. Rice seedlings were grown in liquid culture for 14 days with 2.88 mM KNO_3 as the sole nitrogen source and then N starved for 3 d. CK, continuous N; −N, starved for 3 d; +KNO_3, resupplied with 2.88 mM nitrate; +NH_4Cl resupplied with 2.88 mM NH_4^+; and +Gln, resupplied with 2.88 mM glutamine; +NH_4NO_3, resupplied with 1.44 mM NH_4NO_3. The value of related genes were normalized to its CK control respectively. The mRNA of *OsActin* was used as the reference. Error bars represent SE. LSD values were calculated at the probability of either 5% (*, p<0.05) or (**, P<0.01).

nutrient solution contained: 1.44 mM NH_4NO_3, 0.32 mM NaH_2PO_4, 0.5 mM K_2SO_4, 1 mM $CaCl_2 \cdot 2H_2O$, 1.6 mM $MgSO_4 \cdot 7H_2O$, 50 μM Fe-EDTA, 15 μM H_3BO_3, 9 μM $MnCl_2 \cdot 4H_2O$, 0.12 μM $CuSO_4 \cdot 5H_2O$, 0.12 μM $ZnSO_4 \cdot 7H_2O$, 40.5 μM citric acid and, 0.39 μM $Na_2MoO_4 \cdot 2H_2O$ [38]. 1 M HCl or NaOH was added to adjust pH to 5.5 and the nutrient solution was replaced every 2 d.

Nitrogen treatments

For the analysis of gene expression in response to nitrate, rice seedlings were grown in liquid culture for 14 d with 1.44 mM NH_4NO_3 as the N source. The solution was changed every 2 d. After 10 d, the medium was changed to 2.88 mM KNO_3 as the sole N source for 4 d. Then, the seedlings were deprived of N for 3 d before being re-supplied with KNO_3 or KCl to a final concentration of 2.88 mM. The control plants were grown in continuous 2.88 mM KNO_3 as the sole nitrogen source. The plants from different treatments were harvested at 4 h, 6 h and 8 h after re-supply separately. The roots were frozen in liquid N_2 and stored at −80°C for later analysis.

For analyses of gene expression in response to different N sources, the plants were grown in complete nutrient solution with 2.88 mM KNO_3 as the sole nitrogen source for 14 d [39]. Then, the plants were starved for N for 3 d before being transferred to fresh nutrient solution with the same concentration of different N sources. Roots were harvested 3 h later. For the control treatment, the plants were continuously supplied with 2.88 mM KNO_3 as the only N source. All the nutrient experiments were repeated at least twice with similar results.

P and S treatments

To initiate different P and S treatments, two-week-old seedlings grown in complete nutrient solution as described above. For the P and S starvation treatments, the complete nutrient solution was replaced with nutrient solution lacking P or S for 3 d with PO_4^- or

SO_4^{2-} being replaced by chloride. For the control plants, the seedlings grown in continuous complete nutrient solution were transferred to fresh complete nutrient solution at the same time. For re-supply, the appropriate nutrient, 0.32 mM PO_4^- or 2.1 mM SO_4^{2-} was added in the light period for 3 h before the roots were harvested and frozen in liquid N_2 and stored at −80°C for gene expression analyses [38].

RNA extraction and qPCR

Primers for *OsMADS23*, *OsMADS25*, *OsMADS27*, *OsMADS57* and *OsMADS61* for qPCR (see Table 1) were designed using Primer Premier 5. Total RNA was extracted using the RNAiso Plus reagent (Takara) according to the manufacturer's instructions. The first-strand cDNA was synthesized using Prime-Script RT reagent Kit with gDNA Eraser (Takara). The qPCR was performed in 96-well plates using SYBR Premix Ex Taq II (Takara) according to the manufacturer's instructions. The qPCR was conducted in the following cycling conditions: 95°C for 30 s, followed by 40 cycles of 95°C for 10 s and 60°C for 30 s. Melt curve analysis was used to confirm the absence of non-specific amplification products. Relative expression levels were calculated by subtracting the threshold cycle (Ct) values for *OsActin* (Os03g0718100) from those of the target gene (to give ΔCt) and then calculating 2−ΔCt as we described before [40–42]. RT-PCR experiments were performed with three biological replicates with the representative being shown.

Statistics

The results were analyzed for variance by IBM SPSS Statistics 20. Student's t-test was calculated at the probability at either at 1% (P<0.01 with significant level **) or 5% (P<0.05 with significant level *) as described before [43,44].

A

B

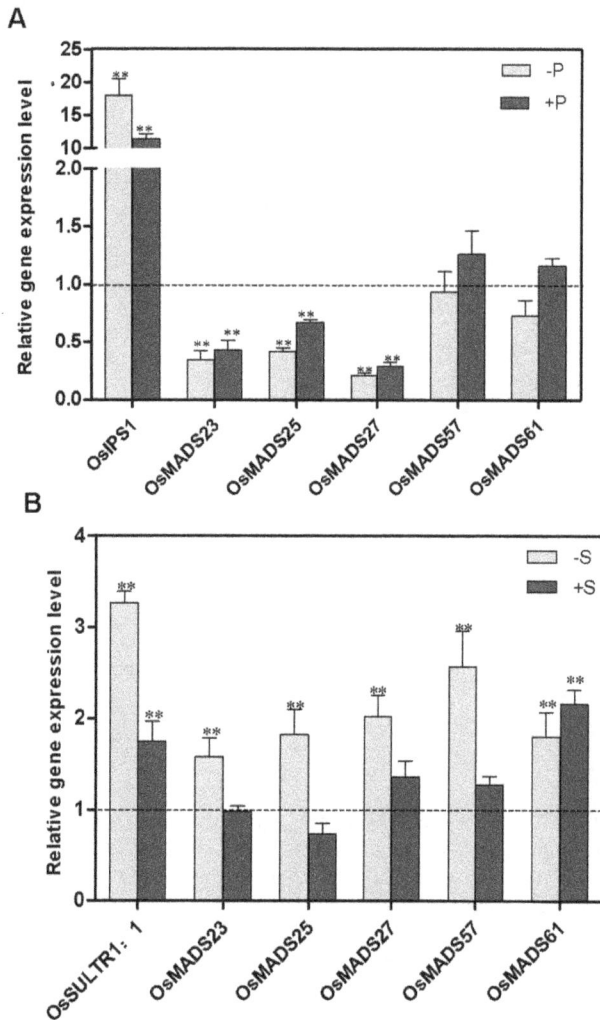

Figure 3. Effect of deprivation and re-supply of phosphate (P) and sulfate (S) on the expression of five *ANR1*-related genes in rice. Two-week old rice seedlings grown hydroponically in complete nutrient solution were deprived of P or S or were maintained on complete nutrient supply for 3 d. In the light period on the day of the experiment, one set of the P-starved and S-starved plants were re-supplied with 0.32 mM $H_2PO_4^-$ and 2.1 mM SO_4^{2-} respectively. Roots were harvested 3 h later from controls: continuous nutrient supply (CK); P-deprived (−P); P-resupply (+P); S-deprived (−S); S-resupply (+S). Total RNA was extracted from roots and qPCR reactions were performed in triplicate for each RNA sample. The mRNA of *OsActin* was used as the reference. The value of related genes were normalized to its CK control respectively. A Student's t-test was calculated at the probability of either 5% (*, p<0.05) or (**, P<0.01).

Results

Five *ANR1*-like genes have different expression patterns in response to nitrate

The *Arabidopsis ANR1* gene was identified as a key gene controlling lateral root growth through NO_3^- signaling [26]. To understand whether these five *ANR1* homologous genes in rice could play similar roles to *ANR1* in *Arabidopsis*, we begin by investigating their expression patterns and levels in response to nitrate. Rice seedlings were grown in complete nutrient solution for 14 d with 2.88 mM KNO_3 as the sole N source They were then deprived of N for 3 d before being re-supplied with 2.88 mM

KNO_3 (or 2.88 mM KCl as control). The rice nitrate transporter gene *OsNAR2.1*, which is known to be nitrate-inducible [45], was used as a positive control. As shown in Figure 1, the expression of *OsNAR2.1* was very strongly upregulated by nitrate re-supply as expected. Each of the five *ANR1*-related genes was found to have a different expression pattern in response to nitrate starvation and re-supply (Figure 1). Similar to *ANR1* gene in *Arabidopsis*, the expression of *OsMADS61* was significantly induced by nitrate starvation at 4 h, 6 h and 8 h and significantly suppressed by nitrate re-supply at 6 h in comparison to the seedlings continuously supplied with nitrate. In contrast, the expression of *OsMADS25* was not significantly affected by nitrate starvation but was induced by nitrate re-supply at 4 h, 6 h and 8 h in comparison to the continuous nitrate treatment. Similarly, the expression of *OsMADS27* was not significantly affected by nitrate starvation but was induced by nitrate re-supply at 4 h and 8 h in comparison to the continuous nitrate treatment. The expression of *OsMADS57* was downregulated by nitrate starvation at 4 h, 6 h and 8 h and was upregulated by nitrate re-supply at 4 h in comparison to the continuous nitrate treatment (Figure 1). *OsMADS23* was the only gene to show no significant response to either nitrate starvation or nitrate re-supply (Figure 1).

Five *ANR1*-like genes respond differently to different N sources

To investigate whether these five *ANR1* homologous genes in rice were also regulated by the other N sources, we further investigated their expression patterns in response to different N sources. The plants were grown in the complete nutrient solution for 10 d before changing media with 2.88 mM KNO_3 as the sole nitrogen source for 4 days. Then, the plants were starved for N nutrient for 3 d followed by re-supply with the same concentrations of different N sources. As shown in Figure 2, the expression of rice transporter *OsNAR2.1* was down–regulated in the roots of N-starved rice plants and then rapidly upregulated when supplied with 2.88 mM KNO_3 for 3 h which was consistent with a previous report [39]. *OsMADS61* and *OsMADS57* were the only genes that were affected by nitrate not by any other N sources, whereas *OsMADS23* was the only gene regulated by various different N sources but not by nitrate. In contrast, *OsMADS25* and *OsMADS27* were upregulated by both nitrate and ammonium nitrate but downregulated by ammonium chloride.

Effect of P and S deprivation and re-supply on the expressions of five *ANR1*-like genes in rice roots

To analyze whether *ANR1*-like homologous genes are involved in the regulation of gene expression in response to other nutritional stresses, we investigated the effect of P and S deprivation and re-supplementation. As shown in Figure 3, in contrast to the phosphate transporter *OsIPS1*, *OsMADS23*, *OsMADS25* and *OsMADS27* were all downregulated by P starvation and P re-supply whereas *OsMADS57* and *OsMADS61* were not significantly affected. For the S treatment, *OsMADS23*, *OsMADS25*, *OsMADS27* and *OsMADS57* were all upregulated by S starvation but not by S re-supply whereas *OsMADS61* was upregulated by both S starvation and re-supply in comparison to the continuous S nutrient treatment.

Effect of N-deprivation and re-supply on the expressions of *OsNRT2.1*, *OsNAR2.1*, *OsIPS1* and *OsSULTR1;1* in rice roots

We have investigated whether there is crosstalk between the N, P and S regulatory pathways in the regulation of expression of the

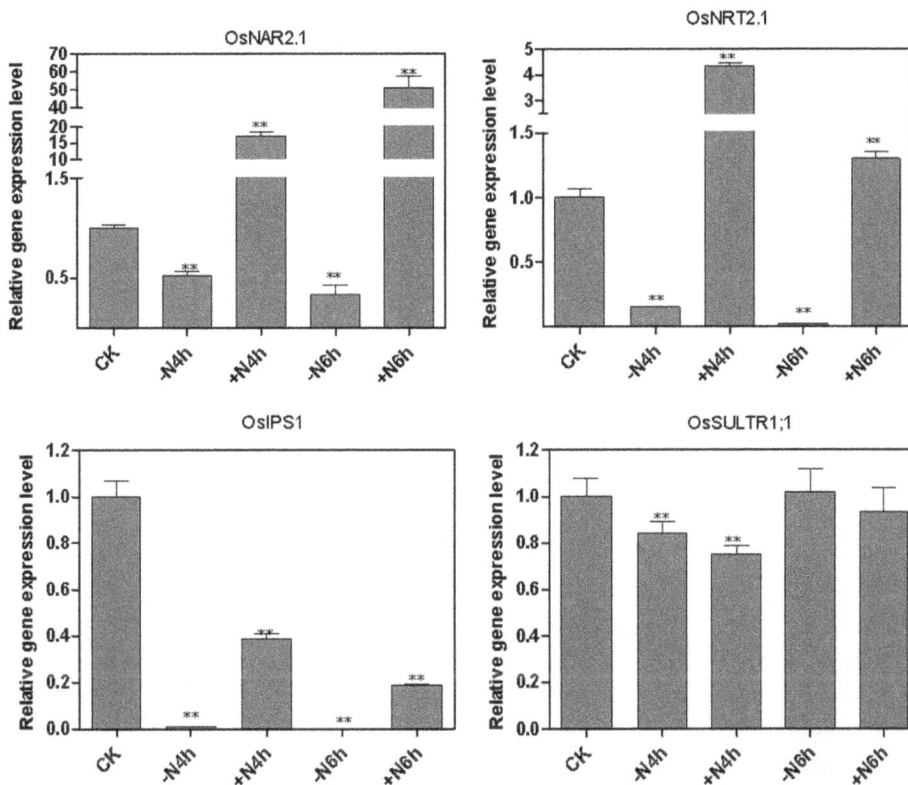

Figure 4. Effect of N-deprivation and re-supply on expression of *OsNRT2.1*, *OsNAR2.1*, *OsIPS1* **and** *OsSULTR1;1* **in rice roots.** Rice seedlings were grown hydroponically in a growth cabinet. Nitrogen treatments were as described in Fig. 1. CK: continuous KNO_3; −N4h: starved of N for 3 d and resupplied with KCl for 4 h; +N4h: resupplied with KNO_3 for 4 h; −N6h: starved of N for 3 d and resupplied with KCl for 6 h; +N6h: resupplied with KNO_3 for 6 h. Total RNA was extracted from roots and qPCR reactions were performed in triplicate for each RNA sample. The mRNA of *OsActin* was used as the reference. A Student's t-test was calculated at the probability of either 5% (*, $p < 0.05$) or (**, $P < 0.01$).

the *OsNRT2.1*, *OsNAR2.1*, *OsIPS1* and *OsSULTR1;1* genes. We first investigated whether the expression of phosphate transporter *OsIPS1* and sulphate transporter *OsSULTR1;1* was regulated in response in nitrate starvation and nitrate re-supply. The *OsNRT2.1* and *OsNAR2.1* genes, which encode components of the high affinity nitrate transport system (HATS) in rice, were used as the positive controls *OsNRT2.1* and *OsNAR2.1*, were previously shown to be upregulated by nitrate and suppressed by NH_4^+. As expected, *OsNRT2.1* and *OsNAR2.1* were downregulated by nitrate starvation and rapidly upregulated by nitrate re-supply (Figure 4), confirming the results of previous studies [45,46]. However, the expression of phosphate transporter *OsIPS1* was significantly downregulated by both nitrate starvation and nitrate re-supply at both 4 h and 6 h time points, whereas *OsSULTR1;1* was only significantly downregulated by both nitrate starvation and re-supply at 4 h and not at 6 h.

Effect of P, S deprivation and re-supply on the expression of *OsNRT2.1*, *OsNAR2.1*, *OsIPS1* and *OsSULTR1;1* in rice roots

In this experiment two-week old rice seedlings grown in complete nutrient solution were deprived of phosphate or sulfate for 3 d and re-supplied with phosphate or sulfate for 3 h. *OsIPS1* was used as P control gene and *OsSULTR1;1* was used as S control gene. As shown in Figure 5, the expression of *OsIPS1* was notably upregulated by P deprivation and downregulated by P re-

supply, which was consistent with previous study [14]. Surprisingly, the gene expression patterns of *OsNRT2.1* and *OsNAR2.1* is very similar to the expression pattern of *OsIPS1*, which were upregulated by both starvation and P re-supply in comparison to the continuous P treatment (Figure 5). However, the expression of *OsSULTR1;1* was significantly downregulated by P starvation and P re-supply in comparison to the continuously P supply treatment. For the S fluctuation treatment, as shown in Figure 6, the mRNA level of *OsSULTR1;1* was significantly increased by sulfate starvation, which was consistent with previous study [47]. The expression patterns of *OsNRT2.1*, *OsNAR2.1* and *OsIPS1* were the same as *OsSULTR1;1*, being upregulated by both S starvation and S re-supply in comparison to the continuous S treatment (Figure 6).

Discussion

Five *ANR1*-like genes have different expression patterns in response to nitrogen

It was previously demonstrated that expression of *ANR1* in roots of hydroponically grown *Arabidopsis* plants was induced by N deprivation and rapidly downregulated by N re-supply [29]. This pattern of N responsiveness differs from the NO_3^- inducibility of *ANR1* as previously obtained in *Arabidopsis* root cultures [26]. In the present study we have shown that the five *ANR1*-related genes in rice have diverse expression patterns in response to N starvation and N re-supply. The expression of

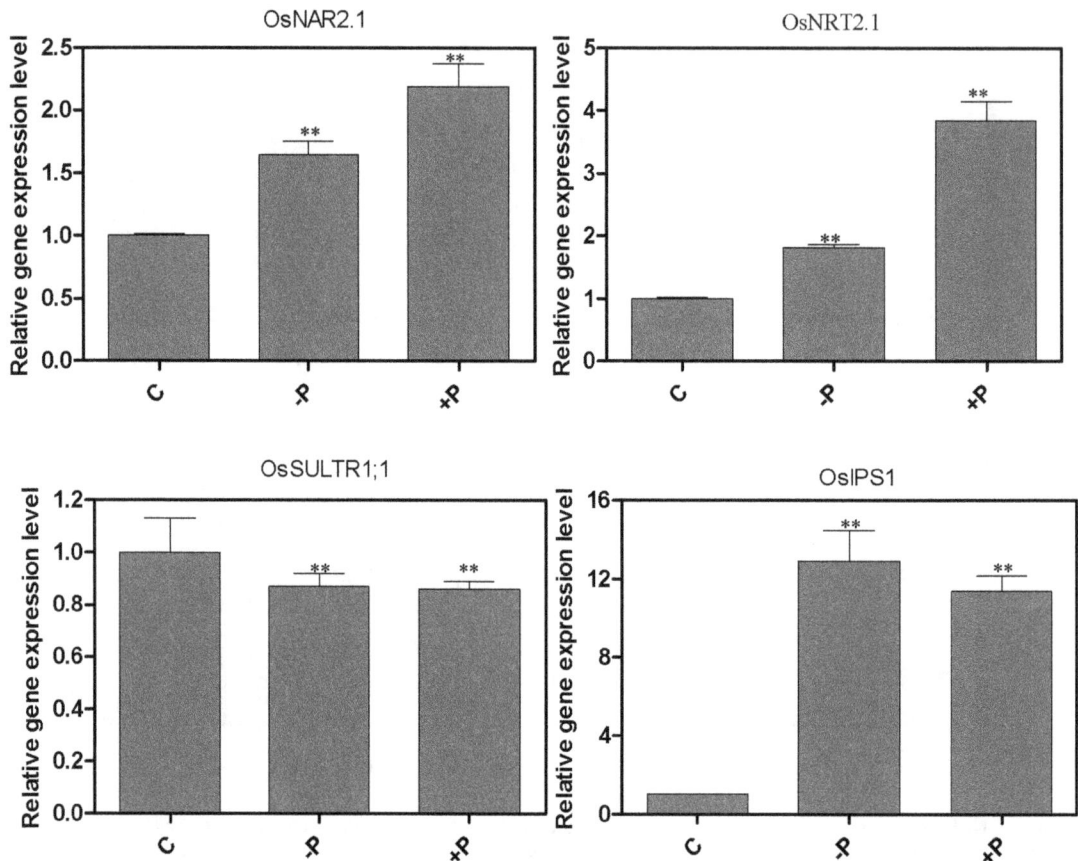

Figure 5. Effect of P-deprivation and re-supply on expression of *OsNRT2.1*, *OsNAR2.1*, *OsIPS1* and *OsSULTR1;1* in rice roots. Rice seedlings were grown hydroponically in a growth cabinet. Phosphorous treatments were as described in Fig. 3. C: continuous complete nutrient supply; −P: starved of P; +P: resupplied with $H_2PO_4^-$ for 3 h. Total RNA was extracted from roots and qPCR reactions were performed in triplicate for each RNA sample. The mRNA of *OsActin* was used as the reference. A Student's t-test was calculated at the probability of either 5% (*, p<0.05) or (**, P<0.01).

OsMADS61 was only significantly induced by N starvation and significantly down regulated by N re-supply, which is very similar to the expression pattern of *ANR1* in *Arabidopsis*. In contrast, the expression of *OsMADS57* was significantly downregulated by N starvation and significantly upregulated by nitrate re-supply. Furthermore, *OsMADS25* and *OsMADS27* were only significantly upregulated by nitrate re-supply not by N starvation. These results were partly consistent with the results from [48], in which the rice seedlings were grown in half-strength liquid Murashige and Skoog medium without N and re-supplied with 3 mM KNO_3. They found that *OsMADS23* and *OsMADS61* were not significantly affected by N fluctuation and that *OsMADS25*, *OsMADS27*, *OsMADS57* were all significantly upregulated by nitrate re-supply [48]. They therefore suggested that the *AGL17*-like related genes in rice had specific functions differing from those of their *A. thaliana* homologs. However, our finding found that the expression pattern of *OsMADS61* is very similar to that of *ANR1* pattern in *Arabidopsis*.

It has recently been shown that microRNA444 (miR444), which targets three OsMADS-box genes (*OsMADS23*, *OsMADS27*, *OsMADS57*) in rice, has multiple roles in the NO_3^- signaling pathway [49]. In this study, *OsMADS23*, *OsMADS27* and *OsMADS57* were downregulated under conditions of N-deprivation and were unaffected following 0.5, 1, or 2 h of 5 mM KNO_3 supplementation [49]. The different sampling times may account for the partial differences between their results and ours.

The phenotype of transgenic lines overexpressing miR444 provided evidence that *OsMADS23*, *OsMADS27* and *OsMADS57* could have a role in regulating the lateral response to localised nitrate in rice [49], as *ANR1* does in *Arabidopsis* [26]. In addition, there was evidence that this group of genes is involved in controlling the root architecture in response to P starvation [49]. However, it has been noted that miR444 has additional target genes (non-MADS box) in rice and there could be other unknown genes involved in the root developmental responses to the nutrient supply [50].

The five *ANR1*-like homologous gene in rice are regulated by P and S fluctuations

Previous results from *Arabidopsis* had demonstrated that expression of *ANR1* was not regulated by fluctuations in P and S supplies and that and *SUPPRESSOR OF OVEREXPRESSION OF CONSTANS 1* (*SOC1*) was the only type-II MADS-box gene that responded to phosphate and sulfate deprivation and re-supplementation [29,41]. A later study found that *AGL12*, *AGL18* and *AGL19* were downregulated by P and S re-supply [41]. In this study, the *OsMADS23*, *OsMADS25* and *OsMADS27* genes were downregulated by both P starvation and P re-supply whereas expression of *OsMADS57* and *OsMADS61* was not significantly regulated by P fluctuations, which was consistent with what was observed for *ANR1* in *Arabidopsis* [29]. These results were partly

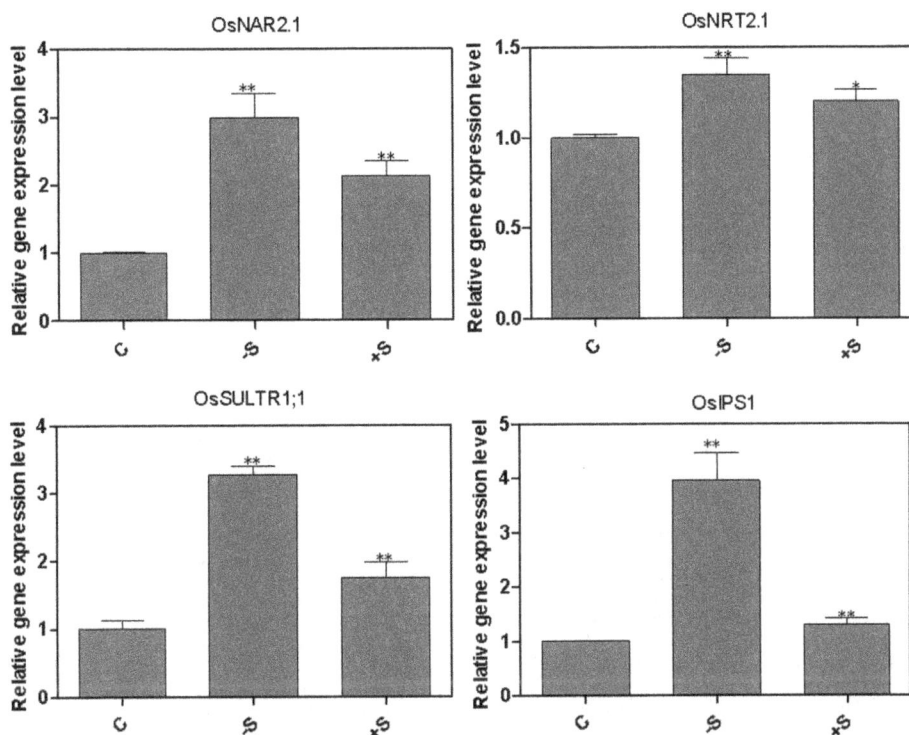

Figure 6. Effect of S-deprivation and re-supply on expression of *OsNRT2.1, OsNAR2.1, OsIPS1* **and** *OsSULTR1;1* **in rice roots.** Rice seedlings were grown hydroponically in a growth cabinet. Sulfur treatments were as described in Fig. 3. C: continuous complete nutrient supply; −S: starved of S; +S: resupplied with SO_4^{2-} for 3 h. Total RNA was extracted from roots and qPCR reactions were performed in triplicate for each RNA sample. The mRNA of *OsActin* was used as the reference. A Student's t-test was calculated at the probability of either 5% (*, p<0.05) or (**, P<0.01).

consistent with the results from [49], in which phosphate starvation reduced the mRNA levels of two miR444 targets (*OsMADS27* and *OsMADS57*). In our result, the expression level of *OsMADS23* and *OsMADS27* were downregulated by P starvation, however in [49], the mRNA abundance of *OsMADS27* and *OsMADS57* were downregulated by phosphate starvation. We used Yoshida [38] rice nutrient solution and Yan chose 1/2 MS culture solution [49] and the time course of phosphate-deprivation was different, which may account for the different gene expression patterns. However, for the S treatment, the expression of all five *ANR1*-related genes in rice was modulated by S starvation, in contrast to what was seen with *ANR1* in Arabidopsis [29], suggesting that these rice genes have specific functions differing from their *Arabidopsis* homologs. Furthermore, the result of the expression of *OsMADS25* regulated by phosphate deprivation was consistent with previous report by [51]. Like *SOC1* in *Arabidopsis*, rice *OsMADS25*, *OsMADS27* and *Os-MADS57* in roots were responsive to nitrate, phosphate and sulfate fluctuations, which suggest that these three genes may be involved in a general stress response pathway to these three macronutrients.

Crosstalk among *OsNRT2.1*, *OsNAR2.1*, *OsIPS1* and *OsSULTR1;1*

As already discussed, depriving seedlings of one mineral nutrient may lead to the disruption of the metabolism of other nutrients [29,52,53]. For example, molybdenum deficiency had positive impacts on genes involved in nitrate and sulfate assimilation and

phosphate transport [54]. Our results show that *OsNRT2.1* and *OsNAR2.1* were regulated by both P and S starvation and re-supply, which are partly consistent with previous research in which the gene expression level of *AtNRT2.1* in *Arabidopsis* was found to be upregulated by P and S re-supply [44]. Our results also indicated that *OsIPS1* was sensitive to N and S fluctuation and that the expression of *OsSULTR1;1* was downregulated by phosphate re-supply, indicating that there is crosstalk between the signaling pathways regulating expression of nitrate, phosphate and sulfate transporters. This is similar to the finding that some nitrate-inducible genes are involved in sulfate metabolism [51,55]. These results suggest that there is a complex regulatory network among these three macronutrients and their transporters. Further work using mutants or other genomic approach needs to be done to identify the mechanisms of the crosstalk among nitrate, phosphate and sulfate transporters.

Acknowledgments

We thank Prof. Hao Yu from the National University of Singapore for critical reading of this manuscript.

Author Contributions

Conceived and designed the experiments: CY SS YG. Performed the experiments: CY SS YX. Analyzed the data: CY SS YZ. Contributed reagents/materials/analysis tools: CY SS LH. Contributed to the writing of the manuscript: CY AY IA YG.

References

1. Rouached H, Stefanovic A, Secco D, Bulak Arpat A, Gout E, et al. (2011) Uncoupling phosphate deficiency from its major effects on growth and transcriptome via *PHO1* expression in *Arabidopsis*. Plant J 65: 557–570.

2. Robertson GP, Vitousek PM (2009) Nitrogen in agriculture: balancing the cost of an essential resource. Annu Rev Environ Resour 34: 97–125.

3. Fan X, Shen Q, Ma Z, Zhu H, Yin X, et al. (2005) A comparison of nitrate transport in four different rice (*Oryza sativa* L.) cultivars. Sci China C Life Sci 48: 897–911.

4. Forde BG, Clarkson DT (1999) Nitrate and ammonium nutrition of plants: physiological and molecular perspectives. Adv Bot Res 30: 1–90.

5. Crawford NM, Forde BG (2002) Molecular and developmental biology of inorganic nitrogen nutrition. The *Arabidopsis* Book. American Society of Plant Biologists. e0011.

6. Vidal EA, Gutiérrez RA (2008) A systems view of nitrogen nutrient and metabolite responses in *Arabidopsis*. Curr Opin Plant Biol 11: 521–529.

7. Krouk G, Ruffel S, Gutiérrez RA, Gojon A, Crawford NM, et al. (2011) A framework integrating plant growth with hormones and nutrients. Trends Plant Sci 16: 178–182.

8. Marschner H (1988) Mineral nutrition of higher plants. Plant Cell Environ 11: 147–148.

9. Schachtman DP, Reid RJ, Ayling SM (1998) Phosphorus uptake by plants: from soil to cell. Plant Physiol 116: 447–453.

10. Raghothama KG (1999) Phospate acquisition. Annu Rev Plant Phys 50: 665–693.

11. Aziz T, Sabir M, Farooq M, Maqsood MA, Ahmad H, et al. (2014) Phosphorus deficiency in plants: responses, adaptive mechanisms, and signaling. In: K. R Hakeem, R. U Rehman and I Tahir, editors. Plant signaling: Understanding the molecular crosstalk. India: Springer. 133–148.

12. Chiou T-J, Lin S-I (2011) Signaling network in sensing phosphate availability in plants. Annu Rev Plant Biol 62: 185–206.

13. Bates TR, Lynch JP (1996) Stimulation of root hair elongation in *Arabidopsis thaliana* by low phosphorus availability. Plant Cell Environ 19: 529–538.

14. Hou XL, Wu P, Jiao FC, Jia QJ, Chen HM, et al. (2005) Regulation of the expression of *OsIPS1* and *OsIPS2* in rice via systemic and local Pi signalling and hormones. Plant Cell Environ 28: 353–364.

15. Sánchez-Calderón L, López-Bucio J, Chacón-López A, Cruz-Ramírez A, Nieto-Jacobo F, et al. (2005) Phosphate starvation induces a determinate developmental program in the roots of *Arabidopsis thaliana*. Plant Cell Physiol 46: 174–184.

16. Svistoonoff S, Creff A, Reymond M, Sigoillot-Claude C, Ricaud L, et al. (2007) Root tip contact with low-phosphate media reprograms plant root architecture. Nat Genet 39: 792–796.

17. Ticconi CA, Delatorre CA, Abel S (2001) Attenuation of phosphate starvation responses by phosphite in *Arabidopsis*. Plant Physiol 127: 963–972.

18. Wu P, Xu G, Lian X (2013) Nitrogen and phosphorus uptake and utilization. In: Q Zhang and R. A Wing, editors. Genetics and Genomics of Rice. New York: Springer. 217–226.

19. Dai X, Wang Y, Yang A, Zhang W-H (2012) *OsMYB2P-1*, an R2R3 MYB transcription factor, is involved in the regulation of phosphate-starvation responses androot architecture in rice. Plant Physiol 159: 169–183.

20. Saito S, Hirai N, Matsumoto C, Ohigashi H, Ohta D, et al. (2004) *Arabidopsis CYP707As* encode (+)-abscisic acid 8′-hydroxylase, a key enzyme in the oxidative catabolism of abscisic acid. Plant Physiol 134: 1439–1449.

21. Tabe LM, Droux M (2002) Limits to sulfur accumulation in transgenic lupin seeds expressing a foreign sulfur-rich protein. Plant Physiol 128: 1137–1148.

22. Robinson D (1994) Tansley review No. 73. the responses of plants to non-uniform supplies of nutrients. New Phytol 127: 635–674.

23. Wang X, Wu P, Xia M, Wu Z, Chen Q, et al. (2002) Identification of genes enriched in rice roots of the local nitrate treatment and their expression patterns in split-root treatment. Gene 297: 93–102.

24. Walch-Liu P, Liu L-H, Remans T, Tester M, Forde BG (2006) Evidence that L-glutamate can act as an exogenous signal to modulate root growth and branching in *Arabidopsis thaliana*. Plant Cell Physiol 47: 1045–1057.

25. Walch-Liu P, Ivanov Ii, Filleur S, Gan Y, Remans T, et al. (2006) Nitrogen regulation of root branching. Ann Bot-london 97: 875–881.

26. Zhang H, Forde BG (1998) An *Arabidopsis* MADS box gene that controls nutrient-induced changes in root architecture. Science 279: 407–409.

27. Linkohr BI, Williamson LC, Fitter AH, Leyser HMO (2002) Nitrate and phosphate availability and distribution have different effects on root system architecture of *Arabidopsis*. Plant J 29: 751–760.

28. Parenicová L, de Folter S, Kieffer M, Horner DS, Favalli C, et al. (2003) Molecular and phylogenetic analyses of the complete MADS-Box transcription factor family in *Arabidopsis*: new openings to the MADS world. Plant Cell 15: 1538–1551.

29. Gan Y, Filleur S, Rahman A, Gotensparre S, Forde BG (2005) Nutritional regulation of *ANR1* and other root-expressed MADS-box genes in *Arabidopsis thaliana*. Planta 222: 730–742.

30. Gan Y, Bernreiter A, Filleur S, Abram B, Forde BG (2012) Overexpressing the *ANR1* MADS-Box gene in transgenic plants provides new insights into its role in the nitrate regulation of root development. Plant Cell Physiol 53: 1003–1016.

31. Messenguy F, Dubois E (2003) Role of MADS box proteins and their cofactors in combinatorial control of gene expression and cell development. Gene 316: 1–21.

32. Alvarez-Buylla ER, Liljegren SJ, Pelaz S, Gold SE, Burgeff C, et al. (2000) MADS-box gene evolution beyond flowers: expression in pollen, endosperm, guard cells, roots and trichomes. Plant J 24: 457–466.

33. Burgeff C, Liljegren S, Tapia-López R, Yanofsky M, Alvarez-Buylla E (2002) MADS-box gene expression in lateral primordia, meristems and differentiated tissues of *Arabidopsis thaliana* roots. Planta 214: 365–372.

34. Filleur S, Walch-Liu P, Gan Y, Forde B (2005) Nitrate and glutamate sensing by plant roots. Biochem Soc Trans 33: 283–286.

35. Lee S, Kim J, Son J-S, Nam J, Jeong D-H, et al. (2003) Systematic reverse genetic screening of T-DNA tagged genes in rice for functional genomic analyses: MADS-box gene as a test case. Plant Cell Physiol 44: 1403–1411.

36. Arora R, Agarwal P, Ray S, Singh A, Singh V, et al. (2007) MADS-box gene family in rice: genome-wide identification, organization and expression profiling during reproductive development and stress. BMC Genomics 8: 242.

37. Meng Y, Shao C, Wang H, Ma X, Chen M (2013) Construction of gene regulatory networks mediated by vegetative and reproductive stage-specific small RNAs in rice (*Oryza sativa*). New Phytol 197: 441–453.

38. Yoshida S, Forno D, Cock J, Gomez K (1976) Routine procedure for growing rice plants in culture solution. In: S Yoshida, D Forno, J Cock and K Gomez, editors. Laboratory manual for physiological studies of rice. Philippines: The International Rice Research Institut. 61–66.

39. Cai C, Wang J-Y, Zhu Y-G, Shen Q-R, Li B, et al. (2008) Gene structure and expression of the high-affinity nitrate transport system in rice roots. J Integr Plant Biol 50: 443–451.

40. Zhou Z, An L, Sun L, Zhu S, Xi W, et al. (2011) Zinc Finger Protein5 Is required for the control of trichome initiation by acting upstream of *Zinc Finger Protein8* in *Arabidopsis*. Plant Physiol 157: 673–682.

41. Gan Y-b, Zhou Z-j, An L-j, Bao S-j, Forde BG (2011) A comparison between northern blotting and quantitative real-time PCR as a means of detecting the nutritional regulation of genes expressed in roots of *Arabidopsis thaliana*. Agric Sci China 10: 335–342.

42. An L, Zhou Z, Sun L, Yan A, Xi W, et al. (2012) A zinc finger protein gene *ZFP5* integrates phytohormone signaling to control root hair development in *Arabidopsis*. Plant J 72: 474–490.

43. Gan Y, Zhou Z, An L, Bao S, Liu Q, et al. (2010) The effects of fluctuations in the nutrient supply on the expression of *ANR1* and 11 other MADS box genes in shoots and roots of *Arabidopsis thaliana*. Botany 88: 1023–1031.

44. Bao S, An L, Su S, Zhou Z,Gan Y (2011) Expression patterns of nitrate, phosphate, and sulfate transporters in *Arabidopsis* roots exposed to different nutritional regimes. Botany 89: 647–653.

45. Feng H, Yan M, Fan X, Li B, Shen Q, et al. (2011) Spatial expression and regulation of rice high-affinity nitrate transporters by nitrogen and carbon status. J Exp Bot 62: 2319–2332.

46. Araki R, Hasegawa H (2006) Expression of rice (*Oryza sativa* L.) genes involved in high-affinity nitrate transport during the period of nitrate induction. Breeding Sci 56: 295–302.

47. Kumar S, Asif MH, Chakrabarty D, Tripathi RD, Trivedi PK (2011) Differential expression and alternative splicing of rice sulphate transporter family members regulate sulphur status during plant growth, development and stress conditions. Funct Integr Genomics 11: 259–273.

48. Puig J, Meynard D, Khong GN, Pauluzzi G, Guiderdoni E, et al. (2013) Analysis of the expression of the *AGL17-like* clade of MADS-box transcription factors in rice. Gene Expr Patterns 13: 160–170.

49. Yan Y, Wang H, Hamera S, Chen X, Fang R (2014) miR444a has multiple functions in the rice nitrate-signaling pathway. Plant J 78: 44–55.

50. Forde BG (2014) Glutamate signalling in roots. J Exp Bot 65: 779–787.

51. Wang R, Okamoto M, Xing X, Crawford NM (2003) Microarray analysis of the nitrate response in *Arabidopsis* roots and shoots reveals over 1,000 rapidly responding genes and new linkages to glucose, trehalose-6-phosphate,iron, and sulfate metabolism. Plant Physiol 132: 556–567.

52. Prosser IM, Purves JV, Saker LR, Clarkson DT (2001) Rapid disruption of nitrogen metabolism and nitrate transport in spinach plants deprived of sulphate. J Exp Bot 52: 113–121.

53. Takehisa H, Sato Y, Antonio B, Nagamura Y (2013) Global transcriptome profile of rice root in response to essential macronutrient deficiency. Plant Signal Behav 8: e24409.

54. Ide Y, Kusano M, Oikawa A, Fukushima A, Tomatsu H, et al. (2011) Effects of molybdenum deficiency and defects in molybdate transporter *MOT1* on transcript accumulation and nitrogen/sulphur metabolism in *Arabidopsis thaliana*. J Exp Bot 62: 1483–1497.

55. Leustek T, Martin MN, Bick J-A, Davies JP (2000) Pathways and regulation of sulfur metabolism revealed through molecular and genetic studies. Annu Rev Plant Phys 51: 141–165.

9

Diapause Induction and Termination in the Small Brown Planthopper, *Laodelphax striatellus* (Hemiptera: Delphacidae)

LiuFeng Wang[1], KeJian Lin[2]*, Chao Chen[1], Shu Fu[1], FangSen Xue[1]*

1 Institute of Entomology, Jiangxi Agricultural University, Nanchang, Jiangxi Province, China, **2** State Key Laboratory for Biology of Plant Diseases and Insect Pests, Institute of Plant Protection, Chinese Academy of Agricultural Sciences, Beijing, China

Abstract

The small brown planthopper, *Laodelphax striatellus* (Fallén) enters the photoperiodic induction of diapause as 3rd or 4th instar nymphs. The photoperiodic response curves in this planthopper showed a typical long-day response type with a critical daylength of approximately 11 h at 25°C, 12 h at 22 and 20°C and 12.5 h at 18°C, and diapause induction was almost abrogated at 28°C. The third stage was the most sensitive stage to photoperiod. The photoperiodic response curve at 20°C showed a gradual decline in diapause incidence in ultra-long nights, and continuous darkness resulted in 100% development. The required number of days for a 50% response was distinctly different between the short- and long-night cycles, showing that the effect of one short night was equivalent to the effect of three long nights at 18°C. The rearing day length of 12 h evoked a weaker intensity of diapause than did 10 and 11 h. The duration of diapause was significantly longer under the short daylength of 11 h than it was under the long daylength of 15 h. The optimal temperature for diapause termination was 26 and 28°C. Chilling at 5°C for different times did not shorten the duration of diapause but significantly lengthened it when chilling period was included. In autumn, 50% of the nymphs that hatched from late September to mid-October entered diapause in response to temperatures below 20°C. The critical daylength in the field was between 12 h 10 min and 12 h 32 min (including twilight), which was nearly identical to the critical daylength of 12.5 h at 18°C. In spring, overwintering nymphs began to emerge in early March-late March when the mean daily temperature rose to 10°C or higher.

Editor: Daniel Doucet, Natural Resources Canada, Canada

Funding: The research was supported by a grant from National Natural Science Foundation of PR China (31272042) and the Sci-Tech Landing Projection of Higher Education of Jiangxi Province (KJLD14030). The funders had no role in study design, data collection and analysis, decision to publish, or preparation of the manuscript.

Competing Interests: The authors have declared that no competing interests exist.

* Email: xue_fangsen@hotmail.com (FX); kjlin@ippcaas.cn (KL)

Introduction

The small brown planthopper, *Laodelphax striatellus* (Fallén) (Hemiptera: Delphacidae) is one of the most serious and destructive pests of agriculture in temperate zones. The planthopper is a plant virus vector that attacks a wide range of economically important crops including rice, wheat, barley, corn and sugarcane causing significant damage [1]. *L. striatellus* is widely distributed in rice-producing regions throughout China and it has the potential to undertake long-distance migration [2,3] and hibernate in temperate regions as 2–5th instar nymphs [4,5]. This is in contrast with the economically important rice planthopper, *Nilaparvata lugens* and *Sogatella furcifera*, which are unable to overwinter in temperate regions in China and instead migrate into these regions every early summer.

Diapause is one of the primary mechanisms whereby insects synchronize their life cycles with local seasonal changes [6]. While undergoing dormancy, insects progress through a series of physiological phases, including diapause induction, maintenance and termination, post-diapause quiescence and post-diapause development. Each of these phases are strongly affected by photoperiod and temperature [7,8]. Diapause in *L. striatellus* has been investigated under laboratory conditions in Japan. The nymphal diapause was induced by short daylengths at low temperatures during the nymphal development and short photoperiod maintained nymphal diapause of *L. striatellus* [9–11]. However, diapause in *L. striatellus* has not been reported in China. For a better understanding of insect life cycles, a detailed understanding of diapause in the planthopper would be desirable because such information is helpful in improving the prediction and management of this pest.

In the present study, diapause induction and termination of the small brown planthopper, *L. striatellus* were systematically investigated under laboratory and field conditions. The purpose of our experiments is to reveal the role of photoperiod and temperature in diapause induction and termination in this planthopper and how the planthopper decides to initiate and terminate diapause under field conditions of changing photoperiod and temperature.

Materials and Methods

Ethics Statement

Because the small brown planthopper *L. striatellus* is a serious and destructive pest of agriculture in the temperate zone in China, no permits were required for collecting the insect and performing the experiments. All experiments were carried out at the Institute of Entomology, Jiangxi Agricultural University, Nanchang, Jiangxi Province (28°46′ N, 115°49′ E).

Experimental Materials and Insect Rearing Conditions

The original colony of *L. striatellus* was collected from the rice fields in the suburbs of Nanchang (28.8° N, 115.9° E), Jaingxi Province. Under local conditions the planthopper exhibits mixed voltinism from four to seven generations per year due to the differences in the diapause intensity of overwintering nymphs and the long ovipositional period of females and overwinters as 1–5 instar nymphs [12]. In the present study, nymphs in the overwintered and the first generations were raised with stems of American sloughgrass, *Beckmannia syzigache*; nymphs in the other generations were raised with rice stems. The planthoppers were reared at 25°C under LD 15:9 (a diapause-averting photoperiod) for several generations before use.

Approximately 50 newly hatched nymphs were put in a glass tube (length: 180 mm; diameter: 32 mm) with stems of American sloughgrass or rice stems, and exposed to different photoperiods and temperatures to observe diapause induction. In the diapause termination experiments, diapausing nymphs were placed in glass tubes with rice stems and they were exposed to different photoperiods and temperatures to observe diapause termination.

All laboratory experiments were performed in illuminated incubators (LRH-250-GS, Guangdong Medical Appliances Plant, Guangdong, China) equipped with four fluorescent 30 W tubes with an irradiance of approximately 1.97 W m^{-2} and the variation in the temperatures was ±1°C.

Diapause Induction Experiments

Photoperiodic Response. The photoperiodic responses for diapause induction in *L. striatellus* were experimentally determined under various photoperiods at constant temperatures of 18, 20, 22 and 25°C. The influence of unnatural photoperiods (including continuous darkness (DD) and continuous light (LL)) on

diapause induction was also examined at 20°C. Diapausing individuals were easy to identify because nymphs entered diapause in 3–4th instars and showed a distinct developmental delay. Thus, the incidence of diapause was determined based on the proportion of nymphs that were still in 3–4th instars within one week after comparable control cultures had completed emergence because the nymphs has stopped development and were almost not feeding during this period. There were 39–129 individuals in each of three replicates for most treatments; a few treatments had two or more than three replicates.

Required Day Number

The required day number (RDN), i.e., the number of light-dark cycles needed to raise the proportion of diapause in a population to 50% [13], was determined by transferring nymphs from a long photoperiod (LD 15:9, a diapause-averting photoperiod) to a short photoperiod (LD 11:13, a diapause-inducing photoperiod) at different times after hatching at 18°C or vice versa. There were 46–198 individuals in each of at least three replicates for all of the treatments.

Detection of Photosensitivity

To determine the greatest sensitivity to diapause-inducing photoperiodic cues, an experiment was performed as described by Spieth [14] involving the periodic interruption of diapause-inducing conditions. Because all nymphs enter diapause at 18°C under the short daylength of LD 11:13 and because five days comprise just one instar time, nymphs reared under a photoperiodic background of LD 11:13 were interrupted by five short nights of LD 15:9 at various nymphal stages at 18°C or vice versa. There were two or three replicates for each treatment.

The Incidence of Diapause under Field Conditions

To understand how the planthoppers decide to diapause under field conditions of changing photoperiod and temperature, nymphs from the autumn generation that hatched at different times from mid-September to mid-November in 2011–2013 were collected daily or every other day then approximately 50 newly hatched nymphs were put in a glass tube with rice stems. There were one to fifteen replicates for each treatment, depending on the hatching amount each day. The nymphs were allowed to develop under outdoor conditions to observe the time course of diapause induction. The incidence of diapause was recorded until all nymphs entered diapause. The total number observed was 7896 nymphs in 2011, 4392 nymphs in 2012 and 6956 nymphs in 2013.

Diapause Termination Experiments

Effect of Diapause-inducing Photoperiod and Temperature on Diapause Intensity. The effect of pre-diapause photoperiod and temperature on diapause intensity (duration) was evaluated by rearing newly hatched nymphs under the diapause-inducing daylengths of 10, 11 and 12 h at the constant temperatures of 18, 20 and 22°C until diapause determination. Diapausing nymphs induced under daylengths of 10, 11 and 12 h at 18, 20 and 22°C were transferred to LD 15:9 and 25°C to test diapause development. The emerged adults were recorded every day until all of the diapausing nymphs emerged.

Effect of Diapause-terminating Photoperiod and Temperature on Diapause Intensity. Diapausing nymphs induced under LD 11:13 at 18°C were divided into two groups. One group was incubated under a long photoperiod of LD 15:9 and a short photoperiod of LD 11:13 at 18°C to test the effect of photoperiod on diapause maintenance and termination. The other group was incubated under LD 15:9 at different temperatures (18,

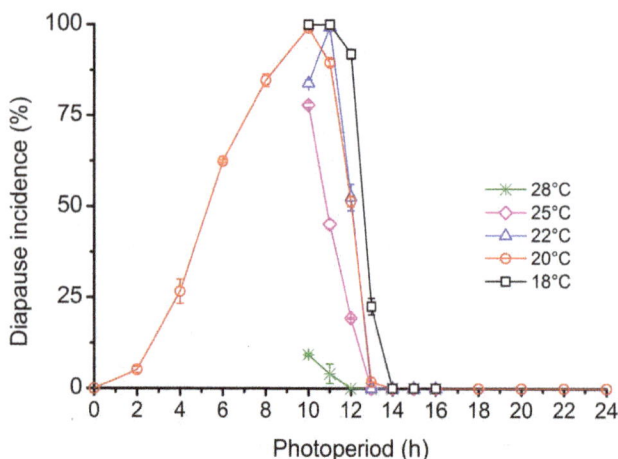

Figure 1. Photoperiodic response curves for diapause induction in *L. striatellus* at constant temperatures of 18, 20, 22 and 25°C. *N*=79–303 for each point.

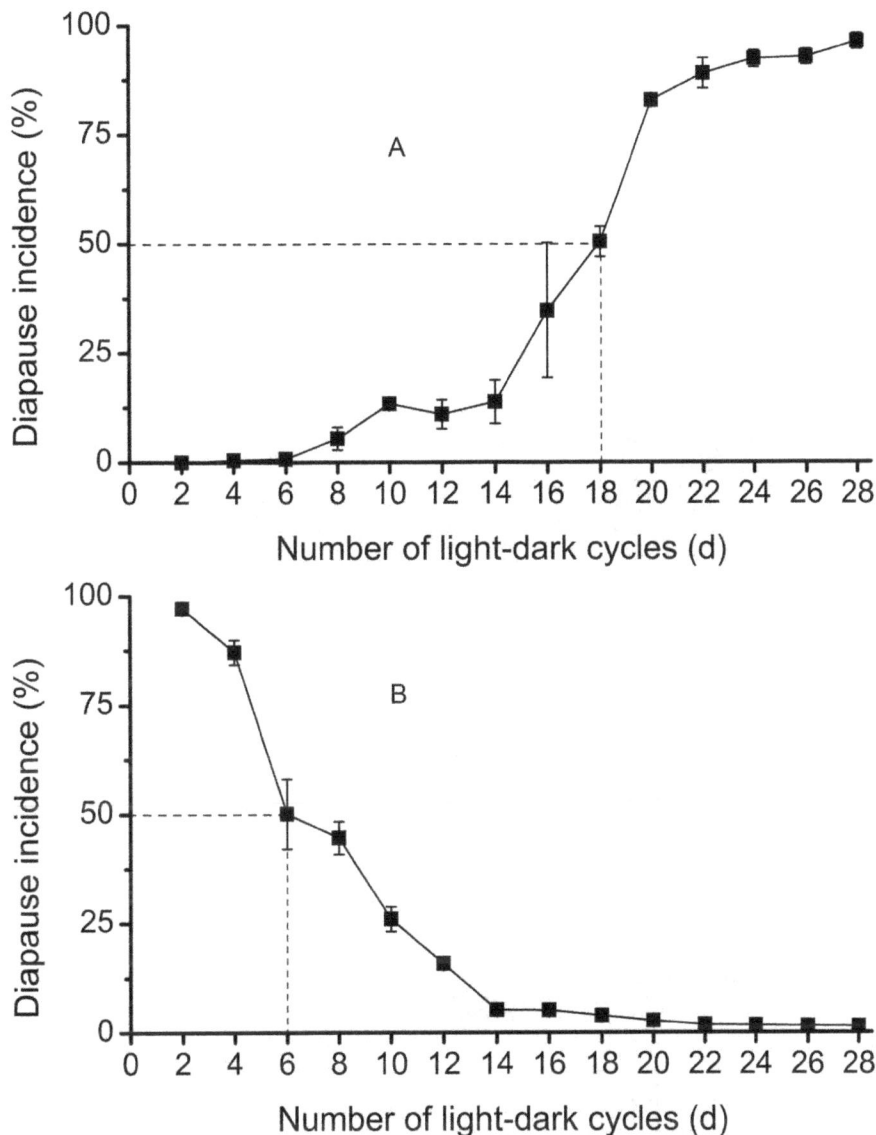

Figure 2. Incidence of diapause in *L. striatellus* at 18°C: (A) when nymphs were exposed to different numbers of long-night cycles (LD11:13) and then moved to short-night cycles (LD 15:9); (B) when nymphs were exposed to different numbers of short-night cycles (LD 15:9) and then moved to long-night cycles (LD 11:13). N = 137–352 for each point.

20, 22, 24, 26, 28, 30 and 32°C) to examine the effect of temperature on diapause maintenance and termination. The emerged adults were recorded every day until all of the diapausing nymphs emerged.

Effect of Chilling on Diapause Termination. To investigate the effect of chilling on diapause development, diapausing nymphs induced under LD 11:13 at 18°C were placed at 5°C for different lengths of time (ranging from 20 to 80 days) in continuous darkness. After chilling, the nymphs were transferred to LD 15:9 and 25°C to terminate diapause. The emerged adults were recorded every day until all of the diapausing nymphs emerged.

Diapause Termination under Field Conditions. The naturally diapausing nymphs that hatched at different times from mid-September to mid-November in 2011 and 2012 were kept under outdoor conditions to observe adult emergence in the next spring. The emerged adults were recorded every day until all overwintering nymphs emerged.

Statistical Analyses

Statistical analyses were conducted using SPSS 19.0 (IBM Inc.). The effects of photoperiod (from 10 h to 16 h), temperature and their interactions on the induction of diapause were tested using a General Linear Model (GLM). The influence of diapause-inducing temperature and photoperiod, diapause-terminating temperature and the chilling period on the duration of diapause were tested using Kruskal–Wallis tests following non-parametric tests. The influence of diapause-terminating photoperiod on the duration of diapause was tested by independent-samples t test. Differences are considered significant if $P < 0.05$.

Results

Photoperiodic Responses for Diapause Induction

Photoperiodic response curves for diapause induction in *L. striatellus* at different temperatures are shown in Fig. 1. The

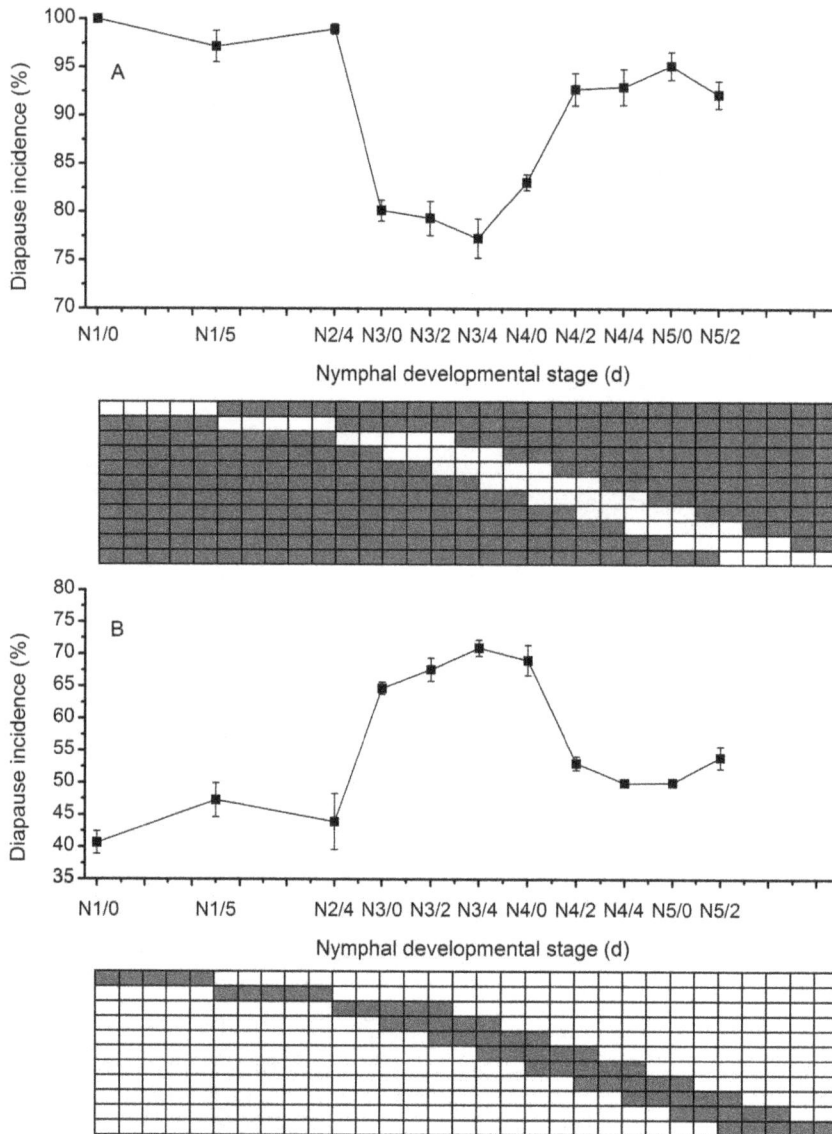

Figure 3. Photosensitivity of diapause during nymphal development at 18°C in *L. striatellus*. (A) when a background period of LD 11:13 was interrupted by five long photoperiods of LD 15:9; (B) when a background period of LD 15:9 was interrupted by five short photoperiods of LD 11:13. White bars represent the long photoperiod (LD15:9), and gray bars represent the short photoperiod (LD11:13). N3/4 means fourth day of third instar. $N = 60$–185 for each point.

photoperiodic response curves showed a typical long-day response type with a critical daylength of approx. 11 h at 25°C, 12 h at 22 and 20°C and 12.5 h at 18°C. The long daylengths of 13–16 h induced 100% development without diapause at 22 and 25°C and 77–100% development at 18 and 20°C. However, the high temperature of 28°C nearly abrogated the diapause-inducing effects of short daylengths; more than 90% individuals developed without diapause under short daylengths. The photoperiod, temperature and their interactions all have a significantly influence on the induction of diapause (Temperature effect: $F_{4,105} = 1064.3$, $P = 0.000$; Photoperiod effect: $F_{6,105} = 3441.8$, $P = 0.000$; Temperature × Photoperiod interactions: $F_{24,105} = 208.5$, $P = 0.000$). The photoperiodic response curve at 20°C showed a gradual decline in diapause incidence in ultra-long nights (from 16 h nightlength to 22 h nightlength), and DD resulted in 100% development.

Required Day Number

The RDN for 50% response was distinctly different between short- and long-night cycles at 18°C. It was 18 days for long-night cycles (Fig. 2A) and 6 days for short-night cycles (Fig. 2B), indicating that the effect of one short-night was equivalent to the effect of three long-nights.

The Most Sensitive Stage to Photoperiod

When the diapause-inducing photoperiod of LD 11:13 was interrupted by five long daylengths at 18°C, the most effective diapause inhibition occurred between the N3/0 (just entered third instar) and N3/4 (the fourth day of third instar) stages (Fig. 3A). Similarly, the diapause-inducing effects were also higher between the N3/0 and N3/4 stages when the diapause-inhibiting photoperiod of LD 15:9 was interrupted by five short daylengths

Figure 5. Diapause duration under LD 15:8 at 25°C in *L. striatellus.* Diapause was induced by the short daylengths of 10, 11 and 12 h at 18, 20 and 22°C. N = 26–48 for each treatment.

11:13 at 20°C and shorter than those of LD 10:14 and LD 11:13 at 22°C ($\chi^2 = 80.39$, $d.f. = 8$, $P<0.05$), whereas diapause-inducing temperature had no significant effect on diapause intensity at different photoperiods ($P>0.05$).

Effect of Diapause-terminating Photoperiod and Temperature on Diapause Intensity

Diapause ended even at a short daylength of LD 11:13; however, the duration of diapause was significantly longer under the short daylength of LD 11:13 (35 days) than it was under the long daylengths of LD 15:9 (26 days) ($t=8.97$, $d.f. = 63$, $P<0.01$) (Fig. 6), indicating that long photoperiods can accelerate diapause development.

In addition, diapause-terminating temperature significantly influenced the diapause intensity ($\chi^2 = 711.42$, $d.f. = 7$, $P<0.05$) (Fig. 7). The duration of diapause was gradually shortened from 25 days at 18 to 11 days at 28°C. Furthermore, the eclosion was less synchronous at 18°C than it was at other temperatures which resulted in changes in the mean, but a substantial proportion of the individuals still eclosed between 10–20 days. However, at the high temperatures of 30 and 32°C, the duration of diapause was significantly longer than the durations at the temperatures of 22, 24, 26 and 28°C. This result indicates that the optimal temperature for diapause termination is in the 26–28°C range.

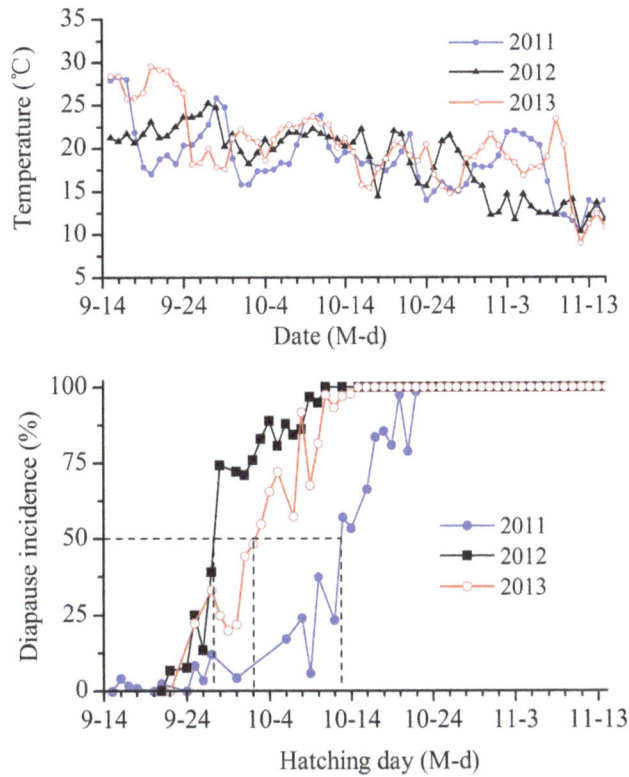

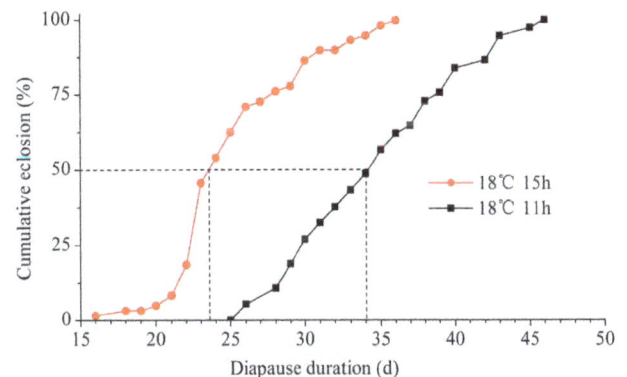

Figure 4. Incidence of nymphal diapause in *L. striatellus* which hatched on different dates under field conditions. N = 59–856 for each point.

(Fig. 3B). The results suggest that the 3rd instar nymph is most sensitive to photoperiod in *L. striatellus*.

The Time Course of Diapause Induction under Field Conditions

Fig. 4 shows the time course of diapause induction under field conditions for three years. Winter diapause had already occurred in some individuals that hatched on 16 September in 2011, 22 September in 2012 and 25 September in 2013. The proportion of diapausing individuals increased with time, and 50% nymphs initiated diapause on 13 October in 2011, 28 September in 2012 and 2 October in 2013. As shown above, the 3rd instar larva was the stage that was most sensitive to photoperiod. The nymphs required approximately 8 days to finish the first and second instars after hatching in late September or mid-October. Therefore, the photosensitive stage started 8 days after hatching, i.e., on 21 October in 2011, 6 October in 2012, and 10 October in 2013. By consulting the astronomical yearbooks [15], the daylength in the Nanchang region between October 6 and October 21 was between 12 h 10 min and 12 h 32 min (including twilight), which was the critical daylength for diapause induction in the field. Nymphs that hatched after mid-October all entered diapause when the mean daily temperature experienced by nymphs was lower than 20°C.

Effect of Diapause-inducing Photoperiod and Temperature on Diapause Intensity

The diapause-inducing photoperiod had a significant effect on diapause intensity (Fig. 5). The duration of diapause induced by a photoperiod of LD 12:12 was significantly shorter than that of LD

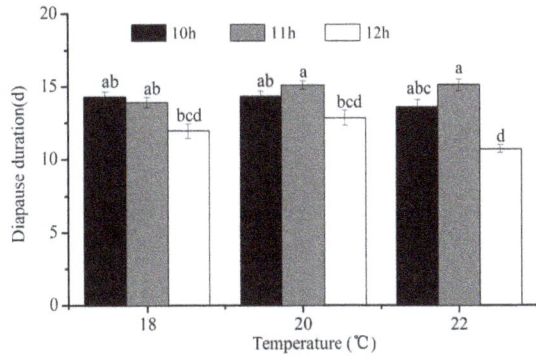

Figure 6. The duration of diapause of *L. striatellus* under the daylengths of 11 and 15 h at 18°C. Diapause was induced under LD 11:13 at 18°C. N = 42–79 for each point.

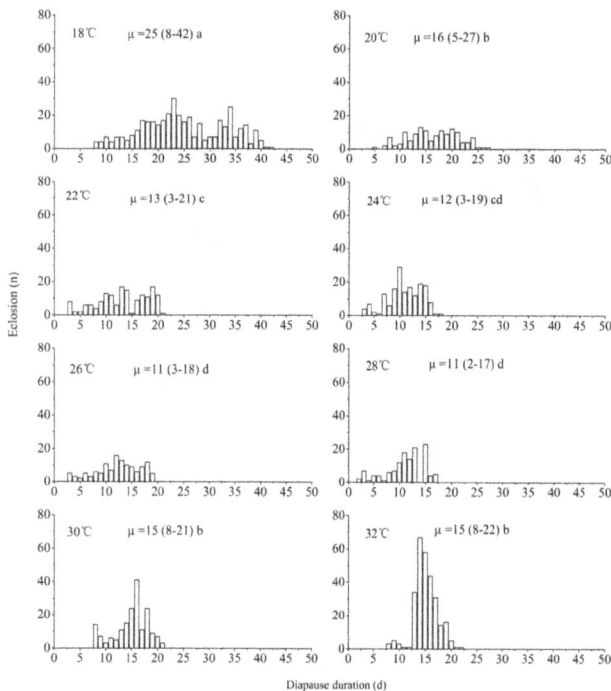

Figure 7. Diapause termination under LD 15:8 at different temperatures in *L. striatellus.* Diapause was induced under LD 11:13 at 18°C. *N* = 127–393 for each point.

Effect of Chilling on Diapause Termination

Fig. 8 shows the cumulative eclosion rate of diapausing nymphs under LD 15:9 at 25°C after chilling at 5°C for 20, 30, 40, 50, 60, 70 and 80 days. The eclosion time was significantly postponed with increasing chilling duration if the time spent under chilling was considered as part of the duration of diapause development ($\chi^2 = 1245.57$, *d.f.* = 7, *P* < 0.01), indicating that chilling at 5°C did not shorten the duration of diapause but lengthened it. In fact, diapausing nymphs without chilling (control) all emerged within 16 days when they were transferred to LD 15:9 at 25°C.

Time Course of Diapause Termination under Field Conditions

The cumulative eclosion rate of the naturally overwintering nymphs is presented in Fig. 9. A few overwintering nymphs began to emerge on 27 March 2012 and 5 March in 2013 when the mean daily temperature rose to 10°C or higher. A proportion of 50% eclosion occurred on 4 April 2012 and on 20 March 2013, showing a 15 day difference. This is most likely because the mean daily temperature in March was 2.8°C lower in 2012 (10.6°C) than it was 2013 (13.4°C). It suggests that the temperatures in March determine the rate of post-diapause development in the planthopper.

Discussion

The photoperiodic response for diapause induction in *L. striatellus* highly depended on temperature. The critical daylength decreased with an increase in temperature from 18 to 25°C (11 h at 25°C, 12 h at 22 and 20°C and 12.5 h at 18°C; Fig. 1). The critical daylength could not be determined at the high temperature of 28°C because this temperature significantly weakened and even completely inhibited the photoperiodic induction of diapause

(more than 90% of the individuals developed without diapause under short daylengths) However, it is important to note that the critical daylengths of 11 h at 25°C and 12 h at 22 and 20°C do not exist in the Nanchang region. In Nanchang, the longest period of daylight in a year is approximately 14 h 56 min (including twilight), the shortest is 11 h 9 min. Therefore, an 11 h daylength occurs in late winter and a 12 h daylength occurs at the end of October. Apparently, the critical daylengths at 20, 22 and 25°C were not suitable to analyze the incidence of diapause in the field. Only the critical daylength of 12.5 h at 18°C was related to diapause induction in the field because this result was consistent with the photoperiod experienced by nymphs entering diapause in the field (12 h 10 min-12 h 32 min; Fig. 4). Our results indicate that the nymphs entered winter diapause in response to short daylengths and low temperatures during autumn, which is consistent with the result reported by Kisimoto [9].

The photoperiodic response curve at 20°C in *L. striatellus* showed a gradual decline in diapause incidence during ultra-long nights (from 16 h nightlength to 22 h nightlength) and DD (Fig. 1), suggesting that different long nights have different inductive effects. This phenomenon has been found in many long-day insect species, such as the large white butterfly, *Pieris brassicae* [16], the Indian meal moth, *Plodia interpunctella* [17], the linden bug, *Pyrrhocoris apterus* [18], the fly, *Chymomyza costata* [19], the spider mite, *Tetranychus urticae* [20], the pine caterpillars species, *Dendrolimus punctatus* and *D. tabulaeformis* [21,22], the endoparasitoid wasp *Microplitis mediator* [23], and the rice stem borer *Chilo suppressalis* [24]. That ultra-long nights and DD result in a decline in diapause incidence presumably reflects the absence of selective pressure, but it may have a physiological significance when one attempts to determine the mechanism of time measurement [25].

One basic concept concerning photoperiodic responses in insects is the RDN [13]. It has been shown that a greater number of exposure days is required for the induction of diapause than are required for its termination [8]. This has been indicated in species as diverse as the mosquito *Aedes atropalpus*, the aphid *Megoura viciae*, the cabbage white butterfly *Pieris rapae*, the European grapevine moth *Lobesia botrana*, the Asian swallowtail *Papilio xuthus* [8], and the zygaenid moth *Pseudopidorus fasciata* [26] and *D. punctatus* [21]. The transfer experiments in *L. striatellus* (Fig. 2) showed that the RDN for short nights was three times less than for that long nights at 18°C (cf. 6 days vs. 18 days), indicating that short nights are photoperiodically more potent.

The intensity of diapause dose not only vary between species but also among individuals of the same species depending on how long they have been exposed to diapause-inducing and terminating conditions [15]. According to the present findings, the intensity of diapause in *L. striatellus* nymphs was affected by diapause-inducing photoperiods. The duration of diapause induced by a photoperiod of LD 12:12 was significantly shorter than those by LD 10:14 and LD 11:13 at 20 and 22°C (Fig. 5). Similar results have been reported for a number of insects, such as the fruit flies, *Drosophila auraria*, *D. subauraria*, *D. triauraria*, the bean bug, *Riptortus clavatus*, the Mediterranean tiger moth, *Cymbalophora pudica* and the cotton bollworm, *Helicoverpa armigera* [27–31], where photoperiods with longer scotophases induced more intense diapause compared with shorter scotophases. Our results further showed that the duration of diapause in *L. striatellus* was significantly affected by the photoperiod during diapause. By transferring diapausing nymphs to a short photoperiod of LD 11:13 and a long photoperiod of LD 15:9 combined with 18°C, diapause was terminated significantly faster at LD 15:9 than it was at LD 11:13 (Fig. 6), suggesting that short daylengths may play a

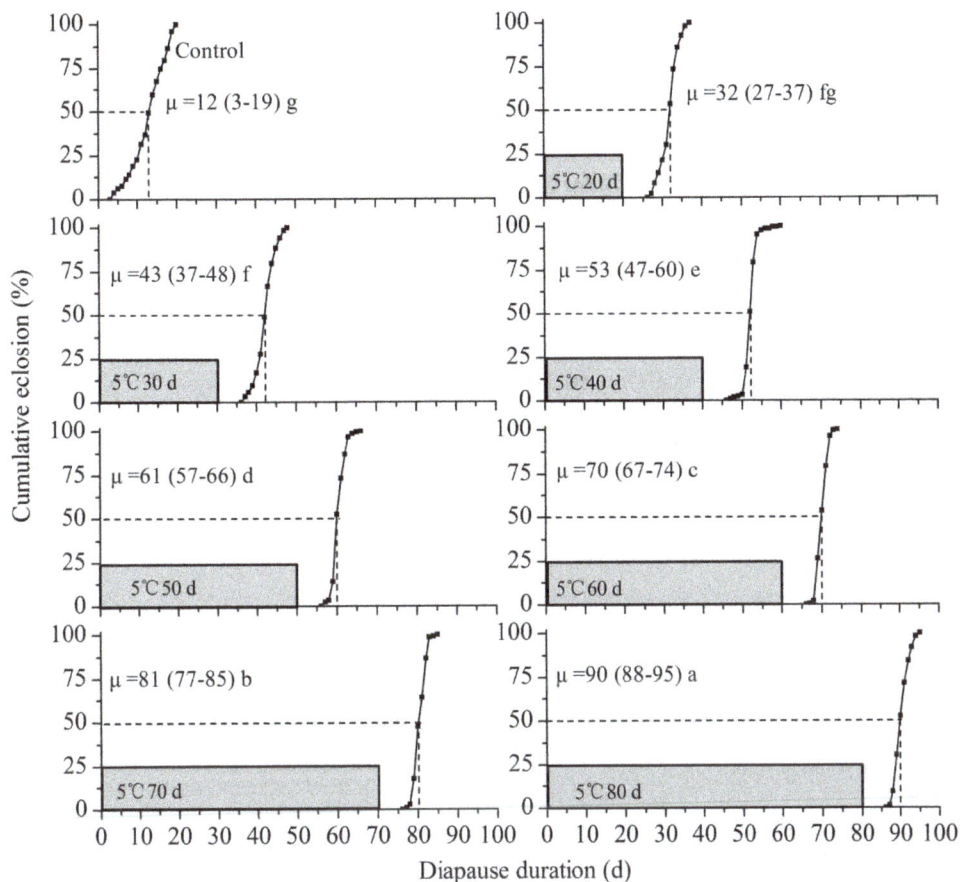

Figure 8. Effect of chilling on diapause termination in *L. striatellus*. Diapausing nymphs induced under LD 11:13 at 18°C were transferred to 5°C DD for 0, 20, 30, 40, 50, 60, 70 and 80 days, and subsequently transferred to LD 15:9 and 25°C. N = 122–200 for each treatment.

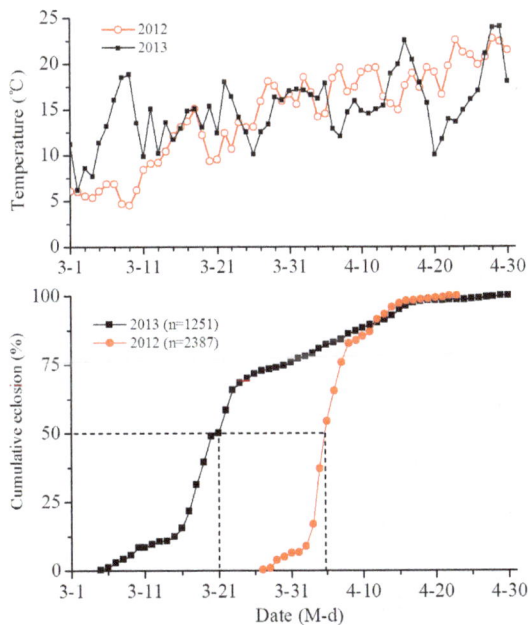

Figure 9. Adult eclosion of overwintering generation of *L. striatellus* under field conditions in spring.

role in maintaining winter diapause. Similar results have also been reported for the mulberry tiger moth *Spilarctia imparilis* [32], the lacewing *Chrysopa downesi* [33] and the cabbage butterfly *Pieris melete*, where long photoperiods during diapause significantly accelerated winter diapause development.

Increasing evidence has shown that chilling is not a prerequisite for the completion of hibernation diapause in many insect species and diapause completion progresses well at intermediate or high temperatures in some insects [24,34–39]. Our data revealed that the diapause could be terminated without exposure to chilling in *L. striatellus*. The rate of diapause completion was positively related to the temperature increase between 18 and 28°C. However, diapause development was delayed when the temperature rose to 30 and 32°C (Fig. 7). This result suggests that the optimal temperature for diapause development is 26 and 28°C in *L. striatellus*. The chilling experiments in Fig. 8 showed that the eclosion time of diapausing nymphs was significantly postponed with an increase in chilling time if the time spent under chilling was considered as part of the duration of diapause development, indicating that chilling at 5°C did not shorten the duration of diapause but lengthened it. Our results suggest that low temperatures during winter may serve primarily to maintain nymphal diapause and prevent the resumption of post-diapause morphogenesis, which in turn synchronizes the adult emergence of the overwintering generation with the availability of host plants [40].

A few studies have examined diapause induction and termination under field conditions of changing photoperiod and temperature [24,41]. In the present study, we systematically investigated the time course of diapause induction and termination of *L. striatellus* under field conditions. Our results reveal that winter diapause had already occurred in some individuals that hatched in mid-September or late September; 100% of the nymphs that hatched after mid-October entered diapause when the mean daily temperature experienced by nymphs decreased to 20°C or lower (Fig. 4). This result suggests that the important cue for the initiation of winter diapause depends primarily on temperature, i.e., the temperatures between mid-September and mid-October. Observations of diapause termination under field conditions reveal that a few diapausing nymphs initiated their eclosion on 5 March or 25 March when the mean daily temperature rose to 10°C or higher (Fig. 9), suggesting that an early or late termination of winter diapause primarily depends on the temperatures in March. Therefore, these data indicate that temperature was strongly correlated with the induction and termination in *L. striatellus*. Combining our results with the climatic data from the locality, we can predict the time of diapause initiation in autumn and adult emergence in spring for this insect.

Furthermore, our results emphasize the importance of understanding how insects decide to initiate and terminate diapause under field conditions of changing photoperiod and temperature. As long as we do not understand this, we can never be certain how the findings from the laboratory relate to conditions in the field.

Acknowledgments

We thank Dr. Shaohui Wu from the Department of Entomology in Rutgers University for advice and comments on the manuscript. We also thank the two anonymous reviewers for their valuable comments.

Author Contributions

Conceived and designed the experiments: FSX KJL. Performed the experiments: LFW CC SF. Analyzed the data: LFW CC. Contributed reagents/materials/analysis tools: LFW CC. Contributed to the writing of the manuscript: FSX.

References

1. Liu XD, Zhai BP, Liu CM (2006) Outbreak reasons of *Laodelphax striatellus* population. Chinese B Entomol 43: 141–146 (in Chinese).
2. Otuka A, Matsumura M, Sanada-Morimura S, Takeuchi H, Watanabe T, et al. (2010) The 2008 overseas mass migration of the small brown planthopper, *Laodelphax striatellus*, and subsequent outbreak of rice stripe disease in western Japan. Appl Entomol Zool 45: 259–266.
3. He Y, Zhu YB, Hou YY, Yao ST, Lu ZJ, et al. (2012) Fluctuation and migration of spring population of small brown planthopper (*Laodelphax striatellus*) on wheat in Jiangsu and Zhejiang provinces. Chinese J Rice Sci 26: 109–117 (in Chinese).
4. Cai BH, Huang FS, Feng WX, Fu YR, Dong QF (1964) Study on *Delphacodes striatellus* (Fallén) (Homoptera: Delphacidae) in North China. Acta Entomologica Sinica 13: 552–571 (in Chinese).
5. Lin ZW, Liu Y, Xin HP (2004) A primary study of *Laodelphax striatellus* (Fallén) bio-character in cold region rice. J Heilongjiang Bayi Agr Univ 16: 15–18 (in Chinese).
6. Denlinger DL (2002) Regulation of diapause. Annu Rev Entomol 47: 93–122.
7. Tauber MJ, Tauber CA, Masaki S (1986) Seasonal adaptations of insects: Oxford University Press, USA.
8. Danks HV (1987) Insect dormancy: an ecological perspective: Biological Survey of Canada (Terrestrial Artropods).
9. Kisimoto R (1958) Studies on diapause in the planthopper.1. Effect of photoperiod on the induction and the completion of diapause in the fourth larval stage of the small brown planthopper, *Delphacodes striatella* (Fallén). Japenese J Appl Entomol Zool 2: 128–134.
10. Kisimoto R (1989) Flexible diapause response to photoperiod of a laboratory selected line in the small brown planthopper, *Laodelphax striatellus* (Fallén). Appl Entomol Zool 24: 157–159.
11. Noda H (1992) Geographic variation of nymphal diapause in the small brown planthopper in Japan. Japan Agr Res Q 6: 124–129.
12. Wang LF, Fu S, Xiao L, Chen C, Xue FS (2013) Life history, reproduction and overwintering biology of the small brown planthopper, *Laodelphax striatellus* (Hemiptera: Delphacidae), in Nanchang, Jiangxi, East China. Acta Entomologica Sinica 56: 1430–11439 (in Chinese).
13. Saunders DS (1971) The temperature-compensated photoperiodic clock "programming" development and pupal diapause in the flesh-fly, *Sarcophaga argyrostoma*. J Insect Physiol 17: 801–812.
14. Spieth HR (1995) Change in photoperiodic sensitivity during larval development of *Pieris brassicae*. J Insect Physiol 41: 77–83.
15. Purple Mountain Observatory, Chinese Academy of Sciences, 2011. In 2012 the Chinese Astronomical Calendar. Science Press 595.
16. Danilevskii AS (1965) Photoperiodism and seasonal development of insects (English translation). Edinburgh and London, U.K.: Oliver and Boyd.
17. Masaki S, Kikukawa S (1981) The diapause clock in a moth: response to temperature signals. In: Follett BK, Follett DE, editors. Biological clocks in seasonal reproductive cycles. Bristol: John Wright & Sons Ltd. 101–112.
18. Saunders DS (1983) A diapause induction-termination asymmetry in the photoperiodic responses of the Linden bug, *Pyrrhocoris apterus* and an effect of near-critical photoperiods on development. J Insect Physiol 29: 399–405.
19. Yoshida MT, Kimura MT (1995) The photoperiodic clock in *Chymimyza costata*. J Insect Physiol 41: 217–222.
20. Kroon A, Veenendaal R, Veerman A (1997) Photoperiodic induction of diapause in the spider mite *Tetranychus urticae*: qualitative or quantitative time measurement? Physiol Entomol 22: 357–364.
21. Huang LL, Xue FS, Wang GH, Han RD, Ge F (2005) Photoperiodic response of diapause induction in the pine caterpillar, *Dendrolimus punctatus*. Entomol Exp Appl 117: 127–133.
22. Han RD, Xue FS, He Z, Ge F (2005) Diapause induction and clock mechanism in the pine caterpillar *Dendrolimus tabulaeformis* (Lep., Lasiocampidae). J Appl Entomol 129: 105–109.
23. Li WX, Li JC, Coudron TC, Lu ZY, Pan WL, et al. (2008) Role of photoperiod and temperature in diapause induction of endoparasitoid wasp *Microplitis mediator* (Hymenoptera: Braconidae). Ann Entomol Soc America 101: 613–618.
24. Xiao HJ, Mou FC, Zhu XF, Xue FS (2010) Diapause induction, maintenance and termination in the rice stem borer *Chilo suppressalis* (Walker). J Insect Physiol 56: 1558–1564.
25. Saunders DS (2002) Insect clocks. Amsterdam, The Netherlands: Elsevier.
26. Hua A, Xue FS, Xiao HJ, Zhu XF (2005) Photoperiodic counter of diapause induction in *Pseudopidorus fasciata* (Lepidoptera: Zygaenidae). J Insect Physiol 51: 1287–1294.
27. Kimura MT (1983) Geographic variation and genetic aspects of reproductive diapause in *Drosophila triauraria* and *D. quadraria*. Physiol Entomol 8: 181–186.
28. Kimura MT (1990) Quantitative response to photoperiod during reproductive diapause in the *Drosophila auraria* species-complex. J Insect Physiol 36: 147–152.
29. Nakamura K, Numata H (2000) Photoperiodic control of the intensity of diapause and diapause development in the bean bug, *Riptortus clavatus* (Heteroptera: Alydidae). Eur J Entomol 97: 19–24.
30. Koštál VI, Hodek I (1997) Photoperiodism and control of summer diapause in the Mediterranean tiger moth *Cymbalophora pudica*. J Insect Physiol 43: 767–777.
31. Chen C, Xia QW, Fu S, Wu XF, Xue FS (2014) Effect of photoperiod and temperature on the intensity of pupal diapause in the cotton bollworm, *Helicoverpa armigera* (Lepidoptera: Noctuidae). B Entomol Res 104: 12–18.
32. Sugiki T, Masaki S (1972) Photoperiodic control of larval and pupal development in *Spilarctia imparilis* Butler (Lepidoptera: Arctiidae). Kontyu 40: 269–278.
33. Tauber MJ, Tauber CA (1976) Environmental control of univoltinism and its evolution in an insect species. Canadian J Zool 54: 260–266.
34. Hodek I (2002) Controversial aspects of diapause development. Eur J Entomol 99: 163–174.
35. Takeda M (2006) Effect of temperature on the maintenance and termination of diapause in overwintering females of *Pseudaulacaspis pentagona* (Hemiptera: Diaspididae). Appl Entomol Zool 41: 429–434.
36. Broufas G, Pappas M, Koveos D (2006) Effect of cold exposure and photoperiod on diapause termination of the predatory mite *Euseius finlandicus* (Acari: Phytoseiidae). Environ Entomol 35: 1216–1221.
37. Wang XP, Yang QS, Zhou XM, Xu S, Lei CL (2009) Effects of photoperiod and temperature on diapause induction and termination in the swallowtail, *Sericinus montelus*. Physiol Entomol 34: 158–162.
38. Chen YS, Chen C, He HM, Xia QW, Xue FS (2013) Geographic variation in diapause induction and termination of the cotton bollworm, *Helicoverpa armigera* Hübner (Lepidoptera: Noctuidae). J Insect Physiol 59: 855–862.
39. Johansen TJ, Meadow R (2014) Diapause development in early and late emerging phenotypes of *Delia floralis*. Insect Sci 21: 103–113.
40. Hodek I, Hodková M (1988) Multiple role of temperature during insect diapause: a review. Entomol Exp Appl 49: 153–165.
41. Xue FS, Kallenborn HG, Wei HY (1997) Summer and winter diapause in pupae of the cabbage butterfly, *Pieris melete* Ménétriés. J Insect Physiol 43: 701–707.

10

Sorghum Phytochrome B Inhibits Flowering in Long Days by Activating Expression of *SbPRR37* and *SbGHD7*, Repressors of *SbEHD1*, *SbCN8* and *SbCN12*

Shanshan Yang[1], **Rebecca L. Murphy**[1], **Daryl T. Morishige**[1], **Patricia E. Klein**[2], **William L. Rooney**[3], **John E. Mullet**[1]*

1 Department of Biochemistry and Biophysics, Texas A&M University, College Station, Texas, United States of America, 2 Department of Horticultural Sciences and Institute for Plant Genomics and Biotechnology, Texas A&M University, College Station, Texas, United States of America, 3 Department of Soil and Crop Sciences, Texas A&M University, College Station, Texas, United States of America

Abstract

Light signaling by phytochrome B in long days inhibits flowering in sorghum by increasing expression of the long day floral repressors *PSEUDORESPONSE REGULATOR PROTEIN* (*SbPRR37*, *Ma1*) and *GRAIN NUMBER, PLANT HEIGHT AND HEADING DATE 7* (*SbGHD7*, *Ma6*). *SbPRR37* and *SbGHD7* RNA abundance peaks in the morning and in the evening of long days through coordinate regulation by light and output from the circadian clock. 58 M, a phytochrome B deficient (*phyB-1*, *ma3^R*) genotype, flowered ~60 days earlier than 100 M (*PHYB*, *Ma3*) in long days and ~11 days earlier in short days. Populations derived from 58 M (*Ma1*, *ma3^R*, *Ma5*, *ma6*) and R.07007 (*Ma1*, *Ma3*, *ma5*, *Ma6*) varied in flowering time due to QTL aligned to *PHYB*/*phyB-1* (*Ma3*), *Ma5*, and *GHD7*/*ghd7-1* (*Ma6*). *PHYC* was proposed as a candidate gene for *Ma5* based on alignment and allelic variation. *PHYB* and *Ma5* (*PHYC*) were epistatic to *Ma1* and *Ma6* and progeny recessive for either gene flowered early in long days. Light signaling mediated by PhyB was required for high expression of the floral repressors *SbPRR37* and *SbGHD7* during the evening of long days. In 100 M (*PHYB*) the floral activators *SbEHD1*, *SbCN8* and *SbCN12* were repressed in long days and de-repressed in short days. In 58 M (*phyB-1*) these genes were highly expressed in long and short days. Furthermore, *SbCN15*, the ortholog of rice Hd3a (FT), is expressed at low levels in 100 M but at high levels in 58 M (*phyB-1*) regardless of day length, indicating that PhyB regulation of *SbCN15* expression may modify flowering time in a photoperiod-insensitive manner.

Editor: Nicholas S. Foulkes, Karlsruhe Institute of Technology, Germany

Funding: This research was supported by Pioneer Hi-Bred International, Inc., Ceres Inc., and the Perry Adkisson Chair in Agricultural Biology. The funders had no role in study design, data collection and analysis, decision to publish, or preparation of the manuscript.

Competing Interests: The authors declare that no competing interests exist. The research was funded in part by two commercial sources (Pioneer Hi-Bred International, Inc., Ceres Inc.). These commercial sources of funding, along with all other potential competing interests related to employment, consultancy, patents, products in development, marketed products, did not influence the conduct of this research and created no competing interests.

* Email: jmullet@neo.tamu.edu

Introduction

Flowering time has a significant impact on plant adaptation to agro-ecological environments, biomass accumulation and grain yield [1]. Floral initiation is regulated by plant development, photoperiod, shading, temperature, nutrient status, and many other factors [2–5]. Signals from many input pathways are integrated in the shoot apical meristem (SAM) through regulation of the meristem identity genes *LEAFY* (*LFY*) and *APETALA1* (*AP1*), which are activated during transition of the SAM from a vegetative meristem to a floral meristem. Long day (LD) plants, such as *Arabidopsis*, flower earlier in LD compared to short days (SD). In contrast, SD plants, such as rice and sorghum, show delayed floral initiation under LD conditions. Photoperiod regulated flowering is mediated by light signaling from photoreceptors and output from the endogenous circadian clock consistent with external coincidence models of flowering time regulation [6].

Photoperiod sensitive *Sorghum bicolor* genotypes delay floral initiation when grown under LD conditions. Sorghum genotypes with reduced photoperiod sensitivity have been identified and used by breeders because they flower early and at similar times in both long and short days, enhancing grain production [7]. In contrast, bioenergy sorghum is highly photoperiod sensitive, flowering in long day environments only after an extended phase of vegetative growth, thereby increasing biomass accumulation and nitrogen use efficiency [1,8].

Photoperiod regulated flowering requires perception of light and signaling by plant photoreceptors such as the red/far-red light sensing phytochromes (Phy), blue light/ultraviolet wavelength sensing cryptochromes (Cry), phototropins, and Zeitlupes [9,10]. Phytochromes play an important role in flowering time regulation in most plants including rice [11], barley [12], and sorghum [13]. The sorghum genome encodes three phytochrome genes, *PHYA*, *PHYB* and *PHYC*. Quail et al. (1994) established a standard nomenclature for phytochrome where PHY corresponds to phytochrome apoproteins, while phytochrome or phy indicates presence of the holoprotein, the fully assembled chromoprotein with chromophore covalently attached to the apoprotein [14]. Since all phytochrome proteins referred in this study are presumed to be holoproteins, Phy is used to represent wild type holoprotein, while phy is used to represent mutant versions of the holoprotein. Inactivation of PhyB results in early flowering in long days [13]. Phytochromes are soluble chromoproteins that contain an N-terminal photosensory domain and a C-terminal dimerization moiety. There are three sub-domains in the N-terminal moiety: PAS (PER, ARNT and SIM), GAF (cGMP phosphodiesterase, adenylate cyclase, Fh1A) and PHY (phytochrome-specific GAF-related), which form a unique structure, the "light-sensing knot" [15]. The PAS/GAF domains transduce light signals and the C-terminal domain, consisting of two PAS and HKRD (histidine-kinase-related domain), is responsible for dimerization and nuclear localization.

The central oscillators of the plant circadian clock are encoded by *TIMING OF CAB EXPRESSION 1 (TOC1), CIRCADIAN CLOCK ASSOCIATED 1 (CCA1)* and *LATE ELONGATED HYPOCOTYL (LHY)* [16]. Rhythmic expression of these central oscillators modulates the expression of *GIGANTEA (GI)*, an output gene of the circadian clock. GI, in concert with other factors, activates expression of *CONSTANS (CO)*, a zinc-finger transcription factor that plays an essential role in photoperiod regulation of flowering time in Arabidopsis [17], rice [18] and sorghum [19]. In *Arabidopsis*, CO is stabilized and accumulates during the evening of long days through the action of Cry1, Cry2 and PhyA, where it activates expression of *FT* and flowering. In SD, CO is not stabilized during the evening because *CO* expression occurs in darkness [20]. FT is produced in leaves and translocated to the SAM where it binds to FD. In Arabidopsis, FT together with *SUPPRESSOR OF OVEREXPRESSION OF CONSTANS (SOC1)*, promotes expression of meristem identity gene *LFY* and *AP1*, leading to floral transition [20].

The core of photoperiod regulatory pathway GI-CO-FT is present in *Arabidopsis*, a LD plant, and the SD plants rice and sorghum. In rice, *OsGI, HEADING DATE 1 (Hd1)*, and *HEADING DATE 3a (Hd3a)* are orthologs of GI, CO, and FT, respectively [21]. Hd1 *(OsCO)* delays flowering time in LD in rice and activates flowering in SD. In addition, Itoh et al. [22] identified a pair of genes in rice, *EARLY HEADING DATE 1 (EHD1)* and *GRAIN NUMBER, PLANT HEIGHT AND HEADING DATE 7 (GHD7)* that regulate flowering in response to day length by modifying expression of *Hd3a* (florigen). *EHD1* activates *Hd3a* expression and induces floral transition. In contrast, *GHD7*, a homolog of wheat *VRN2* [23], represses flowering in LD by down-regulating *EHD1* and *Hd3a*. In maize, 25 FT-like homologs were identified and designated as *Zea mays CENTRORADIALIS (ZCN)* genes. *ZCN8* was identified as a source of florigen [24]. *SbCN8* (ortholog of *ZCN8*) and *SbCN12* (ortholog of *ZCN12*) have been proposed to encode florigens in sorghum [19,25,26]. In sorghum, CO activates flowering in SD by inducing expression of *SbEHD1, SbCN8* and *SbCN12,* whereas in LD, CO activity is inhibited by SbPRR37 [19].

More than 40 flowering time QTL have been identified in sorghum [27] and maturity loci *Ma1–Ma6*, modify photoperiod sensitivity [7,28,29]. Dominance at *Ma1–Ma6* delays floral initiation in long days. *Ma3* encodes phytochrome B, indicating that light signaling through this photoreceptor is required for photoperiod sensitive variation in flowering time [13]. *Ma6* was identified as *SbGHD7*, a repressor of flowering in long days [26]. In LD, SbGhd7 increases photoperiod sensitivity by inhibiting

expression of the floral activators *SbEHD1, SbCN12* and *SbCN8*. *Ma1* was identified as *SbPRR37*, a floral repressor that acts in LD [25]. The orthologs of *SbPRR37* in wheat and barley, *PHOTO-PERIOD 1 (Ppd1, Ppd-H1, Ppd-D1a)* [30,31] and rice *OsPRR37* [32], also modulate flowering time in response to photoperiod. In LD, SbPRR37 inhibits expression of *SbEHD1, SbCN12,* and *SbCN8,* resulting in repression of flowering [25]. Moreover, *SbPRR37* modulates photoperiod sensitivity and floral repression in an additive fashion together with *SbGHD7* [26]. Expression of *SbPRR37* and *SbGHD7* is regulated by the circadian clock and light, suggesting common upstream regulation [26].

The current study focused on elucidating how phytochrome B regulates flowering time in response to day-length in sorghum. We report that *PHYB* is required for light activation of *SbPRR37* and *SbGHD7* expression in the evening of long days, resulting in repression of *SbEHD1, SbCN12, SbCN8* and floral initiation.

Materials and Methods

Phenotypic analysis of sorghum flowering time

The maturity loci and flowering dates of all sorghum lines used in this study are listed in Table S1. To characterize the difference in flowering time between different genotypes and day-length, 100 M and 58 M were planted in Metro-Mix 200 (Sunshine MVP; Sun Gro Horticulture) and grown in a greenhouse in LD (14 h light/10 h dark) and SD (10 h light/14 h dark) conditions. Days to mid-anthesis were recorded and plants were photographed. 100 M plants (n = 5) and 58 M plants (n = 9) were grown in LD and phenotyped for days to anthesis (Figure 1A). The mean days to flowering for 100 M was 126 days (± 4 days) and 62 days (± 3 days) for 58 M, a significant difference in flowering times for these genotypes (p-value$\ll 0.001$, Welch two sample t-test). Under SD, 100 M plants (n = 7) and 58 M plants (n = 5) were used for analysis of flowering time (Figure 1B). The mean days to flowering for 100 M was 59 days (± 4 days) and for 58 M, 48 days (± 1 days), a significant differences in days to flowering (p-value$\ll 0.001$). To establish the interaction between PhyB and photoperiod, factorial ANOVA was run with photoperiod and PhyB alleles as factors. The significance of the effects of PhyB alleles, day-length and PhyB:day-length interaction were detected (p-value$\ll 0.001$). All statistics were run in R 3.1.0. The two-way interaction graphs were plotted using the "HH" package in R.

Sequencing of *PHYB* alleles

To identify coding alleles in the *PHYB* gene, the full-length genomic sorghum *PHYB* genes from historical sorghum cultivars were amplified as three overlapping segments by PCR (Phusion High-Fidelity DNA polymerase, New England BioLabs, Inc). The amplified PCR products were cleaned and concentrated (QIA-quick PCR Purification kit, QIAGEN). PCR products were separated by electrophoresis on 1% agarose gels. Specific PCR products were excised and purified (QIAquick Gel Extraction Kit, QIAGEN). The purified PCR products were sequenced using the BigDye Terminator v3.1 Cycle Sequencing Kit (Applied Biosystems) and the Applied Biosystems 3130xl Genetic Analyzer. All primers used for sequencing were designed using PrimerQuest[SM] software (Integrated DNA Technologies, Inc) and are shown in Table S2. Sequencher v4.8 (Gene Codes) was used for sequence assembly and alignment with the BTx623 whole genome sequence of *Sorghum bicolor* (version 1.4) downloaded from Phytozome v8.0 (http://www.phytozome.net/). The SIFT (sorting intolerant from tolerant) program (http://sift.jcvi.org/) was utilized to predict whether an amino acid substitution affects protein function, based

Figure 1. Photographs of the sorghum lines 100 M and 58 M for flowering time phenotype. (A) Photograph of 100 M (left) and 58 M (right) grown for 109 days in LD (14 h light/10 h dark). 100 M and 58 M flowered after 126 days and 62 days respectively. (B) Photograph of 100 M (left) and 58 M (right) grown in a greenhouse in SD for 53 days (10 h light/14 h dark). 100 M flowered after 59 days and 58 M flowered after 48 days. LD: long days. SD: short days. DTF = number of days to flowering time. Scale bar is 8.6 cm.

on the degree of conservation of amino acid residues in sequence alignments derived from closely related sequences.

QTL analysis of *PHYB* action

The sorghum cultivar 58 M (*Ma1Ma2ma3^{R}Ma4Ma5ma6*) was crossed to R.07007 (*Ma1ma2Ma3Ma4ma5Ma6*) to generate a population for QTL analysis. F1 generation plants were self-pollinated to produce F2 populations from which F3 populations were derived by self pollination. F2 and F3 populations were planted in the greenhouse and grown under long day conditions (14 h light/10 h dark). Days to mid-anthesis of panicles of plants from the F2 and F3 populations were recorded. The median, standard error, and range of Days to Flowering and the number of plants of each genotype analyzed from the F2 and F3 populations are shown in Table3. For analysis of epistatic interaction, three-way ANOVA was run to detect the effect of allelic variation in

three maturity genes (*Ma3*, *Ma5* and *Ma6*) and three two-way interactions (*Ma3:Ma5*, *Ma3:Ma6*, *Ma5:Ma6*). The significance of the effects of single genes and genetic interactions were detected (p-value<<0.001). All statistics were run in R 3.1.0. The two-way interaction graph was plotted using the "HH" package in R.

For genotyping, genomic DNA of 86 F2 individuals and 132 F3 individuals was extracted from leaf tissue using the FastDNA Spin Kit (MP Biomedicals). Template for sequencing on an Illumina GAIIx sequencer was generated following the standard Digital Genotyping (DG) protocol [33]. Genotypes of all individuals from both populations were identified. The genetic map was constructed using the Kosambi mapping function in MAPMAKER v3.0 with 285 markers from the F2 population and 653 markers from the F3 population. QTL were mapped using the genetic map and the Composite Interval Mapping (CIM) function in WinQTL Cartographer v2.5 [34]. Significant LOD thresholds for QTL

Table 1. Sequence analysis of *PHYB* coding alleles in different sorghum lines.

	Exon 1	Exon 1	Exon 3	Exon 4	Sorghum Genotypes
Nucleotide Variation	CAC>...	A>G	A>.	C>G	
Protein Modification	His>...	Asp>Gly	Premature stop codon	Leu>Val	
Mutation Position (AA #)	31	308	1023	1113	
Alignment with PHYB in *Arabidopsis* (AA #)	32	293	1007	1096	
Phytochrome Domain		GAF(N)			
PHYB (*Ma3* or *ma3*)	−	−	−	−	BTx623, 100 M, 90 M, R.07007, Hegari, Tx7000, BTx642, SC56, Shallu, BTx3197
phyB-1 (*ma3^R*)	−	−	+	−	58 M
phyB-2	+	+	−	+	IS3620C

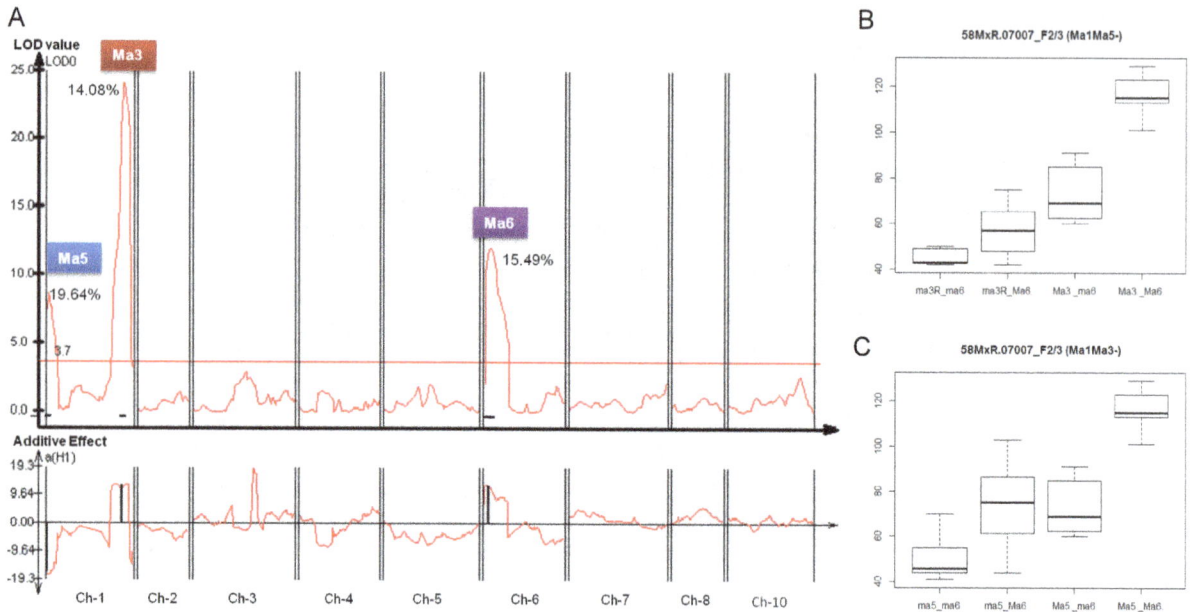

Figure 2. Flowering time QTL and analysis of epistasis in populations derived from 58MxR.07007. (A) Flowering time QTL labeled *Ma3*, *Ma5* and *Ma6*, were identified through analysis of flowering time variation in LD in the F2 population derived from 58MxR.07007. LOD values are shown on the Y-axis and sorghum chromosome numbers on the X-axis. The percent of the variance explained by each QTL is noted. The additive plot is shown in the lower portion of 2A where a positive value corresponds to alleles from R.07007 that delay flowering time. (B) Boxplot of flowering time distribution in the subset of the population with *Ma1Ma5-* genotypes but varying for alleles of *Ma3/ma3^R* and *Ma6/ma6*. (C) Boxplot of flowering time distribution in the subset of the population having *Ma1Ma3-* genotypes but varying for *Ma5/ma5* and *Ma6/ma6*. Median values for flowering time are represented by horizontal lines within boxes.

detection were calculated based on experiment specific permutations with 1000 permutations and $\alpha = 0.05$ [35].

Gene Expression Assays

Sorghum genotypes 100 M and 58 M were planted and grown in a greenhouse under long day conditions (14 h light/10 h dark) for 32 days and then transferred to growth chambers under either LD (14 h light/10 h dark) or SD (10 h light/14 h dark) conditions for seven days for entrainment prior to collection of leaf tissue. In the growth chamber, daytime (lights on) temperature was set at 30°C with a light intensity of ~300 $\mu mol \cdot s^{-1} \cdot m^{-2}$ and night (lights off) temperature was set at 23°C. Relative humidity was ~50% throughout the experiment. At day 39, leaf segments from the top three expanded leaves from three individual plants of each genotype and treatment were collected every 3 hours through one 24 h light-dark cycle and 48 h of continuous light. The leaf tissues at each time point were subjected to total RNA extraction using TRI Reagent (MRC) with the protocol for samples with high levels of polysaccharides. RNA was further purified using the RNeasy Mini kit (QIAGEN), including removal of DNA contamination by on-column DNase I digestion before reverse transcription. RNA integrity was examined on 1% MOPS gels. First-strand cDNA synthesis was performed using the SuperScript III First-Strand Synthesis System (Invitrogen) with oligo dT and random hexamer primer mix. After first-strand cDNA synthesis, the reactions were diluted to 10 ng/μl of the initial total RNA. Gene-specific qPCR reactions were carried out using Power SYBR Green PCR Master Mix (Applied Biosystems). 18S rRNA was selected as the internal control reference and the reactions were performed using the TaqMan Universal PCR Master Mix Protocol with rRNA Probe (VIC Probe) and rRNA Forward/Reverse Primer. All reactions were run on the 7900HT Fast Real-Time PCR System with SDS v2.3 software (Applied Biosystems). The specificity of each gene

specific primer set was validated by melting temperature curve analysis. Amplification efficiency of each primer sets was determined by the serial dilution method [36] (Table S3). Relative expression was determined by the comparative cycle threshold ($\Delta\Delta$Ct) method [36] with calibration from most highly expressed samples. The calculated primer efficiencies were used to adjust data for relative quantification by the efficiency correction method [37]. Each relative expression value was derived from an average of three technical replicates and three biological replicates. The individual expression data points presented as $2^{-\Delta Ct}$ [38]. The significance (p-values) of the difference in expression between genotypes were detected using Welch two sample t-test in R 3.1.0 based on three technical replicates and three biological replicates. P-values were calculated either for certain time points of the day or all time points of the day.

Results

PHYB alleles in diverse sorghum lines

Sorghum genotype 58 M, a photoperiod insensitive early flowering line, has the genotype $ma3^R ma3^R$, corresponding to the *phyB-1* allele [13]. This allele contains a frame shift mutation that results in a prematurely terminated PhyB lacking regions of the protein necessary for dimerization and biological activity. To confirm and extend prior analysis of *PHYB* diversity in sorghum, alleles from several sorghum lines that vary in photoperiod sensitivity were sequenced and compared. The coding sequence of *PHYB* from BTx623 and 100 M (both *Ma3*) was 7285 bp in length consisting of four exons encoding a protein with 1178 amino acid residues. *PHYB* sequences from R.07007, Hegari, Tx7000, BTx642, SC56, Shallu and BTx3197 were identical to BTx623 and 100 M (*Ma3*). The *PHYB* sequence from 58 M ($ma3^R$), referred to as *phyB-1* (Table 1), contains a mutation that

Table 2. Information on flowering time QTL identified in the 58MxR.07007 F2 population.

QTL	Maturity Locus	Chromosome Number	Position (cM)[a]	LOD score	Physical Interval[b]	Additive Effect[c]	Dominant Effect[d]	R²[e]
1	Ma5	Ch_1	1.8	8.66	6139583–9077991	–17.09	18.19	0.1964
2	Ma3	Ch_1	99.4	24.21	60402909–61604749	12.55	16.09	0.1408
3	Ma6	Ch_6	7.2	12.09	203707–1716581	12.83	5.81	0.1549
Total								49.21%

[a]Position of likelihood peak (highest LOD score).
[b]Physical Interval: physical coordinate interval spanning 1 LOD interval across the likelihood peak.
[c]Additive Effect: A positive value means the delay of flowering time due to R.07007 allele. A negative value means the delay of flowering time due to 58 M allele.
[d]Dominant Effect: A positive value means dominance for the delay of flowering time.
[e]R² (coefficient of determination): percentage of phenotypic variance explained by the QTL.

renders the gene inactive [13]. No coding mutations were identified in 90 M, a line that encodes the weak allele *ma3* [28]. IS3620C encodes a different allele, designated *phyB-2*, which differs from *PHYB* by one INDEL and two SNPs, resulting in one amino acid deletion and two amino acid substitutions (Table 1). The first substitution in *phyB-2* could alter function because it produces an Asp^{308}Gly change in the GAF domain of PhyB. The SIFT prediction score of this Asp^{308}Gly substitution is 0.1, indicating moderate intolerance.

PhyB affects flowering time in LD and SD

The sorghum maturity standards, 100 M and 58 M, were constructed from Milo genotypes that contain alleles of *Ma1* and *Ma3* that modify flowering time [28]. The sorghum maturity standard 100 M is photoperiod sensitive with a maturity genotype *Ma1Ma2Ma3Ma4Ma5ma6* [26]. The genotype 58 M is photoperiod insensitive, flowers early in LD and SD, and has the genotype *Ma1Ma2ma3^{R}Ma4Ma5ma6* [26]. Genotype 58 M contains null alleles of *Ma3* (*ma3^R*, *phyB-1*) and *Ma6* (*ghd7-1*). When grown in a greenhouse under 14 h LD during the summer, 58 M plants were spindly and flowered in ~62 days (±3 days), whereas 100 M flowered in ~126 days (±4 days) due to the repressing action of SbPRR37 (*Ma1*) (Figure 1A). This result confirmed that loss of PhyB activity in 58 M reduces the ability of *Ma1* to inhibit flowering in LD (p-value<<0.001) [13]. When grown in a greenhouse in 10 h SD during December–February at lower light intensity, 100 M flowered in ~59 days (±4 days) while 58 M flowered in ~48 days (±1 days) (Figure 1B). Therefore in sorghum, PhyB has a smaller but still significant effect on flowering time in SD (p-value<<0.001). The factorial ANOVA with photoperiod and *PHYB* alleles as factors indicated the effects of PhyB, day-length and PhyB:day-length interaction are all significant (p-value<<0.001) (Figure S1-A).

PHYB is epistatic to Ma1 (SbPRR37) and Ma6 (SbGHD7)

In sorghum, SbPRR37 (*Ma1*) and SbGHD7 (*Ma6*) are primary determinants of photoperiod sensitivity in *Ma3* backgrounds acting in an additive fashion to inhibit flowering in LD [26]. Expression of both genes is induced by light, although the photoreceptor or photoreceptors that mediate light signaling were not known prior to the current study [25,26]. To examine how *PHYB* (*Ma3*), SbPRR37 (*Ma1*), and SbGHD7 (*Ma6*) co-regulate the timing of floral initiation, F2 and F3 populations were derived from a cross of R.07007 (*Ma1Ma3ma5Ma6*) and 58 M (*Ma1-ma3^{R}Ma5ma6*). These populations segregated for a wide range of flowering times (~85 days) when planted in July and grown in a greenhouse in 14 h LD. Digital genotyping [33] was employed to generate DNA markers for genetic map construction. The genetic map spanned all of the ten sorghum chromosomes, although the long arms of SBI02 and SBI09 in its entirety were deficient in DNA markers. QTL analysis identified three significant QTL (LOD score>3.7) for days to anthesis in LD using the F2 population (n = 86), which together explained ~50% of the phenotypic variance for flowering time (Figure 2A). The QTL with the highest LOD score (LOD = 24.2), spanned DNA on chromosome 1 from 60,402,909–61,604,749 bp which encompasses *PHYB* (chromosome_1:60,915,677–60,917,553) (Table 2). Recessive *ma3^R* alleles from 58 M associated with this QTL caused early flowering time phenotypes. The flowering time QTL on chromosome 6 spanning a physical interval from 203,707–1,716,581 bp (1 LOD interval) aligned with SbGHD7 [26]. The recessive *ghd7-1* null allele from 58 M was associated with early flowering in LD. The third flowering time QTL near the proximal end of chromosome 1 (chromosome 1:6,139,583–9,077,991) had a

Table 3. Flowering time of F2/F3 progeny from 58MxR.07007 in LD.

Genotype (All plants = *Ma1Ma1*)			Days to Flowering: median (±SE)	Days to Flowering: range	Number of plants
*Ma3_*	*Ma5_*	*Ma6_*	115 (±5)	101–129	42
*Ma3_*	*Ma5_*	*ma6ma6*	69 (±8)	60–91	19
ma3^{R}ma3^R	*Ma5_*	*Ma6_*	57 (±8)	42–75	15
ma3^{R}ma3^R	*Ma5_*	*ma6ma6*	43 (±2)	42–50	6
*Ma3_*	*ma5ma5*	*Ma6_*	75 (±12)	44–103	52
*Ma3_*	*ma5ma5*	*ma6ma6*	46 (±6)	41–70	30
ma3^{R}ma3^R	*ma5ma5*	*Ma6_*	53 (±6)	42–76	24
ma3^{R}ma3^R	*ma5ma5*	*ma6ma6*	44 (±6)	39–68	17

LOD score of 8.7 and explained 19.6% percent of phenotype variance. This QTL was tentatively identified as *Ma5* because R.07007 was reported to be recessive for *Ma5*, a rare allele in sorghum [29]. No QTL aligned with *Ma1* as expected because both 58 M and R.07007 contain dominant alleles of *Ma1* (*SbPRR37*). The three flowering time QTL were also identified in the corresponding F3 population (data not shown).

Plants from the F2/3 population are homozygous for *Ma1*, a repressor of flowering in LD, but varied in alleles of *Ma3*, *Ma5* and *Ma6*. Three-way ANOVA was used to analyze the effect of allelic variation in three maturity genes (*Ma3*, *Ma5* and *Ma6*) on flowering time, and three two-way interactions (*Ma3:Ma5*, *Ma3:Ma6*, *Ma5:Ma6*) showed that allelic variation of the three *Ma* genes and three two-way interactions were significant (p-values<<0.001). The three two-way interaction graphs between *Ma3:Ma5*, *Ma3:Ma6* and *Ma5:Ma6* are shown in Figure S1-B–D. Progeny with the genotypes *Ma3_Ma5_Ma6_* and *Ma3_-Ma5_ma6ma6* flowered later than genotypes that were homozygous recessive for *ma3^R*, showing that *PHYB* is epistatic to the floral repressors encoded by *Ma1* and/or *Ma6* (Figure 2B; Figure S1-B,C). Progeny with the genotype *Ma3_Ma5_Ma6_* (101–129 days) flowered later than plants with the genotype *Ma3_Ma5_-ma6ma6* (60–91 days), consistent with increased floral repression due to *Ma6* in *Ma1* dominant backgrounds. The effect of *Ma6* was delay flowering with varying extents in different genetic backgrounds ranging from 14 days in *ma3^{R}ma3^{R}Ma5_*, ~29 days in *Ma3_ma5ma5*, and ~9 days in *ma3^{R}ma3^{R}ma5ma5*. Furthermore, it was noted that progeny lacking PhyB with a dominant *Ma6* allele showed a significant range of flowering times (42–75

days), suggesting that additional genes and/or environmental factors affect *Ma6* action in this genetic background (Figure 2B; Table 3). A similar wide range of flowering time (59 days) was observed among plants with the genotype *Ma3_ma5ma5Ma6_* (Figure 2C; Table 3). In addition, plants with the genotype *Ma3_Ma5_Ma6_* flowered later in LD than plants with the genotypes *Ma3ma5ma5Ma6_* or *Ma3ma5ma5ma6ma6* (Figure 2C; Figure S1-D). This shows that *Ma5* is also required for late flowering in LD in *Ma1Ma3* backgrounds and that *Ma5* is epistatic to *Ma1* and *Ma6*. Plants with the genotype *ma3^{R}ma3^R-Ma5_ma6ma6* and *Ma3_ma5ma5ma6ma6* flowered early and in a similar number of days as genotypes that are homozygous recessive for both *ma3^R* and *ma5* (*ma3^{R}ma3^{R}ma5ma5ma6ma6*) indicating that the products of both *Ma3* and *Ma5* are required in LD for delayed flowering mediated by *Ma1* (*SbPRR37*).

The requirement for both PhyB and the product of *Ma5* to observe delayed flowering in LD led us to examine the *Ma5* locus for candidate genes that might explain this interaction. The *Ma5* locus is located on SBI-01 and spans a large number of genes including several genes known to affect flowering time in other plants, including *AP1*, *CK2*, and *PHYC*. *PHYC* appeared to be the best candidate gene for *Ma5* because PhyC modifies flowering time in rice specifically in LD, similar to *Ma5* in sorghum [39], PhyB stabilizes PhyC, and PhyB:PhyC act as heterodimers in both Arabidopsis [40,41] and rice [39], consistent with the co-dependence observed between *PHYB* and *Ma5* in this study. Comparison of *PHYC* sequences from BTx623 (*Ma5*), 100 M (*Ma5*), and R.07007 (*ma5*) revealed four differences in PhyC amino acid sequence between BTx623 and R.07007, and two

Table 4. Sequence analysis of *PHYC* coding alleles in different sorghum lines.

	Exon 1	Exon 1	Exon 1	Exon 2	Sorghum Genotypes
Nucleotide Variation	G>T	G>A	T>C	G>T	
Protein Modification	Gly>Val	Gly>Arg	Val>Ala	Glu>Asp	
Mutation Position (AA #)	124	162	190	922	
Alignment with PHYB in *Arabidopsis* (AA #)	160	198	226	954	
Phytochrome Domain	PAS(N)	PAS(N)	PAS-GAF Loop	HKRD(C)	
PHYC-1 (Ma5)	–	–	–	–	BTx623
PHYC-2	–	+	+	–	100 M, 90 M
phyC-1 (ma5)	+	+	+	+	R.07007

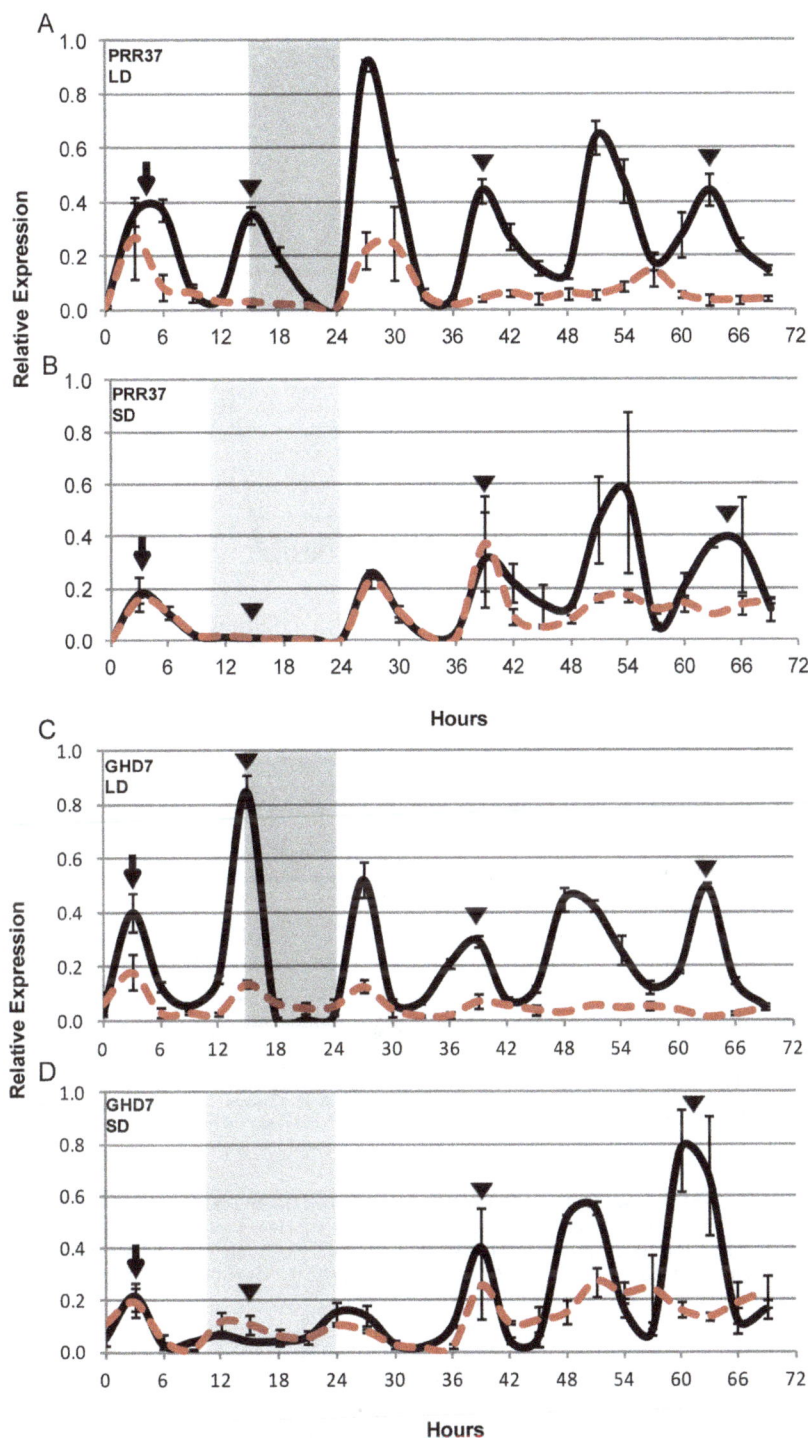

Figure 3. Relative expression of *SbPRR37* **and** *SbGHD7* **in 100 M (***Ma3/PHYB***) and 58 M (***ma3^R/phyB-1***) in LD and SD.** 100 M (solid black line) and 58 M (dashed red line) plants were entrained LD (14 h light/10 h dark) or SD (10 h light/14 h dark) and sampled for one 24 h cycle, followed by 48 h in LL (continuous light and temperature). The grey background corresponds to time when plants are in darkness. Relative gene expression was determined every 3 hours by qRT-PCR. Arrows represent morning peaks of expression and arrowheads represent evening peaks of expression. (A) In LD, the second peak (arrowhead) of *SbPRR37* expression in the evening (~15 h) is missing in the *phyB* deficient line, 58 M. (B) In SD, the second peak (arrowhead) of *SbPRR37* is absent in both 100 M and 58 M. (C) In LD, the second peak (arrowhead) of *SbGHD7* expression in the evening (~15 h) is attenuated in 58 M. (D) In SD, the second peak of *SbGHD7* is attenuated in both 100 M and 58 M. Each data point of relative expression was based on data from three technical replicates and three biological replicates. Error bars indicate SEM.

differences between 58 M/100 M and R.07007 (Table 4). The latter amino acid variants occur in the PAS domain (Gly:Val) and

HKRD domain (Glu:Asp) and SIFT analysis [42] indicated these changes could affect the function of PhyC. These results are

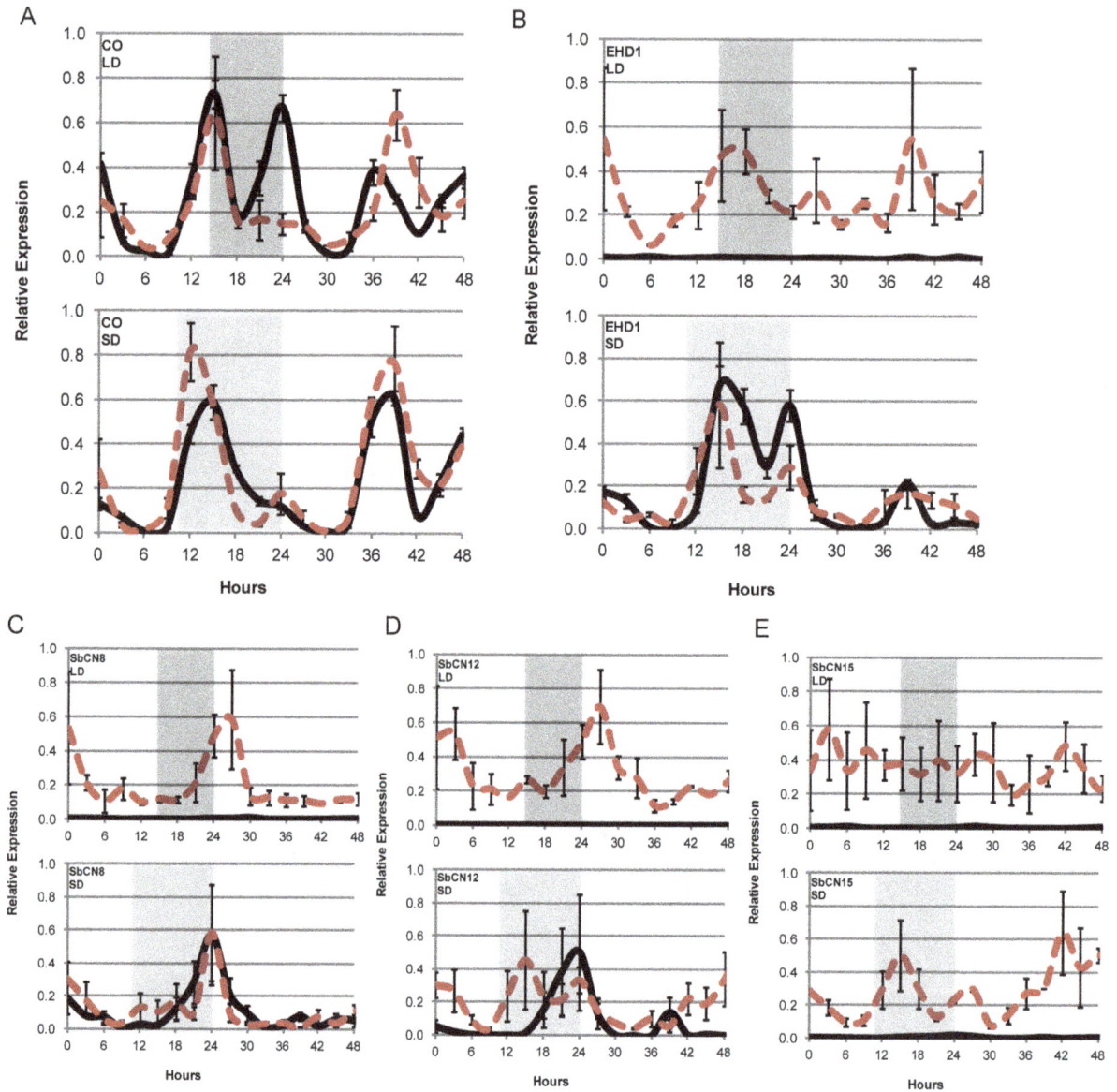

Figure 4. Expression of *SbCO, SbEhd1, SbCN8/12/15* **in 100 M (***Ma3/PHYB***) and 58 M (***ma3^R^/phyB-1***) in LD and SD.** Relative RNA levels in leaves of 100 M (solid black lines) and 58 M (dashed red lines) entrained and sampled in LD (14 h light/10 h dark) or SD (10 h light/14 h dark) for 24 h followed by 24 h in LL (continuous light and temperature). Relative expression levels were determined every 3 hours by qRT-PCR analysis. The gray shaded areas represent the dark periods. (A) *SbCO*, (B) *SbEHD1*, (C) *SbCN8*, (D) *SbCN12*, (E) *SbCN15*. Each data point of relative expression is based on three technical replicates and three biological replicates. Error bars indicate SEM.

consistent with *PHYC* as the candidate gene for *Ma5*. Further analysis is underway to test this assignment.

PhyB modulates expression of *SbPRR37* and *SbGHD7* in long days

Expression of *SbPRR37* and *SbGHD7* in leaves is regulated by light and gating by the circadian clock [25,26]. The influence of PhyB on *SbPRR37* and *SbGHD7* expression was analyzed using 100 M (*PHYB*) and 58 M (*phyB-1*) plants grown for 32 days in LD then entrained for 7 days in LD or SD (Figure 3). Following entrainment, leaf samples were collected from plants for one 24 h LD or SD light-dark cycle, then from plants exposed to continuous light and temperature for an additional 48 h. In leaves of 100 M, *SbPRR37* and *SbGHD7* expression peaked in the morning

(arrow) and evening (arrowhead) in LD as previously reported [25,26] (Figure 3A/C, solid lines). *SbPRR37* and *SbGHD7* RNA abundance continued to oscillate with peaks in the morning and evening when 100 M plants were transferred to continuous light and temperature consistent with regulation by the circadian clock (Figure 3, 24–72 h). In leaves of 58 M in LD (Figure 3A/C, dashed red lines), *SbPRR37* and *SbGHD7* showed an increase in RNA abundance in the morning (arrow) but only a small increase in expression in the evening (arrowhead) compared to 100 M (Figure 3A, p-value<0.1; Figure 3C, p-value<0.05). These results indicate that PhyB is required for elevated evening expression of *SbPRR37* and *SbGHD7* in LD in 100 M.

When 100 M and 58 M plants were entrained and assayed in SD, the morning peak of *SbPRR37* expression was of similar amplitude in both genotypes and expression of *SbPRR37* was low

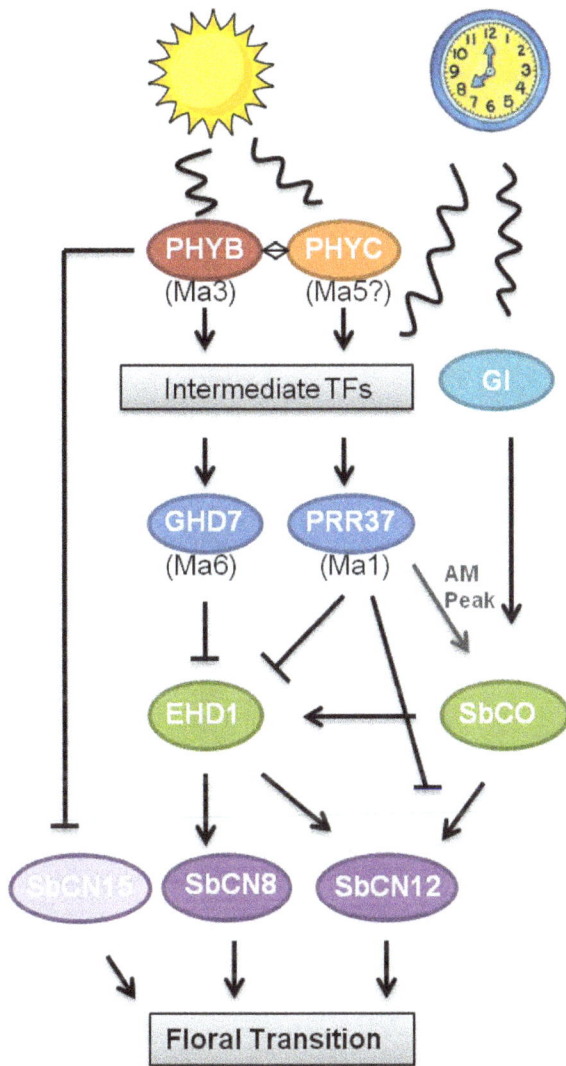

Figure 5. Model of the photoperiod flowering time pathway in sorghum. Phytochrome B (PhyB) is mediates light signaling that modulates flowering time in response to photoperiod in sorghum. In LD, PhyB up-regulates the expression of *PRR37* and *GHD7*, two central floral repressors, during the evening phase of LD but with minimal influence in SD. Induction at this time of day is also dependent on output from the circadian clock. PhyB may stabilize and interact with PhyC, a candidate gene for *Ma5* a locus that also contributes to photoperiod regulation of flowering time. SbPRR37 activates *SbCO* expression peaking at dawn. SbPRR37 and SbGhd7 repress expression of the floral inductors *SbEHD1, SbCN8, SbCN12* and *SbCN15*, leading to delayed flowering in long days. In SD or 58 M (*phyB-1*), expression of the floral repressors SbPRR37 and SbGHD7 is reduced which results in floral initiation once plants have satisfied other requirements for flowering. PhyB was found to mediate repression of *SbCN15* regardless of day length.

during the evening (Figure 3B). Similarly, *SbGHD7* expression in SD was highest in the morning, reaching similar levels in 100 M and 58 M, and lower in the evening when compared to expression levels measured in LD (Figure 3D). These results indicate that in SD, PhyB has a limited effect on *SbPRR37* and *SbGHD7* expression. When 100 M plants entrained in SD were exposed to continuous light, the evening peak of *SbPRR37* and *SbGHD7* expression observed in LD reappeared on the first subjective day

and expression levels were also elevated in the second subjective day (Figure 3B/D). In 58 M, the evening peak of *SbPRR37* and *SbGHD7* reappeared during the first subjective day, however overall expression was attenuated relative to 100 M during the second subjective day.

PhyB modulates expression of *CO, Ehd1, SbCN8, SbCN12* and *SbCN15*

In 100 M entrained to LD, the sorghum ortholog of *CONSTANS* (*SbCO*) shows peaks of expression at dawn (24 h) and in the evening (15 h) that are regulated by *SbPRR37*, the circadian clock, and day length [25]. In 58 M entrained and sampled in LD, the amplitude of the peak of *SbCO* expression at dawn (24 h) was reduced compared to 100 M (Figure 4A, p-value<0.05). The peak of *SbCO* expression at dawn was also reduced and of similar amplitude in plants entrained and sampled in SD (Figure 4A, lower). These results show that the peak of *SbCO* expression at dawn is dependent on PhyB, most likely because expression of *SbPRR37* in the evening of LD is dependent on PhyB (Figure 3A). In contrast, the evening peak (15 h) of *SbCO* expression was similar in both LD and SD in 100 M and 58 M indicating that PhyB does not significantly modulate *SbCO* expression at this time (15 h) of day.

EHD1 is an activator of *Hd3a*, one of the florigens in rice [43]. The sorghum ortholog of *Hd3a* is *SbCN15*. Expression of *SbEHD1* increases when 100 M is transferred from LD to SD in parallel with increased expression of *SbCN8* (ortholog of *ZCN8* [24]) and *SbCN12* (ortholog of *ZCN12*) that have been proposed to encode florigens in sorghum [19,25,26]. *SbPRR37* and *SbGHD7* repress expression of *SbEHD1* in 100 M entrained in LD [25,26]. Therefore *SbEHD1* expression in 58 M and 100 M was quantified and compared to determine if PhyB modulates *SbEHD1* expression. In LD, *SbEHD1* RNA abundance peaked in the evening and was up to ~100-fold higher in 58 M relative to 100 M throughout the time course (Figure 4B, upper; Figure S2-A, p-value<<0.001). In SD, expression of *SbEHD1* was high in both genotypes and peaked during the night (Figure 4B, lower; Figure S2-A).

In 58 M entrained and analyzed in LD, expression of *SbCN8* (Figure 4C, upper) and *SbCN12* (Figure 4D, upper) peaked early in the morning and the relative abundance of RNA derived from these genes was elevated more than ~100-fold relative to their levels in 100 M (Figure S2-B/C, p-values<<0.001). In SD, *SbCN8* (Figure 4C, lower) and *SbCN12* (Figure 4D, lower) expression was similar in both genotypes. Similarly, *SbCN15* (*Hd3a*) expression was increased up to ~60-fold in 58 M compared to 100 M in LD and SD (Figure 4E; Figure S2-D, p-values<<0.001) at all time points assayed, indicating that PhyB mediated repression of *SbCN15* expression occurs regardless of photoperiod.

PhyB could be inducing *SbPRR37* and *SbGHD7* expression directly, and/or indirectly by altering output from the circadian clock. To determine if allelic variation in *PHYB* affected clock gene expression, *TOC1* and *LHY/CCA1*, the central oscillators, and *GI*, a mediator of clock output were examined (Figure S3). In LD and SD, *TOC1, LHY* and *GI* expression in 58 M and 100 M peaked at similar times and most of these genes showed similar amplitude of expression, although expression of GI was approximately 2-fold lower in 58 M. Although three biological replications at the indicated time points may not be sufficient to detect all biologically significant variation present, the small fold differences of circadian clock genes do not appear sufficient to explain the large variation in *SbPRR37* and *SbGHD7* expression observed in

Ma3 vs. *ma3^R* backgrounds. *PHYB* and *PHYC* RNA levels were similar in 100 M and 58 M plants in LD and SD (data not shown).

Discussion

Sorghum genotypes used for grain production are typically photoperiod insensitive and flower in 55–75 days when planted in April in locations such as College Station, Texas where day lengths increase during the early portion of the growing season. Early flowering in grain sorghum helps avoid adverse weather and insect pressure during the reproductive phase, thereby enhancing yield. In contrast, highly photoperiod sensitive energy sorghum genotypes planted in this same location will not initiate flowering for 175 days until mid-September when day lengths decrease to less than 12.2 h [1,29]. Delayed flowering results in long duration of vegetative growth of energy sorghum, increasing biomass yield [8] and nitrogen use efficiency [8]. The importance of optimal flowering time for sorghum productivity led us to investigate the genetic and molecular basis of variation in this trait in sorghum.

Variation of flowering time of sorghum germplasm grown in LD environments is caused principally by differences in photoperiod sensitivity, although shading, GA, temperature, length of the juvenile phase among other factors also affect this trait [7]. A model summarizing information about photoperiod regulation of flowering time in sorghum is shown in Figure 5. In LD, flowering is delayed in photoperiod sensitive sorghum by the additive action of the floral repressors, SbPRR37 (*Ma1*) and SbGhd7 (*Ma6*) [25,26,28,29]. SbPRR37 and SbGhd7 repress expression of the grass specific floral activator, *SbEHD1*. In addition, SbPRR37 inhibits the activity of CO, another activator of flowering in sorghum [19]. The floral activators, SbEhd1 and SbCO, induce expression of *SbCN8* and *SbCN12*, the proposed sources of FT in sorghum. *SbCN15*, the ortholog of *Hd3a* and a source of florigen in rice [21], may also be a source of florigen in sorghum. The circadian clock is shown regulating expression of *SbGI*, *SbCO*, *SbPRR37* and *SbGHD7*, and light regulating expression of *SbGHD7* and *SbPRR37* as shown in previous studies [25,26].

Photoperiod has minimal impact on flowering time in sorghum genotypes such as SM100 that encode null versions of *SbPRR37* and *SbGHD7* [25,26]. Presence of functional alleles of either gene increases photoperiod sensitivity and a further delay in flowering is observed when both genes are present in dominant *Ma3Ma5* backgrounds. Expression of *SbPRR37* and *SbGHD7* is regulated by light and the circadian clock. Both genes show peaks of RNA abundance in the morning and again in the evening in LD and both peaks of RNA are attenuated in darkness. Importantly, the evening peak of expression is attenuated in SD when this phase occurs in darkness, indicating a requirement for light signaling during the evening to maintain sufficiently high levels expression of *SbPRR37* and *SbGHD7* to inhibit flowering. The morning and evening peaks of *SbPRR37* and *SbGHD7* expression observed in sorghum in LD is a pattern of expression first observed in photoperiod versions of this C4 grass. In Arabidopsis, *PRR7*, the ortholog of *SbPRR37*, shows a single peak of clock-regulated expression during the morning [44]. In rice, *SbGHD7* shows a single peak of clock-gated expression in the morning of LD [22]. It will be interesting to determine if the dual peak pattern of *PRR37* and *GHD7* expression observed in sorghum is found in other related C4 grasses such as pearl millet, Miscanthus and sugarcane.

The current study focused on characterizing the light-signaling pathway that regulates *SbPRR37* and *SbGHD7* expression in response to day length. Previous studies showed that sorghum genotypes lacking *PHYB* (58 M, *phyB-1*) flower earlier in LD compared to near isogenic genotypes (100 M) expressing *PHYB*,

demonstrating that light signaling through this photoreceptor is required for photoperiod sensitive variation in flowering time [13]. The current study showed that PhyB (*Ma3*) is epistatic to genes encoding the floral repressors SbPRR37 and SbGhd7 and that PhyB is required for photoperiod-regulated expression of these genes. Moreover, 58 M, a genotype lacking functional PhyB, showed attenuated expression of *SbPRR37* and *SbGHD7* during the evening of LD compared to 100 M (PhyB). In SD, expression of the floral repressors was similar in 58 M and 100 M. Taken together, these results indicate that in sorghum PhyB is required for light signaling in LD that results in elevated expression of *SbPRR37* and *SbGHD7* during the evening.

The molecular basis of PhyB induced expression of *SbPRR37* and *SbGHD7* during the evening of long days is unknown but could involve other photoreceptors and intermediary transcription factors such as PIFs [45]. Detailed studies in rice showed that PhyA, PhyB and PhyC modulate flowering time [39]. PhyC in particular plays a role in natural variation of flowering time in pearl millet [46], Arabidopsis [47], and wheat [48]. In Arabidopsis, a long day plant, PhyB destabilizes CO, an action countered by Cry, PhyA and SPA in LD, leading to floral induction [20]. In rice, *phyB* mutants flower early in LD and SD similar to sorghum. Interestingly, rice *phyC* mutants flower early only in LD [39]. In addition, in rice, both PhyB and PhyC are required to induce *GHD7* expression, where PhyB alone causes some repression of *GHD7* mRNA levels [49]. This indicates that in rice PhyB regulates floral induction in both LD and SD, while PhyC modifies flowering time selectively in LD. The stability of PhyC is reduced in the absence of PhyB in rice and Arabidopsis [40]. PhyB increases PhyC stability, and chromophore-containing PhyB:PhyC heterodimers are required for PhyC activity [41]. Therefore, in sorghum the requirement for PhyB in photoperiod sensitive flowering time may be because PhyB increases PhyC stability and through formation of PhyB:PhyC heterodimers.

Genetic analysis of the role of *PHYB* in sorghum was examined using a population dominant for *Ma1* (*SbPRR37*) and segregating for alleles of *PHYB* (*Ma3*), *Ma5*, and *SbGHD7* (*Ma6*). The presence of *Ma1* in all progeny of the population caused delayed flowering in LD unless the expression or activity of *Ma1* (and in some genotypes *Ma1* and *Ma6*) was altered by recessive alleles of *Ma3* or *Ma5*. The analysis showed that plants homozygous for null alleles of *PHYB* (*phyB-1*) in *Ma5_* backgrounds had reduced photoperiod sensitivity and flowered earlier in LD compared to plants encoding PhyB. Similarly, progeny homozygous for recessive alleles of *Ma5*, in *Ma3_* backgrounds, showed reduced photoperiod sensitivity and flowered earlier in LD. The results indicated that both *PHYB* and *Ma5* are epistatic to *Ma1* and *Ma6*. Progeny recessive for either gene flowered earlier in LD, but showed a range of flowering times, indicating that other genes and/or environmental factors affected flowering time in these backgrounds, although with reduced response to photoperiod. Interestingly, *PHYB* and *Ma5* appear to be co-dependent or acting at a similar point in the regulatory pathway because allelic differences at *Ma5* did not affect flowering time significantly in *phyB-1* backgrounds and vice versa. R.07007 (*Ma3ma5*) and 58 M (*ma3^R Ma5*) show attenuated expression of *SbPRR37* and *SbGHD7* in the evening of LD ([25] and this study) indicating that both *Ma3* (PhyB) and *Ma5* are required for elevated expression of the sorghum floral repressors during the evening of LD. In searching for an explanation for this co-dependence, we found the *Ma5* locus spans several genes known to affect flowering time including *PHYC* and that the sequence of PhyC in R.07007 (*ma5*) contained amino acid changes that could potentially modify the function of this protein. The hypothesis that *Ma5* corresponds to

PHYC is consistent with studies showing that PhyC modifies flowering in an LD specific manner in rice, similar to *Ma5* [39]. In addition, PhyC stability is dependent in part on PhyB and PhyC activity requires the formation of functional heterodimers with PhyB (and other phytochromes) [41]. If sorghum PhyC is regulated by PhyB in a manner similar to their counterparts in rice, this would explain why *Ma5* (presumptive *PHYC*) activity is not observed in *phyB-1* backgrounds. Experiments designed to test this hypothesis are currently underway.

In Arabidopsis, *CO* expression peaks once per day in the evening and the amplitude of *CO* expression is regulated by blue light/GI-FKF1-ZTL mediated turnover of CDF1, a repressor of *CO* expression [50]. PRR7 also modifies *CO* expression through repression of *CDF1* expression [51]. In sorghum, *SbCO* expression peaks twice each day, at dawn and again in the evening in LD. The peak of *SbCO* expression at dawn is attenuated in SD ([25] and this study) and in genetic backgrounds lacking SbPRR37 [19]. It is possible that SbPRR37 modulates *SbCO* expression by repressing sorghum orthologs of *CDF1* as occurs in Arabidopsis [51]. The peak of *SbCO* expression at dawn in LD was not observed in the sorghum genotype lacking PhyB (58 M). Since PhyB is required for elevated *SbPRR37* expression in the evening of LD, and SbPRR37 has been shown to induce elevated expression of *SbCO* at dawn, it is likely that lack of PhyB induced expression of *SbPRR37* during the evenings of LD explains the observed expression of *SbCO* in 58 M.

In rice, *Hd3a*, a member of the PEBP gene family, encodes an FT protein that acts as a florigen [52]. In maize, *ZCN8* and possibly *ZCN12* are sources of florigen [24,53]. Sorghum encodes orthologs of *Hd3a* (*SbCN15*), *ZCN8* (*SbCN8*) and *ZCN12* (*SbCN12*). *SbCN8* and *SbCN12* expression is regulated by day length and by alleles of *SbPRR37*, *SbGHD7*, and *PHYB* in a manner consistent with these genes being sources of florigen in sorghum. In prior studies, *SbCN15* expression was modulated to only a small extent by variation in photoperiod and in mutants of *SbPRR37* and *SbGHD7* that affect flowering time, suggesting that this gene was not an important target of photoperiod regulation [25,26]. In the current study, expression of *SbCN15* was found to be ~60-fold higher in leaves of 58 M (*phyB-1*) compared to 100 M (*PHYB*) in both LD and SD. If *SbCN15* functions as a source of florigen as in rice, photoperiod independent repression of *SbCN15* expression by PhyB suggests that this gene may be responsible for early flowering induced by shading [7]. 58 M plants exhibit shade avoidance responses including longer leaf blades and sheaths, fewer tillers, narrower leaf blades, less leaf area, and more rapid stem elongation [7]. In Arabidopsis, light signaling through PhyB represses shade avoidance responses, and PhyB deficient mutants have elongated stems and an early flowering phenotype associated with "constitutive shade avoidance" [54]. Information on photoperiod regulated flowering time

in sorghum described in this paper will hopefully facilitate analysis of flowering time variation caused by shading and other environmental factors.

Supporting Information

Figure S1 ANOVA interaction graphs showing (A) Day-length:PhyB (Day:Genotype) interaction. (B–D) Three two-way interactions (*Ma3:Ma5*, *Ma3:Ma6*, *Ma5:Ma6*) in the 58MxR.07007 F2/F3 population.

Figure S2 Fold differences of *SbEHD1*, *SbCN8*, *SbCN12* and *SbCN15* RNA abundance at peaks of expression in 100 M and 58 M grown in LD (14 h light/10 h dark) or SD (10 h light/14 h dark). Positive fold difference values indicate higher mRNA levels detected in 58 M. (A) *SbEHD1*, (B) *SbCN8*, (C) *SbCN12*, (D) *SbCN15*. The time point corresponding to peak expression is shown below each graph.

Figure S3 Relative expression levels of circadian clock genes and *GI* in 100 M (black solid line) and 58 M (red dashed line) under either LD (14 h light/10 h dark) or SD (10 h light/14 h dark) conditions. The gray shaded area represents the dark period. The first 24 h covers one light-dark cycle, followed by 24 h of continuous light. (A) GI. (B) TOC1. (C) LHY. Each data point of relative expression corresponds to three technical replicates and three biological replicates. Error bars indicates SEM.

Table S1 Genotypes and flowering dates of sorghum lines.

Table S2 Primer sequences used for *PHYB* alleles amplification and sequencing.

Table S3 Primer sequences and amplification efficiency for qRT-PCR.

Acknowledgments

The authors would like to thank Susan Hall for help with plant management.

Author Contributions

Conceived and designed the experiments: SY JEM. Performed the experiments: SY RLM DTM. Analyzed the data: SY RLM. Contributed reagents/materials/analysis tools: PEK WLR. Contributed to the writing of the manuscript: SY DTM JEM.

References

1. Rooney WL, Blumenthal J, Bean B, Mullet JE (2007) Designing sorghum as a dedicated bioenergy feedstock. Biofuels, Bioproducts and Biorefining 1: 147–157.

2. Srikanth A, Schmid M (2011) Regulation of flowering time: all roads lead to Rome. Cell Mol Life Sci 68: 2013–2037.

3. Welch SM, Dong Z, Roe JL (2004) Modelling gene networks controlling transition to flowering in Arabidopsis; 2004 26 Sep–1 Oct Brisbane, Australia. CDROM. 1–20.

4. Greenup A, Peacock WJ, Dennis ES, Trevaskis B (2009) The molecular biology of seasonal flowering-responses in Arabidopsis and the cereals. Ann Bot 103: 1165–1172.

5. Andres F, Coupland G (2012) The genetic basis of flowering responses to seasonal cues. Nat Rev Genet 13: 627–639.

6. Nozue K, Covington MF, Duek PD, Lorrain S, Fankhauser C, et al. (2007) Rhythmic growth explained by coincidence between internal and external cues. Nature 448: 358–361.

7. Morgan PW, Finlayson SA (2000) Physiology and Genetics of Maturity and Height. In: Smith CW, Frederiksen RA, editors. Sorghum: Origin, History, Technology, and Production. New York: Wiley Series in Crop Science. 240–242.

8. Olson SN, Ritter K, Rooney W, Kemanian A, McCarl BA, et al. (2012) High biomass yield energy sorghum: developing a genetic model for C4 grass bioenergy crops. Biofuels, Bioproducts and Biorefining 6: 640–655.

9. Jiao Y, Lau OS, Deng XW (2007) Light-regulated transcriptional networks in higher plants. Nat Rev Genet 8: 217–230.

10. Kami C, Lorrain S, Hornitschek P, Fankhauser C (2010) Light-Regulated Plant Growth and Development. Current Topics in Developmental Biology: Elsevier Inc. 29–66.

11. Izawa T, Oikawa T, Sugiyama N, Tanisaka T, Yano M, et al. (2002) Phytochrome mediates the external light signal to repress FT orthologs in photoperiodic flowering of rice. Genes Dev 16: 2006–2020.

12. Hanumappa M, Pratt LH, Cordonnier-Pratt MM, Deitzer GF (1999) A photoperiod-insensitive barley line contains a light-labile phytochrome B. Plant Physiol 119: 1033–1040.

13. Childs KL, Miller FR, Cordonnier-Pratt MM, Pratt LH, Morgan PW, et al. (1997) The sorghum photoperiod sensitivity gene, Ma3, encodes a phytochrome B. Plant Physiol 113: 611–619.

14. Quail PH, Briggs WR, Chory J, Hangarter RP, Harberd NP, et al. (1994) Spotlight on Phytochrome Nomenclature. Plant Cell 6: 468–471.

15. Nagatani A (2010) Phytochrome: structural basis for its functions. Curr Opin Plant Biol 13: 565–570.

16. Pruneda-Paz JL, Kay SA (2010) An expanding universe of circadian networks in higher plants. Trends Plant Sci 15: 259–265.

17. Robson F, Costa MM, Hepworth SR, Vizir I, Pineiro M, et al. (2001) Functional importance of conserved domains in the flowering-time gene CONSTANS demonstrated by analysis of mutant alleles and transgenic plants. Plant J 28: 619–631.

18. Yano M, Katayose Y, Ashikari M, Yamanouchi U, Monna L, et al. (2000) Hd1, a Major Photoperiod Sensitivity Quantitative Trait Locus in Rice, Is Closely Related to the Arabidopsis Flowering Time Gene CONSTANS. Plant Cell 12: 2473–2483.

19. Yang S, Weers B, Morishige D, Mullet J (2014) CONSTANS is a photoperiod regulated activator of flowering in sorghum. BMC Plant Biology 14: 148.

20. Turck F, Fornara F, Coupland G (2008) Regulation and identity of florigen: FLOWERING LOCUS T moves center stage. Annu Rev Plant Biol 59: 573–594.

21. Tsuji H, Taoka K, Shimamoto K (2011) Regulation of flowering in rice: two florigen genes, a complex gene network, and natural variation. Curr Opin Plant Biol 14: 45–52.

22. Itoh H, Nonoue Y, Yano M, Izawa T (2010) A pair of floral regulators sets critical day length for Hd3a florigen expression in rice. Nat Genet 42: 635–638.

23. Yan L, Loukoianov A, Blechl A, Tranquilli G, Ramakrishna W, et al. (2004) The wheat VRN2 gene is a flowering repressor down-regulated by vernalization. Science 303: 1640–1644.

24. Meng X, Muszynski MG, Danilevskaya ON (2011) The FT-like ZCN8 Gene Functions as a Floral Activator and Is Involved in Photoperiod Sensitivity in Maize. Plant Cell 23: 942–960.

25. Murphy RL, Klein RR, Morishige DT, Brady JA, Rooney WL, et al. (2011) Coincident light and clock regulation of pseudoresponse regulator protein 37 (PRR37) controls photoperiodic flowering in sorghum. Proc Natl Acad Sci U S A 108: 16469–16474.

26. Murphy RL, Morishige DT, Brady JA, Rooney WL, Yang S, et al. (2014) Ghd7 (Ma6) Represses Flowering in Long Days: A Key Trait in Energy Sorghum Hybrids. The Plant Genome In press.

27. Mace ES, Hunt CH, Jordan DR (2013) Supermodels: sorghum and maize provide mutual insight into the genetics of flowering time. Theor Appl Genet 126: 1377–1395.

28. Quinby JR (1974) Sorghum improvement and the genetics of growth. College Station: Texas A&M Univ. Press.

29. Rooney W, Aydin S (1999) Genetic control of a photoperiod-sensitive response in Sorghum bicolor (L.) Moench. Crop Science 39: 397–400.

30. Beales J, Turner A, Griffiths S, Snape JW, Laurie DA (2007) A pseudo-response regulator is misexpressed in the photoperiod insensitive Ppd-D1a mutant of wheat (Triticum aestivum L.). Theor Appl Genet 115: 721–733.

31. Turner A, Beales J, Faure S, Dunford RP, Laurie DA (2005) The pseudo-response regulator Ppd-H1 provides adaptation to photoperiod in barley. Science 310: 1031–1034.

32. Koo BH, Yoo SC, Park JW, Kwon CT, Lee BD, et al. (2013) Natural Variation in OsPRR37 Regulates Heading Date and Contributes to Rice Cultivation at a Wide Range of Latitudes. Mol Plant 6: 1877–1888.

33. Morishige DT, Klein PE, Hilley JL, Sahraeian SM, Sharma A, et al. (2013) Digital genotyping of sorghum - a diverse plant species with a large repeat-rich genome. BMC Genomics 14: 448.

34. Wang S, Basten CJ, Zeng Z-B (2012) Windows QTL Cartographer 2.5. Department of Statistics, North Carolina State University, Raleigh, NC.

35. Churchill GA, Doerge RW (1994) Empirical threshold values for quantitative trait mapping. Genetics 138: 963–971.

36. Bookout AL, Mangelsdorf DJ (2003) Quantitative real-time PCR protocol for analysis of nuclear receptor signaling pathways. Nucl Recept Signal 1: 1–7.

37. Pfaffl MW (2001) A new mathematical model for relative quantification in real-time RT-PCR. Nucleic Acids Res 29: 2002–2007.

38. Schmittgen TD, Livak KJ (2008) Analyzing real-time PCR data by the comparative CT method. Nature Protocols 3: 1101–1108.

39. Takano M, Inagaki N, Xie X, Yuzurihara N, Hihara F, et al. (2005) Distinct and cooperative functions of phytochromes A, B, and C in the control of deetiolation and flowering in rice. Plant Cell 17: 3311–3325.

40. Monte E, Alonso JM, Ecker JR, Zhang Y, Li X, et al. (2003) Isolation and characterization of phyC mutants in Arabidopsis reveals complex crosstalk between phytochrome signaling pathways. Plant Cell 15: 1962–1980.

41. Clack T, Shokry A, Moffet M, Liu P, Faul M, et al. (2009) Obligate heterodimerization of Arabidopsis phytochromes C and E and interaction with the PIF3 basic helix-loop-helix transcription factor. Plant Cell 21: 786–799.

42. Kumar P, Henikoff S, Ng PC (2009) Predicting the effects of coding non-synonymous variants on protein function using the SIFT algorithm. Nat Protoc 4: 1073–1081.

43. Doi K, Izawa T, Fuse T, Yamanouchi U, Kubo T, et al. (2004) Ehd1, a B-type response regulator in rice, confers short-day promotion of flowering and controls FT-like gene expression independently of Hd1. Genes Dev 18: 926–936.

44. Nakamichi N, Kiba T, Henriques R, Mizuno T, Chua NH, et al. (2010) PSEUDO-RESPONSE REGULATORS 9, 7, and 5 are transcriptional repressors in the Arabidopsis circadian clock. Plant Cell 22: 594–605.

45. Leivar P, Quail PH (2011) PIFs: pivotal components in a cellular signaling hub. Trends Plant Sci 16: 19–28.

46. Vigouroux Y, Mariac C, De Mita S, Pham JL, Gerard B, et al. (2011) Selection for earlier flowering crop associated with climatic variations in the Sahel. PLoS One 6: e19563.

47. Balasubramanian S, Sureshkumar S, Agrawal M, Michael TP, Wessinger C, et al. (2006) The PHYTOCHROME C photoreceptor gene mediates natural variation in flowering and growth responses of Arabidopsis thaliana. Nat Genet 38: 711–715.

48. Distelfeld A, Dubcovsky J (2010) Characterization of the maintained vegetative phase deletions from diploid wheat and their effect on VRN2 and FT transcript levels. Mol Genet Genomics 283: 223–232.

49. Osugi A, Itoh H, Ikeda-Kawakatsu K, Takano M, Izawa T (2011) Molecular dissection of the roles of phytochrome in photoperiodic flowering in rice. Plant Physiol 157: 1128–1137.

50. Imaizumi T, Schultz TF, Harmon FG, Ho LA, Kay SA (2005) FKF1 F-box protein mediates cyclic degradation of a repressor of CONSTANS in Arabidopsis. Science 309: 293–297.

51. Imaizumi T (2010) Arabidopsis circadian clock and photoperiodism: time to think about location. Curr Opin Plant Biol 13: 83–89.

52. Tamaki S, Matsuo S, Wong HL, Yokoi S, Shimamoto K (2007) Hd3a protein is a mobile flowering signal in rice. Science 316: 1033–1036.

53. Danilevskaya ON, Meng X, Hou Z, Ananiev EV, Simmons CR (2008) A genomic and expression compendium of the expanded PEBP gene family from maize. Plant Physiol 146: 250–264.

54. Franklin KA, Quail PH (2010) Phytochrome functions in Arabidopsis development. J Exp Bot 61: 11–24.

11

Transcriptome Wide Identification and Validation of Calcium Sensor Gene Family in the Developing Spikes of Finger Millet Genotypes for Elucidating Its Role in Grain Calcium Accumulation

Uma M. Singh[1¤], Muktesh Chandra[1], Shailesh C. Shankhdhar[2], Anil Kumar[1]*

1 Department of Molecular Biology and Genetic Engineering, Govind Ballabh Pant University of Agriculture and Technology, Pantnagar, Uttarakhand, India, **2** Department of Plant Physiology, Govind Ballabh Pant University of Agriculture and Technology, Pantnagar, Uttarakhand, India

Abstract

Background: In finger millet, calcium is one of the important and abundant mineral elements. The molecular mechanisms involved in calcium accumulation in plants remains poorly understood. Transcriptome sequencing of genetically diverse genotypes of finger millet differing in grain calcium content will help in understanding the trait.

Principal Finding: In this study, the transcriptome sequencing of spike tissues of two genotypes of finger millet differing in their grain calcium content, were performed for the first time. Out of 109,218 contigs, 78 contigs in case of GP-1 (Low Ca genotype) and out of 120,130 contigs 76 contigs in case of GP-45 (High Ca genotype), were identified as calcium sensor genes. Through *in silico* analysis all 82 unique calcium sensor genes were classified into eight calcium sensor gene family viz., CaM & CaMLs, CBLs, CIPKs, CRKs, PEPRKs, CDPKs, CaMKs and CCaMK. Out of 82 genes, 12 were found diverse from the rice orthologs. The differential expression analysis on the basis of FPKM value resulted in 24 genes highly expressed in GP-45 and 11 genes highly expressed in GP-1. Ten of the 35 differentially expressed genes could be assigned to three documented pathways involved mainly in stress responses. Furthermore, validation of selected calcium sensor responder genes was also performed by qPCR, in developing spikes of both genotypes grown on different concentration of exogenous calcium.

Conclusion: Through *de novo* transcriptome data assembly and analysis, we reported the comprehensive identification and functional characterization of calcium sensor gene family. The calcium sensor gene family identified and characterized in this study will facilitate in understanding the molecular basis of calcium accumulation and development of calcium biofortified crops. Moreover, this study also supported that identification and characterization of gene family through Illumina paired-end sequencing is a potential tool for generating the genomic information of gene family in non-model species.

Editor: Manoj Prasad, National Institute of Plant Genome Research, India

Funding: The authors acknowledge the Department of Biotechnology, Government of India, for providing financial support in the form of Programme Support for research and development in Agricultural Biotechnology (Grant No. BT/PR7849/AGR/02/374/2006-Part II). The authors also acknowledge ICAR, New Delhi for providing funding to Uma M. Singh as a Senior Research Fellowship. The funders had no role in study design, data collection and analysis, decision to publish, or preparation of the manuscript.

Competing Interests: The authors have declared that no competing interests exist.

* Email: anilkumar.mbge@gmail.com

¤ Current address: IRRI-South Asia Rice Breeding Hub, ICRISAT, Patancheru, Andhra Pradesh, India

Introduction

Calcium is an important essential element, acts as secondary messenger in various biological processes. Its deficiency causes low bone density, osteoporosis, colon cancer etc [1]. Milk is the major source of calcium for humans in the world. Unavailability of recommended amount of milk in poor population and lactose intolerance may lead to calcium deficiency. Plant based calcium could serve as an alternate source for calcium. Since, daily consumed cereals are poor in calcium, biofortification of these cereals for calcium and understanding molecular basis of its accumulation is in need. Finger millet, rich in calcium and contains about 5–30 times higher calcium in contrast to rice and wheat [2]. In general, plants absorb calcium through roots and deliver it to shoots through the xylem stream [3]. The process of calcium uptake and transport is genetically and epigenetically determined [4]. Earlier studies suggested the roles of various calcium transporters in the accumulation of calcium in plants [5] but the role of calcium sensor genes in regulating calcium transportation has not been reported till date.

In plants, calcium sensor proteins are categorized into two groups *i.e.*, calcium sensor relay and calcium sensor responder [6].

The calcium relay proteins bind calcium and affect their target protein because they themselves do not have enzymatic activity [6]. In contrast, the calcium sensor responder proteins bind calcium and a change in conformation takes place and hence modulates their own activity by intra-molecular interaction. As per literature, two types of calcium sensor protein *viz.*, Calmodulin (CaM) and Calciuneurin B-like protein (CBL) have been reported likewise calcium responders have been classified into six types *viz.*, (i) Ca^{2+} dependent and CaM independent protein kinases (CDPKs); (ii) SOS3/CBL interacting protein kinases (SIPKs/CIPKs); (iii) CaM dependent protein kinases (CaMKs); (iv) Ca^{2+}/CaM dependent protein kinases (CCaMKs); (v) CDPK related protein kinases (CRKs); (vi) phosphoenolpyruvate (PEP) carboxylase kinase-related kinases (PEPRKs) [7]. The role of the calcium sensor genes are reported in the regulation of calcium transporter proteins [9]. Like calcium dependent protein kinases-1 (*CDPK-1*) was reported in regulation of PM type Ca^{2+}-ATPase2 in Arabidopsis by phosphorylation within their N-terminal regulatory domain [8]. However, the exact nature and role of calcium sensor genes in seed calcium accumulation in finger millet requires detailed investigation.

Earlier some studies have been conducted for understanding the role of these genes in seed calcium accumulation, no fruitful inference could be made due to limited genetic information of finger millet. Considering the problems faced by earlier groups, in present study transcriptomics approach was used to characterize calcium sensor gene family from developing spikes of finger millet. This work will represent the first exhaustive analysis for calcium sensor genes in cereal crops. The characterisation, identification, classification, phylogeny and pathway analysis are therefore important steps in understanding the role of calcium sensor genes in grain calcium accumulation that may further highlight our knowledge on plant calcium signaling. The production of specialised cDNA from spikes was used to determine the expression patterns of all calcium sensor genes and highly expressed genes in contrasting genotypes were validated using qPCR analysis.

Materials and Methods

Tissue collection and RNA isolation

Finger millet genotypes *i.e.*, GP-1 (200 mg/100 g seed) and GP-45 (400 mg/100 g seed) differing in their grain calcium content were obtained from Rani Chauri Hill Campus, G.B. Pant University of Agriculture and technology, Pantnagar (Uttarakhand) India. The spike/panicle samples were selected at four stages *viz.*, S1 (spike emergence); S2 (pollination stage); S3 (dough stage) and S4 (maturation stage) on the basis of its morphology along with stages of ovary and anther. RNA was isolated by using total RNA isolation protocol [7]. Total RNA was treated with RNase- free DNase I (Fermentas, Germany) for 30 min at 37°C to remove residual DNA. The quality of RNA samples was checked both by agarose gel electrophoresis and RNA integrity number (RIN) value estimation. The appearance of two prominent bands of 16S and 28S rRNA in RNA sample and the RIN value of both RNA sample was recorded to be 8, which confirmed good quality and integrity of RNA samples 8.

Preparation of cDNA and transcriptome sequencing

Equal amount of RNA from developing spikes collected at all four stages (S1, S2, S3, S4) of both genotypes (GP-1 & GP-45) were mixed separately for subsequent analysis. Pooled RNA were used to purify poly (A) mRNA using Oligotex mRNA midi prep kit (Qiagen, Germany) followed by fragmentation into 200–500 bp

pieces using divalent cations at 94°C for 5 min. The cleaved RNA fragments were copied into first strand cDNA using Superscript II reverse transcriptase (Life Technologies, Inc.) and random primers. After second strand cDNA synthesis, fragments were end repaired, a-tailed and indexed adapters were ligated. The products were purified and enriched with PCR to create the final cDNA library. The cDNA libraries were used for 2×100 bp paired-end sequencing on a single lane of the Illumina HiSeq 2000 (Genomics Core, UZ Leuven, Belgium). After sequencing, the samples were demultiplexed and the indexed adapter sequences were trimmed using the CASAVA v1.8.2 software (Illumina, Inc.). The sequencing and assembly was done by commercial sequencing service provider (NexGenBio, New Delhi, India). A total of 6.54 GB (Gigabyte) and 7.31 GB sequence data were generated from the GP-1 and GP-45 spikes respectively.

RNA-Seq data filter and *De novo* assembly

The customer Perl script (CONDETRI: http://code.google.com/p/condetri) with parameters (-hq = 20 -lq = 10 -frac = 0.8 -lfrac = 0.1 -minlen = 50 -mh = 5 -ml = 5 -sc = 64) was used to remove the sequencing adaptor and low quality reads. *De novo* assembly of high quality reads was done by Trinity assembler to generate a non-redundant set of transcripts using one k-mer length (25-mer) and group_pairs_distance = 250, path_reinforcement_distance = 70, min_glue = 2, min_kmer_cov = 2 keeping other default parameters [10]. The high quality read obtained after sequencing were assembled *de novo* using the Trinity program [15], which produced 109,218 contigs, with an N50 of 1191 bp in GP-1 and 120,130 contigs, with an N50 of 1450 bp in GP-45 genotype. GP-1 and GP-45 transcripts were deposited at NCBI/Gene Bank as the TSA accession SRR1151079 and SRR1151080 respectively and used for further analysis.

Identification and annotation of calcium sensor gene family from finger millet transcriptome data

To identify Calcium sensor genes from transcriptome of both GP-1 and GP-45 genotypes, Coding DNA Sequence (CDS) of rice calcium sensor genes *viz.*, CaM, CMLs, CBLs, CDPKs, CIPKs, CaMKs, CCaMK, CRKs and PEPRKs were retrieved from various sequence databases. Transcripts of both GP-1 and GP-45 genotypes were used to make offline database separately. Rice calcium sensor genes were searched as query sequences from nucleotide sequence of the full transcriptome data of both genotypes using UGENE (UGENE, UniPro, Russia). Afterwards, the selected calcium sensor genes sequences were performed a BLASTn search.

Each gene gave around 100–200 hits in the contigs, hence an excel sheet of all calcium sensor was made which was a huge file of data influx. Then the filters were applied to specify the data according to the e-value (<1) and large amount of data redundancy was thus removed. Finally, the truncated data was cross checked, with database by using the blast search tool BLASTx and the hits with minimum expected value, maximum identity with maximum query coverage were selected.

Re-assembly of retrieved sequences and ORF identification

Hence, contigs verified were used to select the best non redundant contig using SeqMan Pro gene analysis package (DNASTAR Inc., Madison, WI, USA). An ORF finder (http://www.ncbi.nlm.nih.gov/projects/gorf/) online tool was used for identification of open reading frame (ORF) in annotated sequences.

Classification of calcium sensor genes

The calcium sensor genes retrieved from finger millet transcriptome and rice genome database were aligned and evolutionary analyses using Neighbor-Joining (NJ) algorithm were constructed using MEGA6 (http://www.megasoftware.net) [11]. Following parameters were set during construction of phylogenetic tree *viz.*, substitution, poisson model, complete deletion, replication, bootstrap analysis with 1,000 replicates. The percentage similarities of finger millet sequence with rice sequence were also checked through EBI online software (https://www.ebi.ac.uk/Tools/psa/emboss_needle/). Structural and functional verification of each gene was also predicted using ScanProsite (release 20.83) (http://prosite.expasy.org/scanprosite/) and SMART tools (http://smart.embl-heidelberg.de/).

Properties of calcium sensor genes

Analysis of the functional and physiochemical properties of each protein was done by Protein Identification and Analysis Tools on ExPASy Server (http://web.expasy.org/protparam/) and domains were analysed using ScanProsite detection of PROSITE signature matches and ProRule-associated functional and structural residues in proteins (http://prosite.expasy.org/). Prediction of subcellular localization was done by using Target P1.1 server (http://www.cbs.dtu.dk/services/TargetP/). The molecular weight of the protein and its isoelectric point (pI) were estimated by using online expasy server (http://web.expasy.org/compute_pi/). A well tabulated data sheet was prepared comprising the physiochemical properties such as GRAVY, instability index, number of EF hands and their subcellular localization.

Transcriptome based expression analysis of identified calcium sensor genes

To determine the differentially expressed genes between GP-1 and GP-45, in present analysis, the sequencing reads were mapped to each gene and presented in **F**ragments **P**er **K**ilobase of exon per **M**illion fragments mapped (FPKM). In both GP-1 and GP-45 transcriptome data FPKM value of transcripts were compared. The gene expression were measured as normalized expected fragments, allowing for measurement of read counts from platforms that produce one or more reads per single source molecules [12]. Pathway analysis was also performed using Kyoto Encyclopaedia of Genes and Genomes (KEGG) (http://www.genome.jp/kegg/).

Exogenous calcium application and calcium analysis

Seeds of GP-1 and GP-45 genotypes were sown in kharif season of 2011–12 in the pots. The seedlings were transplanted and grown in individual pots (25 cm upper diameter, 17 cm lower diameter, 25 cm height) filled with acid washed sand and placed in the experimental polyhouse. The plants were provided with half strength of macro and micronutrient in a modified Hoagland nutrient solution [13] containing different concentrations of calcium (0.1 mM, 5.0 mM, 10 mM and 20 mM) in the form of $Ca(NO_3)_2.4H_2O$. The nutrient solution (pH 5.5–6.0) was renewed every three days interval after the sand had been rinsed with distilled water. The experimental design was completely randomized with four treatments, arranged in individual pot with nine pots per treatment, each replicated three times. The spikes were sampled at each stage of spike development (S1, S2, S3 and S4) from each treatment. The samples for calcium estimation by atomic absorption spectroscopy (AAS) (SensAA GBC Scientific Equipment, USA), were prepared by wet digestion method [14]. To detect differences between treatments, analysis of variance (ANOVA) was performed using SPSS 21.0 (SPSS Inc.,Chicago, IL, USA).

RNA isolation and qPCR analysis

Total RNA was isolated from the spike collected at each four stages according to protocol described by Ghawana *et al.* [15]. Isolated RNA was treated with RNase-free DNase I for 30 min at 37°C to remove residual DNA. cDNA was prepared from the purified RNA using Revert Aid H-minus reverse transcriptase cDNA synthesis kit (Fermentas, Germany). Quantitative real-time PCR (qPCR) was used to identify the expression patterns of selected calcium sensor genes in different stages of spikes of both genotypes grown at different concentration of exogenous calcium. The tubulin gene (CX265249) was used as an internal control to normalize the expression level of the target gene. Real-time PCR was performed in the reaction volume of 20 µl containing 2.5× Real Master Mix SYBR ROX/20× SYBR solution, 100 nM of each forward and reverse primers and 100 ng of cDNA. All samples were amplified in triplicate and the mean and standard error values were calculated. Relative expression of all genes were calculated the by $_{\Delta\Delta}CT$ method.

Results

Identification of calcium sensor genes

The rice calcium sensor genes were used as reference for identification of their homologues in finger millet transcriptome. Out of 109,218 contigs (in GP-1) and 120,130 contigs (in GP-45), a total of 138 and 137 contigs respectively representing putative calcium sensor and calcium sensor like protein, were filtered. Only hits with e value of <1.0 were considered for further analysis. The sequence with e value <1.0 were reassembled with SeqMan aligner to reduce redundancy and obtaining maximum length of ORF. The contig sequences which are repeated more than one time were used for open reading frame (ORF) selection. The online tool ORF finder allowed identification of ORFs encoding putative calcium sensor protein in finger millet. Structural and functional verification of calcium binding and related domain were carried out. Finally, we demonstrated 80 and 78 non redundant calcium sensor genes in GP-1 and GP-45 respectively. It includes 17 ORFs encoding for CDPKs, 21 for CIPKs, 2 for CaMKs, 1 for CCaMKs, 2 for PEPRKs, 4 for CRKs, 9 for CBLs and 24 for CaM and CaML proteins in GP-1 transcriptome. Similarly, 16 ORFs encoding for CDPKs, 22 for CIPKs, 2 for CaMK, 1 for CCaMK, 2 for PEPRKs, 4 for CRKs, 8 for CBLs and 23 for CaM and CaML proteins were identified in GP-45 transcriptome.

Designating the analyzed sequences as calcium sensor genes

All calcium sensors were classified into eight group from A to H (in Phylogenetic tree) *viz.*, CaM and CaML, PEPRK, CIPK, CRK, CaMK, CDPK, CCaMK and CBL genes on the basis of sequence features, identity, phylogenetic study and were designated according to the name of their rice homologues. The first initial of genus and species name were used in the naming of all finger millet sequences (**Table 1**).The group A contained sequences of CaM & CaML gene, group B contained sequences of PEPRK gene, group C contained sequences of CIPK gene, group D contained sequences of CRK gene, group E contained sequences of CaMK gene, group F contained sequences of CDPK gene, group G contained sequences of CCaMK gene and group H contained sequences of CBL gene. The genes falling in same group is interpreted as these genes may be diverged from the same

Table 1. List of finger millet Calcium sensor genes and its similarity with its rice orthologs.

S.No.	Ca sensor gene (Oryza)	Accession no. (Oryza)	% identity (Oryza vs Eleusine)
1	OsCaM1	LOC_Os03g20370	EcCaM1 (35.9%)
2	OsCaML1	LOC_Os01g59530	EcCaML1 (71.8%)
3	OsCaML2	LOC_Os11g03980	EcCaML2 (87.4%)
4	OsCaML4	LOC_Os03g53200	EcCaML4 (92.9%)
5	OsCaML5	LOC_Os12g41110	EcCaML5 (80.0%)
6	OsCaML8	LOC_Os10g25010	EcCaML8 (90.8%)
7	OsCaML9	LOC_Os05g41200	EcCaML9 (73.2%)
8	OsCaML10	LOC_Os01g72100	EcCaML10 (87.0%)
9	OsCaML11	LOC_Os01g32120	EcCaM11 (79.9%)
10	OsCaML14	LOC_Os05g50180	EcCaM14 (92.4%)
11	OsCaML17	LOC_Os02g39380	EcCaM17 (57.2%)
12	OsCaML18	LOC_Os05g13580	EcCaML18 (81.1%)
13	OsCaML22	LOC_Os04g41540	EcCaML22 (68.2%)
14	OsCaML23	LOC_Os01g72540	EcCaML23 (75.2%)
15	OsCaML24	Os07g0681400	EcCaML24 (82.6%)
16	OsCaML27	LOC_Os03g21380	EcCaML27 (89.6%)
17	OsCaML28	LOC_Os12g12730	EcCaML28 (83.0%)
18	OsCaML29	LOC_Os06g47640	EcCaML29 (75.5%)
19	OsCaML30	LOC_Os06g07560	EcCaML30 (72.2%)
20	OsCaML31	LOC_Os01g72530	EcCaML31 (86.2%)
21			EcCaML34
22			EcCaML35
23			EcCaML36
24			EcCaML37
25			EcCaML38
26			EcCaML39
27	OsCRK2	BAD54109.1	EcCRK2 (88.1%)
28	OsCRK3	BAC79879.1	EcCRK3 (96.7%)
29	OsCRK5	AAK84452.1	EcCRK5 (93.1%)
30	OsCaMK1	AF368282.1	EcCaMK1 (85.6%)
31	OsCaMK	BAC16472.1	EcCaMK2 (90.9%)
32	OsPEPRK2	BAD17519.1	EcPEPRK2 (72.5%)
33			EcPEPRK3 (84.2%)
34	OsCCaMK1	AAT77292.1	EcCCaMK1 (95.5%)
35	OsCIPK2	AK072868	EcCIPK10 (87.7%)
36	OsCIPK4	Os12g41090	EcCIPK4 (68.9%)
37	OsCIPK5	AK065589	EcCIPK8 (87.7%)
38	OsCIPK6	Os08g34240	EcCIPK13 (80.4%)
39	OsCIPK7	AK111510	EcCIPK18 (79.6%)
40	OsCIPK8	AK120431	EcCIPK3 (90.1%)
41	OsCIPK9	OJ1015F07.8	EcCIPK4 (85.9%)
42	OsCIPK10	AK066541	EcCIPK9 (77.2%)
43	OsCIPK11	AK103032	EcCIPK2 (77.9%)
44	OsCIPK14	Os12g02200	EcCIPK7 (66.7%)
45	OsCIPK16	AK061220	EcCIPK12 (75.0%)
46	OsCIPK17	AK100498	EcCIPK11 (45.2%)
47	OsCIPK18	AK101355	EcCIPK6 (88.4%)
48	OsCIPK19	AK069486	EcCIPK14 (60.1%)
49	OsCIPK21	AK107137	EcCIPK16 (81.6%)
50	OsCIPK23	ACD76983.1	EcCIPK23 (98.0%)

Table 1. Cont.

S.No.	Ca sensor gene (Oryza)	Accession no. (Oryza)	% identity (Oryza vs Eleusine)
51	OsCIPK24	AK102270	EcCIPK11 (88.4%)
52	OsCIPK25	AK065374	EcCIPK19 (61.2%)
53	OsCIPK28	A3B529.2	EcCIPK28 (34.0%)
54	OsCIPK29	AK111746	EcCIPK17 (76.6%)
55			EcCIPK30
56			EcCIPK31
57			EcCIPK32
58			EcCIPK33
59	OsCDPK1	LOC_Os01g43410.1	EcCDPK2 (87.8%)
60	OsCDPK3	LOC_Os01g61590.1	EcCDPK18 (96.3%)
61	OsCDPK4	AK060738.1	EcCDPK14 (94.9%)
62	OsCDPK6	LOC_Os02g58520.1	EcCDPK16 (87.0%)
63	OsCDPK7	LOC_Os03g03660.2	EcCDPK9 (87.8%)
64	OsCDPK8	LOC_Os03g59390.1	EcCDPK11 (89.8%)
65	OsCDPK12	Os04t0584600	EcCDPK5 (89.9%)
66	OsCDPK13	AK061881	EcCDPK1 (79.1%)
67	OsCDPK14	LOC_Os05g41270.1	EcCDPK7 (77.8%)
68	OsCDPK17	Os07g0161600	EcCDPK12 (58.0%)
69	OsCDPK19	AP003954	EcCDPK3 (94.0%)
70	OsCDPK20	AP003866	EcCDPK8 (89.8%)
71	OsCDPK24	Os11g0171500	EcCDPK6 (88.6%)
72	OsCDPK25	Os11g0136600	EcCDPK31 (90.9%)
73	OsCDPK29	LOC_Os12g12860.1	EcCDPK29 (87.8%)
74	OsCBL1	ABA54176.1	EcCBL1 (96.7%)
75	OsCBL2	ABA54177.1	EcCBL2 (98.7%)
76	OsCBL3	ABA54178.1	EcCBL3 (95.4%)
77	OsCBL4	ABA54179.1	EcCBL4 (86.4%)
78	OsCBL5	ABA54180.1	EcCBL5 (90.2%)
79	OsCBL6	ABA54181.1	EcCBL6 (62.0%)
80	OsCBL7	ABA54182.1	EcCBL7 (87.3%)
81	OsCBL9	ABA54184.1	EcCBL9 (87.9%)
82	OsCBL10	ABA54185.1	EcCBL10 (84.2%)

ancestral gene and showing varied function though still maintaining high structural homology.

Structural and functional analysis of calcium sensor genes in finger millet

Domain analysis of calcium sensor genes. The domain analysis revealed that most of CaM and CaML genes contained 4 EF-hand, however, 5 contained 3 EF-hand and another 5 contained 2 EF-hand. The PEPRK gene family comprises of only PKC (Protein kinase) domain in finger millet. The CIPK family in finger millet have no EF hand at the carboxyl terminal, but have auto inhibitory domain also known as NAF domain. CRK genes in finger millet possess only single kinase and single EF-hand with an exception of *EcCRK3* having both kinase as well as an EF-hand while *EcCRK2*, 5 and 7 possess only kinase domain. Similar to rice CaMK, finger millet CaMK also contains one kinase and one EF-hand.

Domain analysis revealed that a conserved protein kinase domain was found in EcCDPKs along with four EF-hand domains, with an exception of *EcCDPK6* which only contain four EF-hand domains and protein kinase domain was absent (**Table 2**). In finger millet CCaMKs have the same general structure of CDPKs, but they have calcium-binding domains with three EF-hands. Structure of CBL are similar to CaM and CaML gene, however unlike CaM, CBL possess only 3 EF-hand, with exception of *EcCBL3* which possess only 2 EF-hand.

Motif analysis of calcium sensor genes. In case of PEPRK, CRK, CaMK and CCaMK the motif 2, 3 and 7 were common for each gene. The PEPRK and CCaMK sequences contain only five motifs, but CRK and CaMK gene sequences contain the entire ten motifs. Most of CaM and CaML comprises of four motifs out of ten rest six were not uniformly present. In case of CIPK, out of ten motifs nearly all motifs are present in each CIPK genes, except in case of *EcCIPK18*, where motif 9 was doubled and motif 10 was absent. Out of ten motifs, all motifs were found in most of CDPKs of finger millet and rice. Motif 4

Table 2. List of domain, sub-cellular localization and instability index analysis of finger millet calcium sensor.

S.No	Gene name	Domain	GRAVY	Instability index	Stability	Localization
A. CaM and CaML proteins						
1	EcCaM1	4 EF-hand	−0.602	23.23	stable	-
2	EcCaML2	4 EF-hand	−0.613	38.96	stable	-
3	EcCaML4	4 EF-hand	−0.449	28.05	stable	-
4	EcCaML5	2 EF-hand	−0.946	33.75	stable	-
5	EcCaML8	4 EF-hand	−0.721	20.32	stable	-
6	EcCaML9	3 EF-hand	−0.531	38.70	stable	-
7	EcCaML10	4 EF-hand	−0.226	60.77	unstable	C
8	EcCaML11	4 EF-hand	−0.226	39.16	stable	-
9	EcCaML14	4 EF-hand	0.120	17.34	stable	M
10	EcCaML18	4 EF-hand	−0.332	30.12	stable	-
11	EcCaML22	4 EF-hand	−0.572	61.46	unstable	C
12	EcCaML23	4 EF-hand	−0.370	41.43	unstable	-
13	EcCaML24	3 EF-hand	−0.254	20.22	stable	C
14	EcCaML27	3 EF-hand	−0.210	29.80	stable	C
15	EcCaML28	4 EF-hand	−0.693	19.61	stable	
16	EcCaML29	3 EF-hand	0.128	33.17	stable	-
17	EcCaML30	2 EF-hand	−0.369	51.99	unstable	S
18	EcCaML31	3 EF-hand	−0.168	27.03	stable	-
19	EcCaML34	2 EF-hand	−0.0397	68.38	unstable	C
20	EcCaML35	2 EF-hand	−0.179	66.71	unstable	C
21	EcCaML36	3 EF-hand	−0.448	46.78	unstable	C
22	EcCaML37	3 EF-hand	−0.954	30.71	stable	M
23	EcCaML38	3 EF-hand	−0.076	24.45	stable	-
24	EcCaML39	4 EF-hand	−0.642	36.45	stable	M
B. EcPEPRK proteins						
1	EcPEPRK2	PKC	0.297	54.25	unstable	S
2	EcPEPRK3	PKC	−0.53	45.09	unstable	-
C. EcCRK proteins						
1	EcCRK2	PKC	−0.395	58.56	unstable	-
2	EcCRK3	PKC	−0.161	40.17	unstable	-
3	EcCRK5	PKC	−0.248	49.56	unstable	S
D. EcCaMK proteins						
1	EcCaMK1	PKC, 1 EF hand	−0.245	44.63	unstable	-
2	EcCaMK2	PKC, 1 EF hand	−0.259	43.88	unstable	-
E. EcCCaMK protein						
1	EcCCaMK1	PKC, 3 EF hand	−0.215	46.23	unstable	-
F. EcCBL proteins						
1	EcCBL1	3 EF hand	−0.241	27.76	stable	-
2	EcCBL2	3 EF hand	−0.216	32.51	stable	S
3	EcCBL3	2 EF hand	−0.017	42.23	unstable	-
4	EcCBL4	3 EF hand	−0.260	46.04	unstable	-
5	EcCBL5	3 EF hand	−0.317	50.54	unstable	-
6	EcCBL6	3 EF hand	−0.331	37.08	stable	M
7	EcCBL7	3 EF hand	−0.229	44.93	unstable	M
8	EcCBL9	3 EF hand	−0.220	35.66	stable	-
9	EcCBL10	3 EF hand	−0.137	44.08	unstable	C
G. EcCDPK proteins						
1	EcCDPK1	PKC, 4 EF hand	−0.524	45.98	unstable	M
2	EcCDPK3	PKC, 4 EF hand	−0.467	38.40	stable	-

Table 2. Cont.

S.No	Gene name	Domain	GRAVY	Instability index	Stability	Localization
3	EcCDPK4	PKC, 4 EF hand	−0.387	45.59	unstable	-
4	EcCDPK6	4 EF hand	−0.350	30.95	stable	-
5	EcCDPK7	PKC, 4 EF hand	−0.322	42.97	unstable	-
6	EcCDPK8	PKC, 4 EF hand	−0.499	36.43	stable	-
7	EcCDPK12	PKC, 4 EF hand	−0.330	33.95	stable	-
8	EcCDPK13	PKC, 4 EF hand	−0.342	38.54	stable	C
9	EcCDPK14	PKC, 4 EF hand	−0.485	42.10	unstable	-
10	EcCDPK17	PKC, 4 EF hand	−0.299	43.63	unstable	C
11	EcCDPK19	PKC, 4 EF hand	−0.443	42.65	unstable	-
12	EcCDPK20	PKC, 4 EF hand	−0.487	34.31	stable	-
13	EcCDPK24	PKC, 4 EF hand	−0.307	35.31	stable	C
14	EcCDPK25	PKC, 4 EF hand	−0.403	22.97	stable	-
15	EcCDPK29	PKC, 4 EF hand	−0.405	36.32	stable	-
H. EcCIPK proteins						
1	EcCIPK2	PKC, NAF	−0.401	30.13	stable	-
2	EcCIPK4	NAF	−0.562	44.22	unstable	M
3	EcCIPK5	PKC, NAF	−0.452	33.39	stable	M
4	EcCIPK6	PKC, NAF	−0.133	31.50	stable	-
5	EcCIPK7	PKC, NAF	−0.146	56.39	unstable	M
6	EcCIPK8	PKC, NAF	−0.289	37.40	stable	-
7	EcCIPK9	PKC, NAF	−0.392	34.93	stable	-
8	EcCIPK10	PKC, NAF	−0.385	34.20	stable	-
9	EcCIPK11	PKC, NAF	−0.494	49.27	unstable	M
10	EcCIPK14	PKC, NAF	−0.370	47.32	unstable	C
11	EcCIPK16	PKC, NAF	0.080	36.17	stable	S
12	EcCIPK18	PKC, NAF	−0.360	30.75	stable	-
13	EcCIPK19	PKC, NAF	−0.384	37.33	stable	C
14	EcCIPK21	PKC, NAF	−0.328	41.36	unstable	-
15	EcCIPK23	PKC, NAF	−0.394	37.56	unstable	-
16	EcCIPK24	PKC, NAF	−0.230	33.20	stable	M
17	EcCIPK25	PKC, NAF	−0.228	49.45	unstable	-
18	EcCIPK28	PKC, NAF	−0.365	37.03	unstable	-
19	EcCIPK29	PKC, NAF	−0.123	40.93	unstable	-
20	EcCIPK31	PKC, NAF	−0.546	48.35	unstable	M
21	EcCIPK32	PKC, NAF	−0.460	39.34	stable	M
22	EcCIPK33	PKC, NAF	−0.373	50.28	unstable	-
23	EcCIPK34	PKC, NAF	−0.329	41.66	unstable	-

C Chloroplast, i.e. the sequence contains cTP, a chloroplast transit peptide;
M Mitochondrion, i.e. the sequence contains mTP, a mitochondrial targeting peptide;
S Secretory pathway, i.e. the sequence contains SP, a signal peptide;
_ any other location.

was absent in case of *EcCDPK4* gene. Multilevel consensus sequence analysis for all 10 motifs of finger millet showed common sequences among gene family (**Table 3**).

Sub-Cellular localization, stability and instability index in calcium sensor genes. Based on the information available in the database it was inferred that a total of 13 calcium sensor genes were chloroplast localized contained cTP (a chloroplast transit peptide), 13 were mitochondria localized contained mTP (a mitochondrial targeting peptide) and 5 were present on secretary pathways contained 'SP' (a signal peptide site) and hence were predicted to be located at chloroplast, mitochondria and at the secretary pathway respectively (**Table 2**). Other calcium sensor sequence residing sites could not be predicted due to limitation of the information available in the database. The stability parameters of calcium sensors suggested that 40 calcium sensor genes were found to be stable and remaining 32 calcium sensor genes were found unstable.

Table 3. Multilevel consensus sequences for the MEME defined motifs of members of different calcium sensor genes.

S.No	E-value	Motif length	Motifs
CaM and CaML			
1	1.4e-652	33	LREAFRVFDKDQNGYISAAELRHVMTNLGEKCT
2	3.1e-612	24	VEECQQMICNVDRDGDGQINYHEF
3	2.4e-468	21	LRRVFRLFDKNGDGCITTKEL
4	5.9e-317	41	QNPTEDELQAMINEVDTNGNGCIDFHEFVNLYCRKMKDHDH
5	2.70E-45	50	AQALDYHGLSADRAGLTATVGAYIPWGAAGLRFEDFESLHRALGDALFGP
6	7.10E-40	21	SLSKKPSPSFRLRNGSLNVVR
7	4.70E-23	11	KCMMQGITVWG
8	2.10E-10	8	CTVMRSLG
9	2.50E-04	49	MHGIAPSSPDRSPQYPSCAPIKGRRRSRLFHRCTAQPKCAKNDYSLCCS
10	1.70E-01	11	HHQLTQQQIKE
CRK			
11	2.2e-368	41	KILKSLSGHNNLVQFYDACEDEDNVYIVMELCEGGELLDRI
12	6.5e-379	35	RCYSTEADMWSIGVITYILLCGSRPFWARTESGIF
13	3.30E-273	50	LKAEPNFNEHPWPTISPEAKDFVKRMLNKDYRKRMTAAQALCHPWIRNYQ
14	6.90E-235	50	ARGGKYSEEDAKCVMVQILSVVSFCHLQGVVHRDLKPENFLFTTKDENSP
15	3.90E-268	50	GEEVGRGHFGYTCSAKAKKGEHKGQDVAVKVIPKAKMTTAIAIEDVRREV
16	3.20E-246	50	FVNTLCNLQYRKMDFEEFCAAAISVYQMEALDTWEQHARTAYEYFEKEGN
17	7.40E-216	29	DFGLSDFVKPDERLNDIVGSAYYVAPEVL
18	1.20E-184	41	EQGMNPSVPLHVVLQDWIRHSDGKLSFLGFIKLLHGMSMRS
19	2.30E-172	41	NSSVASTPARGGFKRPFPPPSPAKHIRALLARRHGSVKPNE
20	1.30E-143	41	DMIIYKLMKAYIRSSSLRKAALRALSKTLTTDQLFYLKEQF
CIPK			
21	2.3e-1717	50	EDEARRYFQQLINAVDYCHSRGVYHRDLKPENLLLDENGNLKVSDFGLSA
22	8.1e-1689	27	YDGAKADIWSCGVILYVLMAGYLPFHD
23	1.4e-1212	50	GMMEQIKREISTMKLVRHPNVVQLHEVMATKTKIYFVMEYVTGGELFNKI
24	2.7e-726	30	RLIRRILDPNPMTRITIAEIMEHPWFKKGY
25	1.1e-722	41	KEGRKGVLAIDAEIFEVTPSFHMVELKKTNGDTLEYQQFCN
26	2.9e-566	29	AQCWRQDGLLHTTCGTPNYVAPEVINNKG
27	3.4e-458	29	RYELGRTLGQGTFAKVYHARNTETGQSVA
28	1.70E-253	21	NLMTMYRKICKAEFRCPPWFS
29	1.70E-278	17	MSQGFDLSGMFEEEQEY
30	5.40E-227	41	KRETRFTSQCPPQEIFSKIEEIATPMGFQVQKQNYKMKLMG
CDPK			
31	1.6e-1779	50	DIVGSPYYMAPEVLKRNYGPEADVWSAGVILYILLCGVPPFWAETEQGIF
32	2.2e-1374	50	CAGGELFDRIIARGHYTERAAAQLCRTIVGVVHMCHSMGVMHRDLKPENF
33	9.4e-1369	50	VLSRMKQFSAMNKFKKMALRVIAENLSEEEIAGLKEMFKMMDTDNSGTIT
34	3.6e-1363	50	NNGTIDYGEFITATMHMNKMEREEHLYKAFQYFDKDGSGYITIDELQQAC
35	1.4e-1151	50	QAILRGHIDFKSEPWPSISESAKDLVRKMLNPDPKKRLTAHQVLCHPWIC
36	4.3e-1061	50	LGQGQFGTTYLCTHRATGCRYACKSISKRKLVTPEDVEDVRREIQIMHHL
37	2.5e-747	29	IKDIIQEVDQDNDGRIDYQEFVAMMQKGN
38	7.9e-623	21	LFANKKEDSPLKAIDFGLSVF
39	4.7e-517	29	EELKEGLRKYGSNLSESEIQQLMEAADID
40	3.5e-397	21	NIVSIRGAYEDAHAVHLVMEL
CBL			
41	1.0e-592	50	NGVIGFGEFVRALSVFHPNAPLEEKIDFAFRLYDLRQTGYIERQEVKQMV
42	2.3e-552	50	DKTFEQADTNHDGKIDQEEWRNFVLRHPSLLKNMTLPYLKDITTTFPSFV
43	4.0e-414	50	ALYELYKKISCSVIDDGLIHKEEFQLALFRTSKKENLFADRVFDLFDTKH
44	3.10E-142	29	MGCCCSKQFKQPPGYEDPQVLARETVFSV
45	1.10E-84	15	ESGMNLSDDIVEAII

Table 3. Cont.

S.No	E-value	Motif length	Motifs
46	2.20E-68	50	MDPSSSSNAFGSRSSLTLGELACAALFPVLAILDAVIFATARCFQKSPPR
47	4.60E-16	21	RRPSGGNLFVDRVFDLFDQKR
48	5.90E-13	6	NSQVDD
49	1.80E-09	11	MVQCLDGVRQL
50	2.00E-01	8	CFTVNEVE

Comparative phylogenetic analysis of calcium sensor gene family in finger millet and rice. Further, to compare finger millet CaM and CaMLs, PEPRKs, CIPKs, CRKs, CaMKs, CDPKs, CCaMK and CBL with those of rice, a phylogenetic tree was generated using full length protein sequences (**Fig. 1**). Phylogenetic analysis of these genes of rice and finger millet differentiated them into different clusters from A to H. Cluster A contained CaM & CaML sequence, while cluster B & C contained sequences of PEPRK and CIPK respectively. It is also noteworthy that cluster D and E contained both the sequences of CRK and CaMK. Cluster F, G and H contained sequences of CDPKs, CCaMK and CBLs respectively.

Differential expression analysis based on transcript abundance of calcium sensor genes

In the present study, expression of total 82 calcium sensor genes were compared by using transcriptome data of both GP-1 and GP-45 genotypes. FPKM values are represented in the form of heat map for each identified calcium sensor genes (**Fig. 2**). The results of transcriptome reveals that the expression of 24 genes was higher in the pooled spike sample of GP-45 genotype while the expression of 11 genes was higher in the pooled spike sample of GP-1 genotype. Out of 24 genes highly expressed in the developing spikes of GP-45 genotype, 7 encoded for CaML, 2 for CRK, 5 for CBL, 7 for CIPK and 4 for CDPK genes. Among them, the *EcCIPK9* gene showed very high expression in GP-45 (FPKM value −55.18) when compared to GP-1 genotype (FPKM value − 0). The number of genes that were highly expressed in the spike of GP-1 genotype includes 5-CaML, 2-CRK, 3-CIPK, and 1-CDPK genes.

Pathways analysis of differentially expressed genes

To estimate the functions of the differentially expressed genes of GP-1 and GP-45, biological metabolic pathways were investigated by KEGG pathway analysis. Out of 35 differentially expressed genes, only 10 genes could be assigned to three documented pathways (**Table 4**). These pathways mainly related to stress adaption, hormonal, biotic and abiotic changes. The pathway with the greatest numbers of unique genes was for proteins involved in plant-pathogen interaction. The four genes which have less expression value in GP-45 genotype (*EcCaM1*, *EcCaML4*, *EcCaML14* and *EcCDPK14*) belonged to pathways involved in plant-pathogen interaction and phosphatidylinositol signaling. Out of six genes highly expressed in GP-45 genotype, five (*EcCaML11*, *EcCaML18*, *EcCDPK4*, *EcCDPK13* and *EcCDPK17*) were involved in plant-pathogen interaction and remaining one (*EcCIPK31*) in abiotic stress pathway.

Quantitative real-time PCR validation

To validate the results obtained from differential expression analysis through RNA sequencing, a total of four calcium responder genes having FPKM more than 10 times higher in GP-45 than in GP-1 genotype, were selected and 4 sets of primers were designed for qPCR study. The details of primer sequences, gene name and Tm are given in **Table 5**. The influence of calcium treatment on the expression of four genes was analysed by the real-time PCR and subsequently a heat map was generated using the R- software package ver. 1.12 (**Fig. 3**). According to heat map, the overall expression of *EcCDPK3*, *EcCIPK2*, *EcCIPK9* and *EcCDPK11* was higher in GP-45 as compared to the GP-1 genotype. The transcript level of *EcCDPK3* gene was comparatively higher in calcium deficient conditions (0.1 mM) in the spike of GP-1 genotype at all four stages. Further increase in concentration of exogenous calcium does not show any marked effect on the expression of the genes in four stages of GP-1 genotype. In GP-45, except for S3 stage at 5 mM, increase in exogenous calcium showed no significant change in the *EcCDPK3* gene expression. Although, a slight increase in the expression was detected in the S2 and S4 stages at 20 mM (**Fig. 3**).

There was no marked effect of exogenous calcium on the expression of *EcCIPK2* in GP-1 genotype while a concentration dependent inducibility was observed in case of GP-45. Sharp induction in expression at S3 and S4 stages and S2 and S4 stages was observed at 5 mM and 20 mM concentration of exogenous calcium, respectively (**Fig. 3**). In case of *EcCIPK9* except at S2 stage (0.1 mM) and S3 stage (10 mM) its expression was almost similar in GP-1 genotype at each concentration of exogenously supplied calcium. In GP-45, the expression of *EcCIPK9* was also similar at each concentration of exogenous calcium supplied; however at S3 stage its expression was exceptionally high at sufficient level (5 mM) of calcium (**Fig. 3**). In case of *EcCIPK11* gene except at S3 stage of GP-1 and S4 stage of GP-45 there was almost no effect of exogenously supplied calcium was observed. In GP-1 at S3 stage almost 305 fold expressions was found at 10 mM (Ca excess). However, at S4 stage of GP-45, concentration dependent increase in the expression of *EcCIPK11* gene was observed (**Fig. 3**).

Effects of exogenous calcium on its accumulation in developing spikes

The data (calcium concentration) obtained from all the samples were used to construct graph by STATISTICA v10.0 (StatSoft, Inc.). The increase of calcium accumulation in the spikes occurred progressively up to 20 mM of exogenously supplied calcium in GP-1 genotype. While in case of GP-45, maximum increase was observed at 10 mM of exogenously supplied calcium (**Fig. 4**). Calcium accumulated significantly (P>0.05) in GP-45 genotype up to 10 mM of exogenously supplied calcium, in comparison to

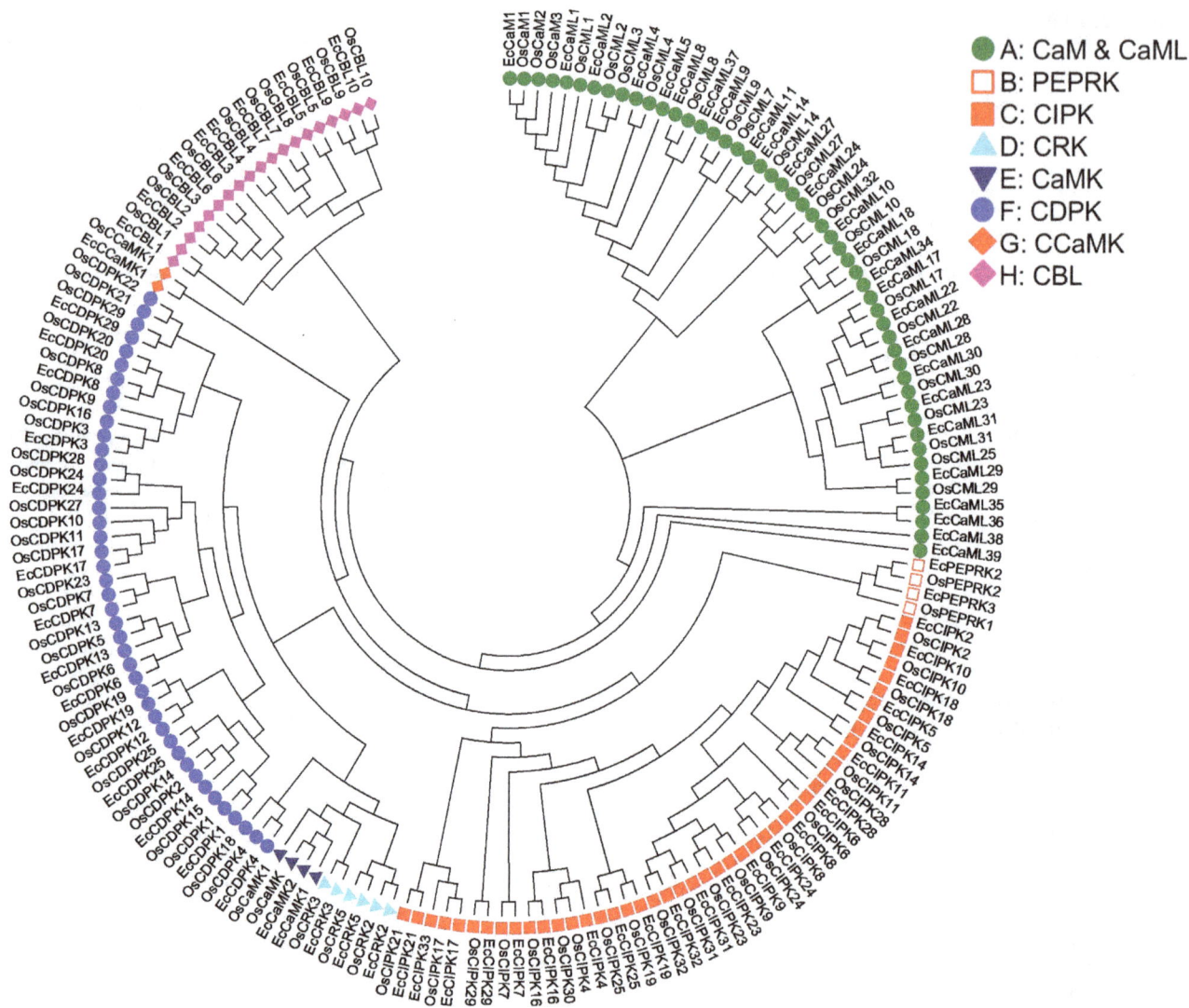

Figure 1. Phylogenetic tree of calcium sensor genes of rice and finger millet. Neighbor-joining tree was created using MEGA6 software with 1000 bootstrap using ORF sequences of rice and finger millet CaM and CaMLs, CBLs, CRKs, PEPRKs, CaMKs and CCaMK proteins. Eight groups were labelled as A, B, C, D, E, F, G and H.

GP-1 genotype which accumulated calcium even at toxic concentration (20 mM). Similar trends of calcium accumulation in the spikes were observed at all four stages of spikes (**Fig. 4**).

Discussion

This is the most comprehensive study of finger millet transcriptome data till date. GP-1 and GP-45 are the finger millet genotypes varying in the content of seed calcium. The sequence data generated from the transcriptome of both genotypes have been analysed to understand the type and function of calcium sensor gene family by comparative phylogeny and digital gene expression profiling.

Identification, classification and designation of calcium sensor genes

Identification of any of the gene families in an organism is a very daunting task especially when genome sequence information is not available. In this situation, either domain features of that gene

family or sequences characterized from related organisms are used as reference for identification of genes in non-sequenced organisms. In present study, the rice calcium sensor genes were used as query for identifying their homologs in finger millet transcriptome. High level of colinearity of rice genome with finger millet genome [16] and the presence of well characterized sequences of calcium sensor genes in rice are the reasons of its selection as a reference sequence in this study. The calcium sensor genes comprise one of the largest families among plant signaling gene. The member of calcium sensor genes showed varied diversity among the species. In our study, we demonstrated 82 calcium sensor proteins in the transcriptome of finger millet developing spike. The abundance of calcium sensor genes in a species may be related with genome duplications (segmental/tandem), rather than the genome size. Rice genome (0.4 GB) include only (110) sensor proteins [17–19]. Further, it is interesting that in some known species (eg. *Arabidopsis thaliana*; genome size ~125 Mb; Calcium sensor genes 134) [17–20] the number of calcium sensor gene family was found to be more than that of large genome size species. However,

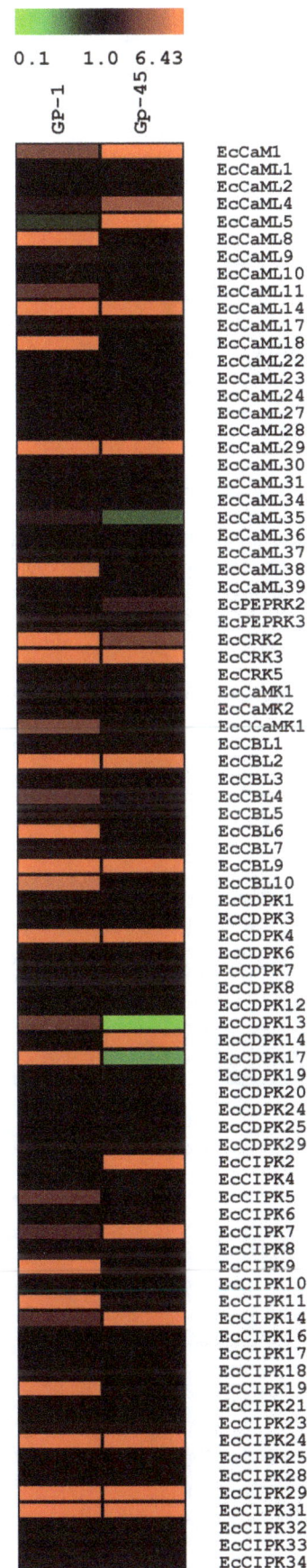

Figure 2. Expression of 82 Calcium sensor genes in pooled spikes of GP-1 (Low calcium) and GP-45 (High calcium) genotype. The number indicated on each cell represents the log2 calculated FPKM values. FPKM values smaller than 1 were not calculated due to negative logarithm and they were stated as in the original data.

it should be noticed that the total number of family members can be altered by the type of data source used in the study. Whole genome sequencing studies might be useful to obtain precise protein abundances compared to the transcriptome-wide data sources.

Correct classification of genes into gene families is important for understanding gene function and evolution. Genes of the same family usually share similar sequences, functional domains and even interacting partners. While some gene families are more dynamic in evolution and show species-specific gene members, others are more conserved and found in distantly related species or even across complete kingdoms of life. Gene family classification, *i.e.*, the grouping of genes or proteins into families, often yields important insights into gene function and gene evolution [21]. Many sequence-based methods for automated gene family classification have been developed within the last 20 years.

In present study a total of 82 genes of calcium sensor family identified from transcriptome data were classified that includes 25-CaM & CaML, 9-CRK, 9-CBL, 23-CIPK and 14-CDPK genes. This includes the gene sequences identified from developing spikes of finger millet and may have failed to spot many other genes of this family which are not expressed. Therefore, only phylogenetic approach of computational phylogeny cannot be applied with these gene sequences. Presence of high level of colinearity and sequence similarity between finger millet and rice genomes could be utilised in this situation [16]. The identified finger millet calcium sensors were verified for conserved sequences, motifs and domains alongside maximum sequence identity and comparative phylogeny.

Structural and functional analysis of calcium sensor genes

Domain and motif analysis of calcium sensor genes. Calcium sensor gene family in plants comprises of protein kinase (PKC), EF- hand and NAF- domain [22–24]. The CaM and CaM-like gene family in rice contains two pairs of EF-hand domain [25]. However, its number varies in case of finger millet and ranges from one pair of EF- hand to two pair of EF-hand (**Table 3**). The variation in EF hands in CaM and CaML genes generally, related with calcium binding capacity as one EF hand interact with one calcium ion. Accordingly, their function might also be modulated in different biological processes [26]. The gene family PEPRK and most of the CRK comprises of only protein kinase domain in plants. However, in phylogenetic tree the PEPRK and CRK are entirely apart from each other, this is due to variation in the amino acid sequences of these proteins other than at protein kinase region. The rice CaMK, CCaMK and CDPK gene family comprises of protein kinase domain and EF-hand domain [24]. The number of protein kinase domain is one in all three gene family but the number of EF-hand varies from one (as in case of CaMK), two (as in case of CCaMK) and four (as in case of CDPK). Same trend of number of EF-hand was also recorded in case of CaMK and CCaMK of finger millet, but variation was recorded in case of CDPK genes. The wide variability of EF-hand domains proteins shows the diversity of the processes in which Ca^{2+} is involved [26]. The CIPK gene family in rice have no EF-hand, but have kinase and NAF domain [27]. Similarly in finger millet, CIPK consisted of NAF-domain that interacts with Calcenurin B-like sensor proteins (CBLs). Whereas N-terminal part

Table 4. Functional category of differentially expressed calcium sensor gene in finger millet.

Calcium sensor proteins	Functional pathway
EcCaM1	Phosphatidylinositol signaling system, Plant-pathogen interaction
EcCaML4	Phosphatidylinositol signaling system, Plant-pathogen interaction
EcCaML11	Plant-pathogen interaction
EcCaML14	Plant-pathogen interaction
EcCaML18	Plant-pathogen interaction
EcCDPK4	Plant-pathogen interaction
EcCDPK13	Plant-pathogen interaction
EcCDPK14	Plant-pathogen interaction
EcCDPK17	Plant-pathogen interaction
EcCIPK31	Abiotic stresses, Cold stress tolerance

of CIPKs comprises of a conserved catalytic domain typical of Ser-threonine kinases. In plant CBL have same general structure of CaM but they have calcium binding domains with three EF-hand. The finger millet CBL also contains same number of EF-hand. As like similarity in domain feature of calcium sensor proteins of finger millet with their rice orthologs, similarity in motif and conserved sequence analysis of both organisms was also recorded. Similarity in the conserve region of finger millet and rice calcium sensor genes imply their correct identification and classification. Regardless of similarity at conserve region of finger millet with rice calcium sensor genes, variation in number of domains and motifs were also recorded. Variation in number of domain among members of rice and finger millet, calcium sensor genes might be due to incomplete sequence of finger millet genes or due to segmental duplication/deletion during the course of evolution among these genes.

Comparative phylogenetic analysis of calcium sensor gene family in finger millet and rice. To classify and predict the biological role of finger millet calcium sensors, phylogenetic distances were computed comparatively with rice calcium sensor proteins (**Fig. 1**). By referring to the rice calcium sensor protein characterization, most of the finger millet calcium sensor proteins were found within same group with rice. However, twelve calcium sensor genes were found diverse from the rice orthologs. This diversity in the structure of some sensor genes of finger millet with

rice might contribute some role in the distinctness of calcium content in finger millet and rice.

Sub-Cellular localization and stability analysis in calcium sensor genes. Many of the calcium sensor genes location could not be traced because of limitation of information available in the database (**Table 3**). However, an inference can be made from all above experimental analysis that cytoplasm and mitochondria of the cell encase most of the proteins and only a few are present in the secretory pathways. Another parameter studied was of stability and most of the calcium sensor proteins are found unstable according to their instability index parameters. Earlier report also suggests that most of CDPKs proteins are unstable and unstable short-lived protein often comprise regulatory functions [28]. Further wet lab experimentation is needed to be done to verify and validate these *in silico* results.

Differential expression of calcium sensor genes

Unlike most animals, plants are sessile and cannot migrate from poor-quality environment therefore have developed mechanisms like efficient uptake of minerals in order to adapt to their environment. Thus, they need to tolerate the particular conditions they encounter to survive. This makes plants an ideal system for the study of adaptive variation, and this is particularly true for finger millet, which shows substantial natural variation in terms of

Table 5. List of Primers designed from highly expressed genes from transcriptome data of high Ca containing genotype for qPCR analysis.

S.No	Primer code	Primer Seq (5'–3')	Primer length	Amplicon size	Tm (°C)
1	EcTubulinF	CTCCAAGCTTTCTCCCTCCT	20	207	58
	EcTubulinR	GCATCATCACCTCCTCCAAT	20		
2	EcCDPK3F	ATGTGCGTTCCGTGTACTCC	20	169	60
	EcCDPK3R	ATCTGGATCTCCCTGCGAAT	20		
3	EcCIPK2F	CGATGAGAACAGCAACCTGA	20	152	58
	EcCIPK2R	CCTTTGCACCGTCATAACCT	20		
4	EcCIPK9F	CCGTACGAGCTGGGGAAGA	19	148	61
	EcCIPK9R	CCCGCTTTATCTGCTCGACC	22		
5	EcCIPK11F	AACTTCTGGCTCAGCAGACT	20	157	61
	EcCIPK11R	GGTAAATCGTTCTTCCCGGC	20		

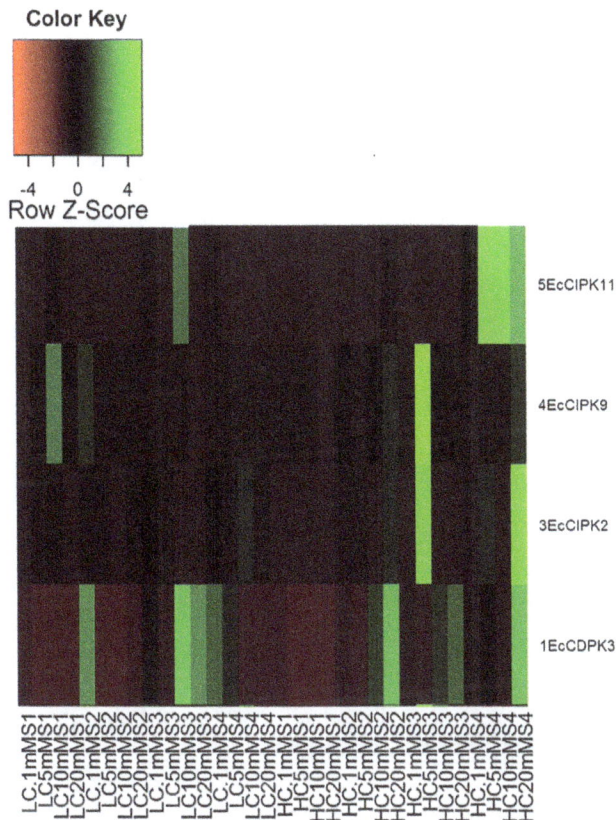

Figure 3. Expression profiles of selected Calcium sensor genes in developing spikes of GP-1 (LC) and GP-45 (HC) exposed to different doses of calcium (0.1 mM, 5.0 mM, 10 mM & 20 mM) as indicated by qPCR analysis. The scale representing the relative signal intensity values is shown above. The different stages of developing spikes are S1 (spike emergence); S2 (pollination stage); S3 (dough stage) and S4 (maturation).

mineral accumulation due to genetic variability. The genetic variability in the distribution of minerals in different plants of same species and within the edible tissues has long been thought to be utilized in biofortification strategies [29]. However, the distinctive patterns in the accumulation of minerals in plant tissues, cell types and sub-cellular compartments are the product of selective transport processes catalyzing their short distance as well as long distance movement [30]. As like in finger millet seed which contain different concentration of calcium in its different layers might be the outcome of efficient calcium transport machinery [31].

In general short distance ion movement depends upon the membrane transporters while the long distance ion movement utilizes the xylem and phloem pathways. The movement of most of the mineral elements in phloem fed tissue (like seeds) occurs predominantly through phloem tissue [30]. The immobility of calcium in phloem tissue [32], makes this study very interesting that how calcium could get accumulated in seed. Efforts were made in present investigation to study the role of calcium sensor genes in seed calcium accumulation by differential expression analysis in two selected genotypes.

The Ca^{2+}- transporters $viz.$, Ca^{2+}-ATPases, Ca^{2+}/Cation exchangers and calcium channels are the main class of calcium transporting proteins [33] and calcium sensors $viz.$, CaM & CaML, CBL, CIPK, CDPK, CRK, CaMK, CCaMK and PEPRK are regulatory proteins which play important role in calcium homeostasis [8,34]. Calcium transporters can directly regulate the trafficking of calcium within cell or tissue while calcium sensors indirectly involved in this process by regulating the activity of calcium transporters [35]. To study the role of calcium sensor gene in seed calcium accumulation, the expression level of these genes were detected by comparing the FPKM value of 82 genes expressed in the spike transcriptome of both the genotypes (**Fig. 2**). FPKM is defined as a quantification method for gene expression by the data obtained from RNA sequencing, to normalize the total read length and the number of sequencing reads [36]. The results of transcriptome reveals that the expression of 24 genes was found higher in the spike of GP-45 genotype and the expression of 11 genes was found higher in the spike of GP-1 genotype. The remaining genes were not assigned any value (0) even though their transcripts were detected. The 0 FPKM values in these genes are due to low expression of these genes and the value smaller than 1 were not calculated due to negative logarithm [37].

In a similar study on different layer of barley leaf transcriptome the transcript of calcium binding and transporter genes was found almost seven times more in calcium rich epidermal cell compared to other cell. Among these were included homologs of the auto-inhibited Ca^{2+}-ATPase (ACA) family, the cation exchanger (CAX) family, the annexins and numerous calcium-dependent/calmodulin interacting protein kinases (CDPK/CIPK) and calmodulin [38]. Homologs have also been found to be enriched in the mesophyll (where Ca is high) in the eudicot *Arabidopsis* by transcriptomic [39] and proteomic methods [40]. The tonoplast proteome of Arabidopsis mesophyll cells (the high Ca accumulating cell type in dicots), identified various calcium sensors, Ca^{2+}-ATPase, Ca^{2+}-Exchanger and two pore calcium channels genes [40]. The abundance of calcium sensor genes in the spike of high calcium containing genotype, as evident from FPKM value as with low calcium containing genotype might be most possible reasons of high seed accumulation in finger millet. The results again indicate the polygenic nature of calcium accumulation, as many genes are differentially expressed in both low and high calcium containing genotypes.

In pathway analysis, out of 35 differentially expressed genes, only 10 can be assigned to three pathways, were detected between GP-1 and GP-45. The majority of the differentially expressed genes belong to pathways involved in plant-pathogen interaction and different kind of biotic and abiotic stresses. During these pathways various calcium mediated signaling are operational. In one of the study it was reported that Arabidopsis AtCIPK24 regulates the activity of vacuolar Ca^{2+}/H^+ exchanger (VCaX1) during salt stress [41]. Similarly, high expression of EcCIPK31 gene in GP-45 and its involvement in salt stress might be linked with activation of any calcium exchanger and in turn high calcium accumulation. However, due to limited knowledge of pathway about differentially expressed gene further study is needed.

Quantitative real-time PCR validation

The calcium responsive study depicts that the availability of calcium in rhizosphere significantly (P<0.05) increased the calcium content in two genotypes of finger millet (**Fig. 4**). The results clearly revealed differential responsiveness of finger millet genotypes towards exogenously supplied calcium. Through qPCR analysis it was found that the expression of EcCDPK11 gene was significantly higher in high calcium containing genotype (GP-45) than low calcium containing genotype (GP-1) (**Fig. 3**). Its expression was also high at high concentration of exogenous calcium application as well as at later stages of spike development

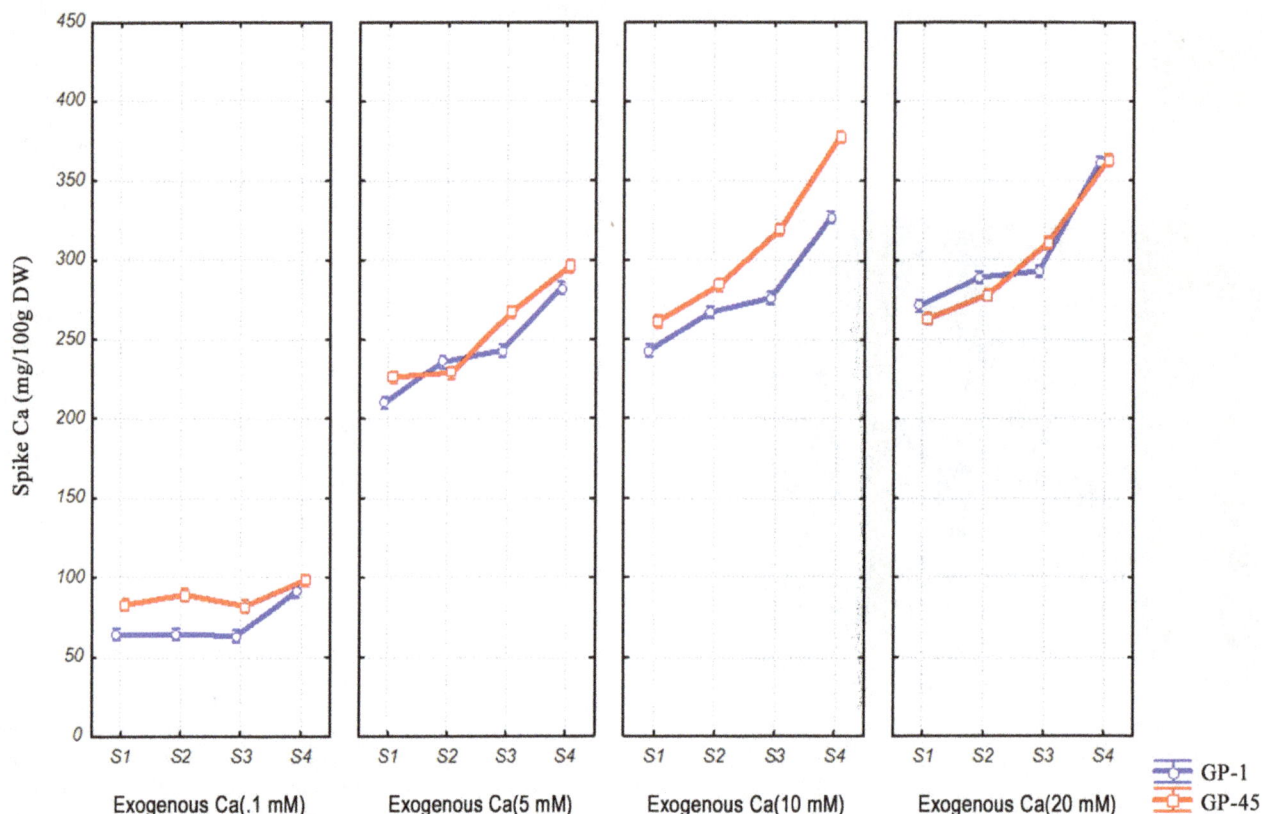

Figure 4. Pattern of calcium accumulation in developing spikes of GP-1 (Low calcium) and GP-45 (High calcium) grown under different concentration of exogenous calcium (0.1 mM, 5.0 mM, 10 mM & 20 mM). The different stages of developing spikes are S1 (spike emergence); S2 (pollination stage); S3 (dough stage) and S4 (maturation).

(**Fig. 3**). The higher accumulation of these genes might be associated with high seed calcium accumulation in finger millet by regulating the activity of Ca^{2+}- ATPase. Recently, in one of the study co-expression of Ca^{2+}-ATPase and CDPK was reported in finger millet [42]. The role of AtCDPK1 in regulation of calmodulin activated Ca^{2+}-ATPase2 (ACA2) was well studied in *Arabidopsis thaliana*. In that AtCDPK1 bind at N-terminal auto-inhibitory region of ACA2 and inhibit their activity [8].

The expression of EcCIPK2, EcCIPK9 and EcCIPK11 genes was comparatively higher in high calcium containing genotype (GP-45) in comparison to low calcium containing genotype (GP-1) (**Fig. 3**). Its expression increases as the amount of exogenous calcium in nutrient medium was increased (**Fig. 4**). Comparatively higher expression of these genes was obtained at later stages of seed development. However, the expression of EcCIPK19 gene was not corroborated with the calcium accumulation study and hence it might have no role in seed calcium accumulation. Increased expressions of rice OsCIPK2 gene was reported during K^+ deficient conditions and suspected that it may regulate activity of the K^+ channel, enhancing K^+ uptake under K^+ -deficient conditions [43]. The role of these genes in high seed calcium accumulation has not yet investigated. In one study, it was demonstrated that Arabidopsis SOS2 (AtCIPK24) regulates the activity of vacuolar Ca^{2+}/H^+ exchanger (VCaX1) [41]. Interaction of CIPK24 with CBL10 at vacuolar membranes was also reported [24] because binding of CBL10 is essential for the full functioning of CIPK24 gene. The higher accumulation of these gene transcripts might be well correlated with the increasing seed

calcium content by activating vascular CaX protein in finger millet.

Conclusions

In this work, we characterize calcium sensor gene families from the transcriptome of developing spikes of two finger millet genotypes differing in their grain calcium contents for ascertaining their role in grain calcium accumulation. To our knowledge, this is first attempt to assemble and characterize the calcium sensor gene family from the developing spike of finger millet transcriptome using Illumina paired-end sequencing methods. Based on the transcriptome assembly, calcium sensor gene family were isolated, classified and characterized. The 82 sequences were predicted as calcium sensor genes in finger millet. Majority of differentially expressed genes (24) among total calcium sensor genes (82) in GP-45 and 11 in GP-1 genotype were found higher FPKM value. Through KEGG pathway analysis out of 35 differentially expressed genes only ten were assigned to stress related pathways. For the validation of the results of transcriptome sequencing, some of the calcium sensor genes (especially responder) having 10 times higher FPKM value in GP-45 than in GP-1 genotype, was used for qPCR analysis under the influence of different concentration of exogenous calcium. Significant correlations with expression value and amount of calcium accumulated were observed. These results fully demonstrate that Illumina paired-end sequencing is a fast and cost effective approach for gene discovery and their characterization.

Author Contributions

Conceived and designed the experiments: AK SCS. Performed the experiments: UMS. Analyzed the data: MC UMS. Contributed reagents/materials/analysis tools: AK. Contributed to the writing of the manuscript: UMS AK.

References

1. Ross AC, Manson JE, Abrams SA, Aloia JF, Brannon PM, et al. (2011) The 2011 report on dietary reference intakes for calcium and vitamin D from the institute of medicine: what clinicians need to know. Journal of Clinical Endocrinology and Metabolism 96: 53–58.
2. National Research Council (1996) Lost crops of Africa. Vol. 1: Grains. National Academy Press, Washington, DC.
3. Clarkson DT (1984) Calcium transport between tissues and its distribution in the plant. Plant, Cell and Environment 7: 449–456.
4. Singh UM, Pandey D, Kumar A (2013) Determination of Calcium responsiveness towards exogenous application in two genotypes of *Eleusine coracana* L. differing in their grain calcium content. Acta Physiologia Plantarum (In Press).
5. Dayod M, Tyerman S, Leigh R, Gilliham M (2010) Calcium storage in plants and the implications for calcium biofortification. Protoplasma 247: 215–231.
6. Sanders D, Pelloux Jrm, Brownlee C, Harper JF (2002) Calcium at the Crossroads of Signaling. The Plant Cell Online 14: S401–S417.
7. Harmon AC, Gribskov M, Gubrium E, Harper JF (2001) The CDPK superfamily of protein kinases. New Phytologist 151: 175–183.
8. Hwang I, Sze H, Harper JF (2000) A calcium-dependent protein kinase can inhibit a calmodulin-stimulated Ca^{2+} pump (ACA2) located in the endoplasmic reticulum of Arabidopsis. Proceedings of the National Academy of Sciences 97: 6224–6229.
9. Ghawana S, Paul A, Kumar H, Kumar A, Singh H, et al. (2011) An RNA isolation system for plant tissues rich in secondary metabolites. BMC Research Notes 4: 85.
10. Liu S, Li W, Wu Y, Chen C, Lei J (2013) De Novo Transcriptome Assembly in Chili Pepper (*Capsicum frutescens*) to Identify Genes Involved in the Biosynthesis of Capsaicinoids. PLoS ONE 8: 1371.
11. Tamura K, Stecher G, Peterson D, Filipski A, Kumar S (2013) MEGA6: Molecular Evolutionary Genetics Analysis Version 6.0. Molecular Biology and Evolution 30: 2725–2729.
12. Trapnell C, Williams B, Pertea G, Mortazavi A, Kwan G, et al. (2010) Transcript assembly and quantification by RNA-Seq reveals unannotated transcripts and isoform switching during cell differentiation. Nature Biotechnology 28: 511–515.
13. Hoagland DR, Arnon DI (1950) The water-culture method for growing plants without soil. California Agricultural Experiment Station 2.
14. Barbeau WE, Hilu KW (1993) Protein, calcium, iron, and amino acid content of selected wild and domesticated cultivars of finger millet. Plant Foods Hum Nutr 43: 97–104.
15. Grabherr MG, Haas BJ, Yassour M, Levin JZ, Thompson DA, et al. (2011) Full-length transcriptome assembly from RNA-Seq data without a reference genome. Nat Biotech 29: 644–652.
16. Dida MM, Gale MD, Devos KM (2007) Comparative analyses reveal high levels of conserved colinearity between the finger millet and rice genomes. Theoretical and Applied Genetics 115: 489–499.
17. Asano T, Tanaka N, Yang G, Hayashi N, Komatsu S (2005) Genome-wide identification of the rice calcium-dependent protein kinase and its closely related kinase gene families: comprehensive analysis of the CDPKs gene family in rice. Plant and Cell Physiology 46: 356–366.
18. Boonburapong B, Buaboocha T (2007) Genome-wide identification and analyses of the rice calmodulin and related potential calcium sensor proteins. BMC Plant Biology 7: 4.
19. Kolukisaoglu Ã, Weinl S, Blazevic D, Batistic O, Kudla J (2004) Calcium sensors and their interacting protein kinases: genomics of the Arabidopsis and rice CBL-CIPK signaling networks. Plant Physiology 134: 43–58.
20. Hrabak EM, Chan CWM, Gribskov M, Harper JF, Choi JH, et al. (2003) The Arabidopsis CDPK-SnRK superfamily of protein kinases. Plant Physiology 132: 666–680.
21. Wu CH, Huang H, Yeh L-SL, Barker WC (2003) Protein family classification and functional annotation. Computational Biology and Chemistry 27: 37–47.
22. Kudla Jr, Xu Q, Harter K, Gruissem W, Luan S (1999) Genes for calcineurin B-like proteins in Arabidopsis are differentially regulated by stress signals. Proceedings of the National Academy of Sciences 96: 4718–4723.
23. Snedden WA, Fromm H (2001) Calmodulin as a versatile calcium signal transducer in plants. New Phytologist 151: 35–66.
24. Klimecka M, Muszynska G (2007) Structure and functions of plant calcium dependent protein kinases. Acta Biochimica Polonica 54(2): 219–233.
25. Gifford JL, Walsh MP, Vogel HJ (2007) Structures and metal-ion-binding properties of the Ca^{2+}-binding helix-loop-helix EF-hand motifs. Biochem J 405: 199–221.
26. Day IS, Reddy VS, Ali GS, Reddy AS (2002) Analysis of EF-hand-containing proteins in Arabidopsis. Genome Biology 3: 1–0056.
27. Albrecht V, Weinl S, Blazevic D, D'Angelo C, Batistic O, et al. (2003) The calcium sensor CBL1 integrates plant responses to abiotic stresses. The Plant Journal 36: 457–470.
28. Lyzenga WJ, Liu H, Schofield A, Muise-Hennessey A, Stone SL (2013) Arabidopsis CIPK26 interacts with KEG, components of the ABA signalling network and is degraded by the ubiquitin–proteasome system. Journal of Experimental Botany. doi:10.1093/jxb/ert123
29. White PJ, Broadley MR (2009) Biofortification of crops with seven mineral elements often lacking in human diets – iron, zinc, copper, calcium, magnesium, selenium and iodine. New Phytologist 182: 49–84.
30. Karley AJ, White PJ (2009) Moving cationic minerals to edible tissues: potassium, magnesium, calcium. Current Opinion in Plant Biology 12: 291–298.
31. Nath M, Roy P, Shukla A, Kumar A (2013) Spatial distribution and accumulation of calcium in different tissues, developing spikes and seeds of finger millet genotypes. Journal of Plant Nutrition 36: 539–550.
32. Busse JS, Palta JP (2006) Investigating the *in vivo* calcium transport path to developing potato tuber using ^{45}Ca: a new concept in potato tuber calcium nutrition. Physiologia Plantarum 128: 313–323.
33. Goel A, Taj A, Pandey D, Gupta S, Kumar A (2011) Genome wide comparative in silico analysis of calcium transporter of Rice and Sorghum. Genomic Proteomics and Bioinformatics 9: 138–150.
34. Sanders D, Brownlee C, Harper JF (1999) Communicating with Calcium. The Plant Cell Online 11: 691–706.
35. Conn SJ, Gilliham M, Athman A, Schreiber AW, Baumann U, et al. (2011) Cell-specific vacuolar calcium storage mediated by *CAX1* regulates apoplastic calcium concentration, gas exchange, and plant productivity in Arabidopsis. The Plant Cell Online 23: 240–257.
36. Mortazavi A, Williams BA, McCue K, Schaeffer L, Wold B (2008) Mapping and quantifying mammalian transcriptomes by RNA-Seq. Nat Meth 5: 621–628.
37. Tombuloglu H, Kekec G, Sakcali M, Unver T (2013) Transcriptome-wide identification of R2R3-MYB transcription factors in barley with their boron responsive expression analysis. Molecular Genetics and Genomics 288: 141–155.
38. Richardson A, Boscari A, Schreiber L, Kerstiens G, Jarvis M, et al. (2007) Cloning and expression analysis of candidate genes involved in wax deposition along the growing barley (*Hordeum vulgare*) leaf. Planta 226: 1459–1473.
39. Yang Y, Costa A, Leonhardt N, Siegel RS, Schroeder JI (2008) Isolation of a strong Arabidopsis guard cell promoter and its potential as a research tool. Plant Methods 4(1): 6.
40. Carter C, Pan S, Zouhar J, Avila EL, Girke T, et al. (2004) The vegetative vacuole proteome of *Arabidopsis thaliana* reveals predicted and unexpected proteins. The Plant Cell Online 16: 3285–3303.
41. Qiu Q-S, Guo Y, Quintero FJ, Pardo JM, Schumaker KS, et al. (2004) Regulation of vacuolar Na^+/H^+ exchanger in *Arabidopsis thaliana* by the salt-overly-sensitive (SOS) Pathway. Journal of Biological Chemistry 279: 207–215.
42. Kumar A, Mirza N, Charan T, Sharma N, Gaur V (2014) Isolation, characterization and immunolocalization of a seed dominant CaM from finger millet (*Eleusine coracana* L. Gartn.) for studying Its functional role in differential accumulation of calcium in developing grains. Applied Biochemistry and Biotechnology: 1–19.
43. Ma T-L, Wu W-H, Wang Y (2012) Transcriptome analysis of rice root responses to potassium deficiency. BMC Plant Biology 12: 161.

12

Comparative Metabolite Profiling of Two Rice Genotypes with Contrasting Salt Stress Tolerance at the Seedling Stage

Xiuqin Zhao[9], Wensheng Wang[9], Fan Zhang, Jianli Deng, Zhikang Li, Binying Fu*

Institute of Crop Sciences, National Key Facility for Crop Gene Resources and Genetic Improvement, Chinese Academy of Agricultural Sciences, Beijing, China

Abstract

Background: Rice is sensitive to salt stress, especially at the seedling stage, with rice varieties differing remarkably in salt tolerance (ST). To understand the physiological mechanisms of ST, we investigated salt stress responses at the metabolite level.

Methods: Gas chromatography-mass spectrometry was used to profile metabolite changes in the salt-tolerant line FL478 and the sensitive variety IR64 under a salt-stress time series. Additionally, several physiological traits related to ST were investigated.

Results: We characterized 92 primary metabolites in the leaves and roots of the two genotypes under stress and control conditions. The metabolites were temporally, tissue-specifically and genotype-dependently regulated under salt stress. Sugars and amino acids (AAs) increased significantly in the leaves and roots of both genotypes, while organic acids (OAs) increased in roots and decreased in leaves. Compared with IR64, FL478 experienced greater increases in sugars and AAs and more pronounced decreases in OAs in both tissues; additionally, the maximum change in sugars and AAs occurred later, while OAs changed earlier. Moreover, less Na^+ and higher relative water content were observed in FL478. Eleven metabolites, including AAs and sugars, were specifically increased in FL478 over the course of the treatment.

Conclusions: Metabolic responses of rice to salt stress are dynamic and involve many metabolites. The greater ST of FL478 is due to different adaptive reactions at different stress times. At early salt-stress stages, FL478 adapts to stress by decreasing OA levels or by quickly depressing growth; during later stages, more metabolites are accumulated, thereby serving as compatible solutes against osmotic challenge induced by salt stress.

Editor: Jauhar Ali, International Rice Research Institute, Philippines

Funding: This work was supported by the Bill & Melinda Gates Foundation Project (Grant No. OPP51587); The Program of Introducing International Super Agricultural Science and Technology (Grant No. 2011-G2B) and The Program of International Science and Technology Cooperation (Grant No. 2012DFB32280). The funders had no role in study design, data collection and analysis, decision to publish, or preparation of the manuscript.

* Email: fubinying@caas.cn

[9] These authors contributed equally to this work.

Introduction

Rice (*Oryza sativa* L.) is the staple food for nearly half of the world's population. However, its growth and yield are greatly influenced by soil salinity [1–2], especially at the seedling stage [3–4]. An understanding of the physio-biochemical attributes that enable plants to survive under saline conditions is beneficial for improving rice production.

A high cellular salt concentration leads to ionic and osmotic imbalance that disrupts plant ion homeostasis and water potential, resulting in metabolic damage, growth arrest and even death [5]. Significant differences in salt tolerance (ST) between varieties are mainly related to the ability of plants to regulate the amount of sodium (Na^+) and chloride reaching leaves or to metabolic regulation when roots are presented with saline conditions [6–8]. Higher tolerance to elevated Na^+ is also sometimes associated with these differences [9]. Compared with the effects of ion toxicity, rice is reported to suffer greater osmotic stress under saline conditions. The contents of several metabolites, including amino acids (AAs), sugars and polyols, are increased under saline conditions, thereby serving as a defense against osmotic challenge by acting as compatible solutes [10–15].

Metabolites are the end products of cellular regulatory processes. Their levels can be regarded as the ultimate response of biological systems to genetic or environmental changes and are closely associated with phenotype [16]. In rice, metabolic analysis of ST has been very limited relative to substantive transcriptomics and/or proteomics research [17–21]. To date, no reports have appeared on metabolic response to salt stress over time in varieties with different ST, even though the systematic analysis of metabolic snapshots is considered a valid approach for quantitative description of cellular regulation and control [16,22]. Furthermore, contradictory views exist regarding the functions of some metabolites during stress response. For example, proline appears to be a preferred organic osmoticum in many plants [10,12,23–26], whereas proline accumulation in rice is considered to be a symptom of injury rather than an indicator of ST [27–29].

In the current study, we used GC-MS to conduct metabolic profiling analysis on leaves and roots of two rice varieties, FL478 and IR64, under control and saline conditions. We also simultaneously investigated morphological and physiological traits. The two varieties differ greatly in ST and in their metabolic responses to the imposed salt stress. Our results provide insights into the specific adaptive response of rice to salt stress.

Material and Methods

Two contrasting rice genotypes, FL478 and IR64, were used as research materials. FL478 is a salt-tolerant recombinant inbred line developed at the International Rice Research Institute using Pokkali (a salt-tolerant donor) and IR29. Salt-sensitive IR64 is a commercial variety from Asia.

Plant growth conditions

Pot experiments, designed to investigate the different metabolic, morphological and physiological responses to salt stress between FL478 and IR64, were conducted in a greenhouse at the Chinese Academy of Agricultural Science, Beijing, China, in May 2012. After establishing water screens, the average daytime temperature in the greenhouse was about 25–29°C, which was appropriate for rice growth. The rice plants were cultivated in nutrient solution [30]. Salt treatments were conducted as reported previously [31]. Briefly, rice plants were cultivated to the four-leaf stage, and stress was then applied by adding NaCl to a final concentration of 100 mM and an electrical conductivity (EC) of 12 dS m^{-1}. The EC of the control tanks was around 1.0 dS m^{-1}. Three replicates were performed per treatment.

Physiological trait analysis

Salt score (SS). Salt scoring was conducted 7 d after the stress treatment. The salt stress symptoms of 10 plants per line were scored according to a modified standard evaluation system [32] for ST at the seedling stage, which ranged from 1 (normal growth) to 9 (almost all plants dead or dying).

Relative water content (RWC). The youngest fully expanded leaves were harvested at 7 d for RWC analysis. RWC (%) was calculated as $100 \times (FLW - DLW)/(TLW - DLW)$, where FLW, TLW and DLW are fresh leaf weight, turgid leaf weight after saturation and dry leaf weight, respectively.

Na$^+$ and K$^+$ in shoots and roots. Na$^+$ and K$^+$ contents in both shoots and roots under control and saline conditions at 7 d were estimated using a flame photometer (S2; Thermo Finnigan, Waltham, MA, USA). After washing with distilled water to remove surface Na$^+$ contamination, shoot and root samples were left to dry at 50°C for 4 d. The dried tissue was ground into a powder in liquid nitrogen. The resulting powder was acid digested by suspending in 5 ml of concentrated nitric acid overnight. Five milliliters of a 10:4 diacid mixture of nitrate and perchlorate was added to the partially digested tissue powder, which was then incubated for 2 h on a sand bath to allow complete digestion. The digested solution was diluted to 25 ml with double-distilled water. Na$^+$ and K$^+$ levels in the acid-digested samples, representing total

Na$^+$ and K$^+$ in the tissue samples, were estimated using the flame photometer.

Length and dry weight of shoots and roots. Plants grown under stress and control conditions were harvested at 7 d. Plant height, root length, and dry weight of the aboveground biomass and roots were measured.

Metabolite extraction and identification

The topmost leaves and roots of plants cultivated under both control and stress conditions were collected simultaneously at 1, 3 and 7 d after salt treatment. The samples were flash-frozen in liquid nitrogen and stored at −70°C until metabolite extraction.

The samples were ground in liquid nitrogen with a pestle and mortar. Separate aliquots of the frozen, ground sample were used for metabolite extraction as described previously [33–34]. The extracted samples were then derivatized and analyzed by gas chromatography-mass spectrometry (GC-MS). The GC-MS system comprised an AOC 5000 auto-sampler, GC 2010 gas chromatograph and Voyager quadrupole mass spectrometer (Shimadzu, Kyoto, Japan). The mass spectrometer was tuned using tris-(perfluorobutyl)-amine. GC was performed on a 30-m Rtx-5 MS column with a 0.25-μm film thickness. A mixture of leaves and roots of both genotypes under stress and control conditions at 1, 3 and 7 d was extracted in bulk and used as a reference. Reference samples were run once every 10 samples. N-methyl-N-[trimethylsilyl] trifluoroacetamide (MSTFA) was run once every five samples to clean the injection surface of potential pollution.

Chromatograms and mass spectra were processed using the find algorithm implemented in GC-MS Postrun Analysis software (Shimadzu). Specific mass spectral fragments were detected in defined retention time windows using the mass spectral library NIST (http://www.nist.gov/mml/chemical_properties/data/) and the public domain mass spectral library of the Max Planck Institute for Plant Physiology, Golm, Germany (http://csbdb.mpimp-golm.mpg.de/). Most AAs, organic acids (OAs) and sugars were verified by performing standard addition experiments using pure authenticated compounds.

Data analysis

The denominator of the quotient was the average response of the reference sample. The sample responses were volume-corrected with ribitol to compensate for errors during sample preparation or GC injection and normalized using sample fresh weight.

Analysis of variance (ANOVA) was performed in SAS v6.12 (SAS Institute Inc., Cary, NC, USA, 1996) was performed to determine the significance of trait differences between genotypes (G), tissues (T), sampling time points (t) and treatments (S). Specifically, metabolite differences associated with G and T were analyzed using data obtained at 1 d under control conditions, and metabolite differences related to t were analyzed with data from each genotype at three sampling time points under control conditions. Finally, metabolite differences due to S were analyzed using data from both control and stress treatments of each genotype at each time point. Differentially changed metabolites were defined as those showing significant concentration increases or decreases relative to their respective controls at $P \leq 0.05$ in ANOVA. Hierarchical cluster analysis of metabolites in roots and leaves was carried out using R software (http://www.r-project.org/).

Results

Salt stress effects on morphological and physiological traits of FL478 and IR64

The two genotypes, FL478 and IR64, were evaluated for growth parameters after the time-series salt treatment. As shown in Figure 1, the salinity score (SS) values of both genotypes altered under salt stress. FL478 had much better ST than IR64, as confirmed by a lower SS ($SS_{FL478} = 2.67$; $SS_{IR64} = 5$) (Figure 1A) and reduced repression of growth. For example, plant height and biomass respectively decreased by 27.6% and 46.7% in FL478 and 45.1% and 56.8% in IR64 (Figure 1B–C). Although salt stress had relatively less effect on root length and root dry weight (Figure 1D–E).

Physiological analysis revealed that RWC decreased by 29.9% in IR64 but only 11.7% in FL478 under salt stress (Figure 2A). Salt treatment drastically decreased the concentration of K^+ and increased that of Na^+ in both leaves and roots. Na^+ concentration increased by 34.8- and 46.2-fold in leaves of FL478 and IR64, respectively (Figure 2B), and 7.7- and 9.1-fold in roots of FL478 and IR64, respectively (Figure 2C). The most significant Na^+ increase was observed in leaves of IR64, with the Na^+ concentration in leaves of IR64 about twice that of FL478 under saline conditions. Additionally, under salt stress, leaves of IR64 displayed higher Na^+/K^+ ratio levels than those of FL478; in roots, similar ratios were observed between the two genotypes (Figure 2D).

Metabolite profiles of FL478 and IR64 under control and salt stress conditions

Using GC-MS, 89 and 86 metabolites were identified in leaves and roots, respectively, of both FL478 and IR64 (Table S1) under control and salt stress conditions.

Differences between genotypes (G), tissues (T) and sampling time points (t) were examined with ANOVA (Table S2). Out of all metabolites identified from both tissues, 30–32 metabolites (35–40%) showed significant G differences in different tissues at 1 d under control conditions; these differences, on average, explained 85.1–87.8% of the total phenotypic variation in these metabolites (Table S2.1). Additionally, 68–70 metabolites (74–76%) showed significant T differences in FL478 and IR64, respectively, which, on average, explained 92–95% of the total phenotypic variation (Table S2.2). Finally, 36–44 (44–49%) metabolites in leaves and 54–56 (63–65%) metabolites in roots were greatly influenced by

sampling time, which, on average, explained 82.3–88.0% of the total phenotypic variation of the measured metabolites (Table S2.3). Because of the greater sensitivity of the metabolites to t and T, we focused our comparative study on individual metabolites showing significant differences under stress treatment in roots and leaves of both genotypes.

An overview of metabolite changes in roots and leaves

All metabolite change data from roots and leaves of both cultivars were analyzed by hierarchical clustering to provide a global view of metabolite changes in response to salt stress (Figure 3). Leaf and root samples were clustered into distinct groups. Clear separations were observed between root samples collected at different time points. In contrast, the clustering of leaf samples was more complicated: IR64 at 1 d and FL478 at 7 d clustered into one group, and IR64 at 3 d and FL478 at 3 d clustered into another, with IR64 at 7 d separated significantly from the other samples. These results indicate that differences in the metabolic phenotypes of sensitive and tolerant rice cultivars were more pronounced in leaves than in roots.

Differentially changed metabolites in leaves of IR64 vs. FL478

To determine the responses of each genotype to salt stress, we compared levels of each metabolite in the stressed plants to levels in the control plants at similar time points (Table S3).

AA regulation under salt stress. Salt stress generally increased the levels of all 15 AAs in FL478 and IR64 (Table S3). Of the total 90 stress vs. control comparisons, 55 (61.1%) cases showed increases, whereas two (2.2%) decreased significantly. Compared with IR64, FL478 had more AAs showing increased levels. Moreover, the maximum number of measured AAs showing significantly increased levels occurred from 3–7 d in FL478, but occurred only at 3 d in IR64. The highest change levels occurred at 1 d in both genotypes. Only phenylalanine consistently increased in both genotypes over the course of the treatment. Five AAs (lysine, valine, proline, isoleucine and threonine) consistently increased in FL478 during salt treatment, with significant 3.25–6.91-fold increases observed; these same AAs were decreased or unchanged in IR64 at 7 d of salt treatment.

Sugar regulation under salt stress. Concentrations of most sugars were increased by salt treatment, indicating that increased sugar is beneficial to ST (Table S3). Out of 186 stress vs. control comparisons, 67 (36%) and 11 (5.9%) involved significant

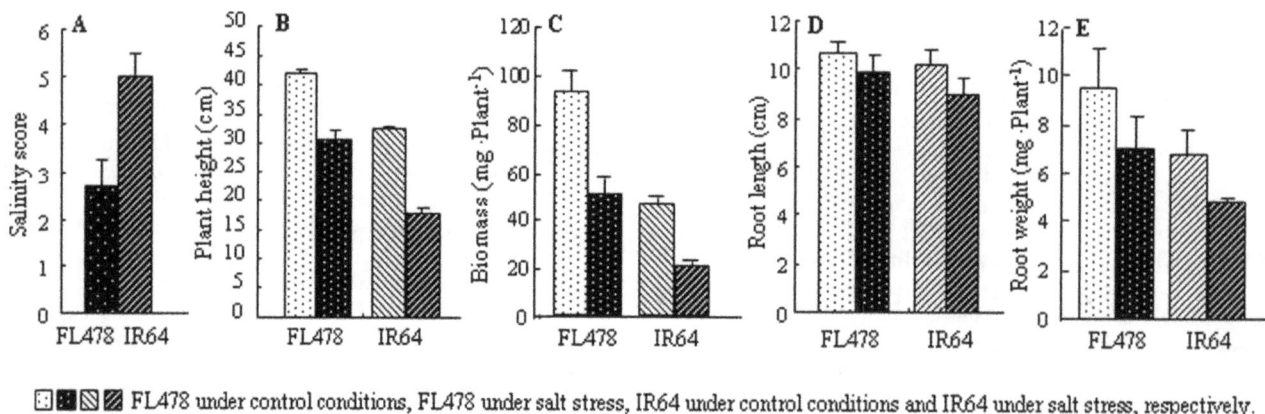

▨▧▨▨ FL478 under control conditions, FL478 under salt stress, IR64 under control conditions and IR64 under salt stress, respectively.

Figure 1. Morphological traits measured in both FL478 and IR64 under control and/or salt stress conditions. A: salinity score under salt conditions. B: plant height. C: plant biomass. D: root length. E: root dry weight.

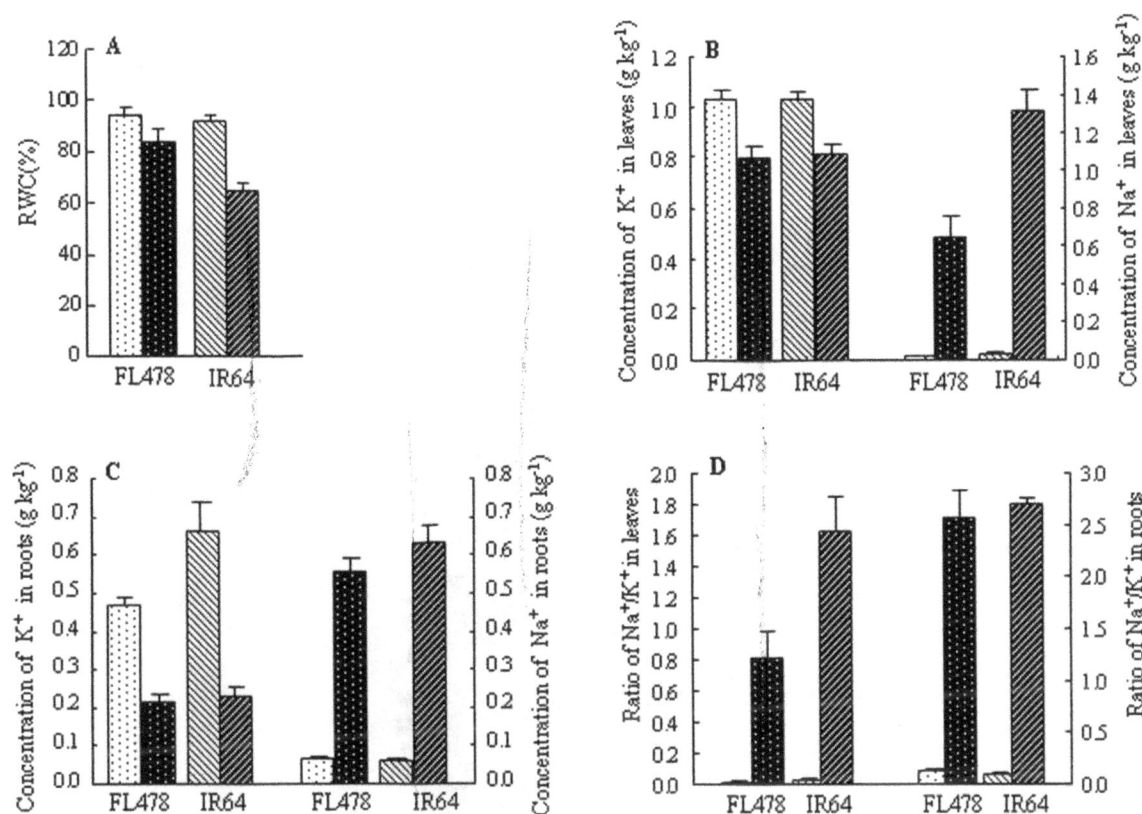

□ ■ ▨ ▧ FL478 under control conditions, FL478 under salt stress conditions, IR64 under control conditions and IR64 under stress conditions, respectively.

Figure 2. Physiological traits measured in both FL478 and IR64 under salt stress and control conditions. A: relative water content (RWC). B: K^+ and Na^+ concentration in leaves. C: K^+ and Na^+ concentration in roots. D: ratio of Na^+/K^+ in leaves and roots.

increases and decreases, respectively. Compared with IR64, FL478 had significantly more sugars showing increased levels. Additionally, the maximum number of sugars showing significantly increased levels occurred at 3–7 d in FL478 and 1–3 d in IR64. The highest change levels occurred at 7 d in FL478 and 3 d in IR64. Only melicitose and raffinose showed consistently increased levels across all time points of the stress treatment in both genotypes. Four sugars (sucrose, lactose, sorbitol and mannitol) were consistently increased over the course of the stress treatment only in FL478.

OA regulation under salt stress. Interestingly, in contrast to the change patterns observed for AAs and measured sugars, salt stress caused a decrease in most OAs in both IR64 and FL478 (Table S3). Of 156 stress vs. control comparisons, 21 (13.5%) involved increases while 31 (19.91%) exhibited significant decreases. Compared with IR64, FL478 had more OAs showing significantly decreased levels, especially at 1–3 d. OAs fumaric acid, succinic acid, malic acid and oxalic acid, involved in the tricarboxylic acid (TCA) cycle, decreased significantly in FL478, but increased or showed no change in IR64 at 1 d. Isocitric acid was specifically increased in FL478 over the course of the treatment.

Differentially changed metabolites in roots of FL478 vs. IR64

We identified 86 compounds in roots. Table S4 shows the range of metabolites detected and how their levels changed after different stress treatment durations. Eighty of the 86 metabolites identified in roots were differentially regulated in both genotypes at least one salt stress time point. Similar to the change pattern of metabolites in leaves, most of the significantly changed metabolites were induced in roots. This result confirms that the metabolite increases were favorable for plant growth under saline conditions.

AA regulation in roots under salt stress. A number of AAs increased under salt stress. Out of 90 stress vs. control comparisons, 28 (31.1%) and 12 (13.3%) cases increased and decreased significantly, respectively (Table S4). Threonine showed consistently increased levels across all stress time points in both genotypes, while five AAs (phenylalanine, tryptophan, glutamic acid, aspartic acid and proline) showed increased levels at two time points in one or both genotypes. Compared with IR64, FL478 had a slightly larger number of increased AAs and a higher average change level, especially at 7 d. Proline levels increased only during stress treatment in FL478.

Sugar regulation in roots under salt stress. More sugars were observed to increase than to decrease in roots of both genotypes under saline conditions (Table S4). Out of 180 stress vs. control comparisons, 67 (37.2%) and 28 (15.6%) cases were

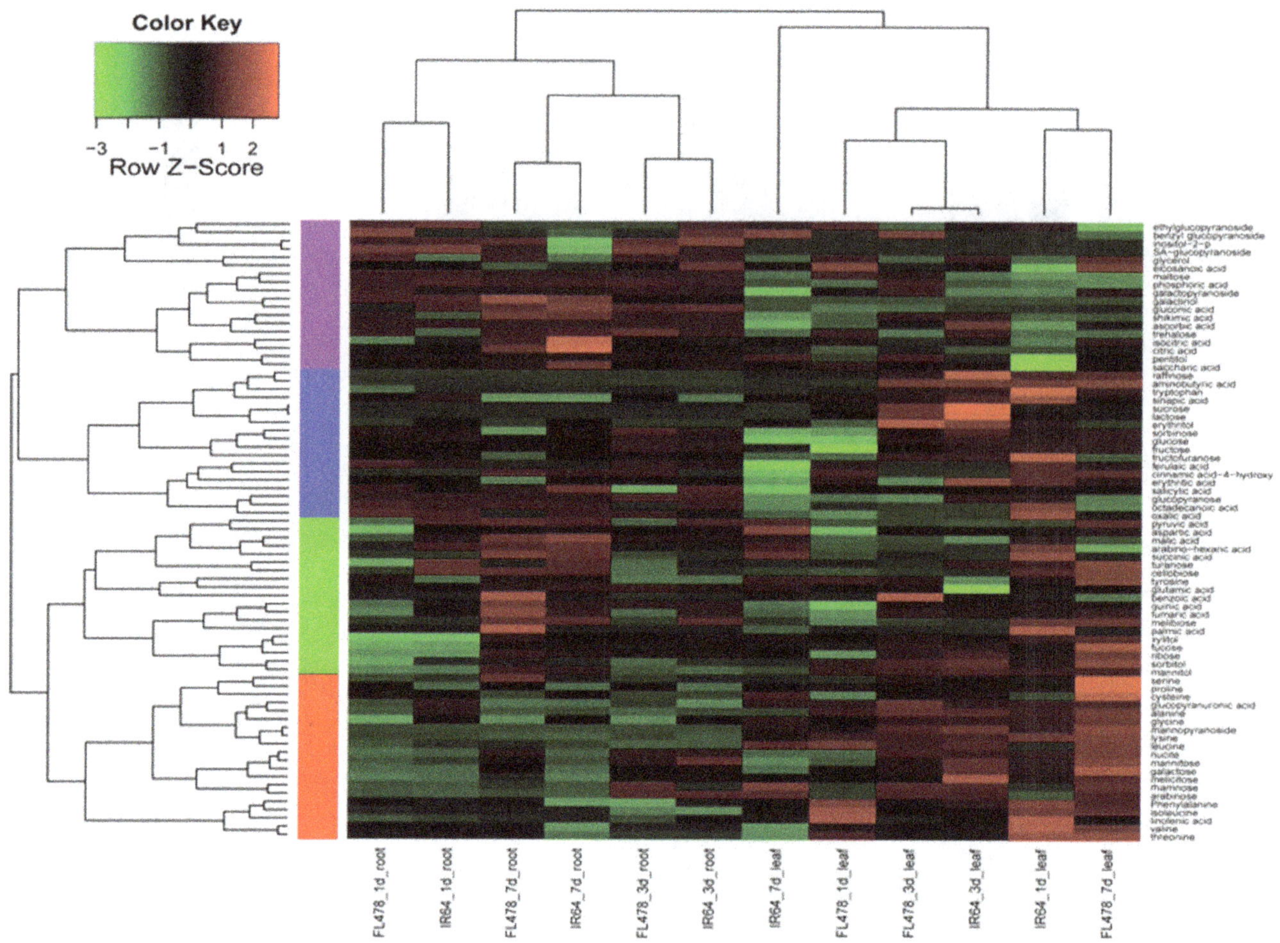

Figure 3. Results of hierarchical cluster analysis of changed metabolite pools. Hierarchical trees were drawn based on detected changed metabolites in leaves and roots of FL478 and IR64 under 1-, 3- and 7-d salt stress treatments. Columns correspond to genotypes at different time points, while rows represent different metabolites. Red and green colors indicate increased and decreased metabolite concentrations, respectively.

associated with significant increases and decreases, respectively. Overall, similar change patterns were uncovered between the roots of the two genotypes, including similar change numbers, change levels and times associated with the maximum number of sugars showing significant variation (3 d in both rice genotypes). Nevertheless, change levels of sugars were generally lower in roots than in leaves over the course of the stress treatment. Only galactinol was commonly increased in both genotypes. Galacto-pyranoside was increased specifically in FL478 (Table S4).

OA regulation in roots under salt stress. Most OAs were differentially regulated in roots at at least one salt stress sampling point (Table S4). Interestingly, unlike the change pattern in leaves in which more OAs were decreased, a larger number of OAs increased in roots under salt stress. Out of 150 stress vs. control comparisons, 53 (35.3%) and 23 (15.3%) cases involved significant increases and decreases in roots, respectively. The maximum number of OAs showing significantly increased levels occurred at 7 d in both FL478 and IR64. Compared with IR64, FL478 had a greater number of decreased OAs across the three stress time points. These decreased OAs included two participants in the TCA cycle: fumaric and succinic acid. Only citric acid was commonly increased in both genotypes across the three sampling time points; however, FL478 had a much lower change level than

IR64, with citric acid levels increasing 1.5–6.8-fold in FL478 and 2.0–12.3-fold in IR64.

Discussion

Rice genotypes vary considerably in salt stress tolerance. We analyzed two rice genotypes, FL478 and IR64, with contrasting ST. Under high salt conditions, these two genotypes performed differently over the 7-d treatment. Compared with IR64, FL478 was less affected, as evidenced by a smaller reduction in biomass, plant height and root length, and a lower SS value. Shoot growth was more extensively affected than root growth in both genotypes, which is consistent with a previous report that leaf growth was more seriously affected by salt stress than root growth [35]. Moreover, a higher RWC was observed in FL478 under salt stress, a condition that may help rice plants maintain cell turgidity and normal metabolism under stress conditions.

Although salt stress caused remarkably decreased potassium and increased sodium levels in both genotypes, the extent of change was much lower in FL478. Under saline conditions, FL478 had a much lower Na^+ concentration in leaves, i.e., nearly half of that found in IR64. At the same time, the Na^+/K^+ ratio was significantly lower in leaves of FL478 than in IR64. Similar Na^+/K^+ ratios were found between roots of the two genotypes

(Figure 2D), demonstrating that less sodium was translocated from roots into FL478 leaves. All these results imply that the ST of FL478 is due to the capacity of this genotype to maintain a lower Na^+/K^+ ratio in leaves. This conclusion is in accord with previous studies [6–9].

GC-MS analysis revealed that most AAs and sugars increased significantly in leaves and roots of both genotypes under salt stress. Compared with IR64, the salt-tolerant variety FL478 had a later and more prolonged increase in AAs and sugars (Tables S3 and S4). This result is consistent with a previous study of barley under salt stress, in which an increase in AA levels occurred later in a salt-tolerant variety than in a sensitive variety [9]. AAs have been observed to generally increase in many plant species, where they act as osmolytes in response to abiotic stress [36] or are considered to be an indicator of general stress [9]. These increased AAs may be the result of *de novo* synthesis, reduced protein synthesis or general protein breakdown during stress [9,36–39]. Nevertheless, the manner in which increased AA levels contribute directly or indirectly to salt stress tolerance is still unclear. Soluble sugars not only function as metabolic resources and structural constituents of cells, but also act as compatible solutes under various stress conditions [37,40–42]. The salt-tolerant variety FL478 had a greater number of increased sugars, whose levels were highest at 7 d, whereas almost no sugars experienced increases at the later stress time in IR64. The heavier accumulation of metabolites at this late stage suggests their possible function as compatible solutes against osmotic challenge, a consequence of higher Na^+ accumulation in plants after long-time exposure to salt stress.

In contrast to the consistent change patterns of AAs and sugars in both tissues, OAs decreased in leaves and increased in roots under salt stress. This opposite change pattern may be related to the different functions of leaves and roots. First, the fact that OAs decreased in leaves indicates that energy production or plant growth was repressed by salt stress. Second, the degree of cation–anion imbalance is one of the key factors determining OA levels in plants. In situations where roots take up an excess of cations, the negative charge required to restore the charge balance is often provided by OAs, such as malate, malonate, citrate and aconitate [43–44]. Thus, the increased OAs observed in roots may function to compensate for a charge imbalance [14], or operate as metabolically active solutes for osmotic adjustment [45]. Importantly, the larger number of more strongly and exclusively decreased OAs, including those participating in the TCA cycle, observed at early stress times (1–3 d) in FL478 compared with IR64, indicate that the growth of FL478 was depressed earlier. This result is consistent with previous observations that stomatal conductance and transpiration rate experienced greater decreases in leaves of FL478 than of sensitive variety IR29 under salt stress [18].

Several metabolites either had a common response in both genotypes or were specifically changed in FL478 during salt treatment, indicating their possible relationship to ST. In total, 6 metabolites, including two AAs (phenylalanine and threonine), one OA (citric acid) and three sugars (raffinose, melicitose and galactinol) were commonly increased in the leaves or roots of both genotypes, while 11 metabolites, including five AAs (lysine, threonine, isoleucine, proline, valine), one OA (isocitric) and five sugars (sucrose, lactose, sorbitol, mannitol and galactopyranoside), were greatly and specifically induced in leaves or roots of FL478 over the course of the treatment. Most of these metabolites are

proposed to be related to stresses [40–41,46–52]. To our knowledge, however, an association between lactose, sorbitol and melicitose and stress tolerance has not been previously reported.

Finally, the debate regarding the function of proline in stress tolerance [23,25–26] can be partially resolved on the basis of our observations that leaf proline contents increased in both genotypes under stress, whereas proline content in roots decreased in IR64 but increased in FL478. We therefore conclude that proline accumulation is an indicator of stress tolerance, not an injury system. Additional longer time-series research using contrasting materials might aid assessment of conflicting opinions regarding the functions of proline in stress tolerance.

Conclusions

We used GC-MS to comparatively analyze metabolite changes in two contrasting rice genotypes under salt stress treatment. As a result, genotype- and time-dependent metabolite profiles in response to salt stress were uncovered. First, the levels of most AAs and sugars increased consistently in both leaves and roots. In contrast, OA levels showed opposite changes between roots and leaves because of the different adaptive mechanisms of the two tissues and the differential functions of OAs against salt stress. Second, the superior ST of FL478 was attributed to its lower Na^+ absorption and higher RWC under stress. At the metabolic level, the salt stress response of FL478 involved a greater decrease in OAs during early stress stages, with a higher accumulation of AAs and sugars, which function as effective compatible solutes against osmotic challenge, during later stress stages. Third, 11 metabolites were exclusively increased in FL478 under stress, implying their positive roles in ST. Three of these compounds—lactose, sorbitol and melicitose—are newly recognized as metabolites related to ST.

Supporting Information

Table S1 Metabolite comparison between leaves and roots of FL478 and IR64 under control and salt stress conditions.

Table S2 S2.1: Results of ANOVA for genotypic differences (G) of metabolites in leaves and roots between FL478 and IR64 under normal growth conditions. S2.2: Results of ANOVA for tissue differences (T) between leaves and roots of IR64 and FL478 under normal growth conditions. S2.3: Results of ANOVA for the stability of metabolites (t) in leaves and roots of FL478 and IR64 under normal growth conditions.

Table S3 Metabolite ratios in salt-treated FL478 and IR64 leaves compared with controls at different time points.

Table S4 Metabolite ratios in salt-treated FL478 and IR64 roots compared with controls at different time points.

Author Contributions

Conceived and designed the experiments: XQZ BYF. Performed the experiments: XQZ WSW JLD. Analyzed the data: XQZ FZ. Contributed reagents/materials/analysis tools: FZ JLD. Wrote the paper: XQZ ZKL BYF.

References

1. Khatun S, Flowers TJ (1995) Effects of salt on seed set in rice. Plant, Cell and Environ 18: 61–67.

2. Akbar M, Yabuno T (1974) Breeding for saline-resistant varieties of rice. II. Comparative performance of some rice varieties to salt during early developing stage. Jpn J Breed 24: 176–181.

3. Pearson GA, Ayers AD, Eberhard DL (1966) Relative salt tolerance of rice during germination and early seedling development. Soil Sci 102: 151–156.

4. Maas EV, Hoffman GJ (1977) Crop salt tolerance-current assessment. J Irrig Drain Div 103: 115–134.

5. Pandit A, Rai V, Sharma TR, Sharma PC, Singh NK (2011) Differentially expressed genes in sensitive and tolerant rice varieties in response to salt-stress, J Plant Biochem Biotechnol 20: 149–154.

6. Flowers TJ, Yeo AR (1981) Variability in the resistance of sodium chloride salt within rice (Oryza.sativa L.). New Phytogist 88: 363–373.

7. Flowers XJ, Salama FM, Yeo AR (1988) Water-use efficiency in rice {Oryza sativa L.} in relation to resistance to salt. Plant Cell Environ 11: 453–459.

8. Yeo AR, Flowers TJ (1983) Varietal differences in the toxicity of sodium ions in rice leaves. Physiol Plantarum 59: 189–195.

9. Widodo, Patterson JH, Newbigin E, Tester M, Bacic A, et al. (2009) Metabolic responses to salt stress of barley (Hordeum vulgare L.) cultivars, Sahara and Clipper, which differ in salt tolerance. J Exp Bot 60: 4089–4103.

10. Liu T, Van SJ (2000) Selection and characterization of sodium chloride-tolerant callus of Glycine max (L.) Merr cv. Acme. Plant Growth Regul 31: 195–207.

11. Gong Q, Li P, Ma S, Rupassara SI, Bohnert H (2005) Salt stress adaptation competence in the exptremophile Thellungiella halophila in comparison with its relative Arabidopsis thaliana. Plant Physiol 44: 826–839.

12. Ghoulam C, Foursy A, Fares K (2002) Effects of salt stress on growth, inorganic ions and proline accumulation in relation to osmotic adjustment in five sugar beet cultivars. Environ Exp Bot 47: 39–50.

13. Sanchez DH, Siahpooh MR, Roessner U, Udvardi M, Kopka J (2008) Plant metabolomics reveals conserved and divergent metabolic responses to salt. Physiol Plantarum 132: 209–219.

14. Zuther E, Koehl K, Kopka J (2007) Comparative metabolome analysis of the salt response in breeding cultivars of rice. In: Jenks MA, Hasegawa PM, Jain SM, eds. Advances in molecular breeding toward drought and salt tolerance crops. Berlin.

15. Wu DZ, Cai SG, Chen MX, Ye LZ, Chen ZH, et al. (2013) Tissue Metabolic Responses to Salt Stress in Wild and Cultivated Barley. PLOS ONE, HTTP://www.plosone.org, 8:e55431.

16. Fiehn Oliver (2002) Metabolomics – the link between genotypes and phenotypes. Plant Mo Biol 48: 155–171.

17. Walia H, Wilson C, Zeng L, Ismail AM, Condamine P, et al. (2007) Genome-wide transcriptional analysis of salt-stressed japonica and indica rice genotypes during panicle initiation stage. Plant Mol Biol 63: 609–623.

18. Walia H, Wilson C, Condamine P, Liu X, Ismail AM, et al. (2005) Comparative transcriptional profiling of two contrasting rice genotypes under salt stress during the vegetative growth stage. Plant Physiol 139: 822–835.

19. Kumari S, Sabharwal VP, Kushwaha HR, Sopory SK, Singla-Pareek SL, et al. (2009) Transcriptome map for seedling stage specific salt stress response indicates a specific set of genes as candidate for saline tolerance in Oryza sativa L. Funct Integr Genomic 9: 109–123.

20. Nohzadeh MS, Habibi RM, Heidari M, Salekdeh GH (2007) Proteomics reveals new salt responsive proteins associated with rice plasma membrane. Biosci Biotech Bioch 71: 2144–2154.

21. Zang X, Komatsu S (2007) A proteomics approach for identifying osmotic-stress-related proteins in rice. Phytochemistry 68: 426–437.

22. Teusink B, Baganz F, Westerhoff HV, Oliver SG (1998) Metabolic control analysis as a tool in the elucidation of the function of novel genes. Meth Microbiol 26: 297–336.

23. Basu S, Gangopadhyay G, Poddar R, Gupta S, Mukherjee BB (1999) Proline enigma and osmotic stress-tolerance in rice (Oryza sativa L.) In: Kavikishor PB (ed) Plant Tissue Culture and Biotechnology Emerging trends (pp 275–281). Universities Press, Hyderabad, India.

24. Basu S, Gangopadhyay G, Mukherjee BB (2002) Salt tolerance in rice in vitro: Implication of accumulation of Na$^+$, K$^+$ and proline. Plant Cell Tiss Org Cul 69: 55–64.

25. Jain S, Nainawatee HS, Jain RK, Chowdhury JB (1991) Proline status of genetically stable salt-tolerant Brassica juncea L. somaclones and their parent cv. Prakash. Plant Cell Rep 9: 684–687.

26. Kavi KPB, Hong Z, Miao GH, Hu CAA, Verma DPS (1995) Overexpression of $^{\Delta}$1-pyrroline-5-carboxylate synthetase increases proline production and confers osmotolerance in transgenic plants. Plant Physiol 108: 1387–1394.

27. Lin CC, Kao CH (1996) Proline accumulation is associated with inhibition of rice seedling root growth caused by NaCI. Plant Sci 114: 121–128.

28. Garcia AB, Engler J, Iyer S, Gerats T, Van MM, et al. (1997) Effects of osmoprotectants upon NaCl stress in Rice. Plant Physiol 115: 159–169.

29. Hoai NTT, Shim IS, Kobayashi K, Kenji U (2003) Accumulation of some nitrogen compounds in response to salt stress and their relationships with salt tolerance in rice (Oryza sativa L.) seedlings. Plant Growth Regul 41: 159–164.

30. Yoshida S, Forna DA, Cock JH, Gomez KA (1976) Laboratory manual for physiological studies of rice. International Rice Research Institute, Los Baños, Philippines.

31. Wang WS, Zhao XQ, Pan YJ, Zhu LH, Fu BY, et al. (2011) DNA methylation changes detected by methylation-sensitive amplified polymorphism in two contrasting rice genotypes under salt stress. J Genet Genomics 38: 419–424.

32. Gregorio GB, Senadhira D, Mendoza RD (1997) Screening rice for salinity tolerance. IRRI Discussion Paper Series Number 22 International Rice Research Institute, Manila, Philippines.

33. Bowne JB, Erwin TA, Juttner J, Schnurbusch T, Langridge P, et al. (2012) Drought responses of leaf tissues from wheat cultivars of differing drought tolerance at the metabolite level. Mol Plant 5: 418–429.

34. Zhao XQ, Wang WS, Zhang F, Zhang T, Zhao W, et al. (2013) Temporal profiling of primary metabolites under chilling stress and its association with seedling chilling tolerance of rice (Oryza sativa L.). Rice, 6: 23 http://www.thericejournal.com/content/6/1/23.

35. Munns R, Termaat A (1986) Whole-plant responses to salinity. Aust J Plant Physiol 13: 143–160.

36. Joshi V, Joung JG, Fei ZJ, Jander G (2010) Interdependence of threonine, methionine and isoleucine metabolism in plants: accumulation and transcriptional regulation under abiotic stress. Amino Acids 39: 933–947.

37. Fougere F, Rudulier DL, Streeter JG (1991) Effects of salt stress on amino acids, organic acids, and carbohydrate composition of roots, bacteroids, and cytosol of alfalfa (Medicago sativa L.). Plant Physiol 96: 1228–1236.

38. Delauney AJ, Verma DPS (1993) Proline biosynthesis and osmoregulation in plants. Plant J 4: 215–223.

39. Good AG, Zaplachinski ST (1994) The effects of drought stress on free amino-acid accumulation and protein-synthesis in Brassicanapus. Physiol Plant 90: 9–14.

40. Nishizawa A, Yabuta Y, Shigeoka S (2008), Galactinol and raffinose constitute a novel function to protect plants from oxidative damage. Plant Physiol 47: 1251–1263.

41. Kaplan F, Kopka J, Haskell DW, Zhao W, Schiller KC, et al. (2004) Exploring the Temperature-Stress Metabolome of Arabidopsis. Plant Physio 136: 4159–4168.

42. Rosa M, Prado C, Podazza G, Interdonato R, González JA, et al. (2009) Soluble sugars—Metabolism, sensing and abiotic stress, a complex network in the life of plants. Plant Signal Behavior, 4: 388–393.

43. Chang KJ, Roberts JKM (1991) Cytoplasmic malate levels in maize root-tips during KC ion uptake determined by ^{13}C-NMR spectroscopy. Biochim Biophys Acta 1092: 29–34.

44. Jones DL (1998) Organic acids in the rhizosphere – a critical review, Plant and Soil 205: 25–44.

45. Yang CW, Guo WQ, Shi DC (2010) Physiological roles of organic acids in alkali-tolerance of the alkali-tolerant halophyte Chlori virgate. Agron J, 102: 1081–1089.

46. Patonnier MP, Peltier JP, Marigo G (1999) Drought-induced increase in xylem malate and mannitol concentration and closure of Fraxinus excelsior L. stomata. J Exp Bot 50: 1223–1229.

47. Loester WH, Tyson RH, Everard JD, Redgwell RJ, Bieleski RL (1992) Mannitol synthesis in higher plants: evidence for the role and characterization of a NADPH-dependent mannose-6-phosphate reductase. Plant Physiol 98: 1396–1402.

48. Thomas JC, Sepahi M, Arendall B, Bohnert HJ (1995) Enhancement of seed germination in high salt by engineering mannitol expression in Arabidopsis thaliana. Plant Cell Environ 18: 801–806.

49. Gupta AK, Kaur N (2005) Sugar 16ignaling and gene expression in relation to carbohydrate metabolism under abiotic stresses in plants. J Bio Sci 30: 761–776.

50. Chiou TZ, Bush DR (1998) Sucrose is a signal molecule in assimilate partitioning. Proc Natl Acad Sci USA 95: 4784–4788.

51. Roitsch T (1999) Source-sink regulation by sugar and stress. Curr Opin Plant Biol 2: 198–206.

52. Smeekens S (2000) Sugar-induced signal transduction in plants. Annu Rev Plant Physiol Plant Mol Biol 51: 49–81.

13

Expression of Wheat High Molecular Weight Glutenin Subunit 1Bx Is Affected by Large Insertions and Deletions Located in the Upstream Flanking Sequences

Yuke Geng[1,2,9], Binshuang Pang[3,9], Chenyang Hao[1], Saijun Tang[2], Xueyong Zhang[1]*, Tian Li[1]*

1 Key Laboratory of Crop Gene Resources and Germplasm Enhancement, Ministry of Agriculture/Institute of Crop Science, Chinese Academy of Agricultural Sciences, Beijing, China, 2 College of Biological sciences, China Agricultural University, Beijing, China, 3 Beijing Engineering and Technique Research Center of Hybrid Wheat, Beijing Academy of Agricultural and Forestry Sciences, Beijing, China

Abstract

To better understand the transcriptional regulation of high molecular weight glutenin subunit (HMW-GS) expression, we isolated four *Glu-1Bx* promoters from six wheat cultivars exhibiting diverse protein expression levels. The activities of the diverse *Glu-1Bx* promoters were tested and compared with β-glucuronidase (*GUS*) reporter fusions. Although all the full-length *Glu-1Bx* promoters showed endosperm-specific activities, the strongest GUS activity was observed with the *1Bx7^{OE}* promoter in both transient expression assays and stable transgenic rice lines. A 43 bp insertion in the *1Bx7^{OE}* promoter, which is absent in the *1Bx7* promoter, led to enhanced expression. Analysis of promoter deletion constructs confirmed that a 185 bp MITE (miniature inverted-repeat transposable element) in the *1Bx14* promoter had a weak positive effect on *Glu-1Bx* expression, and a 54 bp deletion in the *1Bx13* promoter reduced endosperm-specific activity. To investigate the effect of the 43 bp insertion in the *1Bx7^{OE}* promoter, a functional marker was developed to screen 505 Chinese varieties and 160 European varieties, and only 1Bx7-type varieties harboring the 43 bp insertion in their promoters showed similar overexpression patterns. Hence, the *1Bx7^{OE}* promoter should be important tool in crop genetic engineering as well as in molecular assisted breeding.

Editor: Leandro Peña, Instituto Valenciano De Investigaciones Agrarias, Spain

Funding: National Transgenic Research Project (ZX08002-004) and National Natural Science Foundation of China (30671293). The funders had no role in study design, data collection and analysis, decision to publish, or preparation of the manuscript.

Competing Interests: The authors have declared that no competing interests exist.

* Email: zhangxueyong@caas.cn (XZ); litian@caas.cn (TL)

9 These authors contributed equally to this work.

Introduction

Hexaploid wheat (*Triticum aestivum* L.) is one of the most important human food sources. Its complex genetic background leads to great diversity in nutritional and processing qualities among cultivars. High molecular weight glutenin subunits (HMW-GSs) are the main grain storage proteins in the endosperms of wheat and related species [1,2]. Although HWM-GSs grain storage proteins account for only about 12% of the total protein [3], they play a key role in wheat gluten as the skeletal network that to a large extent determines its structure and formation [4]. The compositions and quantities of allelic variation in HMW-GS genes substantially affect the taste and appearance of dough products, such as Chinese noodles and European bread [5]. Therefore, improvement of flour quality based on superior HWM-GS alleles is necessary to meet changing consumer demands.

Both qualitative and quantitative effects of HMW-GS subunits are important for flour quality [6,7]. In the process of breeding, high dough strength is used as a predictor of good-quality bread wheat; and overexpression of Glu-1Bx7 by way of allele *1Bx7^{OE}* makes an important contribution to high dough strength in some cultivars [8,9]. Expression of HMW-GS is regulated by three major factors, which are at the genomic level (gene duplication), transcriptional level and translational level [10–13]. Transcriptional regulation driven by *Glu-1* 5′-upstream flanking regions might provide strategies for improving grain quality in wheat breeding programs [14]. A number of crucial *cis*-acting elements from HMW-GS promoters of various wheat cultivars have been investigated and characterized, as these could affect tissue specificity or expression activity, including conservative endosperm-specific motifs, such as the GCN4 motif [15], the prolamin box [16], AACA/TA motif [17], RY repeat motif [18], and Skn-1 [19,20], each of which is capable of exerting temporal expression [21,22]. However, the basis of transcriptional regulation of divergence caused by large insertion and deletion (InDel) alterations in HMW-GS promoter regions is still not clear. As reported earlier, a tandem 54 bp duplication, known as the "cereal-box" located at −400 bp in the *1Bx* promoter may enhance endosperm-specific expression [1], suggesting this duplicated region might be a key region for control of gene expression [23,24]. There is a 185 bp MITE insertion in the promoters of *1Bx14* and *1Bx20*, but functional verification indicated that this insertion had little effect on gene expression [25–27]. A 43 bp insertion found at −1000 bp in the $1Bx7^{OE}$ promoter was significantly associated with the overexpression phenotype. It was speculated that the overexpression was brought about by gene duplication mediated by the insertion of a retroelement [13], and there was no further study concerning the 43 bp InDel effect on protein expression. Therefore, more experimental data are needed to clarify the effect of InDels in HMW-GS promoters.

Highly active endosperm-specific promoters serve as an important genetic resource for high-quality and high-yield wheat breeding. Use of seed storage protein gene promoters is an attractive strategy for obtaining target gene products exclusively from crop kernels. A number of seed-specific promoters from barley, rice, maize and other species have been investigated functionally [15,21,28]. Transgenic crops with favorable gene stacking require different tissue-specific promoters from various cereals, as this is helpful to reduce homology-based transcriptional gene silencing [29,30]. HMW-GS promoters from wheat, although containing endosperm-specific motifs, may not be spatially controlled in the same way as in their original genetic backgrounds due to subtle differences in respective regulation systems [31]. Hence, further research of key motifs from tissue-specific promoters would boost applicability in genetic engineering.

Among hexaploid wheat HMW-GSs, Glu-1Bx often shows the highest level of expression [32]. We therefore set out to analyse $1Bx$ promoter sequence characteristics to uncover the transcriptional regulation mechanism. Based on diverse protein expression levels in six wheat cultivars, we isolated four Glu-1Bx promoters in approximately 2.2 kb of length and further validated their functions. By comparison with these upstream sequences, several large InDels such as a 43 bp InDel, a 54 bp duplication and a 185 bp MITE resulted in major divergences among the four promoters, including the $1Bx7$ promoter (Pro-1Bx7), $1Bx7^{OE}$ promoter (Pro-$1Bx7^{OE}$), $1Bx13$ promoter (Pro-1Bx13) and $1Bx14$ promoter (Pro-1Bx14). The promoter sequence variation was shown to be an important factor causing differential expression in transient expression systems and in transgenic rice plant assays. Notably, Pro-$1Bx7^{OE}$ is a highly active endosperm-specific promoter that can be made available for crop improvement by transgenic methods. Moreover, we developed a new specific molecular marker in terms of the 43 bp insertion residing in the $1Bx7^{OE}$ promoter, with which we screened 505 Chinese and 160 European cultivars [33]. We found that this functional marker is significantly associated with $1Bx7$ overexpression. Our results further showed that transcriptional regulation might be responsible for $1Bx$ expression diversity to a larger extent than initially expected.

Materials and Methods

Plant materials

Hexaploid wheat (*Triticum aestivum* L.) cultivars (cv.) Yanzhan 1, Atlas 66, Jimai 20, Xiaoyan 54, Yunmai 33 and Chinese Spring were grown in the field. Endosperm of Xiaoyan 54 was prepared for transient expression assays, and rice (*Oryza sativa* L. ssp *japonica*) cv. Kita-ake was used to produce stable transformants. Materials used for molecular marker screening included 505 Chinese and 160 European cultivars [33].

SDS-PAGE and quantification of HMW-GSs

Protein fractions were extracted from single wheat kernels using a previously reported HMW-GS extraction protocol [34]. Identical amounts of protein extracted from seeds of different

varieties were separated by SDS-PAGE and visualized by Coomassie Blue staining as described by Zhang et al. [35]. Densitometric analyses of 1Bx subunits were carried out by Quantity One software (Bio-rad, USA). The value of the optical density multiplication area was used to quantify HMW-GS expression.

Promoter isolation and cis-element prediction

DNA extraction was performed as previously described [36]. Using a pair of specific primer sets, 1Bx2258F/R (Table S1), four full-length $1Bx$ promoters, ~2.3 kb in size, from six wheat varieties were isolated, gel purified and sequenced. Putative regulatory elements within the $1Bx$ promoter were predicted using the Plant Cis-acting Regulatory DNA elements (PLACE) database [37] combined with a previous report [38].

Construction of promoter-GUS chimeric genes and subsequent transformation

Several full-length and truncated $1Bx$ promoters were obtained by PCR amplification with primers introducing DNA restriction enzyme sites for convenient subcloning (Table S1), and cloned into a modified vector PAHC25 [39] for transient expression, and then subcloned into binary vector pCAMBIA1391z, containing the reporter gene *GUS* under the control of different $1Bx$ promoters. In transient expression experiments, immature embryos harvested at 12–14 days post anthesis (DPA) were used for bombardment as described by Ortiz et al. [40]. Different $1Bx$ promoter-GUS constructs were tested in transient expression assays as described previously [41]. For stable transformation, the binary vector constructs were first introduced into *Agrobacterium tumefaciens* strain EHA105, and then rice transformation was carried out as described by Cho et al. [42]. Transgenic rice plants were selected on medium containing 50 mg L^{-1} of hygromycin, and positive lines were grown in the field for further analysis.

PCR and Southern blot analyses of transformed rice plants

Genomic DNA was isolated from leaf tissues of transformed rice plants as previously described [36]. PCR analysis for molecular identification of transgenic rice plants was performed using a set of specific primers for the $1Bx$ promoter and *GUS* gene (listed in Table S1).

For Southern blot analysis, genomic DNA (10 µg) from different transgenic lines were digested with *Bam*HI or *Hind*III (New England Biolabs, USA). Digested DNA was separated by electrophoresis in 0.8% (w/v) agarose gels, and then transferred to Hybond-N$^+$ membranes (Amersham Biosciences, USA) and hybridized with a *GUS* gene fragment labeled with [α-^{32}P]dCTP as described previously [36].

Histochemical GUS assay

Wheat endosperms undergoing transient expression and different tissues of T_3 transgenic rice were used for histochemical GUS assays. GUS staining was performed as described by Kosugi et al. [43]. Images of stained samples were captured using an MZ16 High-tech Stereomicroscope (Leica, Germany). Stained GUS spots were counted, and for statistical comparison, data of each sample was expressed as the mean number of blue spots per endosperm.

Quantification of expression of the *GUS* gene under control of *1Bx* promoters in transgenic rice plants

Total RNA was extracted from kernels of 3 independent positive T_3 transgenic rice lines at 10–18 DPA. Transcriptional levels of *GUS* in all stable transgenic lines were quantified by quantitative real-time PCR (qRT-PCR) with a 7300 Real-time PCR system (Applied Biosystems, USA) using Power SYBR Green PCR Master Mix (Applied Biosystems, USA). Details of primer pairs used for qRT-PCR are given in Table S1. The specificity of the primer sets was assured by confirmation that the resulting products appeared as single peaks in real-time melting temperature curves and as single fragments after separation by agarose gel electrophoresis. To confirm adequate amplification PCR efficiency was assessed using a sample dilution series as templates [44]. Amplification plots and predicted threshold cycle values were obtained from three independent biological replicates with SDS software version 2.1 (Applied Biosystems, USA). *GUS* gene expression levels were presented as fold-changes calculated using the comparative threshold cycle (CT) method as described [45] with rice *GAPDH* used as the internal control.

Results

Identification of HMW-GSs by SDS-PAGE

HMW-GSs of six wheat cultivars were separated by SDS-PAGE (Figure 1A). Their subunit compositions varied from each other (Table S2). Evidently, expression levels of Glu-1A and Glu-1D are generally lower than Glu-1B. There were four allelic variants of *1Bx* among the six cultivars, namely *1Bx7^OE*, *1Bx14*, *1Bx13* and *1Bx7*. The protein level of *1Bx7^OE* in Yunmai 33 was much higher than that of 1Bx in other cultivars, about 2.2-fold that of 1Bx13 and 1Bx7 and 1.8-fold that of 1Bx14 (Figure 1B).

Comparative analysis of upstream sequences of *Glu-1Bx*

Four types of 5' flanking sequences of *1Bx* alleles were isolated with the specific primer pair, 1Bx2258F/R (Table S1 and Figure S1). They were 2,294, 2,253, 2,185 and 2,433 bp in length for *Pro-1Bx7^OE*, *Pro-1Bx7*, *Pro-1Bx13*, and *Pro-1Bx14*, respectively. The four 5' proximal flanking regions contained five common motifs, including DOF recognition sites, bZIP recognition sites, MYB recognition sites, VP1 recognition sites and basal promoter elements (Figure 2 and Table S3), which are conserved in promoters of genes that encode most seed storage proteins [38]. In addition to several single-base substitutions or small deletions, the presence of sequence insertions or deletions (InDels) constituted the main differences among the four entire promoter regions (Figure S2). By comparison with the *1Bx7* promoter, the *1Bx13* promoter has a 54 bp deletion at −400 upstream from the start

codon (54 bp duplication position), the *1Bx14* promoter contains a 185 bp MITE insertion at −874, consistent with previous reports [26,46], and the *1Bx7^OE* promoter possesses a 43 bp insertion at −1047, which is always associated with the overexpression phenotype [12].

The transient expression results for different *1Bx* promoters

We compared the expression efficiencies of the full-length promoters of *1Bx7*, *1Bx13* and *1Bx7^OE* by means of transient expression assays in wheat endosperms (Figure S3). and *GUS* driven by the *1Bx7^OE* promoter exhibited much higher activity than when driven by the *1Bx7* and *1Bx13* promoters (Figure 3A). Since the 43 bp InDel represents the difference between the *1Bx7* and *1Bx7^OE* promoter sequences, we speculated that the 43 bp insertion enhanced the endosperm-specific expression. In addition, the *1Bx13* promoter activity was lower than that of the *1Bx7* promoter, which lacks the 54 bp duplication present in the *1Bx13* promoter. We further investigated the effect of the 185 bp MITE on *1Bx14* expression. Two truncated *1Bx14* promoters were fused to *GUS*, and transient expression results showed that *GUS* expression driven by a 1,192 bp *Pro-1Bx14-D2* was slightly higher than that driven by the 873 bp *Pro-1Bx14-D1* (Figure 3B), confirming that the MITE might positively but weakly affected transcription of *1Bx14* [26].

Histochemical and quantitative assays in transgenic rice

Chimeras were constructed using different *1Bx* promoters fused to *GUS* and then transformed into rice. Through GUS staining assays in the T_3 generation of stable transgenic plants, we detected *GUS* activity driven by the three full-length *1Bx* promoters only in the seeds and not in stems or leaves collected at 15 DPA. These results indicated that the promoters are endosperm-specific (Figure 4A). In contrast, GUS staining was observed in all tissues of transgenic rice carrying the *Ubiquitin* promoter-*GUS* construct (Figure 4A). Therefore, the full-length *1Bx* promoters contained necessary *cis*-elements that specify endosperm-specific regulation in both wheat and rice. Consistent with the transient expression results, the full-length *1Bx7^OE* promoter with the 43 bp insertion exhibited much higher *GUS* activity than either the full-length *1Bx7* or *1Bx13* promoters (Figure 4A). Southern blot analysis confirmed that the transgenic rice lines had single copies of the *GUS* gene (Figure S4); therefore the comparative results of promoter activities were convincing.

We also investigated the activities of different truncated *1Bx* promoters in transgenic rice. Except for the *Pro-1Bx13-D* promoter, other truncated promoters, including *Pro-1Bx14-D1/D2* and *Pro-1Bx7-D*, retained endosperm-specific expression

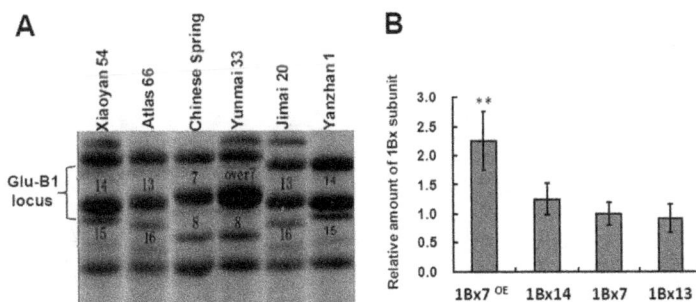

Figure 1. HMW-GSs from six common wheat cultivars. (A) SDS-PAGE profiles of HMW-GSs. 1Bx7 (Chinese Spring), 1Bx13 (Atlas 66 and Jimai 20), 1Bx14 (Xiaoyan 54 and Yanzhan 1) and 1Bx7^OE (Yunmai 33). (B) Relative amounts of four 1Bx subunits estimated by densitometric analysis.

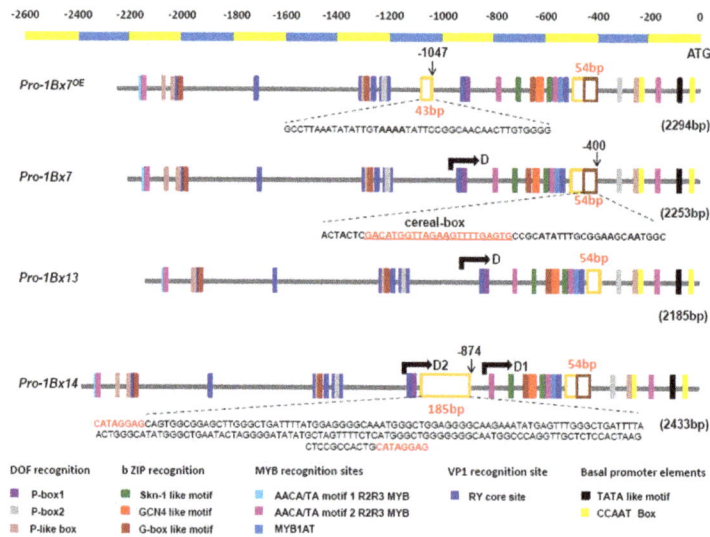

Figure 2. Schematic structure of four *1Bx* promoters. Regulatory elements are indicated by colored rectangles. InDels are labeled with hollow boxes and core sequences are listed in detail under the sketch map. The red underlined sequence shows the "cereal box" in the 54 bp duplication. The 8 bp bases in red at both ends indicate the target site duplication (TSD) of the MITE. Positions of upstream primers used for obtaining truncated promoters are indicated with black arrows.

activity (Figure 4C). Absence of the 54 bp duplication led to loss of endosperm-specific expression controlled by the *Pro-1Bx13-D* promoter, thus indicating that the fragment removed from the *1Bx13* promoter ($-943\sim-2162$) contained necessary *cis*-elements that restricted expression to the endosperm. Like transient expression results, *GUS* expression directed by *Pro-1Bx14-D2* harboring the 185 bp MITE was higher than that directed by *Pro-1Bx14-D1*.

To confirm the results of GUS staining in seeds of transgenic rice, we applied qRT-PCR to determine expression levels of *GUS*. *GUS* expression detected in seeds during 10–18 DPA showed that expression driven by *1Bx* promoters increased rapidly from 10 DPA to 14 DPA, and reached a peak level at 14–16 DPA (Figure 4B and 4D). As the control, *GUS* expression driven by the

Ubiquitin promoter maintained a relatively constant level through 10 to 18 DPA (Figure 4B). The results of *GUS* expression at 14–16 DPA were highly consistent with those of GUS activities based on histochemical staining (Figure 4A and 4C), confirming that the protein expression pattern was similar to the gene expression pattern at the mRNA level.

Phylogenetic analysis of HMW-GS promoters

To address the question of whether these InDels are present in other *HMW-GS* promoters, we analyzed promoters from 14 different wheat *HMW-GS* genes [1,26,47] and identified the regulatory motifs related to endosperm-specific expression in the promoter regions about 1,200 bp upstream of the initiation codon. The numbers of regulatory motifs obviously differed among the

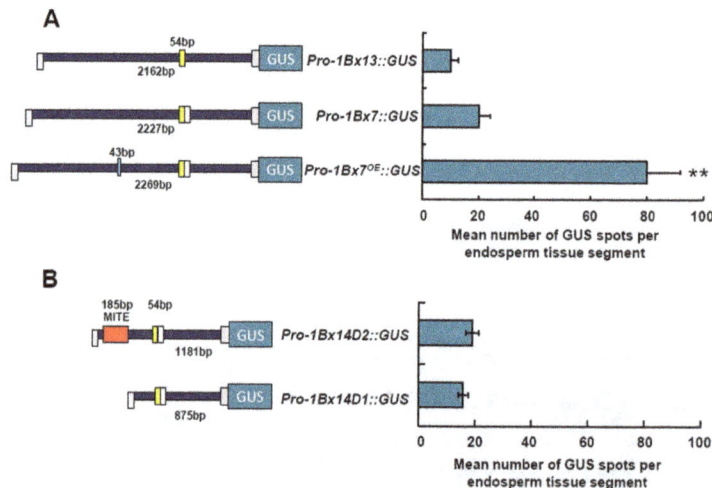

Figure 3. Schematic representation of constructs used for transient expression assays in wheat endosperms and *GUS* activities driven by full-length and truncated *1Bx* promoters. (A) Effects of full-length *1Bx7*, *1Bx13* and *1Bx7^{OE}* promoters on transient expression. (B) Effect of the 185 bp MITE from the *1Bx14* promoter on transient expression. **$P<0.01$ (student's *t*-test) indicates significant difference from others.

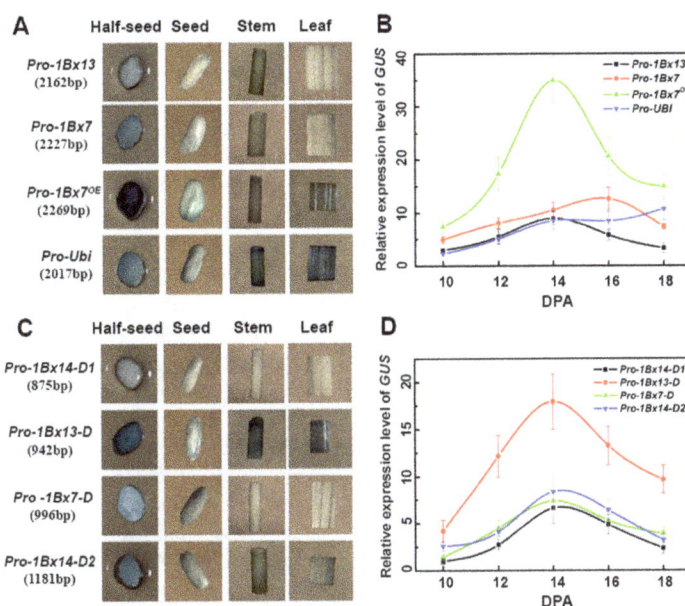

Figure 4. GUS staining of various tissues and quantitative analysis in developing endosperm of transgenic rice. (A) and (C) Histochemical analysis of *GUS* driven by different promoters in transgenic rice tissues collected at 15 DPA; (B) and (D) Relative expression levels of *GUS* in seeds from different transgenic lines during 10 to 18 DPA based on qRT-PCR. Rice *GAPDH* was used as the internal control. Values are shown as means ± s.d (standard deviation) of three independent experiments and three biological replicates. Colored lines at the top right corner represent different transgenic plants.

different HMW-GS promoter sequences (Figure 5). In phylogenetic analysis, all *Glu-1Bx* promoters clustered together in one branch. The 185 bp MITE insertion was present in both the *Glu-1Bx14* and *Glu-1Bx20* promoters. The 54 bp tandem duplication was absent in non-*Glu-1Bx* promoters, but present in all *Glu-1Bx* promoters except the *Glu-1Bx13* promoter. Therefore, InDels contributed to the diversity in HMW-GS promoters, which is an important means for evolution of HMW-GS genes. These large fragment InDels can be used as a potential resource for creating new alleles.

Development of a molecular marker for the 43 bp insertion and its distribution in natural populations

Based on the 43 bp insertion sequence in the *1Bx7^{OE}* promoter, we developed a new molecular marker that differed from those previously reported [34]. Our marker can precisely identify the insertion in HWM-GS promoters among common wheat varieties. PCR amplification resulted in two kinds of bands that distinguish promoters with the 43 bp insertion (a 476 bp fragment) from those without (a 433 bp fragment) (Figure 6A). Among 505 Chinese and 160 European accessions surveyed, we found 3 Chinese and 11 European varieties with the 43 bp insertion when we used this

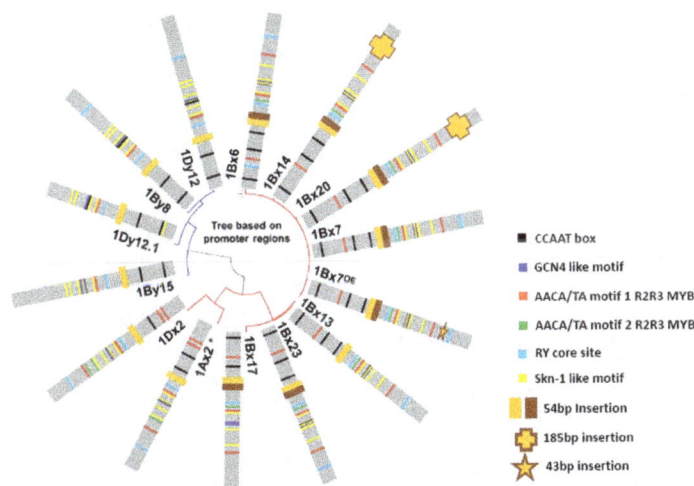

Figure 5. Phylogenetic analysis of 14 *HMW-GS* promoters. Neighbor-Joining Tree of partial length (-1 ~ -1200 bp upstream of the start codon) sequences of *Glu-1* from *Triticum aestivum* L. and other wheat-related grass species. This work was done with the MEGA program (Version 5.2). InDels and conserved cis-elements are shown by markers in different colors.

marker. HMW-GS profiles of accessions containing the 476 bp marker were later obtained by SDS-PAGE (Figure 6B). The presence of particular 1Bx alleles was determined by densitometric analysis; 10 accessions (3 from China and 7 from Europe) exhibited a 1Bx7 overexpression phenotype relative to Chinese Spring used as a control (Figure 6B and Table S4). We also identified some types of *1Bx6* and *1Bx14* with the 43 bp insertion (Figure 6B and Table S4), and confirmed its presence by DNA sequencing. However, these two types of 1Bx did not show overexpression at the protein level. Therefore, the 43 bp insertion in promoters preferentially enhanced 1Bx7 expression although no obvious differences were found among the upstream regions of *1Bx6*, *1Bx14* and *1Bx7^{OE}*.

Discussion

HMW-GSs represent a set of important seed-storage proteins, and both their composition and quantity significantly affect wheat flour quality [7]. Since gene transcriptional regulation is the dominant means of control in production of proteins [48], we isolated four promoter sequences of *Glu-1Bx* and investigated their effects on gene expression. Although the reporter gene driven by all the *1Bx* promoters exhibited an endosperm-specific expression pattern, the *1Bx7^{OE}* promoter from cv. Yunmai 33 produced a markedly stronger activity than other promoters. Previous studies showed that gene duplication was the cause of

1Bx7 overexpression in some wheat cultivars [49]. The connection between the strong activity of the *1Bx7^{OE}* promoter and high protein level produced by the 1Bx7^{OE} subunit clearly indicated that transcription regulation is also a factor in 1Bx7 overexpression. We therefore concluded that multiple factors, including gene duplication and transcriptional regulation determine the expression of *1Bx7*. This work revealed a complex regulatory network of HMW-GS expression in wheat.

The 43 bp insertion at −1047 bp is closely associated with high expression of 1Bx7^{OE}

Large InDels in promoter regions often result in higher rates of transcriptional divergence [50]. In this study, we identified a 43 bp InDel, a 54 bp duplication and a 185 bp MITE in different *Glu-1Bx* promoters, and they accounted for the main differences among *1Bx* promoter sequences. These InDels affected the expression levels of the genes. The presence of the 43 bp insertion at position −1047 upstream of the start codon was shown to be closely associated with high expression levels of the 1Bx7^{OE} subunit [12,13]. We verified that the 43 bp insertion can serve as a strong enhancer to improve expression of the gene by comparing the transcriptional activities between full-length *1Bx7* and *1Bx7^{OE}* promoters (Figure 3A; Figure 4A and 4B). Since there are no known *cis*-elements in the 43 bp insertion, this insertion may facilitate evolutionary tuning of gene expression by affecting local chromatin structure and nucleosome positioning [50].

The 185 bp MITE insertion at −874 bp of *1Bx14* might slightly affect transcription

The 185 bp MITE insertion located at −874 bp in the *1Bx14* promoter may be a remnant of an earlier transposition of a large element or of small, highly repeated elements [51]. In the present study, the *1Bx14* promoter with or without the 185 bp MITE, did not produce a significantly different activity in the transient system or in stable transgenic rice assays, suggesting that it might only slightly affect the transcriptional regulation (Figure 3B; Figure 4C and 4D). The 185 bp MITE exists in both hexaploid and tetraploid wheat, and may be linked to the polyploidization event affecting the constitutions and activities of the genomes of grass species [26].

A 54 bp cereal-box motif is necessary for endosperm-specific expression

The tandem 54 bp duplication at position −400 contains the "cereal-box" implicated in seed-specific expression [1]. Our data demonstrated that the *1Bx13* full-length promoter harboring one 54 bp deletion retains endosperm-specific activity, but a *1Bx13* promoter truncated at −942 bp lacks endosperm-specificity accompanied by increased activity (Figure 4C). We speculate that the 54 bp deletion might complement essential *cis*-elements in the region −940 to −2000 bp of the *1Bx13* promoter to effectively control gene endosperm-specific expression. Without the aid of the *cis*-elements in the region, only one 54 bp cereal-box motif may not be enough to restrict gene expression to endosperm. Based on phylogenetic analysis of HWM-GS promoters, only the *1Bx13* promoter and non-*1Bx* promoters contain a 54 bp deletion (Figure 5). This tandem 54 bp duplication must have occurred before hexaploidization because it is also present in tetraploid wheat. Flanking-sequence divergence was also noted from extensive DNA sequencing analysis of a-gliadin genes [52]. The basis of HMW-GS evolution is repeated sequence events that lead to new alleles [53].

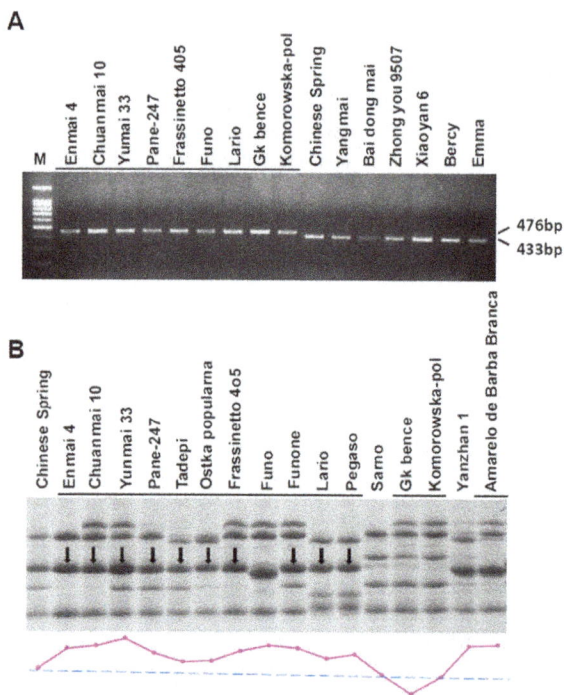

Figure 6. Electrophoretic separation of PCR products from *1Bx* promoters with or without the 43 bp insertion, and SDS-PAGE profiles of HMW-GS of wheat cultivars with the insertion. (A) PCR assays for *1Bx* promoters on a 2% agarose gel. M: 100 bp DNA Ladder. Underlined accessions possess the 43 bp insertion. (B) SDS-PAGE assay of HMW-GS from different accessions containing the 43 bp insertion. Underlined accessions possess the 43 bp insertion. Down black arrows indicate *1Bx7* with 43 bp insertion in accessions from China and Europe. Chinese Spring, Yanzhan 1 and Samo were used as controls. The purple curve represents the relative amounts from different 1Bx subunits.

A simple PCR marker was developed to target high expression of 1Bx7 and 1Bx7^OE

Since previous 43 bp InDel marker covers a region of 1.2~1.3 kb that also contains other InDels such as the 185 bp MITE and 54 bp duplication [34], a new specific marker based on the 43 bp insertion was developed and used effectively in two independent wheat populations combined with SDS-PAGE electrophoresis analysis to identify 1Bx7 overexpressing cultivars. Interestingly, the 43 bp insertion exists not only in the *1Bx7* promoter but also in other *1Bx* promoters such as those of *1Bx14* and *1Bx6* (Figure 6B). Despite harboring the 43 bp insertion in the promoters, the 1Bx14 and 1Bx6 subunits produce no significant increases in protein compared to subunit alleles without the insertion (Figure 6B). The likely reason is that a co-regulatory factor linking the 43 bp insertion to expression efficiency is present in the 1Bx7 alleles or regulation at the translational level might strongly influence the divergence in expression between 1Bx7 and non-1Bx7 subunits.

Putative additive effects of gene duplication and transcriptional regulation on 1Bx7 expression

According to the literature, it is concluded that the 1Bx7 overexpression phenotype is mediated by an LTR retroelement resulting in gene duplication along with the polyploidization event [13]. In the present study, we confirmed that a 43 bp insertion situated in the *1Bx7^OE* promoter is capable of strengthening transcriptional activity markedly through transient expression and transgenic rice assays. By using molecular markers which can be used to indicate *1Bx7* gene duplication [13], we found that only the cultivar Yunmai 33 has both the 43 bp InDel and two 1Bx7 copies (gene duplication), while other 9 cultivars with the 43 bp InDel have only one 1Bx7 copy (Table S4). Although the 1Bx7 subunit of Yunmai 33 is the most abundant in this study (Figure 6B), other cultivars with the 43 bp InDel demonstrate higher 1Bx7 expression than the control, especially Chinese cultivars Enmai 4 and Chuanmai 10 (Figure 6B). So it can be inferred that both gene duplication and transcriptional regulation can lead to 1Bx7 overexpression, and their effects on 1Bx7 expression can be accumulated.

Endosperm is the storage tissue for starch and protein in cereal crops, which are the major sources of carbohydrates and proteins for humans. Improved yield and quality of crops by genetic modification has huge potential, and some significant achievements have already been accomplished [54]. Because continuous high expression of foreign genes in all tissues may cause detrimental effects in host plants [55], identification and application of strong endosperm-specific promoters will attract interest from breeders and biologists. In the current work, we identified a highly active *1Bx7^OE* promoter that can enhance endosperm-specific gene expression at the transcriptional level, and it should be useful for wheat quality improvement by means of genetic transformation and molecular assisted breeding.

Supporting Information

Figure S1   PCR amplification of *1Bx* promoters by using 1Bx1007-F/R (A) and 1Bx2258-F/R (B) primer pairs. Lane 1–4: Chinese Spring (*Pro-1Bx7*); Yunmai 33 (*Pro-1Bx7^OE*); Yanzhan 1 (*Pro-1Bx14*); Atlas 66 (*Pro-1Bx13*); M is a DNA ladder. The PCR products were separated in 1.5% agarose gels.

Figure S2   Alignment of four *1Bx* promoters (*Pro-1Bx7*, *Pro-1Bx7^OE*, *Pro-1Bx13* and *Pro-1Bx14*).

Figure S3   Representative transient expression results of GUS driven by *Pro-1Bx* in wheat endosperms. (A) *Pro-1Bx13*; (B) *Pro-1Bx7*.

Figure S4   Southern blot analysis of transgenic rice lines with full-length *1Bx* promoters (A) or truncated *1Bx* promoters (B). Genomic DNA was digested by *Bam*HI and detected by GUS gene probes.

Table S1   Primers used in this study.

Table S2   HMW-GS compositions of six wheat accessions.

Table S3   Details of 12 known endosperm-specific cis-elements in *1Bx* promoters.

Table S4   Fourteen wheat cultivars harboring the 43 bp insertion in the *1Bx* promoter were identified by marker screening.

Acknowledgments

We are grateful to Dr. Jianmin Wan and Xiuping Guo for help with rice transformation, and gratefully acknowledge help with English editing from Prof. Robert A McIntosh, University of Sydney.

Author Contributions

Conceived and designed the experiments: YG BP XZ. Performed the experiments: YG BP TL. Analyzed the data: CH TL. Contributed reagents/materials/analysis tools: CH ST. Contributed to the writing of the manuscript: YG XZ TL.

References

1. Anderson OD, Greene FC (1989) The characterization and comparative analysis of high-molecular-weight glutenin genes from genomes A and B of a hexaploid bread wheat. Theor Appl Genet 77: 689–700.
2. Reddy P, Apples R (1993) Analysis of a genomic DNA segment carrying the wheat high molecular weight (HMW) glutenin Bx17 subunit and its use as an RFLP marker. Theor Appl Genet 85: 616–624.
3. Halford NG, Forde J, Anderson OD, Greene FC, Shewry PR (1987) The nucleotide and deduced amino acid sequences of an HMW glutenin subunit gene from chromosome 1B of bread wheat (*Triticum aestivum* L.) and comparison with those of genes from chromosomes 1A and 1D. Theor Appl Genet 75: 117–126.
4. Shewry PR, Tatham AS (1990) The prolamin storage proteins of cereal seeds: structure and evolution. Biochem J 267: 1–12.
5. Liu JJ, He ZH, Zhao ZD, Pena RJ, Rajaram S (2003) Wheat quality traits and quality parameters of cooked dry white Chinese noodles. Euphytica 131: 147–154.
6. Barro F, Rooke L, Békés F, Gras P, Tatham AS, et al. (1997) Transformation of wheat with high-molecular-weight subunit genes results in improved functional properties. Nat Biotechnol 15: 1295–1299.
7. Wieser H (2000) Comparative investigations of gluten proteins from different wheat species. I. Qualitative and quantitative composition of gluten protein types. Eur Food Res Technol 211: 262–268.
8. Gupta RB, Paul JG, Cornish GB, Palmer GA, Bekes F, et al. (1994) Allelic variation at glutenin subunit and gliadin loci, *Glu-1*, *Glu-3* and *Gli-1*, of common wheats. I. Its additive and interaction effects on dough properties. J Cereal Sci 19: 9–17.

9. Cornish GB, Békés F, Allen HM, Martin DJ (2001) Flour proteins linked to quality traits in an Australian doubled haploid wheat population. Aust J Agr Res 52: 1339–1348.

10. Harberd NP, Flavell RB, Thompson RD (1987) Identification of a transposon-like insertion in a *Glu-1* allele of wheat. Mol Genet Genomics 209: 326–332.

11. Marchylo BA, Lukow OM, Kruger JE (1992) Quantitative variation in high molecular weight glutenin subunit 7 in some Canadian wheats. J Cereal Sci 15: 29–37.

12. Butow BJ, Gale KR, Ikea J, Juhász A, Bedo Z, et al. (2004) Dissemination of the highly expressed *Bx7* glutenin subunit (*Glu-B1al allele*) in wheat as revealed by novel PCR markers and RP-HPLC. Theor Appl Genet 109: 1525–1535.

13. Ragupathy R, Naeem HA, Reimer E, Lukow OM, Sapirstein HD, et al. (2008) Evolutionary origin of the segmental duplication encompassing the wheat *GLU-B1* locus encoding the overexpressed Bx7 (Bx7^{OE}) high molecular weight glutenin subunit. Theor Appl Genet 116: 283–296.

14. Thomas MS, Flavell RB (1990) Identification of an enhancer element for the endosperm-specific expression of high molecular weight glutenin. Plant Cell 2: 1171–1180.

15. Zheng Z, Kawagoe Y, Xiao S, Li Z, Okita T, et al. (1993) 5′ distal and proximal cis-acting regulator elements are required for developmental control of a rice seed storage protein glutelin gene. Plant J 4: 357–366.

16. Dong G, Ni Z, Yao Y, Nie X, Sun Q (2007) Wheat Dof transcription factor WPBF interacts with *TaQM* and activates transcription of an *alpha-gliadin* gene during wheat seed development. Plant Mol Biol 63: 73–84.

17. Takaiwa F, Yamanouchi U, Yoshihara T, Washida H, Tanabe F, et al. (1996) Characterization of common cis-regulatory elements responsible for the endosperm-specific expression of members of the rice glutelin multigene family. Plant Mol Biol 30: 1207–1221.

18. Fujiwara T, Beachy RN (1994) Tissue-specific and temporal regulation of a β-conglycinin gene: roles of the RY repeat and other cis-acting elements. Plant Mol Biol 24: 261–273.

19. Sha S, Sugiyama Y, Mitsukawa N, Masumura T, Tanaka K (1996) Cloning and sequencing of a rice gene encoding the 13-kDa prolamin polypeptide. Biosci Biotech Bioch 60: 335–337.

20. Wu CY, Suzuki A, Washida H, Takaiwa F (1998) The GCN4 motif in a rice glutelin gene is essential for endosperm-specific gene expression and is activated by *Opaque-2* in transgenic rice plants. Plant J 14: 673–683.

21. Müller M, Knudsen S (1993) The nitrogen response of a barley *C-hordein* promoter is controlled by positive and negative regulation of the GCN4 and endosperm box. Plant J 4: 343–355.

22. Onodera Y, Suzuki A, Wu CY, Washida H, Takaiwa F (2001) A rice functional transcriptional activator, RISBZ1, responsible for endosperm-specific expression of storage protein genes through GCN4 motif. J Biol Chem 276: 14139–14152.

23. Forde J, Malpica JM, Halford NG, Shewry PR, Anderson OD, et al. (1985) The nucleotide sequence of a HMW glutenin subunit gene located on chromosome 1A of wheat (*Triticum aestivum* L.). Nucleic Acids Res 13: 6817–6832.

24. Halford NG, Ford J, Shewry PR, Kreis M (1989) Functional analysis of the upstream regions of a silent and an expressed member of a family of wheat seed-protein genes in transgenic tobacco. Plant Sci 62: 207–216.

25. Anderson OD, Larka L, Christoffers MJ, McCue KF, Gustafson JP (2002) Comparison of orthologous and paralogous DNA flanking the wheat high molecular weight glutenin genes: sequence conservation and divergence, transposon distribution, and matrix-attachment regions. Genome 45: 367–380.

26. Li W, Wan Y, Liu Z, Liu K, Li B, et al. (2004) Molecular charaterization of HMW glutenin subunit allele 1Bx14: further insights into the evolution of Glu-B1-1 alleles in wheat and related species. Theor Appl Genet 109: 1093–1104.

27. Jiang QT, Ma J, Zhao S, Zhao QZ, Lan XJ, et al. (2012) Characterization of HMW-GSs and their gene inaction in tetraploid wheat. Genetica 40: 325–335.

28. Marks MD, Lindell JS, Larkins BA (1985) Quantitative analysis of the accumulation of zein mRNA during maize endosperm development. J Biol Chem 260: 16445–16450.

29. Butaye KMJ, Cammue BPA, Delauré SL, De Bolle MFC (2005) Approaches to minimize variation of transgene expression in plants. Mol breeding 16: 79–91.

30. Qu LQ, Xing YP, Liu WX, Xu XP, Song YR (2008) Expression pattern and activity of six glutelin gene promoters in transgenic rice. J Exp Bot 59: 2417–2424.

31. Furtado A, Henry RJ, Takaiwa F (2008) Comparison of promoters in transgenic rice. Plant Biotechnol J 6: 679–693.

32. Galili G, Feldman M (1983) Genetic control of endosperm proteins in wheat. Theor appl genet 66: 77–86.

33. Su Z, Hao C, Wang L, Dong Y, Zhang X (2011) Identification and development of a functional marker of *TaGW2* associated with grain weight in bread wheat (*Triticum aestivum* L.). Theor Appl Genet 122: 211–223.

34. Radovanovic N, Cloutier S (2003) Gene-assisted selection for high molecular weight glutenin subunits in wheat doubled haploid breeding programs. Mol breeding 12: 51–59.

35. Zhang X, Huang C, Xu X, Hew CL (2002) Identification and localization of a prawn white spot syndrome virus gene that encodes an envelope protein. J Gen Virol 83: 1069–1074.

36. Sambrook J, Fritsch EF, Maniatis T (1989) Molecular Cloning: A Laboratory Manual. Cold Spring Harbor Laboratory Press, New York.

37. Higo K, Ugawa Y, Iwamoto M, Korenaga T (1999) Plant cis-acting regulatory DNA elements (PLACE) database. Nucleic Acids Res 27: 297–300.

38. Juhász A, Makai S, Sebestyén E, Tamás L, Balázs E (2011) Role of conserved non-coding regulatory elements in LMW glutenin gene expression. PloS ONE 6: e29501.

39. Christensen AH, Quail PH (1996) Ubiquitin promoter-based vectors for high-level expression of selectable and/or screenable marker genes in monocotyledonous plants. Transgenic Res 5: 213–218.

40. Ortiz JP, Ravizzini RA, Morata MM, Vallejos RH (1997) A rapid system for studying foreign gene expression in wheat (*Triticum aestivum* L.). Theor Appl Genet 38: 123–130.

41. Oñate L, Vicente-Carbajosa J, Lara P, Diaz I, Carbonero P (1999) Barley BLZ2: a seed-specific bZIP protein that interacts with BLZ1 in vivo and activates transcription from the GCN4-like motif of B-hordein promoters in barley endosperm. J Biol Chem 274: 9175–9182.

42. Cho HJ, Brotherton JE, Widholm JM (2004) Use of the tobacco feedback-insensitive anthranilate synthase gene ASA2 as a selectable marker for legume hairy root transformation. Plant Cell Res 23: 104–113.

43. Kosugi S, Arai Y, Nakajima K, Ohashi Y (1990) An improved assay for β-glucuronidase (GUS) in transformed cells: methanol almost suppresses a putative endogenous *GUS* activity. Plant Sci 70: 133–140.

44. Rasmussen R (2001) Quantification on the Light Cycler instrument. In: Meuer S, Wittwer C, Nakagawara K, eds, Rapid Cycle Real-Time PCR, Methods and Applications. Springer Press, Heidelberg. pp.21–34.

45. Pfaffl MW, Tichopad A, Prgomet C, Neuvians TP (2004) Determination of stable housekeeping genes, differentially regulated target genes and sample integrity: BestKeeper–Excel-based tool using pair-wise correlations. Biotechnol Lett 26: 509–515.

46. Yang ZJ, Li GR, Liu C, Feng J, Zhou JP, et al. (2006) Molecular characterization of a HMW glutenin subunit allele providing evidence for silencing of x-type gene on *Glu-B1*. Acta Genet Sin 33: 929–936.

47. Yan Y, Zheng J, Xiao Y, Yu J, Hu Y, et al. (2004) Identification and molecular characterization of a novel y-type Glu-Dt1 glutenin gene of *Aegilops tauschii*. Theor Appl Genet 108: 1349–1358.

48. Shaw LM, McIntyre CL, Gresshoff PM, Xue GP (2009) Members of the Dof transcription factor family in Triticum aestivum are associated with light-mediated gene regulation. Funct Integr Genomics 9: 485–498.

49. Cloutier S, Banks T, Nilmalgoda S (2005) Molecular understanding of wheat evolution at the *Glu-B1* locus. In: Proceedings of the international conference on plant genomics and biotechnology: challenges and opportunities, Raipur, India, p 40.

50. Vinces MD, Legendre M, Caldara M, Hagihara M, Verstrepen KJ (2009) Unstable tandem repeats in promoters confer transcriptional evolvability. Science 324: 1213–1216.

51. Wessler SR (1998) Transposable elements and the evolution of gene expression. Symposia of the Society for Experimental Biology 51: 115–122.

52. Anderson OD (1997) Applications of molecular biology in understanding and improving wheat quality. In: Steele JL, Chung OK, eds. *Proc Int Wheat Quality Conf*, Grain Industry Alliance, Manhattan, KS. pp. 205–211.

53. SanMiguel P, Tikhonov A, Jin YK, Motchoulskaia N, Zakharov D, et al. (1996) Nested retrotransposons in the intergenic regions of the maize genome. Science 274: 765–768.

54. Bajaj S, Mohanty A (2005) Recent advances in rice biotechnology towards genetically superior transgenic rice. Plant Biol 3: 275–307.

55. Cheon BY, Kim HJ, Oh KH, Bahn SC, Ahn JH, et al. (2004) Overexpression of human erythropoietin (EPO) affects plant morphologies: retarded vegetative growth in tobacco and male sterility in tobacco and Arabidopsis. Transgenic Res 13: 541–549.

14

Bioinformatic Indications That COPI- and Clathrin-Based Transport Systems Are Not Present in Chloroplasts: An Arabidopsis Model

Emelie Lindquist[9], Mohamed Alezzawi[9], Henrik Aronsson*

Department of Biological and Environmental Sciences, University of Gothenburg, Gothenburg, Sweden

Abstract

Coated vesicle transport occurs in the cytosol of yeast, mammals and plants. It consists of three different transport systems, the COPI, COPII and clathrin coated vesicles (CCV), all of which participate in the transfer of proteins and lipids between different cytosolic compartments. There are also indications that chloroplasts have a vesicle transport system. Several putative chloroplast-localized proteins, including CPSAR1 and CPRabA5e with similarities to cytosolic COPII transport-related proteins, were detected in previous experimental and bioinformatics studies. These indications raised the hypothesis that a COPI- and/or CCV-related system may be present in chloroplasts, in addition to a COPII-related system. To test this hypothesis we bioinformatically searched for chloroplast proteins that may have similar functions to known cytosolic COPI and CCV components in the model plants *Arabidopsis thaliana* and *Oryza sativa* (subsp. *japonica*) (rice). We found 29 such proteins, based on domain similarity, in Arabidopsis, and 14 in rice. However, many components could not be identified and among the identified most have assigned roles that are not related to either COPI or CCV transport. We conclude that COPII is probably the only active vesicle system in chloroplasts, at least in the model plants. The evolutionary implications of the findings are discussed.

Editor: Steven M. Theg, University of California - Davis, United States of America

Funding: This work was supported by Olle Engkvist Byggmästare Foundation (to H.A.), and a PhD student fellowship from the Libyan Higher Education (to M.A.). The funders had no role in study design, data collection and analysis, decision to publish, or preparation of the manuscript.

Competing Interests: The authors have declared that no competing interests exist.

* Email: henrik.aronsson@bioenv.gu.se

[9] These authors contributed equally to this work.

Introduction

Chloroplasts, the most fully characterised plastids, contain photosynthetically active thylakoids located in an aqueous stroma, surrounded by a double membrane. In addition to the stroma they have two other aqueous compartments: the intermembrane space between the double membrane's outer and inner envelopes, and the lumen enclosed by the thylakoids. Some chloroplast-localized proteins are encoded by the chloroplast genome. However, most (ca. 95%) are encoded by the nuclear genome, processed in the cytoplasm then transferred to chloroplasts [1]. These proteins are translocated across the outer and inner envelope membranes to the stroma via two translocons, designated TOC and TIC, respectively, mostly aided by cleavable transit peptides [2]. However, some non-canonical proteins may enter the chloroplast without a transit peptide. After entering the chloroplast, proteins are further targeted to specific sub-compartments. Thylakoid targeted proteins are transferred from the stroma via one of four pathways: the Secretory (Sec) pathway, the Signal Recognition Particle (SRP) pathway, the Twin Arginine Translocation (Tat) pathway, or the spontaneous pathway. Proteins transported across the thylakoid membrane into the lumen are using the Sec or the Tat pathway, whereas integral thylakoid membrane proteins are using the SRP or the spontaneous pathway [3,4]. All of these pathways are energy-dependent and mediated by specific combi-nations of proteins except the spontaneous pathway, which requires no energy inputs or specific proteins for protein transport [5].

Although thylakoid membranes contain proteins their main components are lipids, transferred to the thylakoids after synthesis in the envelope [6,7]. Several studies indicate that the lipids could be transported by vesicles [8–10], but as yet there is no clear evidence of protein transport via vesicles in chloroplasts. In contrast, three coated vesicle transport systems have been characterized in the plant cytosol: the COPII (coat protein complex II), COPI (coat protein complex I) and CCV (clathrin coated vesicle) systems, all similar to corresponding systems in yeast and mammals [11–13]. Cytosolic vesicles are known to deliver both soluble and membrane-bound proteins to target membranes, leading to the hypothesis that the vesicle system in chloroplasts may deliver not only lipids, but also proteins [14]. If so, it would represent an uncharacterized fifth pathway for thylakoid-targeted proteins, in addition to the four already identified.

Vesicle transport in chloroplasts has been observed mainly at low temperatures in *Pisum sativum* (pea), *Glycine max* (soybean), *Spinacia oleracea* (spinach) and *Nicotiana tabacum* (tobacco) [15,16]. Proteins required for vesicle transport in the chloroplast are so far suggested to be similar to those of the well-characterized COPII vesicle transport system in the cytosol [14].

COPII, COPI vesicles and CCV in the cytosol have similar functions, but distinct protein and lipid compositions, and recognize different sets of cargo, which make each transport specific [17,18]. COPII-coated vesicles appears to be involved exclusively in transport from ER to Golgi [19,20]. The COPII coat comprises five subunits: Sec23/24, Sec13/31 and Sar1 [11,21]. Formation of a vesicle starts with activation and recruitment of the small GTPase Sar1 to the donor membrane with the help of Sec12p acting as a guanine nucleotide exchange factor (GEF) at ribosome-free ER membranes in the cytosol [22,23]. Subsequently coat proteins are gathered and the vesicle is formed. Most cytosol localized coat subunits of COPII have predicted homologs in chloroplasts [14,24].

Homologues of two important proteins for vesicle transport in the cytosol, RabA5e and Sar1, respectively named CPRabA5e and CPSAR1 (CP = chloroplast localized), have been identified in the chloroplast [9,14,25]. CPSAR1 (which has been detected in the envelopes, stroma and stromal vesicles) is required for thylakoid biogenesis, and is more abundant in the envelopes than the stroma at low temperature (4°C), supporting the hypothesis that it participates in a chloroplast vesicle transport system similar to the cytosolic COPII system [9]. CPRabA5e was subsequently identified in chloroplasts showing an attenuation of vesicles and alteration of thylakoid morphology, under oxidative stress [25].

COPI vesicles primarily mediate transport within the Golgi and between the Golgi and ER [11,26]. The COPI coat (sometimes called coatomer) consists of two main subcomplexes: a cargo-selective F-COPI subcomplex (with β, δ, γ and ζ subunits), and B-COPI subcomplex (with α, β' and ε subunits) [11,26]. The active form of the GTPase ADP-ribosylation factor 1 (Arf1) is needed to initiate coatomer recruitment to Golgi membranes, similarly to the Sar1 requirement for initiation of COPII coat recruitment. Thus, Arf1 and Sar1 act as triggers for COPI- and COPII-coated vesicle maturation, respectively [27].

CCVs play a key role in membrane and protein transport between the trans-Golgi network, plasma membrane and endo-somes [26,28] through the endocytic and late secretory pathways [29]. Their coats consist of clathrin triskelions, structures composed of three "legs" consisting of three heavy chains (each ~190 kDa) and three light chains (each ~25 kDa). They form a basket-like lattice of pentagons and hexagons [30,31] assembled in coordination with other proteins and Arf1. In contrast to COPII and COPI vesicles, adaptor proteins (APs) — including five AP complexes (designated AP1–5), various monomeric adaptors (GGAs) and cargo-specific adaptors — rather than the coat *per se*, are the cargo selectors in CCV vesicles [11,26,29]. They bind to membranes and collect cargo to be transported with the vesicles, sometimes forming networks enabling different kinds of cargo to be transported simultaneously [32,33]. Here we focus on the AP complexes. AP1 and AP2 are dependent on clathrin for vesicle formation, whereas AP3 and AP4 appear to be clathrin-independent [32]. The fifth adaptor protein complex, AP5, was recently discovered in human (HeLa) cells, where it localizes to the endosome and is believed to act independently of clathrin. In this paper vesicles containing these components are collectively referred to as clathrin coated vesicles (CCVs), regardless of their clathrin dependence/independence.

Land plants are known to possess AP1–4, and recent homology analysis suggests they also have AP5 [33]. AP complexes generally consist of four subunits: two large, one medium, and one small [11]. One of the large subunits is called γ, α, δ, ε or ζ, depending on the associated AP complex. The second large subunit is called β and numbered 1–5 depending on the AP complex. Similarly, the medium and small subunits are named μ1–5 and σ1–5,

respectively [33]. However, in plants a single subunit called β1/2 probably functions in both AP1 and AP2 complexes, whereas there are distinct β1 and β2 proteins in mammals [34,35] and σ5 is predicted to be missing from AP5 in Arabidopsis [33]. Like COPI, all AP complexes need Arf proteins for recruitment to membranes [32,36].

There are numerous similarities in the three vesicle systems, e.g. the requirement for activation of small GTPases (Sar1 in the COPII system, Arf1 in the COPI and CCV systems) for recruitment of the coat and additional proteins [11]. There are also similarities in structural architecture of the coats and the domains they possess. For example, α and β' subunits of the B-COPI subcomplex form a triskelion similar to clathrin, generating a curved structure. In terms of domain configuration, the β' subunit of COPI has high similarities to Sec13/31 of COPII, indicating that COPI has similarities to the coats of both CCVs and COPII vesicles [37]. Further similarities include the presence of N-terminal β propellers (enabling binding to AP complexes) and α solenoid legs in the heavy chains of clathrin triskelions of CCVs [37–39], Sec13/31 of the COPII coat [40,41] and the B-COPI subcomplex [26,37]. In addition, γ and β subunits of the F-COPI subcomplex [26] have similarities to "appendages" of the AP complexes of CCVs, and thus are considered to be cargo-binding [11,37].

The similarities in, and differences between, the vesicle systems pose intriguing questions about their origin and evolution. COPII is hypothetically the most ancestral system, since it is an essential biosynthetic pathway in all investigated organisms [26], while COPI (which has strong similarities with both COPII and CCV in domain organization and coat structure, respectively) is putatively an intermediate system [37]. If so, the clathrin system evolved most recently.

Current knowledge of chloroplast vesicles indicates that they are most strongly related to the putatively ancestral COPII system [9,14]. However, the possibility that homologues to cytosolic COPI and CCV systems may be present in chloroplasts has not been systematically explored previously. Thus, we addressed this possibility using the model plant *Arabidopsis thaliana* and yet another model plant *Oryza sativa* (subsp. *japonica*) (rice) to support our findings in Arabidopsis.

Methods

Multiple *in silico* approaches were used to search for proteins in *Arabidopsis thaliana* chloroplasts that could have homologous functions to COPI and CCV proteins in the cytosol of various organisms. The workflow is presented in Figure 1 and described below.

Identifying domains, patterns and motifs

Protein sequences matching cytosolic COPI and CCV subunits in *Arabidopsis thaliana*, *Saccharomyces cerevisiae* (Baker's yeast), *Homo sapiens* (human) and *Mus musculus* (mouse) were retrieved from literature and Uniprot (http://www.uniprot.org) (Figure 1, Tables S1, S2, S3, S4, S5, S6, S7, S8, S9). COPII-related proteins were omitted since they were recently investigated [14].

The collected proteins were compiled and searched for characteristic domains, patterns or motifs, using Prosite release 20.95 (http://prosite.expasy.org) [42] and the Pfam database 26.0 (http://pfam.sanger.ac.uk) [43] (Tables S1, S2, S3, S4, S5, S6, S7, S8, S9). Each identified domain, pattern or motif is denoted by either a PS (Prosite) or PF (Pfam) entry.

After converting the dataset of Arabidopsis chloroplast proteins (GO:0009507) (retrieved from TAIR version 10, www.arabidopsis.

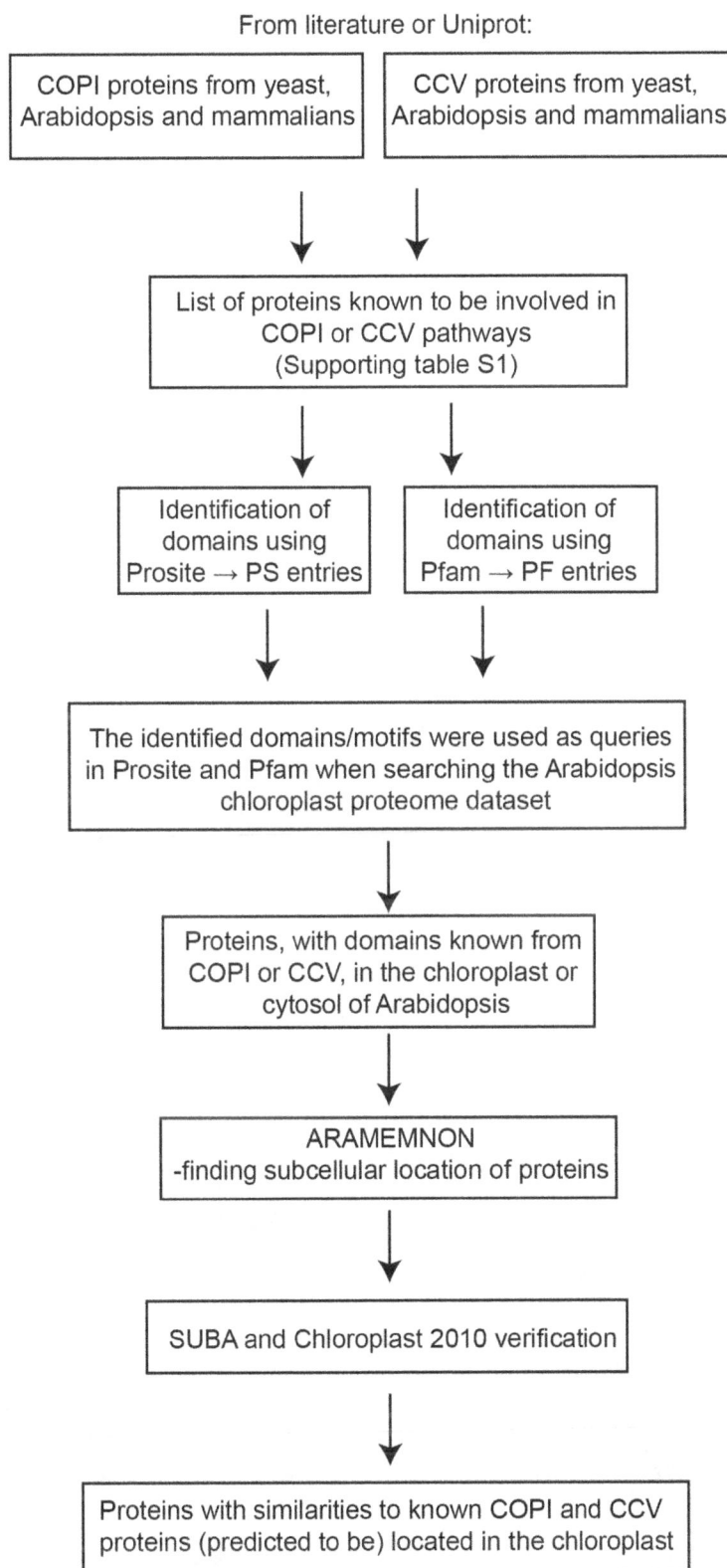

Figure 1. Identification of putative chloroplast COPI and CCV transport components in Arabidopsis. Schematic work flow of the bioinformatics methods used to find putative COPI- and CCV-related transport proteins in chloroplasts. Cytosolic COPI and CCV proteins were retrieved from the literature or Uniprot, and their characteristic domains were identified using Prosite or Pfam. Identified domains were used to search a database of chloroplast-localized proteins, and the localization of proteins found in the chloroplast with relevant domains was further checked using ARAMEMNON, SUBA and Chloroplast 2010. Finally, a list of chloroplast proteins with similarities to known COPI and CCV proteins was compiled.

org) [44] into fasta file format the dataset was searched for the identified Prosite entries using ScanProsite (http://prosite.expasy.org/scanprosite). The Pfam database does not offer corresponding tools so proteins containing the requested Pfam entries were sought manually. The searches generated a list of proteins in Arabidopsis chloroplasts with identical combinations of domains to proteins known to participate in cytosolic COPI or CCV pathways.

As mentioned above, the CCV AP5 complex was first identified in human (HeLa) cells, and subsequently predicted to be present in Arabidopsis, although the degree of conservation is low [33]. Thus, for AP5 we used both human proteins and the predicted proteins in Arabidopsis to identify characteristic domains, which were later used to search the chloroplast dataset.

Subcellular localization of identified proteins

To further check that identified proteins are localized in chloroplasts we used the ARAMEMNON plant membrane protein database [45] to retrieve their names and predict their subcellular localizations, applying 17 tools provided by the host website (http://aramemnon.uni-koeln.de): BaCelLo [46], ChloroP_v1.1 [47], iPSort [48], Mitopred [49], Mitoprot_v2 [50], MultiLoc [51], PA-SUB_v2.5 [52], PCLR_v0.9 [53], PProwler_v1.1 [54], PrediSi [55], Predotar_v1 [56], PredSL [57], SignalP_HMM_v3 [58,59], SignalP_NN_v3 [60], SLP-Local [61], TargetP_v1 [60,62], and WoLF PSort [63]. Using these tools, a Bayesian consensus (SigConsens) score was obtained from ARAMEMNON for each protein with patterns of interest. A score ≥10 was considered reliable, providing a strong prediction of subcellular location [64]. To corroborate the findings we used the SUBA database version 2.21 (http://suba.plantenergy.uwa.edu.au) [65], which in addition to bioinformatics predictions contains information from proteomic and GFP experiments on subcellular localizations of Arabidopsis proteins. Finally, the Chloroplast 2010 database (http://bioinfo.bch.msu.edu/2010_LIMS) [66] was used to confirm the validity of predictions and/or experiments that identified proteins are present in chloroplasts (Figure 1).

Complementary studies using rice

Complementary investigations of rice (*Oryza sativa* subsp. *japonica*) proteins were conducted to assess the validity and generality of the findings from the Arabidopsis analysis. Domains of relevant Arabidopsis, yeast, mouse and human proteins were retrieved from the Prosite and Pfam websites (Tables S1, S2, S3, S4, S5, S6, S7, S8, S9). Proteins with corresponding combinations of domains were identified in the rice subsp. *japonica* dataset of the National Center for Biotechnology Information (NCBI, TaxID39947) using ScanProsite release 20.102, and their subcellular localizations were predicted using Target P 1.1.

Entries were also searched using Pfam, and proteins with domains of interest were identified manually using a rice dataset downloaded from Phytozome v9.1 (http://www.phytozome.net). To ensure that each hit was from rice subsp. *japonica* they were checked using the Rice genome annotation project (http://rice.plantbiology.msu.edu) or RiceChip Annotation Site (http://www.ricechip.org). Subcellular locations of the hits were then identified using Target P 1.1. Names of identified proteins were retrieved from Uniprot, or the Rice genome annotation project if not identified in Uniprot.

Results

To find putative components of a hypothetical COPI or CCV system in Arabidopsis chloroplasts, known COPI or CCV proteins

from the cytosol of various organisms were retrieved and analysed to identify characteristic domains (Figure 1, Tables S1, S2, S3, S4, S5, S6, S7, S8, S9). The domains were used to search a dataset of protein sequences of chloroplast-localized proteins to identify chloroplast proteins with COPI or CCV domains. Diverse tools, based on differing principles, for predicting the likelihood of proteins having transit peptides were used to strengthen the localization. Several proteins identified in the chloroplast dataset were identical to proteins known to act in the cytosol, raising doubts about their true locations. Occasionally, an identified domain or a domain combination was found in several of the cytosolic proteins, and subsequently in several different chloroplast proteins, hence generating chloroplast proteins which could function as several of the cytosolic subunits. These chloroplast proteins are described below as having commonly occurring domains.

Putative Clathrin triskelion related chloroplast proteins in Arabidopsis

A protein named Putative heavy chain of clathrin complex (AtCHC2)/At3g08530 was found in the TAIR chloroplast dataset (Table 1) with a Clathrin propeller repeat (PF01394), a Clathrin, heavy-chain linker (PF09268), a Clathrin-H-link (PF13838), a Region in Clathrin and VPS domain (PF00637), and a Clathrin heavy-chain (CHCR) repeat profile (PS50236) as identified in yeast (Table S1). The chloroplast localization of AtCHC2 was also supported by SUBA and Chloroplast 2010 (Table 1).

Similarly, Putative light chain of clathrin complex (AtCLC1/At2g40060; Table 1) was found to have a Clathrin light chain domain (PF01086), identified with known vesicle proteins from both yeast and Arabidopsis (Table S1). Chloroplast localization for this protein was supported by SUBA and Chloroplast 2010 (Table 1).

Putative Clathrin AP1–5 related chloroplast proteins in Arabidopsis

In clathrin-coated vesicles five AP complexes are known, designated AP1–5. Five proteins similar to the AP1 complex γ subunit were found in the chloroplast dataset: Putative ascorbate peroxidase/At1g07890, Putative thylakoid-bound ascorbate peroxidase (AttAPX)/At1g77490, RNase E/G-type endoribonuclease (AtRNEE/G)/At2g04270, Stromal ascorbate peroxidase (AtsAPX)/At4g08390, and Putative peroxisomal ascorbate peroxidase (AtAPX3)/At4g35000 (Table 2). These five proteins all have the same Peroxidases proximal heme-ligand signature domain (PS00435) as the γ subunit of AP1 in yeast (Table S2). Chloroplast localization was supported for the Putative ascorbate peroxidase and AtAPX3 by Chloroplast 2010, and for AtAPX3 also by SUBA. The other three identified proteins (AttAPX, AtRNEE/G, and AtsAPX) had ARAMEMNON consensus scores >10, indicating a chloroplast location, supported by SUBA and Chloroplast 2010 (Table 2).

Considering AP2 homologues, five proteins similar to the β2 subunit in yeast were identified: Putative large subunit of carbamoyl phosphate synthetase VEN3 (AtCarB)/At1g29900, Putative H-protein of glycine decarboxylase/At1g32470, Acetyl-CoA carboxylase (AtACC2)/At1g36180, Biotin carboxylase subunit of plastidic acetyl-coenzyme A carboxylase complex (AtCAC2)/At5g35360, and Putative RimM-like protein involved in 16S rRNA processing/At5g46420 (Table 3). These five proteins all had a Carbamoyl-phosphate synthase subdomain signature 2 (PS00867) identified using Prosite (Table S3).

Table 1. Putative chloroplast localized CCV triskelion components identified using characteristic domains in searches of the TAIR chloroplast dataset.

Name (ARAMEMNON), Accession No	Role of chloroplast protein (TAIR)	SigConsens (ARAMEMNON)			SUBA	Chloroplast 2010
		CP	MT	SEC		
Putative clathrin heavy chain						
AtCHC2, At3g08530	Protein binding, vesicle transport, endocytosis	0.0	0.0	2.7	Yes (MS/MS)	Yes
Putative clathrin light chain						
AtCLC1, At2g40060	Vesicle transport	0.0	0.0	2.0	Yes (MS/MS)	Yes

CP, chloroplast; MT, mitochondria; SEC, secretory pathway.

Of the five proteins predicted to be chloroplast localized AP2 $\beta2$ subunits, three (AtCarB, AtCAC2 and the Putative RimM-like protein involved in 16S rRNA processing) had ARAMEMNON consensus scores >10 and support for this localization from both SUBA and Chloroplast 2010. ARAMEMNON also strongly predicted chloroplast localization for AtACC2, but a mitochondrial location for the Putative H-protein of glycine decarboxylase, although chloroplast localization for the latter was supported by SUBA and Chloroplast 2010 (Table 3).

For AP3, AP4 and AP5 only subunits with commonly occurring domains were identified (Table 4, Tables S4, S5, S6). Further details regarding these proteins are presented below in a separate paragraph.

Putative B-COPI subcomplex related chloroplast proteins in Arabidopsis

COPI vesicle coats consist of a B-COPI subcomplex and an F-COPI subcomplex, both composed of several subunits (Tables S7, S8). Our searches detected eight proteins similar to the β' subunit of the B-COPI subcomplex, with a Trp-Asp (WD) repeats circular profile (PS50294) and a Trp-Asp (WD) repeats profile (PS50082), which identifiey the β' subunit in both Arabidopsis and human cytosol: Receptor for activated C kinase (AtRACK1A)/At1g18080, Putative U-box-type E3 ubiquitin ligase (AtPUB60)/At2g33340, Putative Cdc20-like mitotic specificity factor for anaphase-promoting complex (AtFZR2/AtCCS52A1)/At4g22910, Putative Cdc20-like mitotic specificity factor for anaphase-promoting complex (AtFZR3/AtCCS52B)/At5g13840, WD40 repeat protein, functions in chromatin assembly (AtMSI1)/

Table 2. Putative chloroplast localized CCV AP1 complex components identified using characteristic domains in searches of the TAIR chloroplast dataset.

Name (ARAMEMNON), Accession No., subunit	Role of chloroplast protein (TAIR)	SigConsens (ARAMEMNON)			SUBA	Chloroplast 2010
		CP	MT	SEC		
Putative clathrin AP1 complex protein						
Putative gamma subunit of coatomer adaptor complex At4g34450*, β1	Cytoskeleton organization, protein transport, catabolic processes, vesicle transport	0.4	0.0	4.0	No	No
Unknown protein At1g51350*, β1	Unknown	20.4	0.0	3.9	No	Yes
Putative ascorbate peroxidase At1g07890, γ	Golgi organization, glycolysis, hyperosmotic response, photorespiration, protein folding (is a ascorbate peroxidase)	0.8	9.0	0.0	No	Yes
AttAPX At1g77490, γ	Chloroplast-nucleus signalling, thylakoid membrane organization (is a ascorbate peroxidase)	22.8	5.0	4.9	Yes (MS/MS)	Yes
AtRNEE/G At2g04270, γ	Chloroplast mRNA processing, chloroplast organisation, thylakoid membrane organization (is a ribonuclease)	11.2	4.3	0.4	Yes (MS/MS)	Yes
AtsAPX At4g08390, γ	Oxidation-reduction processes (is a ascorbate peroxidase)	17.0	5.3	2.2	Yes (MS/MS and GFP)	Yes
AtAPX3 At4g35000, γ	Oxidation-reduction processes (is a ascorbate peroxidase)	0.0	5.8	0.0	Yes (MS/MS)	Yes

*contains common occurring domain(s); CP, chloroplast; MT, mitochondria; SEC, secretory pathway.

Table 3. Putative chloroplast localized CCV AP2 complex components identified using characteristic domains in searches of the TAIR chloroplast dataset.

Name (ARAMEMNON), Accession No., subunit	Role of chloroplast protein (TAIR)	SigConsens (ARAMEMNON)			SUBA	Chloroplast 2010
		CP	MT	SEC		
Putative clathrin AP2 complex protein						
Putative gamma subunit of coatomer adaptor complex At4g34450*, β2	Cytoskeleton organization, protein transport, catabolic processes, vesicle transport	0.4	0.0	4.0	No	No
Unknown protein At1g51350*, β2	Unknown	20.4	0.0	3.9	No	Yes
Unknown protein At5g57460*, μ2	Unknown	6.6	3.0	4.4	Yes (MS/MS)	Yes
AtCarB, At1g29900, β2	Response to phosphate starvation, chromatin silencing, gluconeogenesis, metabolic processes	20.1	3.9	0.0	Yes (MS/MS)	Yes
Putative H-protein of glycine decarboxylase, At1g32470, β2	Glycine processes, PSII assembly, rRNA processing, biosynthesis of cysteine	8.4	16.2	2.3	Yes (MS/MS)	Yes
AtACC2, At1g36180, β2	Fatty acid and metabolic processes (is a acetyl CoA carboxylase)	17.7	6.9	0.2	No	No
AtCAC2, At5g35360, β2	Fatty acid and metabolic processes, brassinosteroid and polysaccharide biosynthesis (is a acetyl CoA carboxylase)	20.8	0.0	0.0	Yes (MS/MS)	Yes
Putative RimM-like protein involved in 16S rRNA processing, At5g46420, β2	Virus defence, metabolic processes, gene silencing, ribosome biogenesis	14.4	2.7	3.1	Yes (MS/MS)	Yes

*contains common occurring domain(s); CP, chloroplast; MT, mitochondria; SEC, secretory pathway.

At5g58230; and three Unknown proteins/At1g24130/At4g02660/At1g15850 (Table 5). Out of these eight proteins only AtFZR2/AtCCS52A1, AtFZR3/AtCCS52B and one of the Unknown proteins (At1g24130) had scores above 10 using ARAMEMNON, and support by Chloroplast 2010. The other five proteins; AtRACK1A, AtPUB60, AtMSI1, and the other two Unknown proteins (At4g02660 and At1g15850), had scores below

Table 4. Putative chloroplast-localized CCV AP3, AP4 and AP5 complex components identified using characteristic domains in searches of the TAIR chloroplast dataset.

Name (ARAMEMNON), Accession No.	Role of protein (TAIR)	SigConsens (ARAMEMNON)			SUBA	Chloroplast 2010
		CP	MT	SEC		
Putative clathrin AP3 complex protein						
δ and β3 subunit						
Putative gamma subunit of coatomer adaptor complex, At4g34450*	Cytoskeleton organization, protein transport, catabolic processes, vesicle transport	0.4	0.0	4.0	No	No
Unknown protein, At1g51350	Unknown	20.4	0.0	3.9	No	Yes
Putative clathrin AP4 complex protein						
ε subunit						
Unknown protein, At5g57460*	Unknown	6.6	3.0	4.4	Yes (MS/MS)	Yes
Putative gamma subunit of coatomer adaptor complex, At4g34450*	Cytoskeleton organization, protein transport, catabolic processes, vesicle transport	0.4	0.0	4.0	No	No
μ4 and σ4 subunit						
Unknown protein, At5g57460	Unknown	6.6	3.0	4.4	Yes (MS/MS)	Yes
Putative clathrin AP5 complex protein						
μ5 subunit						
Unknown protein, At5g57460	Unknown	6.6	3.0	4.4	Yes (MS/MS)	Yes

*contains common occurring domain(s); CP, chloroplast; MT, mitochondria; SEC, secretory pathway.

Table 5. Putative chloroplast localized B-COPI components identified using characteristic domains in searches of the TAIR chloroplast dataset.

Name (ARAMEMNON), Accession No.	Role of protein (TAIR)	SigConsens (ARAMEMNON)			SUBA	Chloroplast 2010
		CP	MT	SEC		
Putative B-COPI subcomplex protein (β′ subunits)						
AtRACK1A, At1g18080	Response to ABA, GA signalling, glycolysis, translation, salt stress, ribosome biogenesis, seed germination	0.0	0.0	0.0	Yes (MS/MS)	No
AtPUB60, At2g33340	Nucleotide binding	0.0	0.0	0.0	Yes (MS/MS)	Yes
AtFZR2 (AtCCS52A1), At4g22910	Protein binding, cell growth, proteasome assembly, regulation of cell division	11.6	0.0	0.0	No	Yes
AtFZR3 (AtCCS52B), At5g13840	Protein binding, DNA methylation, gamete generation, microtubule organization, proteasome assembly, cell division	18.1	0.0	0.0	No	Yes
Unknown protein, At1g24130	Nucleotide binding	10.8	0.0	0.5	No	Yes
Unknown protein, At4g02660	Signal transduction	0.0	0.0	0.0	Yes (MS/MS)	No
AtMSI1, At5g58230	Protein binding, cell proliferation, chromatin modification, seed development, DNA replication	0.4	0.0	0.0	Yes (MS/MS)	Yes
Unknown protein, At1g15850	Nucleotide binding	9.2	0.0	2.9	No	Yes

CP, chloroplast; MT, mitochondria; SEC, secretory pathway.

10 in ARAMEMNON but were supported as chloroplastic by SUBA and/or Chloroplast 2010 (Table 5).

Putative F-COPI subcomplex related chloroplast proteins in Arabidopsis

For the F-COPI subcomplex ζ subunit two proteins were found in the chloroplast: Component of magnesium-protoporphyrin IX chelatase complex (AtCHLD)/At1g08520, and Unknown protein/At1g67120 (Table 6), both having a VWFA domain profile (PS50234) (Table S8). ARAMEMNON strongly predicted chloroplast localization for AtCHLD, but not for the Unknown protein At1g67120, although it was supported for both of these proteins by SUBA and Chloroplast 2010 (Table S8).

Proteins with commonly occurring domains in Arabidopsis chloroplasts

Seven proteins in the chloroplast dataset (four of which were potential Coat GTPases) were found to be possible homologues of two or more components of the COPI and CCV system, since some vesicle proteins from the cytosol share the same domain(s) (Tables S2, S3, S4, S5, S6, S8, S9), which thus identify the same proteins in the chloroplast dataset (Figure 2, Tables 2–4, 6–7).

The first is the Putative gamma subunit of coatomer adaptor complex/At4g34450, which has domain homology with the following subunits: AP1 β1, AP2 β2, AP3 δ, AP3 β3, AP4 ε, and F-COPI γ. All these subunits have the same identifying domain, the Adaptin N terminal region (PF01602) (Tables S2, S3, S4, S5, S6, S8), except the F-COPI γ subunit, which also contains the Coatomer gamma subunit appendage platform subdomain (PF08752) (Table S8). Chloroplast localization was very weakly predicted for this protein by ARAMEMNON (consensus score 0.4), and not supported by either SUBA or Chloroplast 2010 (Tables 2–4, 6).

The second protein, an Unknown protein/At1g51350 also contains the PF01602 domain and could function homologously to the AP1 β1, AP2 β2, AP3 δ, AP3 β3 and AP4 ε subunits (Tables 2–4, S2, S3, S4, S5). It has strongly predicted chloroplast localization according to ARAMEMNON, supported by Chloroplast 2010 (Tables 2–4).

The third protein found in the chloroplast dataset that could have several functions was another Unknown protein/At5g57460 with a Mu homology domain (MHD) profile (PS51072), an identifier of AP2 μ2, AP4 μ4, AP4 σ4, AP5 μ5 and F-COPI coat δ subunits (Tables S3, S5, S6, S8). The Unknown protein (At5g57460) is located in chloroplasts according to SUBA and Chloroplast 2010 (Tables 3–4, 6).

Putative Coat GTPase related chloroplast proteins in Arabidopsis

The last four proteins with commonly occurring domains were identified as putative homologues to Arf proteins. The Arf proteins used as queries in this search were from yeast and the Arabidopsis cytosol, where the latter are divided into four groups (A, B, D and B2) (Bassham et al, 2008). Regardless of their origin, all identified Arf proteins have a small GTPase Arf family profile (PS51417), and an ADP-ribosylation factor family (PF00025) domain (Table S9). Since there was no distinction in the domains identifying the known proteins, the chloroplast search recognized the same four proteins (At1g09180, At1g05810, At4g35860 and At5g57960), regardless of which Arf protein used as a query (Table 7, Table S9). Three of these proteins are already known to be involved in vesicle systems of the Arabidopsis cytosol. At1g09180 is described as a Secretion-associated RAS 1 protein (AtSARA1A) GTPase functioning in COPII transport [11], whereas At1g05810 and At4g35860 are listed as Rab proteins, namely the putative RAB-A-

Table 6. Putative chloroplast localized F-COPI components identified using characteristic domains in searches of the TAIR chloroplast dataset.

Name (ARAMEMNON), Accession No.	Role of protein (TAIR)	SigConsens (ARAMEMNON)			SUBA	Chloroplast 2010
		CP	MT	SEC		
Putative F-COPI subcomplex protein						
ζ subunit						
AtCHLD, At1g08520*	Chlorophyll biosynthesis, cytokinin metabolic process, photosynthesis	22.8	0.4	0.0	Yes (MS/MS)	Yes
Unknown protein, At1g67120	Cytoskeleton organization, embryo sac development, gluconeogenesis	0.0	1.1	3.2	Yes (MS/MS)	Yes
γ subunit						
Putative gamma subunit of coatomer adaptor complex, At4g34450	Cytoskeleton organization, protein transport, catabolic processes, vesicle transport	0.4	0.0	4.0	No	No
δ subunit						
Unknown protein, At5g57460	Unknown	6.6	3.0	4.4	Yes (MS/MS)	Yes

*contains common occurring domain(s); CP, chloroplast; MT, mitochondria; SEC, secretory pathway.

class small GTPase (AtRabA5e) and the putative RAB-B-class small GTPase (AtRabB1c), respectively (Table 7).

Chloroplast localization was supported for AtSARA1A and AtRabB1c by both SUBA and Chloroplast 2010 (Table 7). For AtRabA5e, a transit peptide directing the protein to the chloroplast has been previously suggested [14], its chloroplast location — supported by ARAMEMNON and Chloroplast 2010 (Table 7) — was recently confirmed and it was renamed CPRabA5e to better reflect its location [25]. The only one of these four proteins not already recorded as part of the secretory system [11] is the putative GTPase of unknown function/ At5g57960, assigned a chloroplast location by ARAMEMNON, supported by SUBA and Chloroplast 2010, rendering it a candidate Arf in chloroplasts (Table 7).

Putative CCV, COPI and Coat GTPase related chloroplast proteins in rice

In total, 15 proteins in *O. sativa* (subsp. *japonica*) chloroplasts were found to have domains, or combinations of domains, characteristic of CCV, COPI including Coat GTPases (Table 8). Nine of these proteins correspond only to a single subunit. Two, Clathrin heavy chain 1/LOC_Os11g01380 and Clathrin heavy chain 2/LOC_Os12g01390, were identified as possible clathrin heavy chain proteins with a predicted chloroplast location (Table 8). Both have Clathrin propeller repeat (PF01394), Clathrin, heavy-chain linker (PF09268), Clathrin-H-link (PF13838), Region in Clathrin and VPS (PF00637), Clathrin heavy-chain (CHCR) repeat profile (PS50236) domains and an Orn/DAP/Arg decarboxylases family 2 pyridoxal-P attachment site (PS00878) (Table S10).

Two other proteins were identified as putative AP1 γ subunits: Probable L-ascorbate peroxidase 7 (APX7)/LOC_Os04g35520 and Probable L-ascorbate peroxidase 8 (APX8)/LO-C_Os02g34810 (Table 8), both of which have a Peroxidases proximal heme-ligand signature (PS00435) (Table S10). Two putative AP2 β2 subunits in rice chloroplasts were also identified: Acetyl-CoA carboxylase 2 (ACC2)/LOC_Os05g22940 and Car-bamoyl-phosphate synthase large chain (CARB)/LO-

C_Os01g38970 (Table 8), both containing a Carbamoyl-phosphate synthase subdomain signature 2 domain (PS00867) (Table S10).

Further, a Regulatory-associated protein of TOR 1 (RAP-TOR1)/LOC_Os12g01922 was found to have the required domains — a Trp-Asp (WD) repeats profile (PS50082) and a Trp-Asp (WD) repeats circular profile (PS50294) —for a functional B-COPI β′ subunit, whereas PPR repeat-containing protein/LOC_Os07g14530 has the Coatomer epsilon subunit domain (PF04733) required for B-COPI ε subunits (Table 8, Table S10). In addition, Magnesium-chelatase subunit ChlD (CHLD)/LOC_Os03g59640 was found as a putative F-COPI ζ subunit, with a VWFA domain profile (PS50234), in rice chloroplasts (Table 8, Table S10).

In contrast, the remaining six proteins have commonly occurring domains identifying them as possible homologues for several subunits (Table S10). The Adaptin N terminal region (PF01602) domain was found in Armadillo/beta-catenin-like repeat family protein/LOC_Os11g41990, Adaptin, putative/ LOC_Os01g43630 and the protein AP3 complex subunit delta/ LOC_Os01g32880, identifying them as candidate AP1 β1, AP2 β2, AP3 δ, AP3 β3 and AP4 ε (Table 8, Table S10). In addition, Adaptin, putative/LOC_Os01g43630 has a Beta2-adaptin appendage C-terminal sub-domain (PF09066), providing the domains needed to be a putative AP4 β4 subunit (Table 8, Table S10). Further, Adaptor complexes medium subunit family protein/LOC_Os12g34370 was identified as a putative AP2 μ2, AP3 μ3, AP5 μ5 and F-COPI δ subunit, having Adaptor complexes medium subunit family domain (PF00928), and the Guanine nucleotide-binding protein subunit beta/LO-C_Os03g46650 was identified as a possible B-COPI α or B-COPI β′ subunit, with a Trp-Asp (WD) repeats profile PS50082, Trp-Asp (WD) repeats circular profile (PS50294) and Trp-Asp (WD) repeats signature (PS00678). Finally, a candidate for all ARF groups was identified: the Mitochondrial Rho GTPase/LO-C_Os03g59590 (Table 8), having an ADP-ribosylation factor family domain (PF00025) (Table 8, Table S10).

a.) Clathrin triskelion

✓ Clathrin light chain
⌡ Clathrin heavy chain

b.) Clathrin coat components

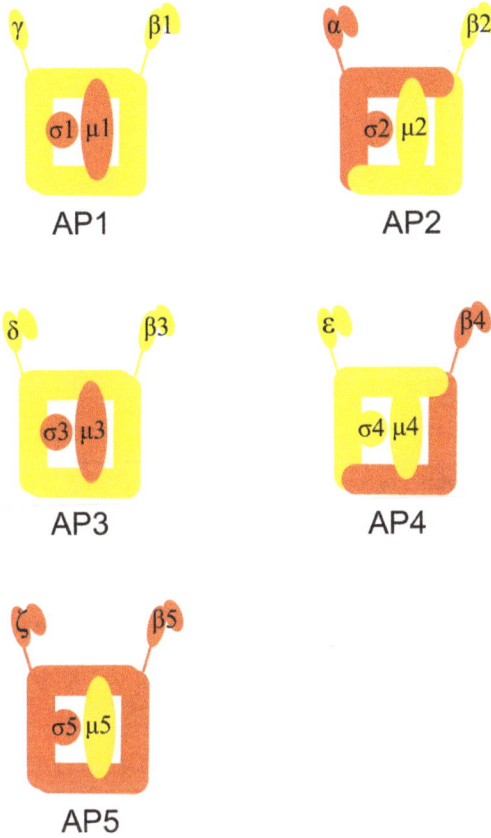

AP1

AP2

AP3

AP4

AP5

c.) COPI coat components and Arf proteins

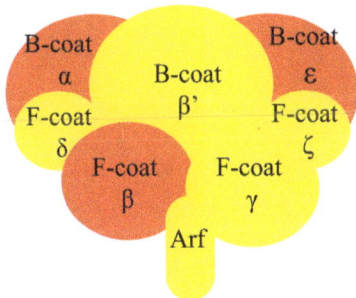

Figure 2. A model of putative CCV and COPI components in Arabidopsis chloroplasts. The figure is based on homologies to components of cytosolic systems (http://www.endocytosis.org/Adaptors/index.html) [130,131]. Red, no proteins with domains of interest detected in chloroplasts. Yellow, proteins with domains of interest identified in chloroplasts, but known to have other roles than vesicle transport and/or the proteins identified had commonly occurring domains, and thus predicted as different subunits in the chloroplasts, and unknown proteins. Green, proteins with domains of interest found in chloroplasts and previously known to have a vesicle transport role.

Discussion

We identified 22 proteins in Arabidopsis that may function as parts of a functional COPI or CCV transport system in chloroplasts: a putative clathrin heavy chain component, clathrin light chain, five AP1 γ subunits, five AP2 β2 subunits, eight B-COPI β' subunits, and two F-COPI ζ subunit proteins (all having characteristic domains, patterns or motifs of cytosolic counterparts; Table 9).

In addition, seven chloroplast proteins with commonly occurring domains were identified as several putative subunits, possibly with multiple functions (Table 9). Four were identified as similar to Arf proteins, while two others could function as an AP1 β1, AP2 β2, AP3 δ, AP3 β3 and/or AP4 ε subunit. One of these two could also function as an F-COPI γ subunit. The seventh protein was found to have a potential function as an AP2 μ2, AP4 μ4, AP4 σ4, AP5 μ5 and/or F-COPI δ subunit (Table 9).

Thus, various possible components of a COPI or CCV system have been detected in chloroplasts, but several of these have other assigned roles, whereas some required components could not be identified at all. Hence, a key question is whether sufficient components are present to form a functional COPI- or CCV-like transport system.

Evidence for clathrin-coated vesicle system components in Arabidopsis chloroplasts

Triskelion proteins. Concerning triskelion proteins AtCHC2 and AtCLC1 were identified as putative clathrin heavy and light chain respectively inside chloroplast, but have previously been assigned same roles in the cytosol [11,67–69] but both SUBA and Chloroplast 2010 indicate a chloroplast location (Table 1). A possible explanation for this apparent discrepancy, supported by mass spectrometry experiments [70–73], is that they have dual locations. However, further tests of this hypothesis are required.

AP1. Known Arabidopsis AP1 complex γ subunit proteins have specific domains, or combinations of domains, detected in none of the chloroplast localized proteins (Table S2). However, the AP1 complex γ subunit in yeast contains a domain called Peroxidases proximal heme-ligand signature (PS00435), which was also found in five proteins in the Arabidopsis chloroplast dataset: Putative ascorbate peroxidase, AttAPX, AtsAPX, and AtAPX3 (all ascorbate peroxidases), and the RNAse AtRNEE/G (Table 2). AttAPX, AtRNEE/G, and AtsAPX were predicted to be chloroplast-targeted, but Putative ascorbate peroxidase and AtAPX3 do not have unambiguously chloroplast locations (Table 2). Regardless of the localization it could be argued that even if having the same domain as the yeast AP1 complex γ subunit the proteins are not likely to act as components in the AP complex since they act as peroxidases rather than as clathrin-related components.

AP1 and AP2 complexes in the Arabidopsis cytosol are believed to share a β1/β2 subunit. Since the constitution of a hypothetical COPI or CCV system in the chloroplast is inevitably unknown, we included the separate β1 subunit of AP1 and β2 subunit of AP2 from yeast, in addition to the β1/β2 subunit from Arabidopsis as queries in our searches. As for AP1, this made no difference since

Table 7. Putative chloroplast localized CCV and COPI Coat GTPase components identified using characteristic domains in searches of the TAIR chloroplast dataset.

Name (ARAMEMNON), Accession No.	Role of chloroplast protein (TAIR)	SigConsens (ARAMEMNON)			SUBA	Chloroplast 2010
		CP	MT	SEC		
Putative Coat GTPases						
AtSARA1A, At1g09180*	Intracellullar transport	0.4	0.0	19.9	Yes (MS/MS)	Yes
AtRabA5e, At1g05810*	Protein transport, GTP mediated signalling	19.2	1.2	2.8	No	Yes
AtRabB1c, At4g35860*	Protein transport, vesicle transport, protein targeting to the vacuole	0.0	0.0	5.9	Yes (MS/MS)	Yes
Putative GTPase of unknown function, At5g57960*	GTP binding	17.3	1.4	0.0	Yes (MS/MS)	Yes

*contains common occurring domain(s); CP, chloroplast; MT, mitochondria; SEC, secretory pathway.

we found no chloroplast proteins similar to either β1/β2 or β1 (Table S2).

AP2. Regarding the AP2 complex, we found a characteristic domain of the β2 subunit (PS00867) in five chloroplast proteins: Putative H-protein of glycine decarboxylase, AtCarB, AtACC2, AtCAC2, and Putative RimM-like protein involved in 16S rRNA processing (Table S3). This indicates that these proteins are more similar to the yeast subunits than those in the Arabidopsis cytosol.

Table 8. Overview of putative chloroplast localized proteins in rice (*subsp. Japonica*) with characteristic domains of CCV and COPI subunit counterparts.

Subunits	Putative chloroplast localized proteins holding domains characteristics for each subunit respectively
CCV	
Clathrin heavy chain	Clathrin heavy chain 1 (LOC_Os11g01380), Clathrin heavy chain 2 (LOC_Os12g01390)
AP1 γ subunit	Probable L-ascorbate peroxidase 7 (APX7) (LOC_Os04g35520), Probable L-ascorbate peroxidase 8 (APX8) (LOC_Os02g34810)
AP1 β1 subunit	Armadillo/beta-catenin-like repeat family protein (LOC_Os11g41990)*, Adaptin, putative (LOC_Os01g43630)*, AP-3 complex subunit delta (LOC_Os01g32880)*
AP2 β2 subunit	Armadillo/beta-catenin-like repeat family protein (LOC_Os11g41990)*, Carbamoyl-phosphate synthase large chain (CARB) (LOC_Os01g38970), Acetyl-CoA carboxylase 2 (ACC2) (LOC_Os05g22940), Adaptin, putative (LOC_Os01g43630)*, AP-3 complex subunit delta (LOC_Os01g32880)*
AP2 µ2 subunit	Adaptor complexes medium subunit family protein (LOC_Os12g34370)*
AP3 δ subunit	Armadillo/beta-catenin-like repeat family protein (LOC_Os11g41990)*, Adaptin, putative (LOC_Os01g43630)*, AP-3 complex subunit delta (LOC_Os01g32880)*
AP3 β3 subunit	Armadillo/beta-catenin-like repeat family protein (LOC_Os11g41990)*, Adaptin, putative (LOC_Os01g43630)*, AP-3 complex subunit delta (LOC_Os01g32880)*
AP3 µ3 subunit	Adaptor complexes medium subunit family protein (LOC_Os12g34370)*
AP4 β4 subunit	Adaptin, putative (LOC_Os01g43630)*
AP4 ε subunit	Armadillo/beta-catenin-like repeat family protein (LOC_Os11g41990)*, Adaptin, putative (LOC_Os01g43630)*, AP-3 complex subunit delta (LOC_Os01g32880)*
AP5 µ5 subunit	Adaptor complexes medium subunit family protein (LOC_Os12g34370)*
COPI	
B-COPI α-subunit	Guanine nucleotide-binding protein subunit beta (LOC_Os03g46650)*
B-COPI β'-subunit	Guanine nucleotide-binding protein subunit beta (LOC_Os03g46650)*, Regulatory-associated protein of TOR 1 (RAPTOR1) (LOC_Os12g01922)
B-COPI ε-subunit	PPR repeat containing protein (LOC_Os07g14530)
F-COPI δ subunit	Adaptor complexes medium subunit family protein (LOC_Os12g34370)*
F-COPI ζ-subunit	Magnesium-chelatase subunit ChlD (LOC_Os03g59640)
Coat GTPases	
ArfA group	Mitochondrial Rho GTPase (LOC_Os03g59590)*
ArfB group	Mitochondrial Rho GTPase (LOC_Os03g59590)*
ArfB2 group	Mitochondrial Rho GTPase (LOC_Os03g59590)*
ArfD group	Mitochondrial Rho GTPase (LOC_Os03g59590)*

* = proteins with several assigned roles, having common occurring domains.

Table 9. Overview of putative chloroplast-localized proteins with characteristic domains of CCV and COPI subunit counterparts.

Subunits	Putative chloroplast localized proteins holding domains characteristics for each subunit respectively
CCV	
Clathrin heavy chain	AtCHC2
Clathrin light chain	AtCLC1
AP1 γ subunit	Putative ascorbate peroxidase, AttAPX, AtRNEE/G, AtsAPX, AtAPX3
AP1 β1 subunit	Putative gamma subunit of coatomer adaptor complex*, Unknown protein* (At1g51350)
AP2 β2 subunit	AtCarB, Putative H-protein of glycine decarboxylase, AtACC2, AtCAC2, Putative RimM-like protein involved in 16S rRNA processing, Putative gamma subunit of coatomer adaptor complex*, Unknown protein* (At1g51350)
AP2 μ2 subunit	Unknown protein* (At5g57460)
AP3 δ subunit	Putative gamma subunit of coatomer adaptor complex*, Unknown protein* (At1g51350)
AP3 β3 subunit	Putative gamma subunit of coatomer adaptor complex*, Unknown protein* (At1g51350)
AP4 ε subunit	Putative gamma subunit of coatomer adaptor complex*, Unknown protein* (At1g51350)
AP4 μ4 subunit	Unknown protein* (At5g57460)
AP4 σ4 subunit	Unknown protein* (At5g57460)
AP5 μ5 subunit	Unknown protein* (At5g57460)
COPI	
B-COPI β' subunit	AtRACK1A, AtPUB60, AtFZR2/AtCCS52A1, AtFZR3/AtCCS52B, AtMSI1, Unknown proteins (At1g24130, At4g02660, At1g15850)
F-COPI γ subunit	Putative gamma subunit of coatomer adaptor complex*
F-COPI δ subunit	Unknown protein* (At5g57460)
F-COPI ζ subunit	AtCHLD, Unknown protein (At1g67120)
Coat GTPases	
ArfA group	AtSARA1A*, AtRabA5e*, AtRabB1c*, Putative GTPase of unknown function*
ArfB group	AtSARA1A*, AtRabA5e*, AtRabB1c*, Putative GTPase of unknown function*
ArfD group	AtSARA1A*, AtRabA5e*, AtRabB1c*, Putative GTPase of unknown function*
ArfB2 group	AtSARA1A*, AtRabA5e*, AtRabB1c*, Putative GTPase of unknown function*

* = proteins with several assigned roles, having common occurring domains.

Furthermore, although they have a predicted chloroplast location they have already been assigned roles that are not related to vesicle transport (Figure 2). The Putative H-protein of glycine decarboxylase is a component of the glycine decarboxylase complex (GDC) that decarboxylates and deaminates glycine, a step in photorespiration occurring in mitochondria [74], despite experimental indications of a chloroplast location [72] (Table 3). AtCarB is part of a Carbamoyl phosphate synthase involved in arginine synthesis, likely in the chloroplast [75]. AtACC2 and AtCAC2 are both acetyl-CoA carboxylases (AACs) [76,77]. The role of AACs is to convert acetyl-CoA to malonyl-CoA during fatty acid synthesis, and plants generally have two types: heterotrimeric and homomeric ACCs. The heterotrimeric ACC in the chloroplast has four subunits: a biotin carboxyl carrier protein, a biotin carboxylase and two carboxyl transferases (α and β) [78]. The homomeric ACC is encoded by two genes, *ACC1* and *ACC2*, and has been considered to be cytosolic, but it was recently shown that the *ACC2* protein product is located in plastids of Arabidopsis [78]. The last putative protein to be discussed as an AP2 β2 subunit is the Putative RimM-like protein involved in 16S rRNA processing, which has been found in the stroma and is involved in RNA processing [70]. Thus, none of the proteins identified as putative AP2 β2 subunits are likely to act in this manner in the chloroplast based on their proposed functions, which are not related to vesicle transport (Figure 2).

In contrast to AP1 and AP2 subunit candidates, putative AP3, AP4 and AP5 subunits identified only had commonly occurring

domains, making them weaker candidates as true vesicle transport system components (Figure 2).

Evidence for COPI vesicle system components in Arabidopsis chloroplasts

B-COPI subcomplex. Eight candidate proteins for the B-COPI coat β' subunit of COPI vesicles with a predicted chloroplast location were detected. The first is AtRACK1 (Table 5), which plays various roles in plants. Plant mutants defective in this protein have reduced sensitivity to various hormones and impairments in developmental processes, including leaf production [79]. AtRACK1A is also a negative regulator of abscisic acid (ABA) responses [80], and recently a number of proteins have been suggested to interact with AtRACK1A, including proteins involved in photosynthesis and stress responses [81]. Its location is ambiguous; chloroplast localization lacks support from ARAMEMNON and Chloroplast 2010, but it has been found experimentally in chloroplasts according to SUBA (Table 5).

The next three identified proteins (AtPUB60, AtFZR2/CCS52SA1 and AtFZR3/CCS52B) have demonstrated roles in protein degradation (Table 5). AtPUB60 is a U-box protein similar to E3 ubiquitin ligases in yeast and humans, involved in plant innate immunity and plant pathogen resistance [82], in addition to its role in the ubiquitin degradation pathway [83]. In Arabidopsis, the cell cycle process is regulated by a number of cyclins, grouped into A, B, D and H cyclins. Some group B cyclins are degraded during mitosis by a specific ubiquitin E3 ligase, known as anaphase promoting complex (APC), following activation by subunits, which include AtFZR2/CCS52A1 and AtFZR3/CCS52B [84,85]. Further, both of these proteins are involved in endoreduplication, which increases ploidy by inhibiting mitosis [86]. Considering the roles of AtFZR2/CCS52A1 and AtFZR3/CCS52B one might assume a cytosolic location, but ARAMEMNON assigns a chloroplast location, but this has not been experimentally proven according to SUBA.

Further, AtMSI1 was identified as a putative β′ subunit (Table 5). Together with FAS1 and FAS2 this is a member of the Chromatin assembly factor-1 (CAF-1) complex in Arabidopsis, which functions as a histone chaperone in chromatin assembly [87,88], and is also important for additional processes such as seed development [89].

The last three proteins identified in this category are unknown and largely uncharacterized (Table 5). Two, Unknown proteins At1g15850 and At1g24130 are only known to be nucleotide binding, having WD40 domains. The third (At4g02660), is a putative transport protein with a BEACH domain, found in trichome cells [90] and chloroplasts [71]. The function of the BEACH domain is unknown, but appears to be crucial for a number of proteins involved in e.g. vesicle transport [90,91]. However this domain could not be identified using Prosite or Pfam. Thus, only the three Unknown proteins can be considered as likely candidates for hypothesized β′ subunits, as the only ones lacking other assigned roles.

F-COPI subcomplex. Two proteins were identified as putative ζ subunits of the F-COPI subcomplex: AtCHLD and an Unknown protein (At1g67120) (Table 6). AtCHLD has already been identified as involved in the secretory system [11]. Closer examination revealed that AtCHLD significantly differs from known ζ subunits in Arabidopsis and yeast secretory systems (Table S8). It has a VWFA domain profile (PS50234), similar to von Willebrand factor type A domain (PF13519), and a Magnesium chelatase subunit ChlI domain (PF01078), which are not present in any other known ζ subunits. Other ζ subunits have a Clathrin adaptor complex small chain domain (PF01217) lacked by AtCHLD. Thus, AtCHLD appears to be the Magnesium-chelatase subunit ChlD (Uniprot) of Magnesium chelatase, a complex with three subunits [92,93]. This complex is involved in chlorophyll biosynthesis, mediating insertion of magnesium ions into protoporphyrin IX, thereby generating Mg-protoporphyrin IX, and is located in chloroplasts [94,95].

The other putative ζ subunit identified in the chloroplast dataset was the Unknown protein (At1g67120) (Table 6), likely to be chloroplastic according to Chloroplast 2010 and SUBA [73].

Given the distinct differences between AtCHLD and other known ζ subunits, previous reports that AtCHLD functions as a magnesium chelatase in the chloroplast [92,93], and the finding that the Unknown protein has the same domains as AtCHLD, there are probably no homologues of the ζ subunit in chloroplasts (Figure 2).

Commonly occurring domains in Arabidopsis chloroplast proteins

Some proteins are reported to perform several roles, such as the AP4 μ4 subunit in Arabidopsis cytosol which has also been noted as the σ4 subunit in the same complex [11] and the newly identified AP5 ζ subunit which has been previously designated a DNA helicase [96]. Thus, the possibility that some of the putative subunits identified here could play several roles and/or other roles than previously reported should not be excluded. We found three proteins that all correspond to several known subunits: Putative

gamma subunit of coatomer adaptor complex, and Unknown proteins At1g51350 and At5g57460 (Table 9). The Putative gamma subunit of the coatomer adaptor complex has been ambiguously called both Sec21 and a COPI γ subunit [24,97–99], but is considered to be a γ subunit located in Golgi and ER membranes in Arabidopsis [99]. It has even been used experimentally as a Golgi marker [98,100] and shown to be involved in cytosolic vesicle transport in Arabidopsis [11], raising doubts about a true chloroplast location, and thus the likelihood of its involvement in vesicle transport in chloroplasts (Figure 2).

It has been suggested that the Unknown protein At1g51350 is a homologue of the human ARMC8α [101], and involved in endosomal sorting and trafficking [102]. However, ARAMEMNON strongly indicates that it is chloroplast localized. The other Unknown protein, At5g57460, has no clear assigned function yet. Thus, the two Unknown proteins could be involved in some of the suggested functions, but further confirmation is needed (Figure 2).

Coat GTPases in Arabidopsis chloroplasts

Four other proteins with commonly occurring domains were found in the Arabidopsis chloroplast, sharing domains with the previously described cytosolic Arf proteins: AtSARA1A, AtRabA5e, AtRabB1c and the Putative GTPase of unknown function (Table 7). All but one of these four proteins has been ascribed other functions, showing that searches for proteins with this domain will not detect only Arf proteins. The Putative GTPase of unknown function, strongly predicted to be chloroplastic by all the databases and experimental data [70,71], is downregulated in a cold-resistant bri1 (brassinosteroid-insensitive 1) Arabidopsis mutant [103]. Thus, its possible involvement in vesicle transport is not clear (Figure 2).

AtSARA1A has both a small GTPase Arf family profile domain (PS51417) and an ADP-ribosylation factor family domain (PF00025) and has already been identified in the secretory system of Arabidopsis as a Sar1 protein [11] (Table S9). It acts as a GTPase, regulating COPII coat assembly in the cytosol [104,105]. However, SARA1A has also been detected in chloroplasts [106], thus it has an ambiguous or possibly dual localization. Interestingly, another Sar1 protein, CPSAR1, identified in the chloroplast has been shown to affect vesicle transport [9].

Two Rab proteins were identified, CPRabA5e and RabB1c (Table 7). CPRabA5E has previously been predicted as an Arf protein [24], but was recently shown to be a Rab protein with a chloroplast location involved in thylakoid biogenesis. It was affected by oxidative stress, accumulating vesicles at the envelope in chloroplasts when incubated at low temperature under oxidative stress [25].

RabB1c is assumed to participate in vesicle transport according to Uniprot, and is a member of the AtRabB family, which is related to human Rab2 GTPases that are involved in COPI transport in mammalian cells [107,108], and may play a similar role in the Arabidopsis secretory system [109]. RabB1c lacks a transit peptide [14], but has been detected in chloroplasts experimentally [71]. Thus, the only plausible candidate Arf in the chloroplast is the Putative GTPase of unknown function, but GTP binding is apparently not sufficient for a functional Arf, thus further confirmation that it acts as one is required (Figure 2).

Evidence for clathrin-coated vesicle system components in rice

Six proteins were identified in rice with CCV relevant domains and chloroplast localization according to Target P. Clathrin heavy chain 1 and Clathrin heavy chain 2 (Table 8) are referred to as clathrin heavy chains [110,111], based on their similarity to other clathrin components (Uniprot) but have not been characterized. If the predicted chloroplast localization is correct further investigation is warranted since their homologies and designations clearly imply a role in vesicle transport.

Two ascorbate peroxidases were identified, APX7 and APX 8 (Table 8). In plants, ascorbate peroxidases use ascorbate as an electron donor to convert H_2O_2 to H_2O. In rice there are eight known *APX* genes, and four of which are believed to be chloroplast localized (*APX5-APX8*) [112,113]. However, they are unlikely to act as AP1 γ subunits in chloroplasts due to their role as peroxidases.

Two AP2 β2 subunit candidates, ACC2 and CARB (Table 8), were identified. However, as discussed above in the Arabidopsis analysis, ACC2 and CARB are involved in fatty acid and arginine synthesis; hence they are unlikely to be subunits of chloroplast vesicles.

Evidence for COPI system components in rice

With COPI relevant domains and chloroplast location according to Target P, three proteins were identified. RAPTOR1 was identified as a putative B-COPI β′ subunit (Table 8). However, in Arabidopsis RAPTOR1 is known as to regulate TOR1 (TARGET OF RAPAMYCIN), a kinase involved in growth signalling pathways, and interacts with a putative substrate of TOR, S6K1, in vivo [114]. The role of RAPTOR1 in the TOR pathway in Arabidopsis makes it an unlikely candidate as possible B-COPI β′ subunit also in rice.

As a putative B-COPI ε subunit, the PPR repeat containing protein was identified (Table 8). Pentatricopeptide repeat proteins (PPR proteins) are RNA-binding proteins involved in various post-transcriptional processes in both mitochondria and chloroplasts [115]. The PPR family is defined by a tandem 35 amino acid motif. The proteins are predicted to have multiple α helices, placing them in the α-solenoid superfamily together with e.g. HEAT domain proteins [115,116]. One of the PPR proteins in rice, OsPPR1, is located in chloroplasts, essential for chloroplast biogenesis, and its suppression results in chlorophyll deficiency [117].

As also found in the Arabidopsis analysis, the only protein corresponding to F-COPI ζ identified in rice chloroplasts was a magnesium chelatase, CHLD [118] (Table 8). This again raises doubts about its function as a COPI component, which was previously indicated [11].

Commonly occurring domains of proteins including Coat GTPases in rice chloroplasts

Six proteins of rice chloroplasts were identified with commonly occurring domains found in multiple subunits, but due to the low specificity of the identifying domains they are less robust candidates. One domain, the Adaptin N terminal region (PF01602), was detected in AP-3 complex subunit delta, Armadillo/beta-catenin-like repeat family protein and Adaptin, putative (Table S10). Little is known about these proteins; AP-3 complex subunit delta has a name implying a role in vesicle transport, but has not yet been characterized. The Armadillo/beta-catenin-like repeat family protein has Armadillo repeats, placing it in the ARM repeat superfamily together with AP-3 complex subunit delta, according to Uniprot and Superfamily 1.75 [119]. Armadillo repeats are found in proteins with various roles, for instance β-catenin [120]. They are about 40 amino acids long and usually tandemly repeated, forming an armadillo domain. Adaptin, putative has an Adaptin N terminal region (PF01602), but also a Beta2-adaptin appendage C-terminal sub-domain (PF09066) (Table S10). The protein has not yet been characterized in rice.

Since two of these proteins have names related to vesicle transport, and all three share domain PF01602, they could potentially all be true subunits.

Adaptor complexes medium subunit family protein/LOC_Os12g34370 has, similarly to AP-3 complex subunit delta and Adaptin putative not either been characterized but a name implying a role in vesicle transport. In Arabidopsis, At1g56590 was annotated as Clathrin adaptor complexes medium subunit family protein [121] and is considered as the AP3 μ3 subunit [11]. Hence, a role in vesicle transport in chloroplasts cannot be excluded.

Guanine nucleotide-binding protein subunit beta was identified as a putative B-COPI α and B-COPI β′ subunit. Its name implies a role as a subunit of a heterotrimeric G-protein, but it has not yet been characterized. The domains identified in this protein (PS50294, PS50082 and PS00678) all refer to WD repeats (Table S10). WD repeat proteins have four of more repetitive subunits, each consisting of about 40–60 amino acids and usually ending with tryptophan (W) and aspartic acid (D) [122]. It has been assumed that all WD repeat proteins form β propellers, and the best characterized is the β subunit of the heterotrimeric G protein [39,122,123]. WD repeat proteins have known importance in various processes, including vesicle transport [122,123], but as shown here simply detecting WD repeats in a protein is not sufficient to elucidate a protein's functions completely.

The Ras superfamily of small GTPases is divided into five families: Rab, Arf/Sar, Ran, Ras and Rho. In plants, no representatives of the Ras family have been found [124]. One protein was identified in a search for proteins with an ADP-ribosylation factor family domain (PF00025): Mitochondrial Rho GTPase. Its Uniprot name indicates a mitochondrial location, but its Rice Genome Annotation project designation is less specific (ATP/GTP/Ca++ binding protein, putative, expressed), and it is located in the chloroplast according to Target P (Table S10). Thus, future experiments are needed to resolve its location.

Conclusion

The acquired data indicate that no transport system resembling cytosolic CCV or COPI systems is present in Arabidopsis chloroplasts. Several putative subunits identified in the chloroplast dataset were shown to be located elsewhere according to previous studies or various tools, having a possible dual location and/or roles unrelated to vesicle transport. Out of 29 proteins identified in Arabidopsis, the majority had either commonly occurring domains, vesicle unrelated or unknown function (Figure 2). Only two proteins among the suggested, Putative heavy chain of clathrin complex (AtCHC2)/At3g08530 and Putative light chain of clathrin complex (AtCLC1)/At2g40060, could be considered likely subunits in the chloroplast, having known roles related to vesicle transport. Several subunits could not be identified at all in the chloroplast, when searching for relevant domains (Figure 2). The findings indicate that if a CCV- or COPI-like vesicle system is present in chloroplasts it probably differs substantially from the cytosolic counterpart. However, the possible presence of a different and/or simplified CCV or COPI system cannot be excluded. The occurrence of a putative AP2 β2 subunit supports the possible presence of a unique system, since this homologue is present in yeast but not Arabidopsis cytosol, and many of the putative subunits identified have greater resemblance to yeast counterparts than Arabidopsis counterparts (Tables S1, S2, S3, S4, S5, S6, S7, S8, S9).

Considering rice, most of the subunits that could be identified are uncharacterized and named by their similarity to other proteins. As in Arabidopsis chloroplasts, many proteins were also found to have commonly occurring domains. Only two proteins (still uncharacterized in rice) have names indicating a role in vesicle transport, a predicted chloroplast location and domains that are not commonly occurring: Clathrin heavy chain 1 and Clathrin heavy chain 2. It is interesting to note that the results in rice support the findings in Arabidopsis i.e. not many proteins can be clearly said to be chloroplast localized and involved in vesicle transport.

No prokaryote vesicle transport system has been reported [16,125,126], but a few examples of prokaryotic structures analogous to vesicles have been observed [35]. The vesicle system in eukaryotes has been hypothesized as a trait that developed soon after the divergence from prokaryotes and thereafter further specialized as adaptations to new environments [35]. Chloroplasts, believed to have resulted via endosymbiosis of early eukaryotes with cyanobacteria, have vesicles with properties resembling other eukaryotic vesicles, including probable regulation of their formation by GTPases, and inhibition of fusion by microcystin LR and low temperature [16]. However, two proteins of prokaryotic origin have suggested involvement in vesicle formation in chloroplasts: CPSAR1 and Vipp1 [9,127]. Vesicles have also been found in representatives of embryophytes, including bryophytes, pteridophytes, spermatophytes (gymnosperms and angiosperms), but not in other groups including cyanobacteria, glaucocystophytes, rhodophytes, chlorophytes and charophytes. Hence, it has been proposed that the vesicles in chloroplasts evolved after the division of embryophytes from charophytes as an adaptation to land colonization [125].

Plastids occur in several forms, in diverse organisms, and their broad variation in thylakoid organization is assumed to have arisen via evolution in different hosts after the ancestral endosymbiosis [126]. Regarding the three known vesicle transport systems, the COPII system is likely to ancestral, since it is used in essential biosynthetic pathways in all eukaryotes, while the COPI and CCV systems could be later specializations involved in recycling resources to the ER and endocytosis [26,128,129].

Taken together, the available evidence indicates that a vesicle system arose in early eukaryotes, COPII is the ancestral machinery, chloroplast vesicles show clear eukaryotic traits and first evolved during land colonization. In addition, we conclude that no COPI- or CCV-like vesicle system is likely to be found in chloroplasts, in contrast to a COPII-like system, for which a chloroplast location has bioinformatic support [14]. Speculatively, early eukaryotes gained a COPII-like vesicle system, engulfed cyanobacteria and developed plastids, to which the system was transferred. If so, since some photosynthetic eukaryotes do not have vesicles in their plastids, a major speciation event was presumably involved, separating those that form COPII-like chloroplast vesicles from others, before all lines continued to develop the COPI and CCV systems in the cytosol. Alternatively, all three vesicle systems may have already developed in the cytosol of the ancestral eukaryotes when cyanobacteria were engulfed, but only the COPII system was transferred to the chloroplast, or the other two were lost during subsequent evolution. Thus, future experimental evidence is needed to solve the intriguing questions how, when and why a suggested COPII system emerged as the sole vesicle system in chloroplasts.

Supporting Information

Table S1 CCV triskelion proteins from Arabidopsis (*A. thaliana*) cytosol (retrieved from Bassham et al, 2008) and yeast (*S. cerevisiae*), mouse (*M. musculus*) and human (*H. sapiens*) cytosol

(retrieved from Uniprot). Domains of these proteins were extracted using Prosite and Pfam, then run against the chloroplast protein dataset to identify proteins putatively involved in vesicle transport in chloroplasts.

Table S2 CCV AP1 complex proteins from Arabidopsis (*A. thaliana*) cytosol (retrieved from Bassham et al, 2008) and yeast (*S. cerevisiae*), mouse (*M. musculus*) and human (*H. sapiens*) cytosol (retrieved from Uniprot). Domains of these proteins were extracted using Prosite and Pfam, then run against the chloroplast protein dataset to identify proteins putatively involved in vesicle transport in chloroplasts.

Table S3 CCV AP2 complex proteins from Arabidopsis (*A. thaliana*) cytosol (retrieved from Bassham et al, 2008) and yeast (*S. cerevisiae*), mouse (*M. musculus*) and human (*H. sapiens*) cytosol (retrieved from Uniprot). Domains of these proteins were extracted using Prosite and Pfam, then run against the chloroplast protein dataset to identify proteins putatively involved in vesicle transport inside chloroplasts.

Table S4 CCV AP3 complex proteins from Arabidopsis (*A. thaliana*) cytosol (retrieved from Bassham et al, 2008) and yeast (*S. cerevisiae*), mouse (*M. musculus*) and human (*H. sapiens*) cytosol (retrieved from Uniprot). Domains of these proteins were extracted using Prosite and Pfam, then run against the chloroplast protein dataset to identify proteins putatively involved in vesicle transport inside chloroplasts.

Table S5 CCV AP4 complex proteins from Arabidopsis (*A. thaliana*) cytosol (retrieved from Bassham et al, 2008) and yeast (*S. cerevisiae*), mouse (*M. musculus*) and human (*H. sapiens*) cytosol (retrieved from Uniprot). Domains of these proteins were extracted using Prosite and Pfam, then run against the chloroplast protein dataset to identify proteins putatively involved in vesicle transport in chloroplasts.

Table S6 CCV AP5 complex proteins from Arabidopsis (*A. thaliana*) cytosol (retrieved from Hirst et al, 2011), and human (*H. sapiens*) cytosol (retrieved from Uniprot). Domains of these proteins were extracted using Prosite and Pfam, then run against the chloroplast protein dataset to identify proteins putatively involved in vesicle transport inside chloroplasts.

Table S7 B-COPI subcomplex proteins from Arabidopsis (*A. thaliana*) cytosol (retrieved from Bassham et al, 2008) and yeast (*S. cerevisiae*), mouse (*M. musculus*) and human (*H. sapiens*) cytosol (retrieved from Uniprot). Domains of these proteins were extracted using Prosite and Pfam, then run against the chloroplast protein dataset to identify proteins putatively involved in vesicle transport inside chloroplasts.

Table S8 F-COPI subcomplex proteins from Arabidopsis (*A. thaliana*) cytosol (retrieved from Bassham et al, 2008) and yeast (*S. cerevisiae*), mouse (*M. musculus*) and human (*H. sapiens*) cytosol (retrieved from Uniprot). Domains of these proteins were extracted using Prosite and Pfam, then run against the chloroplast protein dataset to identify proteins putatively involved in vesicle transport inside chloroplasts.

Table S9 Coat GTPase proteins from Arabidopsis (*A. thaliana*) cytosol (retrieved from Bassham et al, 2008) and yeast (*S. cerevisiae*), mouse (*M. musculus*) and human (*H. sapiens*) cytosol (retrieved from Uniprot). Domains of these proteins were extracted using Prosite and Pfam, then run against the chloroplast protein dataset to identify proteins putatively involved in vesicle transport inside chloroplasts.

Table S10 CCV and COPI proteins from Arabidopsis (*A. thaliana*) cytosol (retrieved from Bassham et al, 2008) and yeast (*S. cerevisiae*), mouse (*M. musculus*) and human (*H. sapiens*) cytosol (retrieved from Uniprot). Domains of these proteins were extracted using Prosite and Pfam, run against the rice (subsp. *japonica*) protein dataset to identify proteins with the same domains, then

those putatively involved in vesicle transport in chloroplasts were identified using Target P, and listed.

Acknowledgments

The authors thank Jenny Carlsson and Sazzad Karim for critical reading of the manuscript, Kemal Sanli for valuable help with organising Pfam data, Nadir Zaman Khan, Selvakumar Sukumuran, and Mageshwaran Rajasekar for initial assistance.

Author Contributions

Conceived and designed the experiments: EL MA HA. Performed the experiments: EL MA HA. Analyzed the data: EL MA HA. Contributed reagents/materials/analysis tools: EL MA HA. Wrote the paper: EL MA HA.

References

1. Abdallah F, Salamini F, Leister D (2000) A prediction of the size and evolutionary origin of the proteome of chloroplasts of Arabidopsis. Trends in plant science 5: 141–142.
2. Aronsson H, Jarvis P (2008) The chloroplast protein import apparatus, its components, and their roles. In: Sandelius AS, Aronsson H (eds) Plant Cell Monograph "Chloroplast - interactions with the environment" Springer Verlag, Berlin, Germany.
3. Jarvis P, Robinson C (2004) Mechanisms of protein import and routing in chloroplasts. Current Biology 14: R1064–R1077.
4. Robinson C, Thompson SJ, Woolhead C (2001) Multiple pathways used for the targeting of thylakoid proteins in chloroplasts. Traffic 2: 245–251.
5. Spetea C, Aronsson H (2012) Mechanisms of Transport Across Membranes in Plant Chloroplasts. Current Chemical Biology 6: 230–243.
6. Kelly AA, Dörmann P (2004) Green light for galactolipid trafficking. Current opinion in plant biology 7: 262–269.
7. Shimojima M, Ohta H, Iwamatsu A, Masuda T, Shioi Y, et al. (1997) Cloning of the gene for monogalactosyldiacylglycerol synthase and its evolutionary origin. Proceedings of the National Academy of Sciences 94: 333–337.
8. Andersson MX, Kjellberg JM, Sandelius AS (2001) Chloroplast biogenesis. Regulation of lipid transport to the thylakoid in chloroplasts isolated from expanding and fully expanded leaves of pea. Plant physiology 127: 184–193.
9. Garcia C, Khan NZ, Nannmark U, Aronsson H (2010) The chloroplast protein CPSAR1, dually localized in the stroma and the inner envelope membrane, is involved in thylakoid biogenesis. Plant Journal 63: 73–85.
10. Räntfors M, Evertsson I, Kjellberg JM, Stina Sandelius A (2000) Intraplastidial lipid trafficking: Regulation of galactolipid release from isolated chloroplast envelope. Physiologia Plantarum 110: 262–270.
11. Bassham DC, Brandizzi F, Otegui MS, Sanderfoot AA (2008) The secretory system of Arabidopsis. The Arabidopsis Book/American Society of Plant Biologists 6.
12. Donaldson JG, Cassel D, Kahn RA, Klausner RD (1992) ADP-ribosylation factor, a small GTP-binding protein, is required for binding of the coatomer protein beta-COP to Golgi membranes. Proceedings of the National Academy of Sciences 89: 6408–6412.
13. Kirchhausen T (2000) Three ways to make a vesicle. Nature Reviews Molecular Cell Biology 1: 187–198.
14. Khan NZ, Lindquist E, Aronsson H (2013) New putative chloroplast vesicle transport components and cargo proteins revealed using a bioinformatics approach: an Arabidopsis model. PLoS One 8: e59898.
15. Morré DJ, Selldén G, Sundqvist C, Sandelius AS (1991) Stromal low temperature compartment derived from the inner membrane of the chloroplast envelope. Plant Physiology 97: 1558–1564.
16. Westphal S, Soll J, Vothknecht UC (2001) A vesicle transport system inside chloroplasts. FEBS Letters 506: 257–261.
17. Rothman JE (1994) Mechanisms of intracellular protein transport. Nature 372: 55–63.
18. Schekman R, Orci L (1996) Coat proteins and vesicle budding. Science 271: 1526–1533.
19. Lee MCS, Miller EA, Goldberg J, Orci L, Schekman R (2004) Bi-directional protein transport between the ER and Golgi. Annual Review of Cell and Developmental Biology 20: 87–123.
20. Bethune J, Wieland F, Moelleken J (2006) COPI-mediated transport. Journal of Membrane Biology 211: 65–79.
21. Bednarek SY, Ravazzola M, Hosobuchi M, Amherdt M, Perrelet A, et al. (1995) COPI-and COPII-coated vesicles bud directly from the endoplasmic reticulum in yeast. Cell 83: 1183–1196.
22. Barlowe C, Schekman R (1993) SEC12 encodes a guanine-nucleotide-exchange factor essential for transport vesicle budding from the ER. Nature 365: 347–349.

23. Yoshihisa T, Barlowe C, Schekman R (1993) Requirement for a GTPase-activating protein in vesicle budding from the endoplasmic reticulum. Science 259: 1466–1468.
24. Andersson MX, Sandelius AS (2004) A chloroplast-localized vesicular transport system: a bio-informatics approach. BMC Genomics 5: 40.
25. Karim S, Alezzawi M, Garcia-Petit C, Solymosi K, Khan NZ, et al. (2014) A novel chloroplast localized Rab GTPase protein CPRabA5e is involved in stress, development, thylakoid biogenesis and vesicle transport in Arabidopsis. Plant Molecular Biology 84: 675–692.
26. McMahon HT, Mills IG (2004) COP and clathrin-coated vesicle budding: different pathways, common approaches. Current Opinion in Cell Biology 16: 379–391.
27. Lee M, Orci L, Hamamoto S, Futai E, Ravazzola M, et al. (2005) Sar1p N-terminal helix initiates membrane curvature and completes the fission of a COPII vesicle. Cell 122: 605–617.
28. Bonifacino JS, Glick BS (2004) The mechanisms of vesicle budding and fusion. Cell 116: 153–166.
29. Dell'Angelica EC (2001) Clathrin-binding proteins: Got a motif? Join the network! Trends in Cell Biology 11: 315–318.
30. Kirchhausen T (2000) Clathrin. Annual Review of Biochemistry 69: 699–727.
31. Wilbur JD, Hwang PK, Brodsky FM (2005) New faces of the familiar clathrin lattice. Traffic 6: 346–350.
32. Robinson MS (2004) Adaptable adaptors for coated vesicles. Trends in Cell Biology 14: 167–174.
33. Hirst J, Barlow LD, Francisco GC, Sahlender DA, Seaman MN, et al. (2011) The fifth adaptor protein complex. PLoS Biology 9: e1001170.
34. Boehm M, Bonifacino JS (2001) Adaptins The Final Recount. Molecular Biology of the Cell 12: 2907–2920.
35. Dacks JB, Field MC (2007) Evolution of the eukaryotic membrane-trafficking system: origin, tempo and mode. Journal of Cell Science 120: 2977–2985.
36. Paleotti O, Macia E, Luton F, Klein S, Partisani M, et al. (2005) The small G-protein Arf6GTP recruits the AP-2 adaptor complex to membranes. Journal of Biological Chemistry 280: 21661–21666.
37. Lee C, Goldberg J (2010) Structure of coatomer cage proteins and the relationship among COPI, COPII, and clathrin vesicle coats. Cell 142: 123–132.
38. Fotin A, Cheng Y, Sliz P, Grigorieff N, Harrison SC, et al. (2004) Molecular model for a complete clathrin lattice from electron cryomicroscopy. Nature 432: 573–579.
39. Ter Haar E, Harrison SC, Kirchhausen T (2000) Peptide-in-groove interactions link target proteins to the β-propeller of clathrin. Proceedings of the National Academy of Sciences 97: 1096–1100.
40. Fath S, Mancias JD, Bi X, Goldberg J (2007) Structure and organization of coat proteins in the COPII cage. Cell 129: 1325–1336.
41. Stagg SM, LaPointe P, Razvi A, Gürkan C, Potter CS, et al. (2008) Structural basis for cargo regulation of COPII coat assembly. Cell 134: 474–484.
42. Gattiker A, Gasteiger E, Bairoch A (2002) ScanProsite: a reference implementation of a PROSITE scanning tool. Applied Bioinformatics 1: 107–108.
43. Ortiz-Zapater E, Soriano-Ortega E, Marcote MJ, Ortiz-Masiá D, Aniento F (2006) Trafficking of the human transferrin receptor in plant cells: effects of tyrphostin A23 and brefeldin A. Plant Journal 48: 757–770.
44. Lamesch P, Berardini TZ, Li D, Swarbreck D, Wilks C, et al. (2012) The Arabidopsis Information Resource (TAIR): improved gene annotation and new tools. Nucleic Acids Research 40: D1202–D1210.
45. Schwacke R, Schneider A, van der Graaff E, Fischer K, Catoni E, et al. (2003) ARAMEMNON, a novel database for Arabidopsis integral membrane proteins. Plant Physiology 131: 16–26.
46. Pierleoni A, Martelli PL, Fariselli P, Casadio R (2006) BaCelLo: a balanced subcellular localization predictor. Bioinformatics 22: e408–e416.

47. Emanuelsson O, Nielsen H, Heijne GV (1999) ChloroP, a neural network-based method for predicting chloroplast transit peptides and their cleavage sites. Protein Science 8: 978–984.

48. Bannai H, Tamada Y, Maruyama O, Nakai K, Miyano S (2002) Extensive feature detection of N-terminal protein sorting signals. Bioinformatics 18: 298–305.

49. Guda C, Fahy E, Subramaniam S (2004) MITOPRED: a genome-scale method for prediction of nucleus-encoded mitochondrial proteins. Bioinformatics 20: 1785–1794.

50. Claros MG, Vincens P (1996) Computational method to predict mitochondrially imported proteins and their targeting sequences. European Journal of Biochemistry 241: 779–786.

51. Höglund A, Dönnes P, Blum T, Adolph H-W, Kohlbacher O (2006) MultiLoc: prediction of protein subcellular localization using N-terminal targeting sequences, sequence motifs and amino acid composition. Bioinformatics 22: 1158–1165.

52. Lu Z, Szafron D, Greiner R, Lu P, Wishart DS, et al. (2004) Predicting subcellular localization of proteins using machine-learned classifiers. Bioinformatics 20: 547–556.

53. Schein AI, Kissinger JC, Ungar LH (2001) Chloroplast transit peptide prediction: a peek inside the black box. Nucleic Acids Research 29: e82–e82.

54. Bodén M, Hawkins J (2005) Prediction of subcellular localization using sequence-biased recurrent networks. Bioinformatics 21: 2279–2286.

55. Hiller K, Grote A, Scheer M, Münch R, Jahn D (2004) PrediSi: prediction of signal peptides and their cleavage positions. Nucleic Acids Research 32: W375–W379.

56. Small I, Peeters N, Legeai F, Lurin C (2004) Predotar: A tool for rapidly screening proteomes for N-terminal targeting sequences. Proteomics 4: 1581–1590.

57. Petsalaki EI, Bagos PG, Litou ZI, Hamodrakas SJ (2006) PredSL: a tool for the N-terminal sequence-based prediction of protein subcellular localization. Genomics, Proteomics and Bioinformatics 4: 48–55.

58. Nielsen H, Brunak S, von Heijne G (1999) Machine learning approaches for the prediction of signal peptides and other protein sorting signals. Protein Engineering 12: 3–9.

59. Nielsen H, Krogh A (1998) Prediction of signal peptides and signal anchors by a hidden Markov model. ISMB-98 Proceedings 6: 122–130.

60. Nielsen H, Engelbrecht J, Brunak S, Heijne GV (1997) A neural network method for identification of prokaryotic and eukaryotic signal peptides and prediction of their cleavage sites. International Journal of Neural Systems 8: 581–599.

61. Matsuda S, Vert JP, Saigo H, Ueda N, Toh H, et al. (2005) A novel representation of protein sequences for prediction of subcellular location using support vector machines. Protein Science 14: 2804–2813.

62. Emanuelsson O, Nielsen H, Brunak S, von Heijne G (2000) Predicting subcellular localization of proteins based on their N-terminal amino acid sequence. Journal of Molecular Biology 300: 1005–1016.

63. Horton P, Park K-J, Obayashi T, Nakai K (2006) Protein Subcellular Localisation Prediction with WoLF PSORT. APBC: 39–48.

64. Schwacke R, Fischer K, Ketelsen B, Krupinska K, Krause K (2007) Comparative survey of plastid and mitochondrial targeting properties of transcription factors in Arabidopsis and rice. Molecular Genetics and Genomics 277: 631–646.

65. Heazlewood JL, Verboom RE, Tonti-Filippini J, Small I, Millar AH (2007) SUBA: the Arabidopsis subcellular database. Nucleic Acids Research 35: D213–D218.

66. Lu Y, Savage LJ, Larson MD, Wilkerson CG, Last RL (2011) Chloroplast 2010: a database for large-scale phenotypic screening of Arabidopsis mutants. Plant Physiology 155: 1589–1600.

67. Holstein SE (2002) Clathrin and plant endocytosis. Traffic 3: 614–620.

68. Chen X, Irani NG, Friml J (2011) Clathrin-mediated endocytosis: the gateway into plant cells. Current Opinion in Plant Biology 14: 674–682.

69. Ito E, Fujimoto M, Ebine K, Uemura T, Ueda T, et al. (2012) Dynamic behavior of clathrin in Arabidopsis thaliana unveiled by live imaging. Plant Journal 69: 204–216.

70. Olinares PDB, Ponnala L, van Wijk KJ (2010) Megadalton complexes in the chloroplast stroma of Arabidopsis thaliana characterized by size exclusion chromatography, mass spectrometry, and hierarchical clustering. Molecular and Cellular Proteomics 9: 1594–1615.

71. Zybailov B, Rutschow H, Friso G, Rudella A, Emanuelsson O, et al. (2008) Sorting signals, N-terminal modifications and abundance of the chloroplast proteome. PLoS One 3: e1994.

72. Kleffmann T, Russenberger D, von Zychlinski A, Christopher W, Sjolander K, et al. (2004) The Arabidopsis thaliana chloroplast proteome reveals pathway abundance and novel protein functions. Current Biology 14: 354–362.

73. Froehlich JE, Wilkerson CG, Ray WK, McAndrew RS, Osteryoung KW, et al. (2003) Proteomic Study of the Arabidopsis t haliana Chloroplastic Envelope Membrane Utilizing Alternatives to Traditional Two-Dimensional Electrophoresis. Journal of Proteome Research 2: 413–425.

74. Maurino VG, Peterhansel C (2010) Photorespiration: current status and approaches for metabolic engineering. Current Opinion in Plant Biology 13: 248–255.

75. Slocum RD (2005) Genes, enzymes and regulation of arginine biosynthesis in plants. Plant Physiology and Biochemistry 43: 729–745.

76. Yanai Y, Kawasaki T, Shimada H, Wurtele ES, Nikolau BJ, et al. (1995) Genomic organization of 251 kDa acetyl-CoA carboxylase genes in Arabidopsis: tandem gene duplication has made two differentially expressed isozymes. Plant and Cell Physiology 36: 779–787.

77. Han X, Yin L, Xue H (2012) Co-expression Analysis Identifies CRC and AP1 the Regulator of Arabidopsis Fatty Acid Biosynthesis. Journal of Integrative Plant Biology 54: 486–499.

78. Sasaki Y, Nagano Y (2004) Plant acetyl-CoA carboxylase: structure, biosynthesis, regulation, and gene manipulation for plant breeding. Bioscience, Biotechnology, and Biochemistry 68: 1175–1184.

79. Wen W, Chen L, Wu H, Sun X, Zhang M, et al. (2006) Identification of the yeast R-SNARE Nyv1p as a novel longin domain-containing protein. Molecular Biology of the Cell 17: 4282–4299.

80. Guo J, Wang J, Xi L, Huang W-D, Liang J, et al. (2009) RACK1 is a negative regulator of ABA responses in Arabidopsis. Journal of Experimental Botany 60: 3819–3833.

81. Kundu N, Dozier U, Deslandes L, Somssich IE, Ullah H (2013) Arabidopsis scaffold protein RACK1A interacts with diverse environmental stress and photosynthesis related proteins. Plant Signaling and Behavior 8: e24012.

82. Monaghan J, Xu F, Gao M, Zhao Q, Palma K, et al. (2009) Two Prp19-like U-box proteins in the MOS4-associated complex play redundant roles in plant innate immunity. PLoS Pathogens 5: e1000526.

83. Wiborg J, O'Shea C, Skriver K (2008) Biochemical function of typical and variant Arabidopsis thaliana U-box E3 ubiquitin-protein ligases. Biochemical Journal 413: 447–457.

84. Fülöp K, Tarayre S, Kelemen Z, Horváth G, Kevei Z, et al. (2005) Arabidopsis anaphase-promoting complexes: multiple activators and wide range of substrates might keep APC perpetually busy. Cell Cycle 4: 4084–4092.

85. Gutierrez C (2009) The Arabidopsis cell division cycle. The Arabidopsis book/ American Society of Plant Biologists 7.

86. Larson-Rabin Z, Li Z, Masson PH, Day CD (2009) FZR2/CCS52A1 expression is a determinant of endoreduplication and cell expansion in Arabidopsis. Plant Physiology 149: 874–884.

87. Kaya H, Shibahara K-i, Taoka K-i, Iwabuchi M, Stillman B, et al. (2001) FASCIATA genes for chromatin assembly factor-1 in Arabidopsis maintain the cellular organization of apical meristems. Cell 104: 131–142.

88. Zhu Y, Dong A, Shen W-H (2012) Histone variants and chromatin assembly in plant abiotic stress responses. Biochimica et Biophysica Acta (BBA)-Gene Regulatory Mechanisms 1819: 343–348.

89. Köhler C, Hennig L, Bouveret R, Gheyselinck J, Grossniklaus U, et al. (2003) Arabidopsis MSI1 is a component of the MEA/FIE Polycomb group complex and required for seed development. EMBO Journal 22: 4804–4814.

90. Wienkoop S, Zoeller D, Ebert B, Simon-Rosin U, Fisahn J, et al. (2004) Cell-specific protein profiling in Arabidopsis thaliana trichomes: identification of trichome-located proteins involved in sulfur metabolism and detoxification. Phytochemistry 65: 1641–1649.

91. Jogl G, Shen Y, Gebauer D, Li J, Wiegmann K, et al. (2002) Crystal structure of the BEACH domain reveals an unusual fold and extensive association with a novel PH domain. EMBO Journal 21: 4785–4795.

92. Eckhardt U, Grimm B, Hörtensteiner S (2004) Recent advances in chlorophyll biosynthesis and breakdown in higher plants. Plant Molecular Biology 56: 1–14.

93. Papenbrock J, Gräfe S, Kruse E, Hänel F, Grimm B (1997) Mg-chelatase of tobacco: identification of a Chl D cDNA sequence encoding a third subunit, analysis of the interaction of the three subunits with the yeast two-hybrid system, and reconstitution of the enzyme activity by co-expression of recombinant CHL D, CHL H and CHL I. Plant Journal 12: 981–990.

94. Masuda T (2008) Recent overview of the Mg branch of the tetrapyrrole biosynthesis leading to chlorophylls. Photosynthesis Research 96: 121–143.

95. Solymosi K, Aronsson H (2013) Etioplasts and Their Significance in Chloroplast Biogenesis. Plastid Development in Leaves during Growth and Senescence: Springer. pp. 39–71.

96. Stabicki M, Theis M, Krastev DB, Samsonov S, Mundwiller E, et al. (2010) A genome-scale DNA repair RNAi screen identifies SPG48 as a novel gene associated with hereditary spastic paraplegia. PLoS Biology 8: e1000408.

97. Gao C, Christine K, Qu S, San MWY, Li KY, et al. (2012) The Golgi-localized Arabidopsis endomembrane protein12 contains both endoplasmic reticulum export and Golgi retention signals at its C terminus. Plant Cell 24: 2086–2104.

98. Kleine-Vehn J, Dhonukshe P, Swarup R, Bennett M, Friml J (2006) Subcellular trafficking of the Arabidopsis auxin influx carrier AUX1 uses a novel pathway distinct from PIN1. Plant Cell 18: 3171–3181.

99. Movafeghi A, Happel N, Pimpl P, Tai G-H, Robinson DG (1999) Arabidopsis Sec21p and Sec23p homologs. Probable coat proteins of plant COP-coated vesicles. Plant Physiology 119: 1437–1446.

100. Vanhee C, Zapotoczny G, Masquelier D, Ghislain M, Batoko H (2011) The Arabidopsis multistress regulator TSPO is a heme binding membrane protein and a potential scavenger of porphyrins via an autophagy-dependent degradation mechanism. Plant Cell 23: 785–805.

101. Kobayashi N, Yang J, Ueda A, Suzuki T, Tomaru K, et al. (2007) RanBPM, Muskelin, p48EMLP, p44CTLH, and the armadillo-repeat proteins ARMC8α and ARMC8β are components of the CTLH complex. Gene 396: 236–247.

102. Tomaru K, Ueda A, Suzuki T, Kobayashi N, Yang J, et al. (2010) Armadillo Repeat Containing 8α Binds to HRS and Promotes HRS Interaction with Ubiquitinated Proteins. The Open Biochemistry Journal 4: 1.

103. Kim SY, Kim BH, Nam KH (2010) Reduced expression of the genes encoding chloroplast–localized proteins in a cold-resistant bri1 (brassinosteroid-Insensitive 1) mutant. Plant Signaling and Behavior 5: 458–463.

104. Robinson DG, Herranz M-C, Bubeck J, Pepperkok R, Ritzenthaler C (2007) Membrane dynamics in the early secretory pathway. Critical Reviews in Plant Sciences 26: 199–225.

105. Cevher-Keskin B (2013) ARF1 and SAR1 GTPases in Endomembrane Trafficking in Plants. International Journal of Molecular Sciences 14: 18181–18199.

106. Joyard J, Ferro M, Masselon C, Seigneurin-Berny D, Salvi D, et al. (2010) Chloroplast proteomics highlights the subcellular compartmentation of lipid metabolism. Progress in Lipid Research 49: 128–158.

107. Tisdale EJ, Bourne JR, Khosravi-Far R, Der CJ, Balch W (1992) GTP-binding mutants of rab1 and rab2 are potent inhibitors of vesicular transport from the endoplasmic reticulum to the Golgi complex. Journal of Cell Biology 119: 749–761.

108. Rutherford S, Moore I (2002) The Arabidopsis Rab GTPase family: another enigma variation. Current Opinion in Plant Biology 5: 518–528.

109. Moore I, Diefenthal T, Zarsky V, Schell J, Palme K (1997) A homolog of the mammalian GTPase Rab2 is present in Arabidopsis and is expressed predominantly in pollen grains and seedlings. Proceedings of the National Academy of Sciences 94: 762–767.

110. Wei Z, Hu W, Lin Q, Cheng X, Tong M, et al. (2009) Understanding rice plant resistance to the brown planthopper (Nilaparvata lugens): a proteomic approach. Proteomics 9: 2798–2808.

111. Park C-J, Sharma R, Lefebvre B, Canlas PE, Ronald PC (2013) The endoplasmic reticulum-quality control component SDF2 is essential for XA21 mediated immunity in rice. Plant Science 210: 53–60.

112. Teixeira FK, Menezes-Benavente L, Margis R, Margis-Pinheiro M (2004) Analysis of the molecular evolutionary history of the ascorbate peroxidase gene family: inferences from the rice genome. Journal of Molecular Evolution 59: 761–770.

113. Teixeira FK, Menezes-Benavente L, Galvão VC, Margis R, Margis-Pinheiro M (2006) Rice ascorbate peroxidase gene family encodes functionally diverse isoforms localized in different subcellular compartments. Planta 224: 300–314.

114. Mahfouz MM, Kim S, Delauney AJ, Verma DPS (2006) Arabidopsis TARGET OF RAPAMYCIN interacts with RAPTOR, which regulates the activity of S6 kinase in response to osmotic stress signals. Plant Cell 18: 477–490.

115. Schmitz-Linneweber C, Small I (2008) Pentatricopeptide repeat proteins: a socket set for organelle gene expression. Trends in Plant Science 13: 663–670.

116. Small ID, Peeters N (2000) The PPR motif–a TPR-related motif prevalent in plant organellar proteins. Trends in Biochemical Sciences 25: 45–47.

117. Gothandam KM, Kim E-S, Cho H, Chung Y-Y (2005) OsPPR1, a pentatricopeptide repeat protein of rice is essential for the chloroplast biogenesis. Plant Molecular Biology 58: 421–433.

118. Zhang H, Li J, Yoo J-H, Yoo S-C, Cho S-H, et al. (2006) Rice Chlorina-1 and Chlorina-9 encode ChlD and ChlI subunits of Mg-chelatase, a key enzyme for chlorophyll synthesis and chloroplast development. Plant Molecular Biology 62: 325–337.

119. Gough J, Karplus K, Hughey R, Chothia C (2001) Assignment of homology to genome sequences using a library of hidden Markov models that represent all proteins of known structure. Journal of Molecular Biology 313: 903–919.

120. Tewari R, Bailes E, Bunting KA, Coates JC (2010) Armadillo-repeat protein functions: questions for little creatures. Trends in Cell Biology 20: 470–481.

121. Niihama M, Takemoto N, Hashiguchi Y, Tasaka M, Morita MT (2009) ZIP genes encode proteins involved in membrane trafficking of the TGN–PVC/vacuoles. Plant and Cell Physiology 50: 2057–2068.

122. Smith TF, Gaitatzes C, Saxena K, Neer EJ (1999) The WD repeat: a common architecture for diverse functions. Trends in Biochemical Sciences 24: 181–185.

123. Li D, Roberts R (2001) Human Genome and Diseases: WD-repeat proteins: structure characteristics, biological function, and their involvement in human diseases. Cellular and Molecular Life Sciences 58: 2085–2097.

124. Vernoud V, Horton AC, Yang Z, Nielsen E (2003) Analysis of the small GTPase gene superfamily of Arabidopsis. Plant Physiology 131: 1191–1208.

125. Westphal S, Soll J, Vothknecht UC (2003) Evolution of chloroplast vesicle transport. Plant and Cell Physiology 44: 217–222.

126. Vothknecht UC, Westhoff P (2001) Biogenesis and origin of thylakoid membranes. Biochimica et Biophysica Acta (BBA)-Molecular Cell Research 1541: 91–101.

127. Kroll D, Meierhoff K, Bechtold N, Kinoshita M, Westphal S, et al. (2001) VIPP1, a nuclear gene of Arabidopsis thaliana essential for thylakoid membrane formation. Proceedings of the National Academy of Sciences 98: 4238–4242.

128. Cavalier-Smith T (2009) Predation and eukaryote cell origins: a coevolutionary perspective. International Journal of Biochemistry and Cell Biology 41: 307–322.

129. Cavalier-Smith T (2000) Membrane heredity and early chloroplast evolution. Trends in Plant Science 5: 174–182.

130. Nickel W, Brugger B, Wieland FT (2002) Vesicular transport: the core machinery of COPI recruitment and budding. Journal of Cell Science 115: 3235–3240.

131. Bonifacino JS, Lippincott-Schwartz J (2003) Coat proteins: shaping membrane transport. Nature Review Molecular Cell Biology 4: 409–414.

15

Genetic Analysis and QTL Detection for Resistance to White Tip Disease in Rice

Tong Zhou[1], Cunyi Gao[1,2], Linlin Du[1], Hui Feng[1], Lijiao Wang[1], Ying Lan[1], Feng Sun[1], Lihui Wei[1], Yongjian Fan[1], Wenbiao Shen[2]*, Yijun Zhou[1]*

1 Institute of Plant Protection, Jiangsu Academy of Agricultural Sciences, Nanjing, China, 2 College of Life Science, Nanjing Agricultural University, Nanjing, China

Abstract

The inheritance of resistance to white tip disease (WTDR) in rice (*Oryza sativa* L.) was analyzed with an artificial inoculation test in a segregating population derived from the cross between Tetep, a highly resistant variety that was identified in a previous study, and a susceptible cultivar. Three resistance-associated traits, including the number of *Aphelenchoides besseyi* (*A. besseyi*) individuals in 100 grains (NA), the loss rate of panicle weight (LRPW) and the loss rate of the total grains per panicle (LRGPP) were analyzed for the detection of the quantitative trait locus (QTL) in the population after construction of a genetic map. Six QTLs distributed on chromosomes 3, 5 and 9 were mapped. *qNA3* and *qNA9*, conferring reproduction number of *A. besseyi* in the panicle, accounted for 16.91% and 12.54% of the total phenotypic variance, respectively. *qDRPW5a* and *qDRPW5b*, associated with yield loss, were located at two adjacent marker intervals on chromosome 5 and explained 14.15% and 14.59% of the total phenotypic variation and possessed LOD values of 3.40 and 3.39, respectively. *qDRPW9* was considered as a minor QTL and only explained 1.02% of the phenotypic variation. *qLRGPP5* contributed to the loss in the number of grains and explained 10.91% of the phenotypic variation. This study provides useful information for the breeding of resistant cultivars against white tip disease in rice.

Editor: Tai Wang, Institute of Botany, Chinese Academy of Sciences, China

Funding: This research was financially supported by grants from the National Key Basic Research and Development Program (973 Program) of China (2013CBA01403), the Special Fund for Agro-Scientific Research in the Public Interest of China (201303021, 201003031), the Jiangsu Province Science and Technology Support Project (BE2012303) and the Jiangsu Agricultural Science and Technology Independent Innovation Fund (cx[12]1003). The funders had no role in study design, data collection and analysis, decision to publish, or preparation of the manuscript.

Competing Interests: The authors have declared that no competing interests exist.

* Email: yjzhou@jaas.ac.cn (YZ); wbshenh@njau.edu.cn (WS)

Introduction

White tip disease of rice (WTDR), which is caused by the rice white tip nematode (*Aphelenchoides besseyi*, *A. besseyi*), is one of the most serious nematode diseases affecting rice worldwide [1–5]. Plants infected with *A. besseyi* exhibit whitening and withering at the tip of the leaf, and the symptoms also include small grains and erect panicles in later growth stages, which can result in large losses in production. The traditional methods used to control *A. besseyi* (including insecticide application and crop rotation) are expensive and cause serious environmental problems [4]. In addition, the wide-scale use of direct-sowing technology for rice, which is not compatible with the treatment of seeds, has recently made this problem more acute in China. Thus, the development of resistant rice cultivars has been considered as the primary strategy for controlling white tip disease [6,7].

Rice varieties that are resistant against WTDR have been reported by researchers, e.g., *cvs Arkansas Fortuna, Nira 43, Asa-Hi, Binam* and *Domsiah*, which were screened from a large number of rice varieties [4,8,9]. However, the genetic mechanism of WTDR resistance is still poorly understood. In addition, some of the resistant varieties exhibited resistance to WTDR only in

particular regions or were highly susceptible to other pathogens [10], which hindered projects aimed at breeding for resistance against WTDR. In addition, many nematode resistance loci, such as *H1*, *GroV1*, *Cre* and *Mi3*, have been identified in tomato, potato, soybean and other crops [11–14], and some have even been cloned and functionally analyzed [15–18]. In addition, a gene (*Has-1^Og*) resistant to the cyst nematode (*Heterodera sacchari*) has been identified in rice [19]. These results suggested that host plants, including rice, harbored defense mechanisms to fight against nematodes, and these mechanisms have developed over long-term co-evolution. The lack of information on the inheritance of the resistance to WTDR may slow the progress of breeding programs. Therefore, elucidating the resistance mechanisms involved will contribute to a better understanding of nematode-plant interactions and assist with the breeding of nematode-resistant cultivars [20].

In our previous study [21], a collection of germplasm resources was screened using an inoculation test, and an *indica* variety, Tetep, showed high resistance to WTDR. Herein, we present the inheritance mode of resistance to WTDR in rice and the quantitative trait loci (QTLs) related to resistance against WTDR.

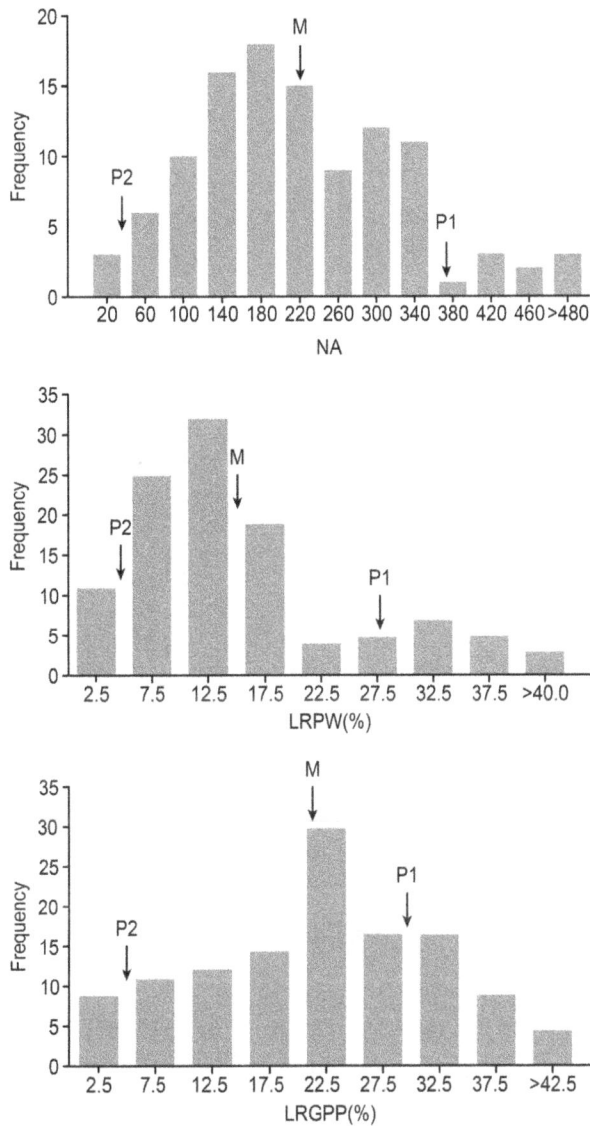

Figure 1. The frequency distributions of three resistance-associated traits in the F$_{2:3}$ lines. P$_1$ represents Huaidao No.5, P$_2$ represents Tetep, and M represents the mean value of the F$_{2:3}$ lines.

Materials and Methods

Plant materials

In 2008, Huaidao No.5, a *japonica* cultivar highly susceptible to WTDR, and Tetep (an *indica* variety) were grown at the experimental station in Jiangsu Academy of Agricultural Sciences. In the same year, F$_1$ was developed from a cross between Huaidao No.5 and Tetep. In 2009, the F$_1$ was grown and self-pollinated at the experimental station to generate F$_2$ lines that were used as mapping populations in Nanjing, Jiangsu Province. In 2009–2010, the F$_2$ lines were grown at the experimental station in Lingshui, Hainan Province, and the leaves of the F$_2$ lines were collected, numbered and stored at −70°C. A total of 138 F$_2$ individuals were selected and self-pollinated to generate 138 F$_{2:3}$ families. The resistance of the two parents, the F$_1$ generation and the F$_{2:3}$ lines were evaluated at the experimental station in Jiangsu Academy of Agricultural Sciences in 2012. Seeds were pretreated by soaking

them in water at 55°C for 15 minutes to ensure there were no live *A. besseyi* before the seeds were sown [4].

Nematode preparation

The seed-borne ectoparasitic *A. besseyi* was initially isolated from infected rice seeds using the Baermann funnel technique and surface-sterilized with 3% H$_2$O$_2$ for 10 minutes. After isolation, *A. besseyi* was cultured on *Botrytis cinerea*, which were grown on potato dextrose agar (PDA) medium for approximately 3 weeks at 25°C [22]. Then, the nematodes were rinsed thoroughly with distilled water and used as inoculation material [23].

Evaluation of resistance

40-µl nematode suspension with 400 juveniles of *A. besseyi* was inoculated between the leaf sheath and the culm with a pipette at the top tillering stage. To keep the micro-environment moist so that the nematodes were able to move and feed easily, a small wad of absorbent cotton was placed in the infection spot [7,21,24]. The mature seeds of each inoculated panicle were harvested. The number of grains per panicle and the weight of the panicle were determined. The number of nematodes in 100 grains from each inoculated panicle was counted using the following protocol. The peeled seed and the shell were soaked in distilled water for 24 hours, after which the free *A. besseyi* were collected from the solution using the Baermann funnel technique, and a microscope was used to count the number of *A. besseyi* [21]. At least 2 panicles of 25 plants from each F$_{2:3}$ line were used for the inoculation. Distilled water without *A. besseyi* was used as the control.

The number of *A. besseyi* individuals in 100 grains (NA), the loss rate of panicle weight (LRPW) and the loss rate of the total grains per panicle (LRGPP) were calculated using the following formulas: NA = (the total number of *A. besseyi* in the counted grains/the total number of counted grains)×100, LRPW = (the panicle weight of the control–the panicle weight of the inoculated plant)/the panicle weight of the control and LRGPP = (the total number of grains per panicle in the control–the total number of grains per panicle in the inoculated plant)/the total number of grains per panicle in the control.

Genotyping, linkage map construction and QTL analysis

The DNA of the two parents, the F$_1$ generation and the 138 F$_2$ lines was extracted using the CTAB method described by Rogers and Bendich [25]. SSR markers (842 pairs), obtained from Gramene (http://www.gramene.org), and polymorphism markers for the two parents were tested. PCR amplification was performed as described by Septiningsih et al. [26], and PCR products were detected using 8% denaturing polyacrylamide gel electrophoresis. The polymorphic markers were used for genotype analysis of the F$_2$ lines to assemble the linkage map. The genetic linkage map was constructed by employing MAPMAKER/EXP 3.0 software [27,28]. Marker distances in centimorgans (cM) were calculated using the Kosambi function.

Composite interval mapping (CIM) was applied to analyze the phenotypic and genotypic data for detecting the QTLs responsible for resistance to *A. besseyi* by using Windows QTL Cartographer 2.5 software [29]. Experiment-wide significance (P<0.05) thresh-old values of the LOD scores for putative QTL detection were determined with 1000 permutations. The threshold value of the LOD was 2.5 at a significance level of P = 0.05 for CIM. The additive effect and the explanation of the phenotypic variance for each QTL were also acquired using this software.

Table 1. Three performance traits of the two parents and the $F_{2:3}$ lines.

Traits	Huaidao No.5	Tetep	$|P_1-P_2|$	$F_{2:3}$ lines	
				Mean±SD	Range
NA	472±80	51±19	421	222±110	24–514
LRPW	0.28±0.19	0.04±0.01	0.24	0.15±0.10	0.01–0.44
LRGPP	0.29±0.09	0.05±0.01	0.24	0.22±0.13	0.00–0.64

P_1 and P_2 represent Huaidao No.5 and Tetep, respectively.

Results

Phenotypic variation

A dramatic difference in NA, LRPW and LRGPP was observed between Huaidao No.5 and Tetep (Table 1). The values for NA, LRPW and LRGPP of Huaidao No.5 were 472, 0.28 and 0.29, respectively. In comparison, the values for Tetep were much lower, at 51, 0.04 and 0.05, respectively. These data supported the hypothesis that Huaidao No.5 was highly susceptible to WTDR, whereas Tetep was highly resistant. The values for NA, LRPW and LRGPP of the $F_{2:3}$ lines ranged from 24 to 514, from 0.01 to 0.44 and from 0 to 0.64, respectively. Each of the frequency distributions for NA, LRPW and LRGPP showed a continuous distribution, which indicated that these traits behaved as quantitative variables (Figure 1).

Correlation coefficients among the three resistance-associated traits

The correlation analysis among the three resistance-associated traits was performed using SPSS ver. 20.0 software (Table 2). The results showed that NA was not significantly correlated with LRPW or LRGPP. However, a significant positive correlation was observed between LRPW and LRGPP at the P<0.01 level.

Genetic linkage map

A total of 160 polymorphic SSR markers were found between the two parents of 842 total markers, for a ratio of 19.01%. A linkage map was constructed, which included 12 linkage groups and spanned a total of 2179.6 cM in genetic distance with an average of 17.16 cM among 127 SSR polymorphic markers. Because these SSR markers were evenly distributed on 12 chromosomes, the linkage map was suitable for QTL detection.

QTL analysis

QTL identification using CIM indicated that a total of six QTLs for resistance to WTDR were found on chromosomes 3, 5 and 9 (Table 3, Figure 2). Two important QTLs (qNA3 and qNA9) responsible for the reproduction numbers of A. besseyi in the panicles were detected at the marker intervals of RM5626–

RM7097 and RM5526–RM3912 on chromosomes 3 and 9, respectively. The LOD values of qNA3 and qNA9 were 3.04 and 2.62, which accounted for 16.91% and 12.54% of the total phenotypic variation, respectively. Three QTLs, i.e., qLRPW5a, qLRPW5b and qLRPW9, which conferred yield loss, were mapped at the marker intervals of RM163–RM18620, RM440–RM161 and RM5526–RM3912, respectively, on chromosomes 5 and 9 (Table 3). qLRPW5a and qLRPW5b, which were located at two adjacent marker intervals on chromosome 5, explained 14.15% and 14.59% of the phenotypic variation and possessed LOD values 3.40 and 3.39, respectively. qLRPW9 was considered as a minor QTL because it only explained 1.02% of the phenotypic variation and had an LOD value of 2.55. We also identified a QTL (qLRGPP5) responsible for the loss in the number of the grains at the marker interval RM18632–RM163 on chromosome 5 (Table 3), which explained 10.91% of the phenotypic variation. Among these QTLs, four (qNA3, qLRPW5a, qLRPW5b and qLRGPP5) were derived from Tetep, and the others (qNA9 and qLRPW9) originated from Huaidao No.5.

Discussion

It has been reported that not all the susceptible rice varieties exhibited symptoms of both white leaf tips and small grains after infection with A. besseyi, although plants without symptoms still showed large losses in production [21]. Therefore, the relative yield loss was preferentially used in this study to evaluate the resistance level of plants against WTDR, which was described by the LRPW. The final number of nematodes in the mature grains was considered as an important assessment index by most researchers because of the potential transmission of the nematodes by seeds and because nematodes can contribute to the loss of the grain, which directly reduces the harvest. The number of A. besseyi in 100 grains was used as another assessment index based on a previous study [21]. We also used the LRGPP to measure the loss in the number of grains, which caused the small grain symptoms. Another symptom, i.e., white leaf tips, was barely seen in our study, which was reported by Feng [21] and Fortuner [30], and

Table 2. Correlation coefficients among the three resistance-associated traits.

Traits	NA	LRPW	LRGPP
NA	1		
LRPW	0.022	1	
LRGPP	−0.059	0.424**	1

**Significant at the 0.01 level.

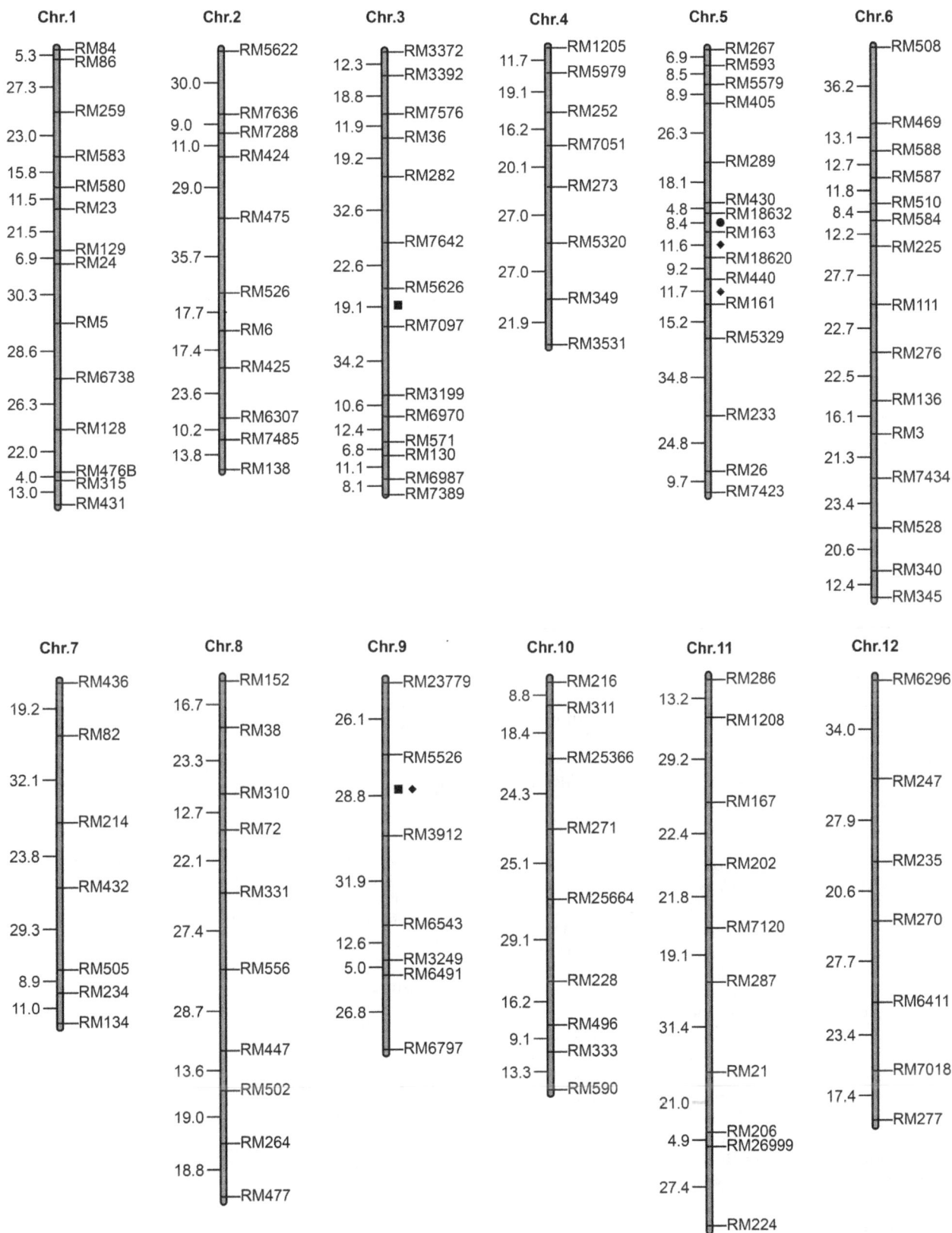

Figure 2. The SSR linkage map and chromosomal distributions of the putative loci responsible for resistance to WTDR. ■, ◆ and ● indicate the resistance QTLs detected using the resistance-associated traits of NA, LRPW and LRGPP, respectively.

Table 3. QTLs for resistance to WTDR detected in the F_2 lines developed from crossing Huaidao No.5 and Tetep.

Traits	QTL	Chr.	Interval	R^2 (%)	A (%)	LOD
NA	qNA3	3	RM5626-RM7097	16.91	71.18	3.04
	qNA9	9	RM5526-RM3912	12.54	−61.08	2.62
LRPW	qLRPW5a	5	RM163-RM18620	14.15	5.43	3.40
	qLRPW5b	5	RM440-RM161	14.59	5.51	3.39
	qLRPW9	9	RM5526-RM3912	1.02	−1.5	2.55
LRGPP	qLRGPP5	5	RM18632-RM163	10.91	1.94	2.83

QTL represents the putative QTL at an LOD level ≥2.5.
Chr. represents Chromosome.
R^2 represents the phenotypic variance of each QTL.
A is an additive effect.

was not investigated. Correlation analysis showed that there was no relationship between NA and LRPW or LRGPP. This finding suggested that the reproduction level of *A. besseyi* in the grains was not entirely responsible for the loss in production and that there may be unknown factors involved in the resistance process. Roberts [31] believed there was tolerance factor involved in reducing the yield loss caused by WTDR. In contrast, LRPW was significantly correlated with LRGPP, which indicated that the decrease in the number of grains in the mature panicles infected by *A. besseyi* caused a direct loss in yield. The mapping results confirmed this hypothesis, i.e., one of the loci for LRPW, *qLRPW5a*, was close to *qLRGPP5*, a locus for LRGPP. *qLRPW9* and *qNA9* were close to each other, but there was no relationship between these two traits. This lack of a relationship might stem from the fact that *qLRPW9* was a minor QTL and only accounted for 1.02% the phenotypic variation.

In this study, we identified 6 QTLs related to resistance against WTDR. Three important QTLs (*qLRPW5a*, *qLRPW5b* and *qLRGPP5*) located on chromosome 5 contributed to a decrease in the yield loss and grain number loss caused by WTDR. There were several genes related to the number, yield and weight of the grain, including *gw-5*, *gw5-1*, *Gwt5a*, *tgwt5* and *qYI-5* in the region of *qLRPW5a* and *qLRGPP5*, which were very close to each other [32–36]. The region containing another QTL (*qLRPW5b*) also included yield-related genes such as *gp5*, *QGwp5*, *gwt5*, *Pdw5*, *ssp-5* and *tgwt5b* [37–41]. This pattern suggests that these genes may be involved in reducing the damage in the grains after infection by *A. besseyi*, i.e., although these genes do not contribute directly to resistance against WTDR, they are involved in yield protection and are useful for resistance breeding projects. The resistance loci responsible for different types of pathogens are thought to be clustered in a chromosomal region in many organisms [42]. We found several resistance genes (e.g., *qBB3-1* and *rbr3*) against pathogens (rice bacterial blight and rice blast) by comparative mapping using a common marker [43,44]. Additionally, the additive effect of *qNA9* and *qLRPW9* were −61.8% and −1.5%, respectively, i.e. the resistance of these two QTLs originated from Huaidao No.5. It suggested that Huaidao No.5, a susceptible variety, also possessed the resistance loci. Lorieux

et al. [19] mapped a resistance gene (*Has-1^{Og}*) to the cyst nematode on rice chromosome 11 between two markers, i.e., RM206 and RM254. In our study, however, no gene was identified on chromosome 11, most likely because different resistance mechanisms occurring in rice depended on the different types of nematodes.

Recently, the *A. besseyi* infestation has caused increasing losses in rice yield in China [21,45–47]. Breeding and utilizing resistant cultivars is considered the most effective strategy for resolving this urgent problem. In this study, we identified the QTLs responsible for resistance to WTDR, which can be further improved by fine mapping to identify the tightly linked markers. Based on these findings, molecular-level breeding efforts target at WTDR can be more efficient, especially for locating both the nematode-resistance genes and the yield-protection genes.

It was reported previously that some resistance genes against nematodes had been cloned. Functional analysis has shown that these resistance genes can be divided into two types. The first type, which possessed a nucleotide binding site-leucine-rich repeat (NBS-LRR) structure, was located in the cytoplasm but lacked a signal sequence, e.g., *Mi-1*, *HeroA*, *Gpa2* and *Gro1-4* [15,17,18,20]. The second type has a transmembrane structure and a signal sequence, e.g., *Hs1^{pro-1}*, *Rhg1*, and *Rhg4* [16,48]. Hosts have evolved various defense pathways to protect themselves against invading nematode diseases. The vast amount of available information can help to rapidly and easily identify the set of resistance genes in this cultivar.

Acknowledgments

We gratefully acknowledge Ying Wang (Nanjing Agricultural University, China) for constructing the segregating population used in this study.

Author Contributions

Conceived and designed the experiments: TZ YZ WS. Performed the experiments: TZ CG LD L. Wang. Analyzed the data: TZ CG L. Wang LD. Contributed reagents/materials/analysis tools: YL HF FS L. Wei YF. Contributed to the writing of the manuscript: TZ CG LD YZ.

References

1. Yoshii H, Yamamoto S (1950) A rice nematode disease, "Senchu Shingare Byo." Symptom and pathogenic nematode. J Fac Agric Kyushu Univ 9: 309–333.
2. Huang CS, Chiang YC (1975) The influence of temperature on the ability of *Aphelenchoides besseyi* to survive dehydration. Nematologica 21: 351–357.
3. Jamali S, Pourjam E, Alizadeh A, Alinia F (2006) Incidence and distribution of *Aphelenchoides besseyi* in rice areas in Iran. J Agric Technol 2: 337–344.
4. Jamali S, Pourjam E, Safai N (2011) Determining the relationship between population density of white tip nematode and rice yield. J Agric Sci Technol 14: 195–203.

5. Nicol JM, Turner SJ, Coyne DL, den Nijs L, Hockland S, et al. (2011) Current nematode threats to world agriculture. Genomics Mol Genet Plant-Nematode Interact 21–43.

6. Hulbert SH, Webb CA, Smith SM, Sun Q (2001) Resistance gene complexes: Evolution and utilization. Annu Rev Phytopathol 39: 285–312.

7. De Waele D (2002) Foliar nematodes: *Aphelenchoides* species. Plant Resist Parasitic Nematodes 141–151.

8. Cralley EM (1949) White tip of rice. Phytopathology 39: 5.

9. Nishizawa T (1953) Studies on the varietal resistance of rice plant to the rice nematode disease "senchu shingare byo" (VI). Bull Kyushu Agric Experiment Station 1: 339–349.

10. Bridge J, Plowright RA, Peng D (2005) Nematode parasites of rice. In: Luc M, Sikora RA, Bridge J, editors. Plant parasitic nematodes in subtropical and tropical agriculture. Wallingford, UK: CAB International. pp. 87–130.

11. Pineda O, Bonierbale MW, Plaisted RL, Brodie BB, Tanksley SD (1993) Identification of RFLP markers linked to the H1 gene conferring resistance to the potato cyst nematode *Globodera rostochiensis*. Genome 36: 152–156.

12. Jacobs JME, van Eck HJ, Horsman K, Arens PFP, Verkerk-Bakker B, et al. (1996) Mapping of resistance to the potato cyst nematode *Globodera rostochiensis* from the wild potato species *Solanum vernei*. Mol Breed 2: 51–60.

13. Williams KJ, Fisher JM, Langridge P (1994) Identification of RFLP markers linked to the cereal cyst nematode resistance gene (*Cre*) in wheat. Theor Appl Genet 89: 927–930.

14. Yaghoobi J, Kaloshian I, Wen Y, Williamson VM (1995) Mapping a new nematode resistance locus in *Lycopersicon peruvianum*. Theor Appl Genet 91: 457–464.

15. Ernst K, Kumar A, Kriseleit D, Kloos DU, Phillips MS, et al. (2002) The broad-spectrum potato cyst nematode resistance gene (*Hero*) from tomato is the only member of a large gene family of NBS-LRR genes with an unusual amino acid repeat in the LRR region. Plant J 31: 127–136.

16. Cai D, Kleine M, Kifle S, Harloff HJ, Sandal NN, et al. (1997) Positional cloning of a gene for nematode resistance in sugar beet. Science 275: 832–834.

17. Vos P, Simons G, Jesse T, Wijbrandi J, Heinen L, et al. (1998) The tomato *Mi-1* gene confers resistance to both root-knot nematodes and potato aphids. Nat Biotechnol 16: 1365–1369.

18. Van Der Vossen EAG, Der Voort V, Rouppe JNAM, Bendahmane A, Sandbrink H, et al. (2000) Homologues of a single resistance-gene cluster in potato confer resistance to distinct pathogens: a virus and a nematode. Plant J 23: 567–576.

19. Lorieux M, Reversat G, Diaz SXG, Denance C, Jouvenet N, et al. (2003) Linkage mapping of *Hsa-1^{Og}*, a resistance gene of African rice to the cyst nematode, *Heterodera sacchari*. Theor Appl Genet 107: 691–696.

20. Paal J, Henselewski H, Muth J, Meksem K, Menéndez CM, et al. (2004) Molecular cloning of the potato *Gro1-4* gene conferring resistance to pathotype Ro1 of the root cyst nematode *Globodera rostochiensis*, based on a candidate gene approach. Plant J 38: 285–297.

21. Feng H, Wei L, Lin M, Zhou Y (2013) Assessment of rice cultivars in China for field resistance to *Aphelenchoides besseyi*. J Integr Agric doi:10.1016/S2095-3119(13)60608-5.

22. Yoshida K, Hasegawa K, Mochiji N, Miwa J (2009) Early embryogenesis and anterior-posterior axis formation in the white-tip nematode *Aphelenchoides besseyi* (Nematoda: Aphelenchoididae). J Nematol 41: 17–22.

23. Sun MJ, Liu WH, Lin MS (2009) Effects of temperature, humidity and different rice growth stages on vertical migration of *Aphelenchoides besseyi*. Rice Sci 16: 301–306.

24. McGawley EC, Rush MC, Hollis JP (1984) Occurrence of *Aphelenchoides besseyi* in Louisiana rice seed and its interaction with *Sclerotium oryzae* in selected cultivars. J Nematol 16: 65–68.

25. Rogers SO, Bendich AJ (1985) Extraction of DNA from milligram amounts of fresh, herbarium and mummified plant tissues. Plant Mol Biol 5: 69–76.

26. Septiningsih EM, Sanchez DL, Singh N, Sendon PM, Pamplona AM, et al. (2012) Identifying novel QTLs for submergence tolerance in rice cultivars IR72 and Madabaru. Theor Appl Genet 124: 867–874.

27. Lander ES, Green P, Abrahamson J, Barlow A, Daly MJ, et al. (1987) MAPMAKER: an interactive computer package for constructing primary genetic linkage maps of experimental and natural populations. Genomics 1: 174–181.

28. Lincoln SE, Lander SL (1993) Mapmaker/exp 3.0 and mapmaker. qtl 1.1. technical report. Cambridge: Whitehead Institute of Medical Research.

29. Wang S, Basten CJ, Zeng ZB (2005) Windows QTL cartographer, version 2.5. Statistical genetics. Raleigh: North Carolina State University.

30. Fortuner R, Williams KJO (1975) Review of the literature on *Aphelenchoides besseyi* hristie, 1942, the nematode causing "white tip" disease in rice. Helminthological Abstracts Series B, Plant Nematology 44: 1–40.

31. Roberts PA (1982) Plant resistance in nematode pest management. J Nematol 14: 24–33.

32. Lu C, Shen L, He P, Chen Y, Zhu L, et al. (1997) Comparative mapping of QTLs for agronomic traits of rice across environments by using a doubled-haploid population. Theor Appl Genet 94: 145–150.

33. Thomson MJ, Tai TH, McClung AM, Lai XH, Hinga ME, et al. (2003) Mapping quantitative trait loci for yield, yield components and morphological traits in an advanced backcross population between *Oryza rufipogon* and the *Oryza sativa* cultivar Jefferson. Theor Appl Genet 107: 479–493.

34. Tan Z, Shen L, Yuan Z, Lu CF, Chen Y, et al. (1996) Identification of QTLs for ratooning ability and grain yield traits of rice and analysis of their genetic effects. Zuo Wu Xue Bao 23: 289–295.

35. Lin HX, Qian HR, Zhuang JY, Lu J, Min SK, et al. (1996) RFLP mapping of QTLs for yield and related characters in rice (*Oryza sativa* L.). Theor Appl Genet 92: 920–927.

36. Cho YC, Suh JP, Choi IS, Hong HC, Baek MK, et al. (2003) QTLs analysis of yield and its related traits in wild rice relative *Oryza rufipogon*. Treat Crop Res 4: 19–29.

37. Yu SB, Li JX, Xu CG, Tan YF, Gao YJ, et al. (1997) Importance of epistasis as the genetic basis of heterosis in an elite rice hybrid. Proc Natl Acad Sci USA 94: 9226–9231.

38. Li Z, Pinson SRM, Park WD, Paterson AH, Stansel JW (1997) Epistasis for three grain yield components in rice (*Oryxa sativa* L.). Genetics 145: 453–465.

39. Zhuang JY, Lin HX, Lu J, Qian HR, Hittalmani S, et al. (1997) Analysis of QTL× environment interaction for yield components and plant height in rice. Theor Appl Genet 95: 799–808.

40. CiXin H, Jun Z, JuQiang Y, Benmoussa M, Ping W (2000) QTL mapping for developmental behaviour of panicle dry weight in rice. Zhongguo Nong Ye Ke Xue 33: 24–32.

41. Lu C, Shen L, Tan Z, Xu Y, He P, et al. (1996) Comparative mapping of QTLs for agronomic traits of rice across environments using a doubled haploid population. Theor Appl Genet 93: 1211–1217.

42. Gebhardt C, Mugniery D, Ritter E, Salamini F, Bonnel E (1993) Identification of RFLP markers closely linked to the H1 gene conferring resistance to *Globodera rostochiensis* in potato. Theor Appl Genet 85: 541–544.

43. Ramalingam J, Vera Cruz CM, Kukreja K, Chittoor JM, Wu JL, et al. (2003) Candidate defense genes from rice, barley, and maize and their association with qualitative and quantitative resistance in rice. Mol Plant Microbe Interact 16: 14–24.

44. Chen H, Wang S, Xing Y, Xu C, Hayes PM, et al. (2003) Comparative analyses of genomic locations and race specificities of loci for quantitative resistance to *Pyricularia grisea* in rice and barley. Proc Natl Acad Sci USA 100: 2544–2549.

45. Lin MS, Ding XF, Wang ZM, Zhou FM, Lin N (2004) Description of *Aphelenchoides besseyi* from abnormal rice with "small grains and erect panicles" symptom in China. Rice Sci 12: 289–294.

46. Wang ZY, Yang HF, Ji MX (2006) The small grains and erect panicle of rice caused by nematode and control the disease in Jiangsu Province. Nanjing Nong Ye Da Xue Xue Bao 29: 54–56.

47. Liu W, Lin M, Li H, Sun MJ (2007) Study on dynamic development of *Aphelenchoides besseyi* on rice plant by artificial inoculations in greenhouse. Zhongguo Nong Ye Ke Xue 12: 14.

48. Hauge BM, Wang ML, Parsons JD, Wang M (2006) Nucleic acid molecules and other molecules associated with soybean cyst nematode resistance: U.S. Patent 7,154,021[P].

16

Determination of Critical Nitrogen Dilution Curve Based on Stem Dry Matter in Rice

Syed Tahir Ata-Ul-Karim, Xia Yao, Xiaojun Liu, Weixing Cao, Yan Zhu*

National Engineering and Technology Center for Information Agriculture, Jiangsu Key Laboratory for Information Agriculture, Nanjing Agricultural University, Nanjing, Jiangsu, P. R. China

Abstract

Plant analysis is a very promising diagnostic tool for assessment of crop nitrogen (N) requirements in perspectives of cost effective and environment friendly agriculture. Diagnosing N nutritional status of rice crop through plant analysis will give insights into optimizing N requirements of future crops. The present study was aimed to develop a new methodology for determining the critical nitrogen (N_c) dilution curve based on stem dry matter (S_{DM}) and to assess its suitability to estimate the level of N nutrition for rice (*Oryza sativa* L.) in east China. Three field experiments with varied N rates (0–360 kg N ha^{-1}) using three Japonica rice hybrids, Lingxiangyou-18, Wuxiangjing-14 and Wuyunjing were conducted in Jiangsu province of east China. S_{DM} and stem N concentration (SNC) were determined during vegetative stage for growth analysis. A N_c dilution curve based on S_{DM} was described by the equation ($N_c = 2.17W^{-0.27}$ with W being S_{DM} in t ha^{-1}), when S_{DM} ranged from 0.88 to 7.94 t ha^{-1}. However, for $S_{DM} < 0.88$ t ha^{-1}, the constant critical value $N_c = 1.76\%$ S_{DM} was applied. The curve was dually validated for N-limiting and non-N-limiting growth conditions. The N nutrition index (NNI) and accumulated N deficit (N_{and}) of stem ranged from 0.57 to 1.06 and 51.1 to -7.07 kg N ha^{-1}, respectively, during key growth stages under varied N rates in 2010 and 2011. The values of ΔN derived from either NNI or N_{and} could be used as references for N dressing management during rice growth. Our results demonstrated that the present curve well differentiated the conditions of limiting and non-limiting N nutrition in rice crop. The S_{DM} based N_c dilution curve can be adopted as an alternate and novel approach for evaluating plant N status to support N fertilization decision during the vegetative growth of Japonica rice in east China.

Editor: Guoping Zhang, Zhejiang University, China

Funding: This work was supported by grants from the National High-Tech Research and Development Program of China (863 Program) (2011AA100703), Special Program for Agriculture Science and Technology from Ministry of Agriculture in China (201303109), Priority Academic Program Development of Jiangsu Higher Education Institutions (PAPD), and Science and Technology Support Plan of Jiangsu Province (BE2011351, BE2012302). The funders had no role in study design, data collection and analysis, decision to publish, or preparation of the manuscript.

Competing Interests: The authors have declared that no competing interests exist.

* Email: yanzhu@njau.edu.cn

Introduction

Estimating nitrogen (N) nutritional status is a key to investigating, monitoring, and managing cropping systems [1]. Conventional farming has led to extensive use of N as a tool for ensuring profitability in the soils with uncertain fertility levels, which has raised the concerns about environmental sustainability. A reliable diagnosis of crop N requirement and nutritional status give insight into optimization of qualitative and quantitative aspects of crop production. It also improve N use efficiency and add to environmental protection [2]. Soil and plant-based strategies are two principle approaches, extensively used to derive information about the N nutrition status of crops, for satisfying their demand for N and to minimize N losses [3]. The former rarely describes the intensity of N release over a longer period, so the latter are widely accepted and adopted. Therefore, the present study investigates a plant-based strategy for an in-season assessment of N nutrition status for rice crop.

In plant-based approaches, the N nutrition status is generally monitored to determine the requirement for top dressing in crops [3]. For this purpose, several plant-based diagnostic tools, such as critical N concentration (N_c) approach, chlorophyll meter, hyper-spectral reflectance and remote sensing, have been successfully used for in-season N management [4]. They differ in scope, in context of reference spatial scale, in terms of monetary and time resources, as well as skills and expertise required for their implementation at field [5]. Despite being simple, chlorophyll meter readings are affected by leaf thickness, abiotic stress and nutrient variability [6]. Canopy reflectance method's accuracy is affected by solar illumination, soil background effects and sensor viewing geometry [4]. However, the concept of N_c can be used as a potential alternate to these techniques, and it can give insight into relative N status of a crop. The present study utilizes this concept for an in-season N fertilizer management in rice crop.

The concept of N_c is crop specific, precise, simple and biologically sound, because it is based on actual crop growth. Whole plant dry matter based N_c approach was successfully applied for N management in winter wheat [7,8], corn [9] and spring wheat [10]. This approach was successfully applied for a Indica rice in tropics and Japonica rice in subtropical temperate region [11,12]. Dry matter partitioning among different plant organs affects the weight/N concentration relationship, and changes the shape of the dilution curve, thus limits its acceptance as a reliable method [13,14]. The concept of N_c for specific plant

Table 1. Changes of stem dry matter (S_{DM}) with time (days after transplantation) under different N rates in two rice cultivars in experiments conducted during 2010 and 2011.

Year	Cultivar	DAT	Sampling date	Stem dry matter/Applied N (kg ha^{-1})					F prob.	LSD
				0	80	160	240	320		
2010	LXY-18	16	07-Jul	0.23	0.27	0.38	0.48	0.49	*	0.028
	LXY-18	26	17-Jul	0.63	0.78	0.95	1.11	1.12	*	0.055
	LXY-18	36	27-Jul	1.04	1.28	1.55	1.77	1.81	*	0.075
	LXY-18	48	08-Aug	2.23	2.73	3.25	3.61	3.51	*	0.226
	LXY-18	60	20-Aug	3.87	4.47	4.94	5.23	5.29	*	0.146
	LXY-18	70	30-Aug	4.72	5.56	6.7	7.01	7.22	*	0.279
	WXJ-14	16	07-Jul	0.22	0.27	0.32	0.36	0.35	*	0.019
	WXJ-14	26	17-Jul	0.39	0.54	0.73	0.9	0.91	*	0.045
	WXJ-14	36	27-Aug	0.55	0.8	1.13	1.38	1.46	*	0.063
	WXJ-14	48	08-Aug	1.22	1.65	1.99	2.18	2.23	*	0.11
	WXJ-14	60	20-Aug	2.77	3.46	3.72	4.19	3.97	*	0.233
	WXJ-14	70	30-Sep	3.69	4.4	5.04	5.81	5.7	*	0.205

Year	Cultivar	DAT	Sampling date	Stem dry matter/Applied N (kg ha^{-1})					F prob.	LSD
				0	90	180	270	360		
2011	LXY-18	18	09-Jul	0.22	0.33	0.4	0.56	0.59	*	0.042
	LXY-18	30	21-Jul	0.67	0.76	0.92	1.19	1.19	*	0.053
	LXY-18	42	02-Aug	1.12	1.28	1.42	1.78	1.76	*	0.132
	LXY-18	54	15-Aug	2.24	2.41	2.76	3.43	3.48	*	0.127
	LXY-18	64	25-Aug	3.59	4.02	4.37	4.75	4.85	*	0.164
	LXY-18	74	04-Sep	5.68	5.91	6.41	7.84	8.04	*	0.172
	WXJ-14	18	09-Jul	0.19	0.28	0.32	0.34	0.36	*	0.02
	WXJ-14	30	21-Jul	0.37	0.6	0.71	0.86	0.9	*	0.06
	WXJ-14	42	02-Aug	0.54	0.96	1.2	1.37	1.47	*	0.128
	WXJ-14	54	15-Aug	1.51	1.84	2.24	2.52	2.68	*	0.137
	WXJ-14	64	25-Aug	2.49	3.07	3.36	4.06	4.04	*	0.169
	WXJ-14	74	04-Sep	4.41	5.04	5.75	6.2	6.27	*	0.178

*: F statistic significant at 0.01 probability level.

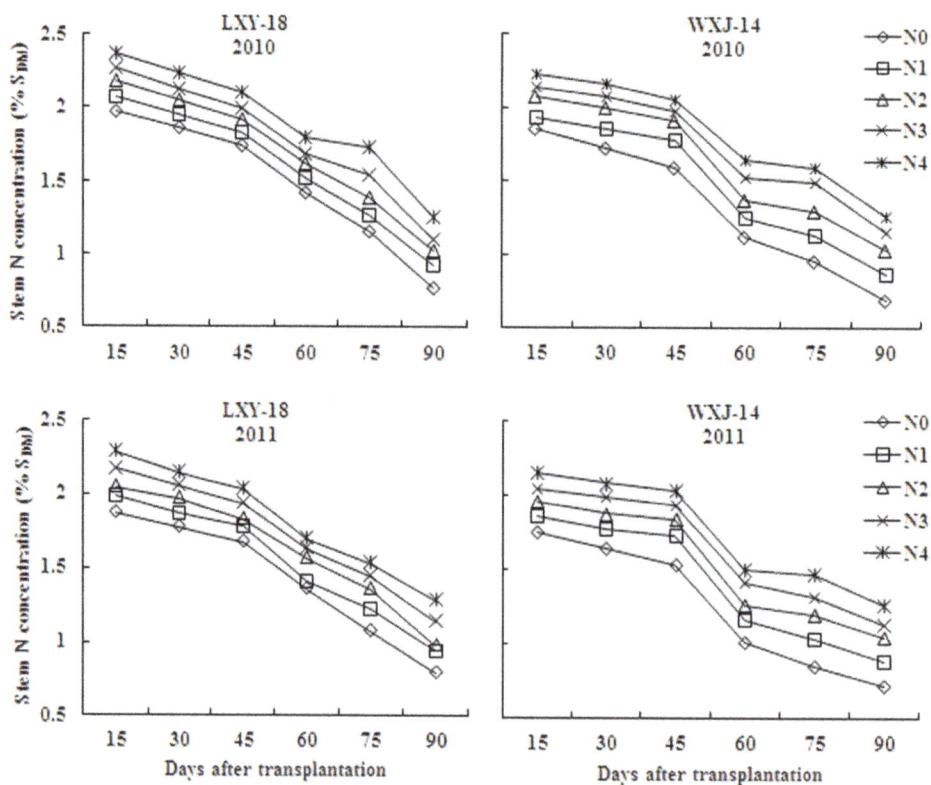

Figure 1. Changes of stem nitrogen concentration (% S_{DM}) with time (days after transplantation) for rice under different N rates in experiments conducted during 2010 and 2011.

organs (e.g., leaves and stem) is similar to that on whole plant basis. Leaf based diagnosis of N status in crops is affected by progressive shading by newer leaves, decline of leaf N concentration due to aging, pest attack, abiotic stresses and increase in the proportion of structural tissues [15]. Stem sap nitrate concentration is influenced by phenological phase, cultivar, temperature and solar radiation [16]. During vegetative phase, the contribution of stem dry matter (S_{DM}) towards total plant dry matter is significantly higher than that of leaf dry matter (L_{DM}), hence it is the most determining factor for N dilution of the whole plant [17]. Thus, the idea of using N_c curve based on S_{DM} over whole plant dry matter and L_{DM} based methods, can be used as an alternate approach for determination of N_c dilution curve.

The objectives of this work were to develop a N_c dilution curve based on S_{DM} and to assess the plausibility of this curve to estimate N nutrition status of Japonica rice. The estimation based on this approach will be more reliable than existing methods due to consistency at different growth stages.

Materials and Methods

Ethics statement

The experiments land is owned and managed by Nanjing Agricultural University, Nanjing, China. Nanjing Agricultural University permits and approvals obtained for the work and study. The field studies did not involve wildlife or any endangered or protected species.

Experimental details

Three field experiments with multiple N rates (0– 360 kg N ha^{-1}) were conducted using three contrasting Japonica

rice hybrids, Lingxiangyou-18 (LXY-18), Wuxiangjing-14 (WXJ-14) and Wuyunjing (WYJ), at Yizheng (32°16′N, 119°10′E) and Jiangning (31°56′N, 118°59′E) located in lower Yangtze River Reaches of east China. The soil was clay loam and was classified as Ultisoles. The rice-wheat cropping system is practiced in the region. The applied N rates varied significantly among different farmers. The average rate of N fertilizer reached 387 kg ha^{-1} during the period of 2004–2008 [18].

The whole experimental area was ploughed and subsequently harrowed before transplanting. All bunds were compacted to prevent seepage into and from adjacent plots. A plastic lining was installed to a depth of 40 cm between drain and the bund of each plot to minimize seepage across the bunds towards the drains. To further minimize seepage of water from control plot (N$_0$), double bunds were constructed separating them and the adjacent plots. Experiments were arranged in a randomized complete block design with three replications. The size of each experimental plot was 8 m by 4.5 m, with planting density of approximately 22.2 hills per m^2. At site 1, soil pH, organic matter, total N, available phosphorous (P), and available potassium (K) were 6.2, 17.5 g kg^{-1}, 1.6 g kg^{-1} 43 mg kg^{-1}, 90 mg kg^{-1}, and 6.4, 15.5 g kg^{-1}, 1.3 g kg^{-1} 38 mg kg^{-1}, and 85 mg kg^{-1} in 2010 and 2011, respectively. The corresponding soil properties were 6.5, 13.5 g kg^{-1}, 1.13 g kg^{-1} 45 mg kg^{-1}, 91 mg kg^{-1} in 2007 at site 2. For experiments conducted at site 1 in 2010 and 2011, treatment consisted of five N rates as 0, 80, 160, 240, and 320 kg N ha^{-1}, and 0, 90, 180, 270, and 360 kg N ha^{-1}, respectively, while for experiment conducted at site 2 in 2007, treatment consisted of three N rates as 110, 220, and 330 kg N ha^{-1}. N in all experiments was distributed as 50% at pre planting, 10% at tillering, 20% at jointing, and 20% at booting, with urea as the N source. Aside from N fertilizer, phosphorus (135 kg ha^{-1}) and potassium (190 kg ha^{-1}) fertilizers were basally incorporated at the last harrowing and leveling in all plots before transplanting as monocalcium phosphate $Ca(H_2PO_4)_2$ and potassium chloride (KCl). Rice seedlings at five leaves stage were transplanted in experimental fields on June 20 (site 1) in 2010 and 2011, and on 29 June (site 2) in 2007, respectively. Pre-emergence herbicides were used to control weeds at early growth stages. Also plots were regularly hand-weeded until canopy was closed to prevent weed damage. Insecticides were used to prevent insect damage. All other agronomic practices were used according to local recommendations to avoid yield loss.

Sample collection and measurement

Rice plants were sampled from each plot at the intervals of 10–12 days from 0.23 m^2 area (5 hills) at active tillering, mid tillering, stem elongation, panicle initiation, booting and heading stages during the period of each experiment for growth analysis. The plants were manually severed at ground level on each sampling date. Fresh plants were divided into green leaf blades and culm plus sheath. Samples were oven-dried at 105°C for half an hour to rapidly stop metabolism and then at 70°C until constant weight to obtain stem dry matter (S$_{DM}$, t ha^{-1}). The dried stem samples were ground and analyzed for total stem N concentration (SNC, %) by Kjeldahl method. Stem N accumulation (SNA, kg N ha^{-1}) was obtained as summed product of the S$_{DM}$ by the SNC. The SNC of whole-plant stem was calculated as SNA divided by S$_{DM}$.

Statistical analysis

The S$_{DM}$ and SNC data for each sampling date, year and variety was separated and subjected to analysis of variance (ANOVA) using GLM procedures in SPSS-16 software package (SPSS Inc., Chicago. IL, USA). The differences among treatment means were measured by using the least significant difference (LSD) test at 90% level of significance, instead of classically used 95% in order to reduce the occurrence of Type II errors that could be high in such field experiments. For each measurement date, year and variety, the variation in the SNC versus S$_{DM}$ across the different N levels was combined into a bilinear relation composed of a linear regression representing the joint increase in SNC and S$_{DM}$ and a vertical line corresponding to an increase in SNC without significant variation in S$_{DM}$. The theoretical N$_c$ points corresponds to the ordinate of the breakout of the bilinear regression. Regression analysis was performed using Microsoft Excel (Microsoft Cooperation, Redmond, WA, USA).

Construction and validation of critical, maximum and minimum N dilution curves

For determination of N$_c$ dilution curve it is necessary to determine the N concentration that did not limit the S$_{DM}$ production either by its excess or deficiency. The data used to construct the N$_c$ dilution curve came from two experiments conducted in 2010 and 2011 by distinguishing the data points for N-limiting and non-N-limiting growth. The N-limiting growth treatment is defined as a treatment for which an additional N application leads to a significant increase in S$_{DM}$. The non-N-

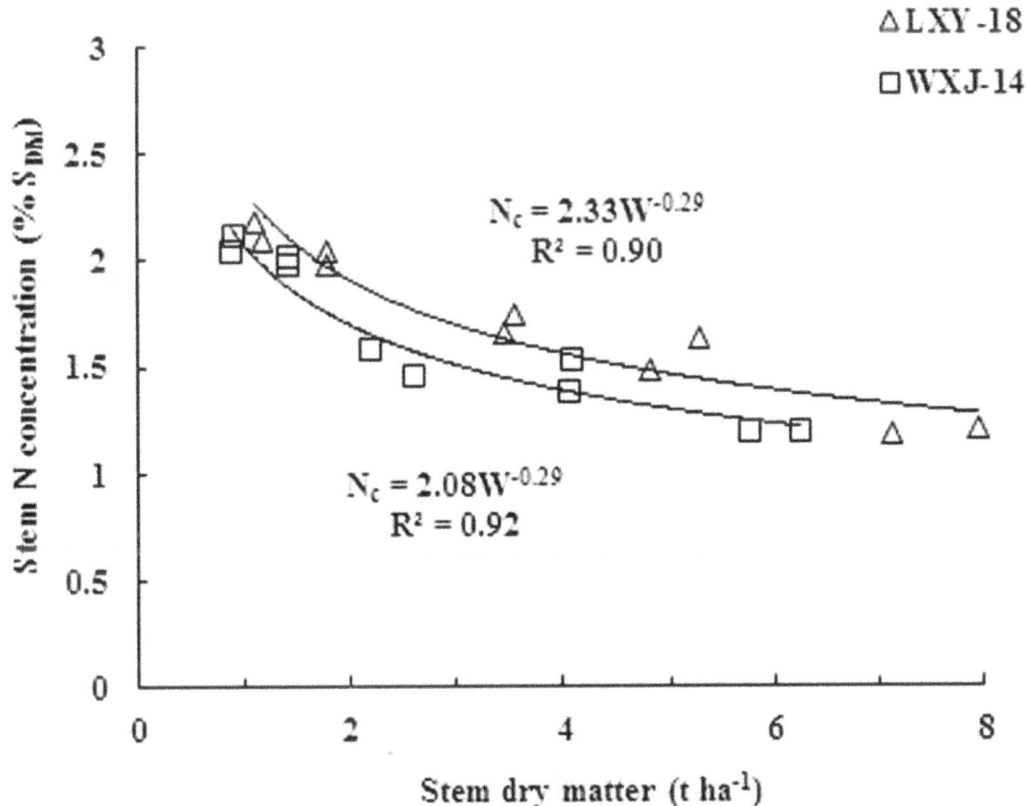

Figure 2. Critical nitrogen data points and N$_c$ dilution curves in stem obtained by non-linear fitting for two rice cultivars (LXY-18, N$_c$ = 2.33W$^{-0.29}$ and WXJ-14, N$_c$ = 2.08W$^{-0.29}$) under different N rates in experiments conducted during 2010 and 2011.

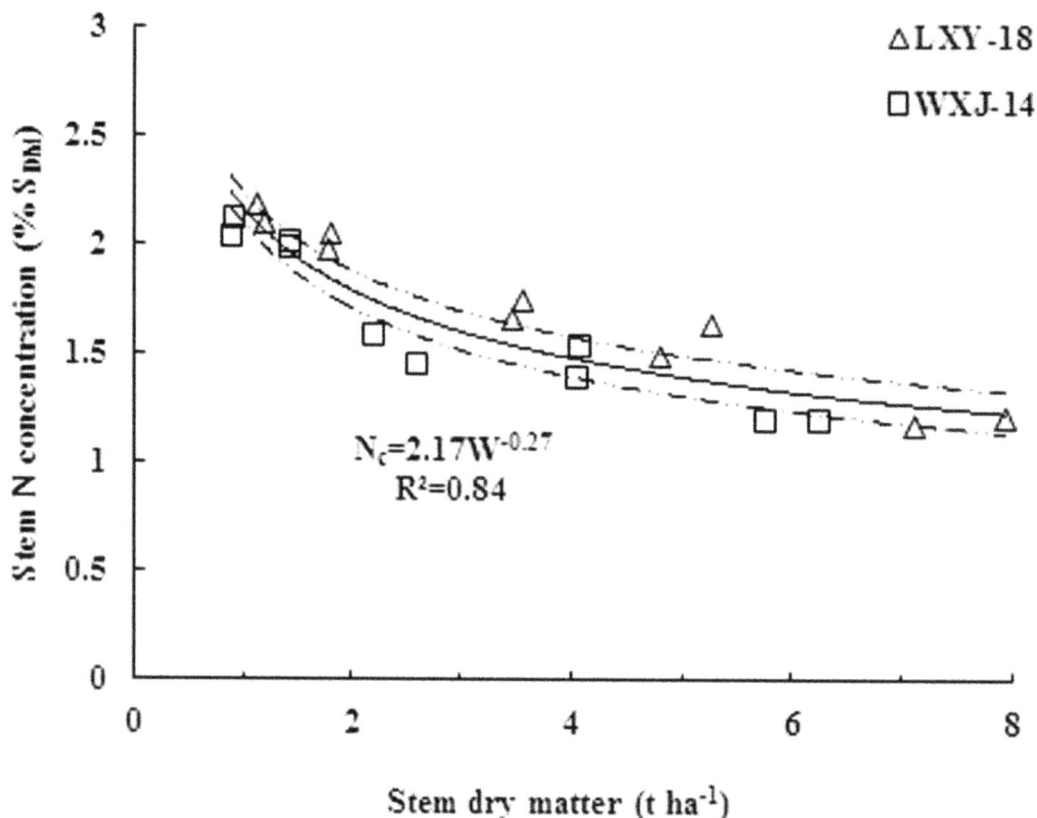

Figure 3. Critical nitrogen data points used to define the N_c dilution curve when data were pooled over for two rice cultivars (LXY-18 and WXJ-14). The solid line represents the N_c dilution curve ($N_c = 2.17W^{-0.27}$; $R^2 = 0.84$) describing the relationship between the N_c and stem dry matter of rice. The dotted lines represent the confidence band (P = 0.95).

limiting growth treatment is defined as a treatment, for which a supplement of N application does not lead to an increase in S_{DM} and, at the same time, exhibits a significant increase in SNC. If at the same measurement date, statistical analysis distinguished at least one set of N-limiting and non-N limiting data point, these data points were used either for construction of the N_c dilution curve or to validate it [7]. Consistent with earlier studies, an allometric function based on power regression (Freundlich model) was used to determine the relationship between the observed decreases in N_c with increasing S_{DM}. The N_c dilution curve was validated first by using the data points not retained for establishing the parameters of the allometric function in 2010 and 2011, and then with independent data set from experiment conducted in 2007.

The data points (n = 13) from most plethoric N treatments (N_4 plots) was assumed to represent the maximum N dilution curve (N_{max}) while the minimum N dilution curve (N_{min}) was determined by using the data points (n = 13) from the most N-limiting treatments for which N application was zero (N_0 check plots).

Calculation of critical N dilution curve based diagnostic tools

To identify the N status in the S_{DM} of rice during vegetative growth, the nitrogen nutrition index (NNI) and accumulated nitrogen deficit (N_{and}) were established for each sampling date, experiment and variety. The NNI value was obtained by dividing the total N concentration of S_{DM} by N_c value determined by critical dilution curve, [9]. The N_{and} value for rice crop on each

sampling date was obtained by subtracting the N accumulation under the N_c condition (N_{cna}) from actual N accumulation (N_{na}) under different N rates [12]. For in-season recommendation of supplemental N application, the difference value of NNI (ΔNNI), N_{and} (ΔN_{and}) and difference value of N application rate (ΔN) between different N treatments was calculated according to the method proposed by Ata-Ul-Karim et al. [12].

Results

Stem dry matter and nitrogen concentration

The S_{DM} production was significantly affected by N fertilization during the growth period of rice. The increase in S_{DM} followed a continuous increasing trend along with sampling dates for both the varieties during each year with increasing N rates from N_0 to N_4; however, there was no significant difference between N_3 and N_4 in all the cases (Table 1). This increase in the S_{DM} production with N fertilization may be linked to a higher absorption of N fertilizer. S_{DM} ranged from minimum 0.22 t ha^{-1} and 0.19 t ha^{-1} (N_0) in WXJ-14 to a maximum of 7.22 t ha^{-1} and 8.04 t ha^{-1} (N_4) in LXY-18 during 2010 and 2011, respectively. The results showed that there was no positive correlation between S_{DM} and N rates, as the S_{DM} tend to decrease when N rate exceeded a critical level. During each experimental year, S_{DM} conferred with the following inequality under different N ratess.

$$S_{DM0} < S_{DM1} < S_{DM2} < S_{DM3} = S_{DM4} \qquad (1)$$

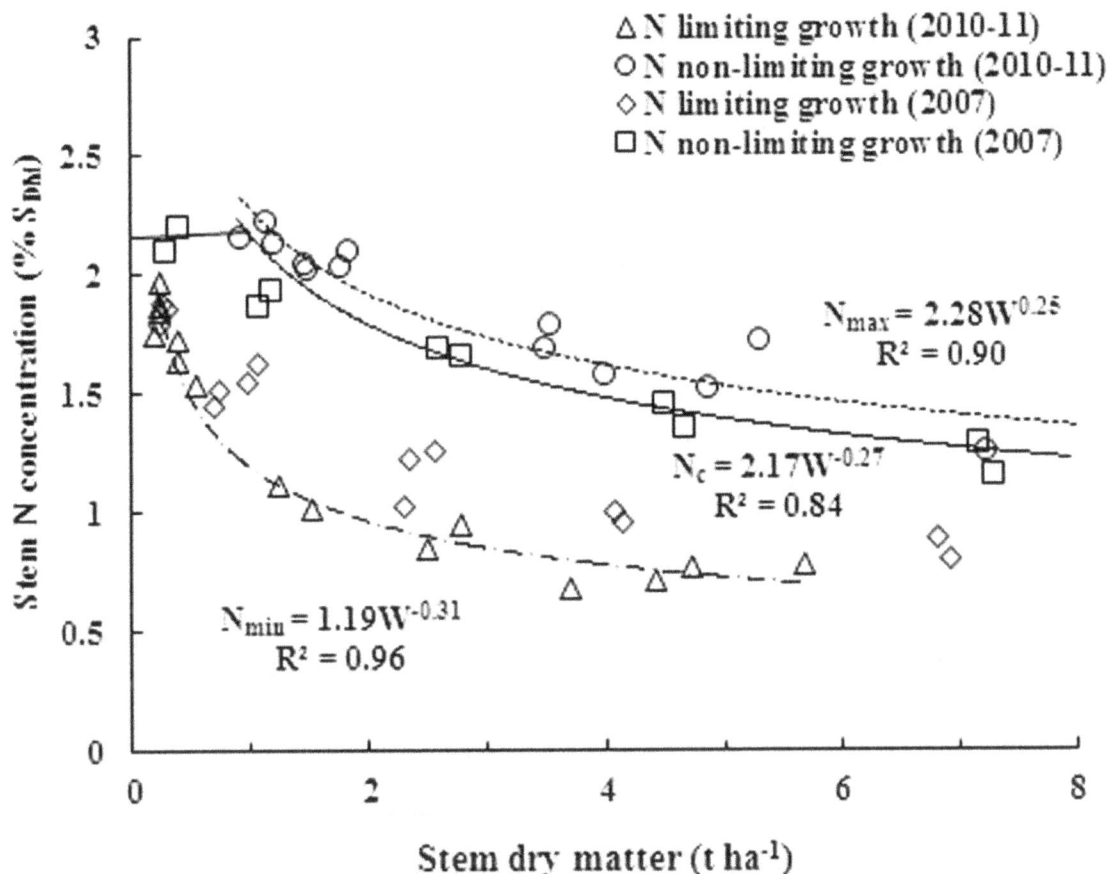

Figure 4. Comprehensive validation of N_c dilution curve using independent data set from experiment conducted in 2007. Data points ($\Diamond$) represent N limiting growth conditions, while ($\Box$) represent N non-limiting conditions. The solid line in the middle represents the N_c curve ($N_c = 2.17W^{-0.27}$) describing the relationship between the N_c and stem dry matter of rice. The data points (Δ) and ($\bigcirc$) not engaged for establishing the parameters of allometric function (2010 and 2011) were used to develop two boundary curves, (–•–•–•) minimum limit curve ($N_{min} = 1.19\ W^{-0.31}$) and (------) maximum limit curve ($N_{max} = 2.27W^{-0.25}$).

where S_{DM0}, S_{DM1}, S_{DM2}, S_{DM3} and S_{DM4} stands for S_{DM} of N_0, N_1, N_2, N_3 and N_4, respectively.

Stem N concentration response to N fertilizer rates was usually linear and a higher rate of N mostly resulted in a higher SNC, hitherto a decline in SNC was observed with increasing S_{DM} from active tillering to heading. Maximum variation in SNC of both cultivars was observed on 16 and 18 DAT, while minimum on 70 and 74 DAT, in years 2010 and 2011, respectively. The SNC ranged from 2.28 to 0.78 for LXY-18 and 2.16 to 0.71 for WXJ-14 during 2010, while 2.36 to 0.77 for LXY-18 and 2.23 to 0.68 for WXJ-14 during 2011 (Fig. 1).

Determination of critical nitrogen dilution curves based on stem dry matter

A set of twenty theoretical data points for both cultivar, obtained from two experiments (10 data points for each cultivar) from active tillering to heading were used to calculate the N_c for a given level of S_{DM}. The S_{DM} data that fit the statistical criteria for establishing N_c dilution curve varied from 0.88 t ha^{-1} to 7.94 t ha^{-1}. A power functions were fitted to the calculated N_c points as equations (2) and (3), the coefficient for which were 0.90 and 0.92 for LXY-18 and WXJ-14, respectively (Fig. 2).

$$N_c = 2.33W^{-0.29} \qquad \left(W \geq 0.88\ t\ ha^{-1}, R^2 = 0.90, n = 10\right) \quad (2)$$

$$N_c = 2.08W^{-0.29} \qquad \left(W \geq 0.88\ t\ ha^{-1}, R^2 = 0.92, n = 10\right) \quad (3)$$

where W is the S_{DM} expressed in t ha^{-1}; N_c is the critical N concentration in stem expressed in % S_{DM}; a and b are estimated parameters. The parameter a represents the N concentration in the S_{DM} when $W = 1$ t ha^{-1}, and b represents the coefficient of dilution describing the relationship between N concentration and S_{DM}.

The F-value (0.72) of two curves was less than the critical value of $F_{(1-18)} = 4.41$ at 5% probability level, showing non-significant difference between the curves [19], thus the data for the two varietal groups were united, and a unified dilution curve was determined as equation 4.

$$N_c = 2.17W^{-0.27} \qquad \left(W \geq 0.88\ t\ ha^{-1}, R^2 = 0.84, n = 20\right) \quad (4)$$

The model accounted for 84% of the total variance. At early growth stages of rice crop, the N_c varied between 2.24% S_{DM} to 2.10% S_{DM} (95% confidence interval) for a S_{DM} of 0.88 t ha^{-1} at the lower end while 7.94 t ha^{-1} at the higher end, respectively (Fig. 3).

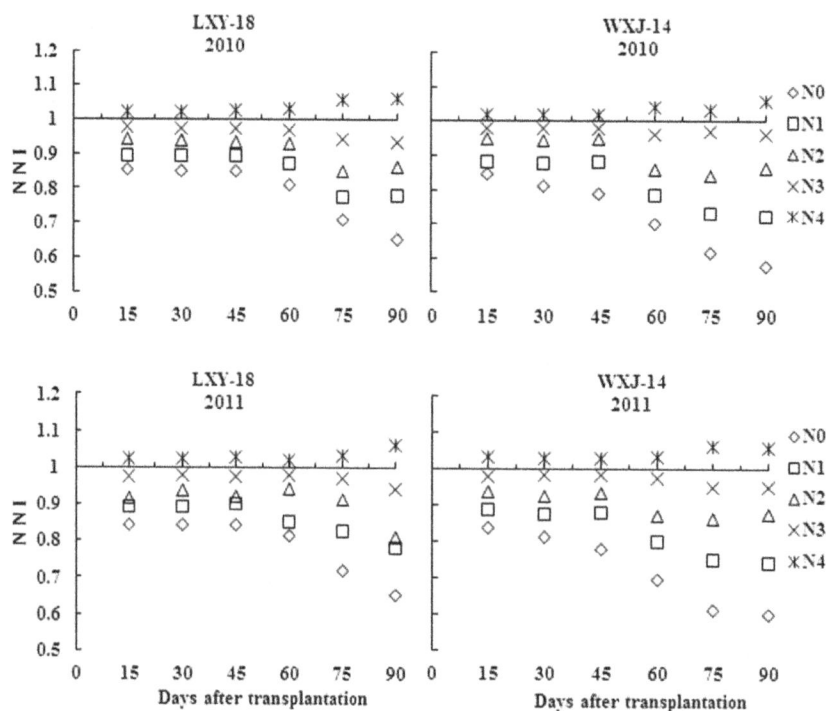

Figure 5. Changes of nitrogen nutrition index (NNI) with time (days after transplantation) for rice stem under different N rates in experiments conducted during 2010 and 2011.

For the S_{DM} range of 0.1 to 0.88 t ha^{-1}, corresponding to early growth stages, increasing N rates at sowing did not significantly affect S_{DM}, because N requirement is relatively low during these early stages. Therefore, the N_c dilution curve cannot be applied to the low $S_{DM} < 0.88$ t ha^{-1} at early growth stages due to relatively smaller decline of N_c with increasing S_{DM}. For these S_{DM}, the N_c could not been determined by the same statistical method because the very high slope of the linear regression resulted in a highly variable estimate [7]. Hence, for the data points of S_{DM} ranging from 0.37 to 0.88 t ha^{-1} a constant N_c (1.76% S_{DM}) was calculated as the mean value between the minimum N concentration of non-limiting N points (2.26% S_{DM}) and the maximum N concentration of limiting N points (1.25% S_{DM}), based on extrapolation of equation 4.

The above S_{DM} based N_c dilution curve was dually validated for N-limiting and non-N-limiting situations within the range for which it was developed. First, the curve was partially validated by combining the data points not engaged for establishing the parameters of the allometric function. In addition, the comprehensive validation of the curve was performed by using the data points from an independent experiment conducted in 2007. The results revealed that the N concentration data that led to the highest significant yields in S_{DM} were positioned close to or above the N_c dilution curve and considered to be non-N-limiting concentrations, whereas the data for the lowest significant S_{DM} yields, were positioned close to or under the N_c dilution curve and classified as N limiting values (Fig. 4). To determine N_{max}, data points were selected only from non-N-limiting treatments (n = 13), and for N_{min}, data points were selected from the treatment without N application (n = 13). Thus, the present N_c dilution curve could well discriminate the N limiting and non-N-limiting growing conditions in this study

Changes of NNI and N_{and}

Nitrogen nutrition index and N_{and} are helpful in determining the crop nutrition status i.e. deficient, optimal or excess of N nutrition. N nutrition is considered as optimum when NNI = 1 and $N_{and} = 0$, while NNI > 1 and $N_{and} < 0$ indicates luxury consumption of N nutrition, values of NNI < 1 and $N_{and} > 0$ represents N shortage. NNI and N_{and} can be used to quantify the intensity of the N stress after the onset of N deficiency. Our results of significant differences in NNI and N_{and} across the growing seasons, N rates, and phenological stages in rice are in agreement with earlier reports for maize and wheat [10]. As seen in Figure 5 and 6, during 2010 and 2011 the NNI ranged from 0.65 to 1.06 for LXY-18 and 0.57 to 1.06 for WXJ-14, while the N_{and} ranged from 51.1 kg ha^{-1} to −7.07 kg ha^{-1} for LXY-18 and 43.3 kg ha^{-1} to −4.5 kg ha^{-1} for WXJ-14. The results showed that NNI amplified while N_{and} declined with increasing N rates, while both intensified steadily with growth of rice crop and reached to peaks at heading stage for N_0, N_1, N_2 and N_3 (N limiting treatments), nevertheless, for N_3 this intensification was minor. In contrast, surplus N nutrition existed till heading stage for N_4 (non-N-limiting treatment). The estimates based on NNI and N_{and} can be used to identify the N nutritional status at any stage of rice growth, allowing us to assess whether the N fertilizer dosage was ample enough to obtain higher yield in practice. These results confirmed the plausibility of using NNI and N_{and} to assess the status of N nutrition in rice plants growing under various conditions and stages.

Figure 7 and 8 showed that ΔN had a positive correlation with ΔNNI and ΔN_{and}. The simple linear regression equation showed non-significant differences between two varieties, although noticeable differences were observed among different phenological stages. Therefore, ΔN during growth period for both varieties could be derived from ΔNNI and ΔN_{and}, respectively, according

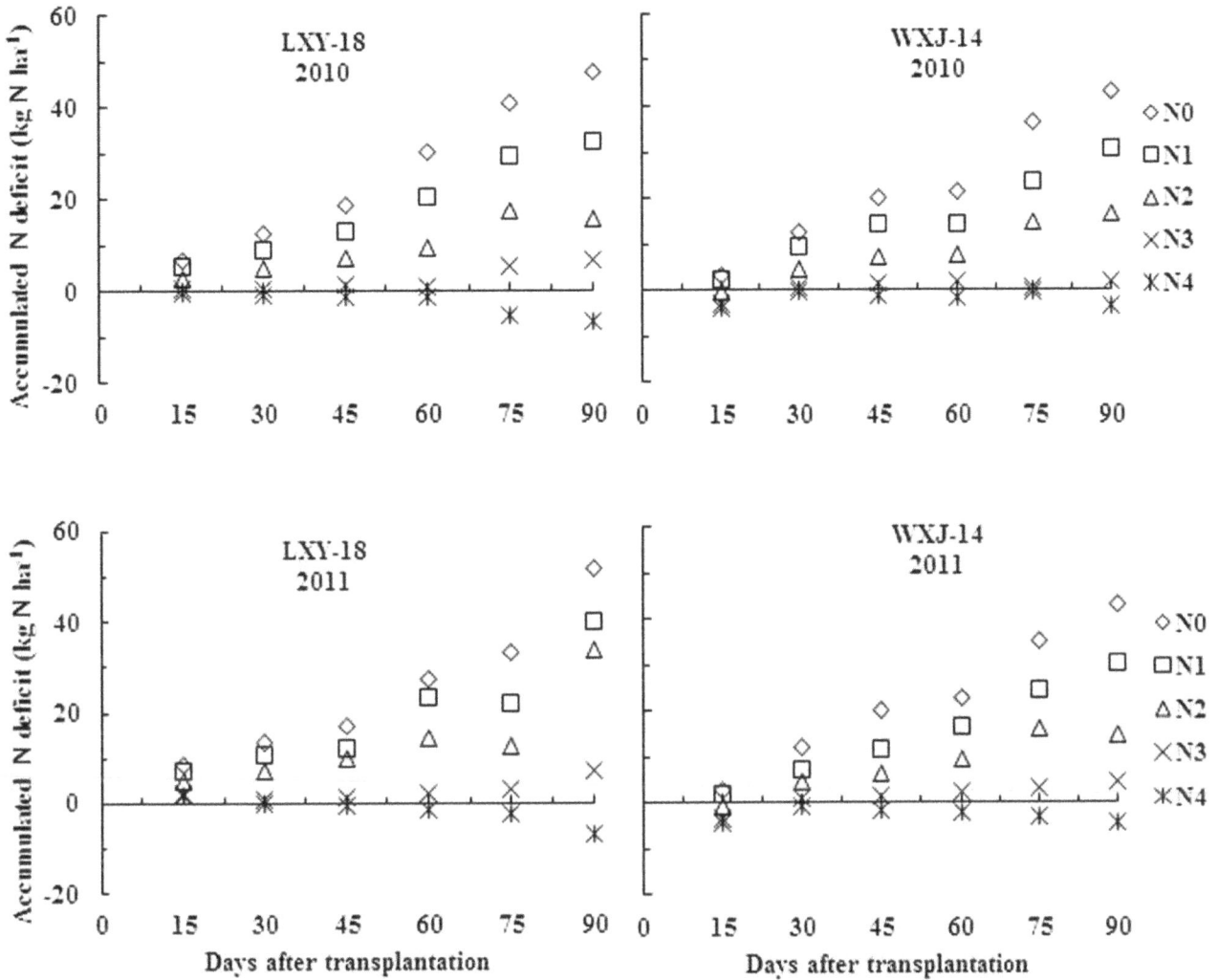

Figure 6. Changes of accumulated N deficit (N_{and}) with time (days after transplantation) for rice stem under different N rates in experiments conducted during 2010 and 2011.

to the equations 5 & 6 as follows:

$$\Delta N = A \times \Delta NNI + B \qquad (5)$$

$$\Delta N = C \times \Delta N_{and} + D \qquad (6)$$

The parameters A, B, C, and D could be calculated from days after transplanting (DAT) using the equations as:

$$A = -16.60 \times DAT + 2101 \qquad (R^2 = 0.95) \quad (7)$$

$$B = -0.024 \times DAT^2 + 2.57 \times DAT - 40.07 \qquad (R^2 = 0.62) \quad (8)$$

$$C = 18.97 ln(DAT) - 89.22 \qquad (R^2 = 0.98) \quad (9)$$

$$D = -9.98 ln(DAT) + 52.76 \qquad (R^2 = 0.19) \quad (10)$$

The ΔN obtained on the basis of relationship between ΔNNI, ΔN_{and} and ΔN, allowed us to make corrective decisions of N dressing recommendation for the precise N management during or even before the period of highest demand of the rice crop.

Discussion

Application of N fertilizer for crop production is an economically viable option in terms of low cost as compared to the value of the marketable agricultural products themselves; however, N usage cannot assure a significant increase in crop productivity due to diminishing returns after certain levels. There is an increasing demand by strategy makers for simple-to-use, technically established and economically viable N indicators, which may allow monitoring and assessment of policy measures and offer tools for farm N management. With the advent of technology, more emphasis should be put on plant-based indicators, which simultaneously reflect the interactions between the plant and the

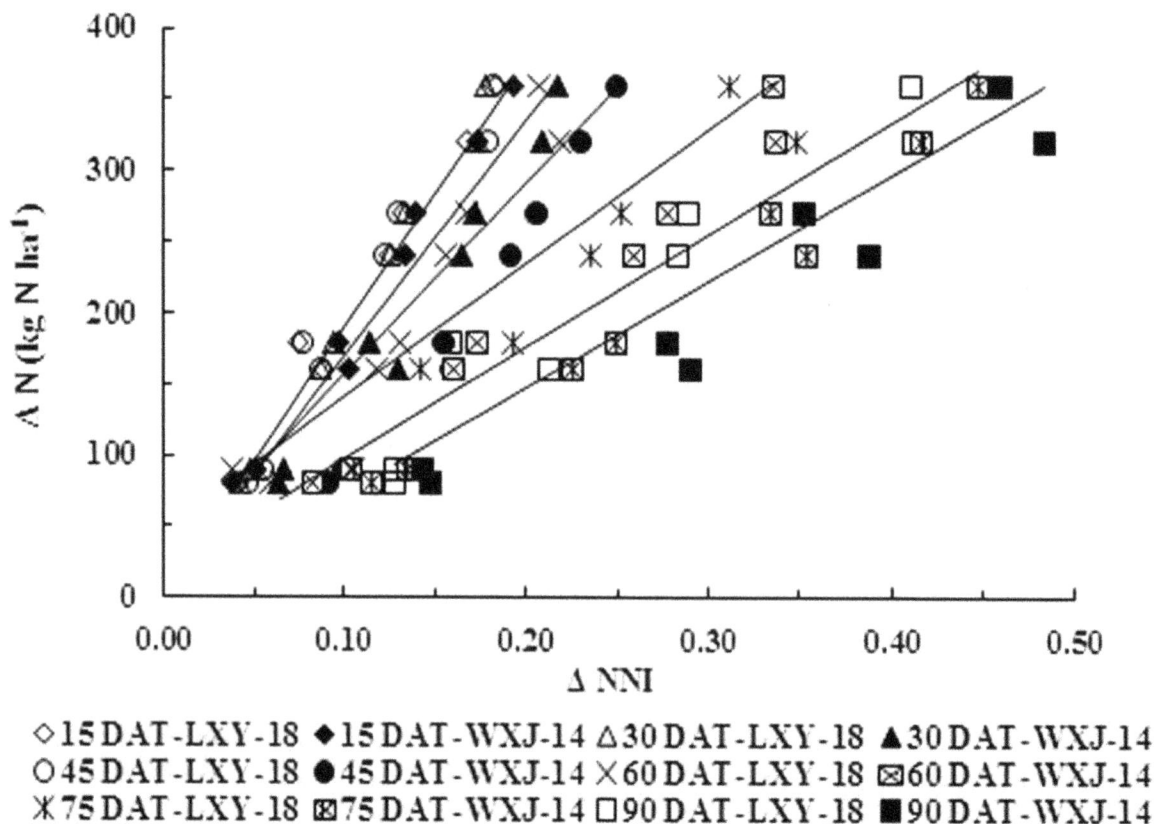

Figure 7. Relationship between changes of nitrogen nutrition index (ΔNNI) and changes of nitrogen application rates (ΔN, kg N ha⁻¹) at different growth stages in experiments conducted during 2010 and 2011. The open symbols represent different growth stages for LXY-18 while filled symbols represent different growth stages for WXJ-14. ($\Delta N = A \times \Delta NNI + B$; $A = -16.60 \times DAT + 2101$, $R^2 = 0.95$; $B = -0.024 \times DAT^2 + 2.57 \times DAT - 40.07$, $R^2 = 0.62$).

soil. So far, there have been several reports on estimating the N_c concentration on whole plant dry matter basis in various crops, including rice [11,12], and on L_{DM} basis in rice [20], yet no attempt was made to determine the N_c dilution curve on S_{DM} basis for any crop including rice. The current study has developed a S_{DM} based N_c dilution curve for rice in east China, thus providing a new approach for diagnosing and regulating N in crop species.

Minimum and maximum nitrogen dilution curves

An obvious variability in SNC for a given range of S_{DM} was observed when all the data from three year experiments were analyzed for interpretation. This variability in SNC towards maturity of rice crop in present study was in agreement with earlier studies on winter wheat [7] and Japonica rice [12], and this variability could be attributed to a decline in the fraction of total plant N associated with photosynthesis [21], change in leaf/shoot ratio and self-shading of leaves [8].

Two boundary curves for N maximum (N_{max}) and minimum (N_{min}) have been determined by using maximum and minimum N concentration in S_{DM} and can be represented as equations:

$$N_{max} = 2.28 W^{-0.25} \qquad (11)$$

$$N_{min} = 1.19 W^{-0.31} \qquad (12)$$

The N_{max} curve corresponding to the maximum N uptake in the S_{DM} without interfering with productivity and it can be considered as the first assessment of a maximum N dilution on S_{DM} basis in crops, and can be obtained with increasing N rates for maximum growth and N accumulation. This curve is an estimate of the maximum N accumulation capacity of stem which is regulated by mechanism associated with the growth and availability of soil N directly or indirectly via N metabolism [22]. The N_{max} curve in the present study shows a luxury consumption of N under N_4 treatment, when N concentration exceeds N_c dilution curve and S_{DM} does not increase with increasing N rate. In contrast, the N_{min} curve is considered as a lower limit at which the N metabolism would soon stop to function. It corresponds to the minimum N taken up by rice plants under N_0 treatment in present study. Thus, the N_{min} were used as the threshold concentration for proper metabolic functionality of the plant.

Moreover, the value of parameter b for the N_{max} was not significantly different from that of N_c dilution curve, which indicate that the partitioning of dry matter remains relatively constant when N uptake exceeds the N_c dilution curve. This is consistent with the concept of N_c dilution curve, which represents the lowest N at which maximum dry matter accumulation occurs. This implies that under luxury consumption of N, when N exceeds N_c dilution curve, dry matter accumulations does not increase with N and hence, dry matter partitioning will have similar value of parameter b. In contrast, for N_{min} curve under N stress, the value for parameter b tended to be slightly lower than the dilution curve.

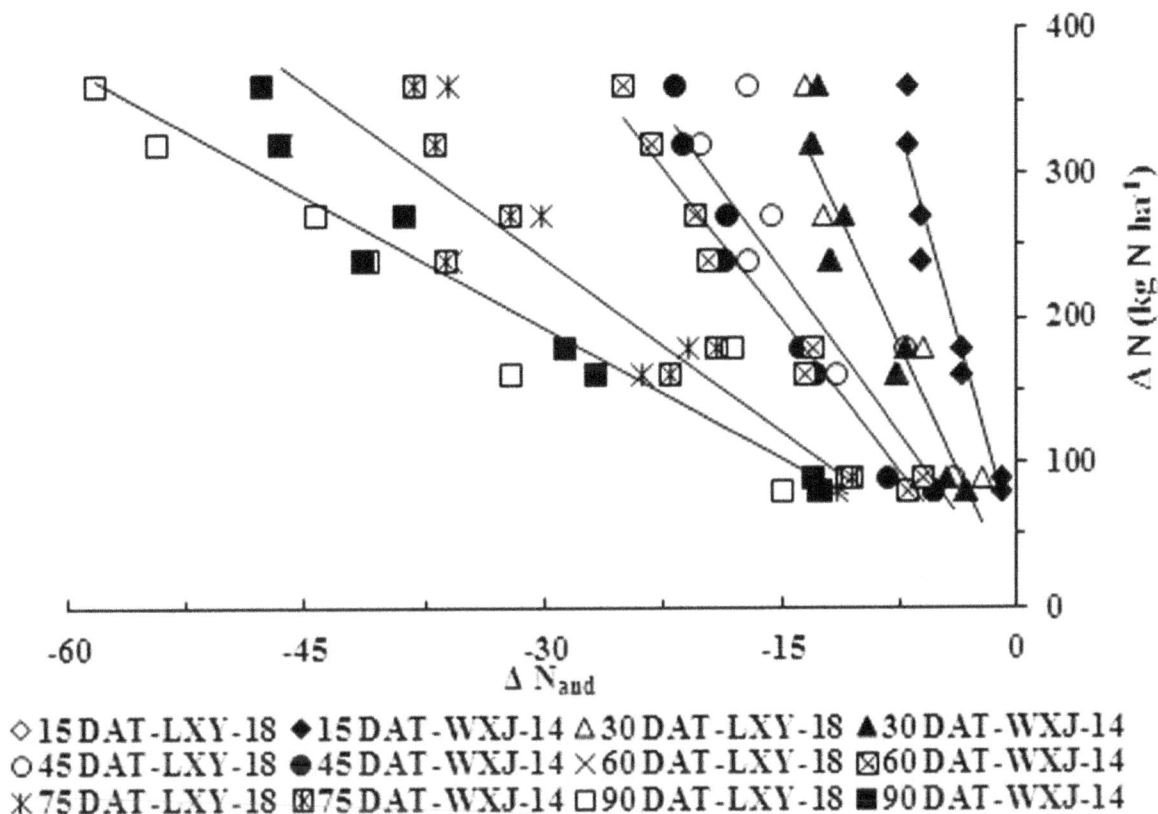

Figure 8. Relationship between changes of accumulated N deficit (ΔN_{and}) and changes of nitrogen application rates (ΔN, kg N ha^{-1}) at different growth stages in experiments conducted during 2010 and 2011. The open symbols represent different growth stages for LXY-18 while filled symbols represent different growth stages for WXJ-14 ($\Delta N = C \times \Delta N_{and} + D$; $C = 18.97$ ln(DAT)-89.22, $R^2 = 0.98$ and $D = -9.98$ ln(DAT)+ 52.76, $R^2 = 0.19$), respectively.

The relatively low value for b was associated with a change in dry matter partitioning.

Comparison with other critical nitrogen dilution curves

The concept of N_c dilution curve on whole plant dry matter and L_{DM} basis have already been successfully implicated for several crops including rice, yet no attempt was made to construct a S_{DM} based N_c dilution curve in any crop including rice. Figure 9 showed that the parameter a of N_c dilution curve on S_{DM} basis with Japonica rice developed in present study (2.17) was lower than the reference curve on whole plant dry matter basis of Indica rice in tropics (5.20) by Sheehy et al. [11] as well as lower than the curves developed with Japonica rice on whole plant dry matter basis (3.53) by Ata-Ul-Karim et al. [12], and L_{DM} basis (3.76) by Yao et al. [20].

The differences observed between the parameter a of dilution curve developed in present study and the curves on whole plant dry matter basis [11,12] were due to morphological aggregation of structural components, which relates to the weight/N concentration in the whole plant [13]. Stress responses may cause differences in the partitioning of dry matter among various plant organs, and thereby affect the shape of the dilution curves. Moreover, dissimilarities in climatic conditions and genetic differences of Indica and Japonica rice contributed to the differences between the curves. The ability of Indica to hold higher plant N content and total N uptake [23–25] and faster growth rate [26], compared with those of Japonica rice, also lead to the differences between N_c dilution curve of Sheehy et al. [11] and that described in the

present study. The differences of S_{DM} based curve with that of L_{DM} based one [20] are mainly attributed to leaf/stem ratio, because decrease in stem N during vegetative phase is related to decline in the metabolic biomass with high N contents, and increase in proportion of structural and non-photosynthetic biomass with low N contents [8]. Thus, higher proportion of structural biomass in stem than in leaves is responsible for the differences between the L_{DM} and S_{DM} based curves of Japonica rice.

The parameter b of the dilution curve indicates the dilution intensity of N during growth and the higher values of b indicate lower N dilutions [17]. The coefficients b were (-0.50, -0.28, -0.22 and -0.274) for N_c dilution curve of Indica rice and for Japonica rice based on whole plant dry matter, L_{DM}, and on S_{DM}, respectively. The observed differences between the coefficients b of Indica rice and current S_{DM} based dilution curve might be explained by the differences in duration of vegetative phase in tropical and subtropical climates, while the differences between coefficients b of the curves of Japonica rice based on L_{DM} and S_{DM} were directly related to the distribution of dry matter between green leaves and the stem [17]. In contrast, the differences between coefficients b of the curves of Japonica rice on whole plant dry matter basis compared with that of S_{DM} basis, are negligible due to the reason that stem have a dilution effect on the N in the above ground tissues, because of their higher weight percentage in the total dry matter [27]. Therefore, the S_{DM} based dilution curve can be used as a potential alternative for in-season estimation of

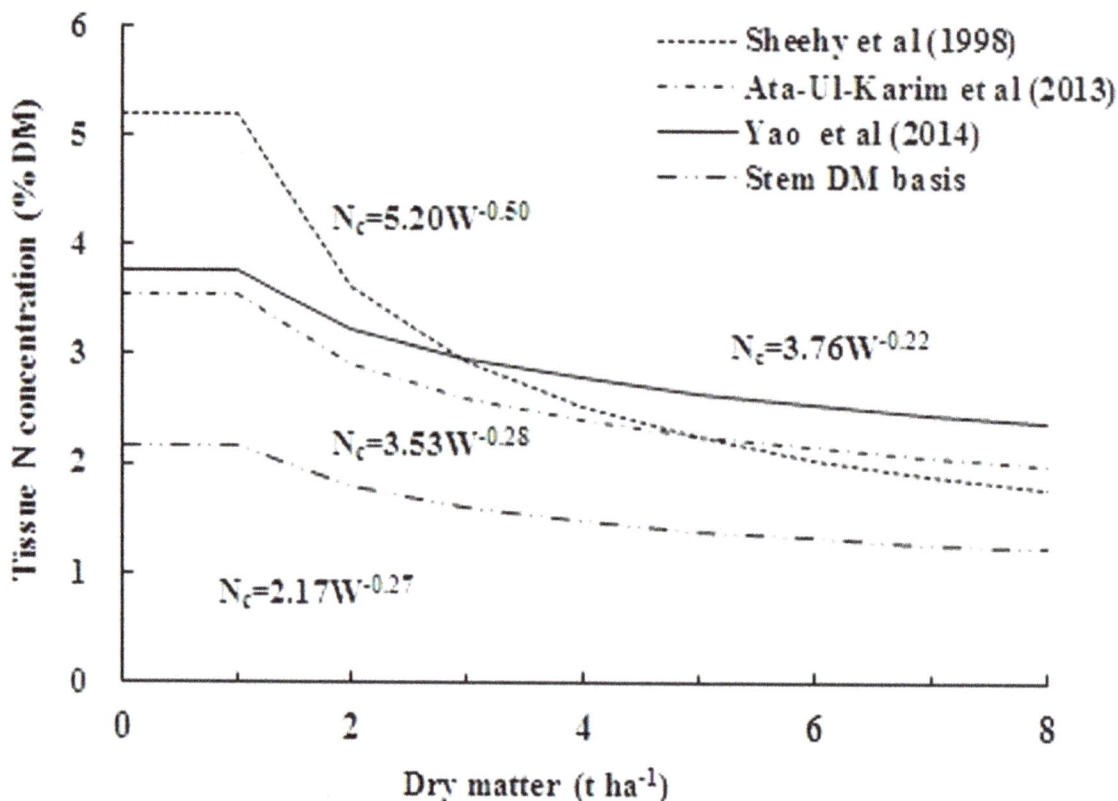

Figure 9. Comparison of different N_c dilution curves. The (------) represents the N_c dilution curve of Sheehy et al. (1998) ($N_c = 5.20W^{-0.50}$) on plant dry matter basis in Indica rice under tropic environment. The (-•-•-•) represents the N_c dilution curve of Ata-Ul-Karim et al. (2013) ($N_c = 3.53W^{-0.28}$) on plant dry matter basis in Japonica rice in Yangtze River Reaches. The (——) line represents N_c dilution curve of Yao et al. (2014) ($N_c = 3.76W^{-0.22}$) on leaf dry matter basis in Japonica rice in Yangtze River Reaches, and the (-••-••-) line represents N_c dilution curve on stem dry matter basis in present study ($N_c = 2.17W^{-0.27}$).

plant N nutrition status, instead of existing whole plant dry matter and L_{DM} based approaches.

Implication for nitrogen diagnosis

The application of the present N_c dilution curve as a diagnostic tool for accurate N management to make corrective decisions of N dressing recommendation during rice production is very interesting. The N_c dilution curve can be used for a priori analysis intended to optimize fertilizer N management or for a posteriori diagnosis intended to detect N limiting nutrition for rice within experimental trials or fields in production. The a priori diagnosis of plant N status consists of timely detection of plant N deficiency during the crop growth cycle to determine the necessity of applying additional N fertilizer. Present study showed that the N_c dilution curve, resulting NNI and N_{and} effectively distinguished conditions of deficient, optimal and surplus N nutrition in rice. The values of ΔN in present study obtained on the basis of relationship between ΔNNI, ΔN_{and} and ΔN, permitted us to make corrective decisions of N dressing recommendation for precise N management during or even before the period of peak demand of the rice crop. The main limitation in using the present NNI and N_{and} directly as diagnostic tools is the need to determine the actual dry matter and N concentration, which can be monitored by the non-destructive means including remote sensing [28–30]. Moreover, a good correlation between these analytical tools and chlorophyll meter readings was previously reported by [9]. These indirect methods could possibly be a substitute for assessing NNI

and N_{and} and portray crops and environments in conditions where they cannot be measured directly [31]. Thus, the models of NNI and N_{and}, based on N_c dilution curve in relation to actual growth status, can be exploited directly for the estimation of crop N status to recommend the necessities of further N application during plant growth. These novel algorithms can also be combined into crop growth and management models to forecast crop N status and quantify N dressing plan. Although, NNI and N_{and} calculated in present study distinguished well the N-limiting and non-N-limiting growth conditions, a more comprehensive validation using different N management practices, N availabilities and cultivars is mandatory to robustly confirm the reliability of NNI and N_{and} usage as an investigative indicators for different ecological regions and rice production systems.

Conclusions

In conclusion, we found that N fertilization endorses increase in the S_{DM}, which was influenced by variations in SNC. A higher rate of N fertilizer generally increased SNC in Japonica rice; however, towards advancing maturity this increase followed a declining trend under different N levels, sampling dates and growing seasons. S_{DM} during vegetative growth period ranged from minimum value of 0.19 (N_0) in WXJ-14 to a maximum value of 8.04 (N_4) in LXY-18, whereas SNC varied from 0.68% in WXJ-14 to 2.36% in LXY-18 on S_{DM} basis under different N rates and growth stages. A new N_c dilution curve on S_{DM} basis for Japonica rice grown in east China was developed and can be described by equation, $N_c = 2.17W^{-0.274}$, when S_{DM} ranges from 0.88 and 7.94 t ha^{-1}, however for $S_{DM} < 0.88$ t ha^{-1}, the constant critical value $N_c = 1.76\%$ S_{DM} was applied, which was independent of S_{DM}. Additionally, the values of NNI and N_{and} at different sampling dates for N limiting condition were generally < 1 and > 0, while > 1 and < 0, respectively for non-N-limiting supply. The values of ΔN derived on the basis of relationship between ΔNNI, ΔN_{and} and ΔN, can be used to make corrective decisions of N dressing recommendation for precise N management, prior to or on the onset of the period of highest demand of the rice crop. We conclude that the S_{DM} based dilution curve developed in the present study offers a new vision into plant N status and can possibly be adopted as an alternate practical tool for reliable diagnosis of plant N status to correct N fertilization decision during the vegetative growth of rice in east China.

Author Contributions

Conceived and designed the experiments: ST AUK XY XL WC YZ. Performed the experiments: ST AUK XY XL. Analyzed the data: ST AUK XY YZ. Wrote the paper: ST AUK YZ.

References

1. Jaggard K, Qi A, Armstrong M (2009) A meta-analysis of sugarbeet yield responses to nitrogen fertilizer measured in England since 1980. J Agric Sci-(Camb) 147: 287–301.
2. Ghosh M, Mandal B, Mandal B, Lodh S, Dash A (2004) The effect of planting date and nitrogen management on yield and quality of aromatic rice (Oryza sativa). J Agric Sci-(Camb) 142: 183–191.
3. Cabangon R, Castillo E, Tuong T (2011) Chlorophyll meter-based nitrogen management of rice grown under alternate wetting and drying irrigation. Field Crops Res 121: 136–146.
4. Lin FF, Qiu LF, Deng JS, Shi YY, Chen LS, et al. (2010) Investigation of SPAD meter-based indices for estimating rice nitrogen status. Compu Electron Agric 71: S60–S65.
5. Confalonieri R, Debellini C, Pirondini M, Possenti P, Bergamini L, et al. (2011) A new approach for determining rice critical nitrogen concentration. J Agric Sci-(Camb) 149: 633–638.
6. Smeal D, Zhang H (1994) Chlorophyll meter evaluation for nitrogen management in corn. Commun Soil Sci Plant Anal 25: 1495–1503.
7. Justes E, Mary B, Meynard JM, Machet JM, Thelier-Huche L (1994) Determination of a critical nitrogen dilution curve for winter wheat crops. Ann Bot 74: 397–407.
8. Yue S, Meng Q, Zhao R, Li F, Chen X, et al. (2012) Critical nitrogen dilution curve for optimizing nitrogen management of winter wheat production in the North China Plain. Agron J 104: 523–529.
9. Ziadi N, Brassard M, Bélanger G, Cambouris AN, Tremblay N, et al. (2008) Critical nitrogen curve and nitrogen nutrition index for corn in eastern Canada. Agron J 100: 271–276.
10. Ziadi N, Belanger G, Claessens A, Lefebvre L, Cambouris AN, et al. (2010) Determination of a critical nitrogen dilution curve for spring wheat. Agron J 102: 241–250.
11. Sheehy JE, Dionora MJA, Mitchell PL, Peng S, Cassman KG, et al. (1998) Critical nitrogen concentrations: implications for high-yielding rice (Oryza sativa L.) cultivars in the tropics. Field Crops Res 59: 31–41.
12. Ata-Ul-Karim ST, Yao X, Liu X, Cao W, Zhu Y (2013) Development of critical nitrogen dilution curve of Japonica rice in Yangtze River Reaches. Field Crops Res 149: 149–158.
13. Kage H, Alt C, Stützel H (2002) Nitrogen concentration of cauliflower organs as determined by organ size, N supply, and radiation environment. Plant Soil 246: 201–209.
14. Vouillot MO, Huet P, Boissard P (1998) Early detection of N deficiency in a wheat crop using physiological and radiometric methods. Agronomie 18: 117–130.
15. Ziadi N, Bélanger G, Gastal F, Claessens A, Lemaire G, et al. (2009) Leaf nitrogen concentration as an indicator of corn nitrogen status. Agron J 101: 947–957.
16. Lemaire G, Jeuffroy MH, Gastal F (2008) Diagnosis tool for plant and crop N status in vegetative stage: Theory and practices for crop N management. Eur J Agron 28: 614–624.
17. Oliveira ECAd, de Castro Gava GJ, Trivelin PCO, Otto R, Franco HCJ (2013) Determining a critical nitrogen dilution curve for sugarcane. J Plant Nutr Soil Sci 176: 712–723.
18. Chen J, Huang Y, Tang Y (2011) Quantifying economically and ecologically optimum nitrogen rates for rice production in south-eastern China. Agric Ecosyst Environ 142: 195–204.
19. Hahn WS (1997) Statistical Methods for Agriculture and Life Science. Seol: Free Academy Publishing Co. 747 p.
20. Yao X, Ata-Ul-Karim ST, Zhu Y, Tian Y, Liu X, et al. (2014) Development of critical nitrogen dilution curve in rice based on leaf dry matter. Eur J Agron 55: 20–28.
21. Bélanger G, Richards JE (2000) Dynamics of biomass and N accumulation of alfalfa under three N fertilization rates. Plant Soil 219: 177–185.
22. Gayler S, Wang E, Priesack E, Schaaf T, Maidl FX (2002) Modeling biomass growth, N-uptake and phenological development of potato crop. Geoderma 105: 367–383.
23. Islam M, Islam M, Sarker A (2008) Effect of phosphorus on nutrient uptake of Japonica and Indica rice. J Agric Rural Dev 6: 7–12.
24. Shan Y, Wang Y, Yamamoto Y, Huang J, Yang L, et al. (2001) Study on the differences of nitrogen uptake and use efficiency in different types of rice. J Yangzhou Univ (Nat Sci Ed) 4: 42.
25. Yoshida H, Horie T, Shiraiwa T (2006) A model explaining genotypic and environmental variation of rice spikelet number per unit area measured by cross-locational experiments in Asia. Field Crops Res 97: 337–343.
26. Ying J, Peng S, He Q, Yang H, Yang C, et al. (1998) Comparison of high-yield rice in tropical and subtropical environments: I. Determinants of grain and dry matter yields. Field Crops Res 57: 71–84.
27. Oliveira ECAd, Freire FJ, Oliveira RId, Freire M, Simoes Neto DE, et al. (2010) Extração e exportação de nutrientes por variedades de cana-de-açúcar cultivadas sob irrigação plena. Rev Bras de Ciênc Solo 34: 1343–1352.
28. Wang W, Yao X, Tian Y, Liu X, Ni J, et al. (2012) Common spectral bands and optimum vegetation indices for monitoring leaf nitrogen accumulation in rice and wheat. J Integr Agric 11: 2001–2012.
29. Zhao B, Yao X, Tian Y, Liu X, Ata-Ul-Karim ST, et al. (2014) New critical nitrogen curve based on leaf area index for winter wheat. Agron J 106: 379–389.
30. Ata-Ul-Karim ST, Zhu Y, Yao X, Cao W (2014) Determination of critical nitrogen dilution curve based on leaf area index in rice. Field Crops Res: In press.
31. Debaeke P, Rouet P, Justes E (2006) Relationship between the normalized SPAD index and the nitrogen nutrition index: Application to durum wheat. J Plant Nutr 29: 75–92.

17

Effects of Winter Cover Crops Straws Incorporation on CH$_4$ and N$_2$O Emission from Double-Cropping Paddy Fields in Southern China

Hai-Ming Tang*, Xiao-Ping Xiao*, Wen-Guang Tang, Ke Wang, Ji-Min Sun, Wei-Yan Li, Guang-Li Yang

Hunan Soil and Fertilizer Institute, Changsha, PR China

Abstract

Residue management in cropping systems is believed to improve soil quality. However, the effects of residue management on methane (CH$_4$) and nitrous oxide (N$_2$O) emissions from paddy field in Southern China have not been well researched. The emissions of CH$_4$ and N$_2$O were investigated in double cropping rice (*Oryza sativa* L.) systems with straw returning of different winter cover crops by using the static chamber-gas chromatography technique. A randomized block experiment with three replications was established in 2004 in Hunan Province, China, including rice–rice–ryegrass (*Lolium multiflorum* L.) (Ry-R-R), rice–rice–Chinese milk vetch (*Astragalus sinicus* L.) (Mv-R-R) and rice–rice with winter fallow (Fa-R-R). The results showed that straw returning of winter crops significantly increased the CH$_4$ emission during both rice growing seasons when compared with Fa-R-R. Ry-R-R plots had the largest CH$_4$ emissions during the early rice growing season with 14.235 and 15.906 g m^{-2} in 2012 and 2013, respectively, when Ry-R-R plots had the largest CH$_4$ emission during the later rice growing season with 35.673 and 38.606 g m^{-2} in 2012 and 2013, respectively. The Ry-R-R and Mv-R-R also had larger N$_2$O emissions than Fa-R-R in both rice seasons. When compared to Fa-R-R, total N$_2$O emissions in the early rice growing season were increased by 0.05 g m^{-2} in Ry-R-R and 0.063 g m^{-2} in Mv-R-R in 2012, and by 0.058 g m^{-2} in Ry-R-R and 0.068 g m^{-2} in Mv-R-R in 2013, respectively. Similar result were obtained in the late rice growing season, and the total N$_2$O emissions were increased by 0.104 g m^{-2} in Ry-R-R and 0.073 g m^{-2} in Mv-R-R in 2012, and by 0.108 g m^{-2} in Ry-R-R and 0.076 g m^{-2} in Mv-R-R in 2013, respectively. The global warming potentials (GWPs) from paddy fields were ranked as Ry-R-R>Mv-R-R> Fa-R-R. As a result, straw returning of winter cover crops has significant effects on increase of CH$_4$ and N$_2$O emission from paddy field in double cropping rice system.

Editor: Dafeng Hui, Tennessee State University, United States of America

Funding: This study was supported by the National Natural Science Foundation of China (No. 31201178), and the Public Research Funds Projects of Agriculture, Ministry of Agriculture of the P.R. China (No. 201103001). The funders had no role in study design, data collection and analysis, decision to publish, or preparation of the manuscript.

Competing Interests: The authors have declared that no competing interests exist.

* Email: hntfsxxping@163.com (XPX); tanghaiming66@163.com (HMT)

Introduction

With the current rise in global temperatures, numerous studies have focused on greenhouse gases (GHG) emissions [1–3]. Agriculture production is an important source of GHG emission [4]. In addition to carbon dioxide (CO$_2$), methane (CH$_4$) and nitrous oxide (N$_2$O) play important roles in global warming. The global warming potentials (GWPs) of CH$_4$ and N$_2$O are 25 and 298 times that of CO$_2$ in a time horizon of 100 years, respectively [5]. The concentrations of CH$_4$ and N$_2$O in the atmosphere are estimated to be increasing at the rates of 1% and 0.2–0.3% per year [6]. In addition to industrial emissions, farmland is another important source of atmospheric GHG [7–10]. Numerous results indicate that rice (*Oryza sativa* L.) paddy field is a significant source of CH$_4$ and N$_2$O emissions [10,11]. The anaerobic conditions in wetland rice field are favorable for fostering CH$_4$ emission [12]. Thus, the characteristics of CH$_4$ and N$_2$O emissions from paddy field and the reduction of emission have received attentions from scientists.

A considerable number of studies have shown that some farm operations can influence CH$_4$ and N$_2$O emission. For example, cropping system, crop type, water and nitrogen (N) management, organic matter application and tillage can regulate CH$_4$ and N$_2$O emission [13–15]. Tillage and crop straws retention have a great influence on CH$_4$ and N$_2$O emission through the changes of soil properties (e.g., soil porosity, soil temperature and soil moisture, etc.) [16–17]. In paddy soils, CH$_4$ is produced by archaea bacteria during the anaerobic degradation of organic matter and oxidized by methanotrophic bacteria [18]. Incorporation of organic material into soil can enhance the number and activity of archaea bacteria [19] and provide large amounts of active organic substrate for CH$_4$ production [20]. Soil amendment with organic material, such as crop straw [21] and green manure incorporation [22], has been well estimated to promote CH$_4$ emission in paddy fields. Biogenic N$_2$O production originates from nitrification and denitrification [23], which are processes involving microorganisms in the soil. N$_2$O flux in paddy fields was small in flooding condition, but peaked after drainage [24]. Some studies have indicated that the cropping system of winter fallow with cover crops has advantages of promoting soil quality, enhancing nutrient utilization, increasing crop yield, reducing soil erosion and chemical runoff, and inhibiting weed growth in paddy field [25–26].

Winter cover crops, which are grown during an otherwise fallow period, are a possible means of improving nutrient dynamics in the surface layer of intensively managed cropping systems. Chinese milk vetch (*Astragalus sinicus* L.) and ryegrass (*Lolium multiflorum* L.) are the main winter cover crops in Southern China. Growing these cover crops with straw mulching in the winter season after late rice harvest and incorporating them into soil as green manure before early rice transplanting next year is a traditional practice as well as rice straw incorporation. Hermawan and Bomke [26] suggested that growing winter cover crops such as annual ryegrass may protect aggregate breakdown during winter and result in a better soil structure after spring tillage, as opposed to leaving soil bare. Other potential benefits of winter cover crops are the prevention of nitrate leaching [27]; weed infestation [28]; and improvement of soil water retention, soil organic matter content and microbial activity [29]. Returning of crop straws have been suggested to improve overall soil conditions, reduce the requirement for N fertilizers and support sustainable rice productivity.

In recent years, many researches have studied the effects of winter cover crops on soil physical properties and crop productivity, methane emission, N availability and nitrogen surplus [30–32]. However, relatively few studies related to CH$_4$ and N$_2$O emissions and yields under different double cropping rice systems with different winter cover crops have been conducted in double-cropping paddy field in Southern China. Monitoring CH$_4$ and N$_2$O emissions of different winter cover crops–double cropping rice cultivation modes is important to maintain soil productivity, increase carbon (C) storage, and regulate the greenhouse effects. Therefore, the objectives of this research were: (1) to quantify CH$_4$ and N$_2$O emissions from paddy field and grain yield under different winter cover crops and double cropping rice systems, (2) to evaluate the GWPs of different winter cover crops–double cropping rice treatments in southern China.

Materials and Methods

Experimental site

The experiment was initiated in winter 2004 at the experimental station of the Institute of Soil and Fertilizer Research, Hunan Academy of Agricultural Sciences, China (28°11′58″ N, 113°04′47″ E). The typical cropping system in this area is double cropping rice. The soil type is a Fe–accumuli–Stagnic Anthrosol derived from Quaternary red clay (clay loam). The characteristics of the surface soil (0–20 cm) in 2004 are as follows: pH 5.40, soil organic carbon (SOC) 13.30 g kg^{-1}, total N 1.46 g kg^{-1}, available N 154.5 mg kg^{-1}, total phosphorous (P) 0.81 g kg^{-1}, available P 39.2 mg kg^{-1}, total potassium (K) 13.0 g kg^{-1}, and available K 57.0 mg kg^{-1}. All these data were tested before the experiment in 2004. This region has the subtropical monsoonal humid climate with a long hot period and short cold period. The average annual precipitation is approximately 1500 mm and the annual mean temperature is 17.1°C, the annual frost-free period is approximately from 270 days to 310 days. The daily precipitation and mean temperature data during the early and late rice growing season during 2012–2013 are presented in Fig. 1. The cropping system was that the early rice rotated with the late rice, and then planted winter cover crops till the next year's early rice transplanting.

Experimental design and field management

A randomized block experiment with three replications was established in 2004, and this study was conducted from 2012 to 2013. The experiment included three cropping systems: rice–rice–ryegrass (Ry-R-R), rice–rice–Chinese milk vetch (Mv-R-R), and rice–rice with winter fallow (Fa-R-R). The plot area was 1.1 m^2 (1 m × 1.1 m). After winter cover crops harvested, a moldboard plow was used to incorporate part of the crop straw into soil: both the ryegrass and Chinese milk vetch straw returned was 22500.0 kg ha^{-1}. All the plots were plowed once to a depth of 20 cm by using a moldboard plow 15 d before rice seedling transplanting. The early rice variety (*Oryza sativa* L.) Lingliangyou 211 and late rice variety (*Oryza sativa* L.) Fengyuanyou 299 were used as the materials in 2012 and 2013. One-month-old seedlings were transplanted with a density of 150,000 plants ha^{-1} (one seed per 16 cm × 16 cm) and 2–3 plants per hill. Gramoxone (paraquat) was applied to control weeds at 2 d before rice transplantation. The basal fertilizer of the early and late rice was applied at the rate of 150.0 kg N ha^{-1} and 180.0 kg N ha^{-1} as urea (60% for basal; 40% for top–dressed at the tillering stage), 75.0 kg P$_2$O$_5$ ha^{-1} as diammonium phosphate and 120.0 kg K$_2$O ha^{-1} as potassium sulfate. The different treatments during early and late rice season and field management were presented in Table 1.

Collection and measurement of CH$_4$ and N$_2$O

CH$_4$ and N$_2$O emitted from paddy field were collected using the static chamber–GC technique at 9:00–11:00 in the morning during the early and late rice growing season. The chamber (50 cm × 50 cm × 120 cm) was made of 5 mm PVC board with a PVC base. The base had a groove in the collar, in which the chamber could be settled. The chamber base was inserted into soil about 5 cm in depth with rice plant growing inside the base. The groove was 1 cm below flooded water, and the chamber was settled into the groove of the collar with water to prevent leakage and gas exchange. The chamber contained a small fan for stirring air, a thermometer sensor, and a trinal–venthole. From the second day after transplanting of early or late rice, gases were sampled weekly. Before sampling, the fan in the chamber started working to allow an even mix of air before extracting the air with a 50 ml injector at 0, 10, 20, and 30 min after closing the box. The air samples were transferred into 0.5 L sealed sample bags by rotating trinal venthole.

The quantities of CH$_4$ and N$_2$O emission were measured with a gas chromatograph (Agilent 7890A) equipped with flame ionization detector (FID) and electron capture detector (ECD). Methane was separated using 2 m stainless-steel column with an inner diameter of 2 mm 13XMS column (60/80 mesh), with FID at 200°C. Nitrous oxide was separated using a 1 m stainless-steel column with an inner diameter 2 mm Porapak Q (80/100 mesh) and ECD at 330°C.

Data analysis

Fluxes of CH$_4$ and N$_2$O were calculated with the following equation [33]:

$$F = ph \times \frac{273}{273+t} \times \frac{dc}{dt}$$

Where, F is the CH$_4$ flux (mg m^{-2} h^{-1}) or N$_2$O flux (μg m^{-2} h^{-1}); T is the air temperature (°C) inside the chamber; ρ is the CH$_4$ or N$_2$O density at standard state (0.714 kg m^{-3} for CH$_4$ and

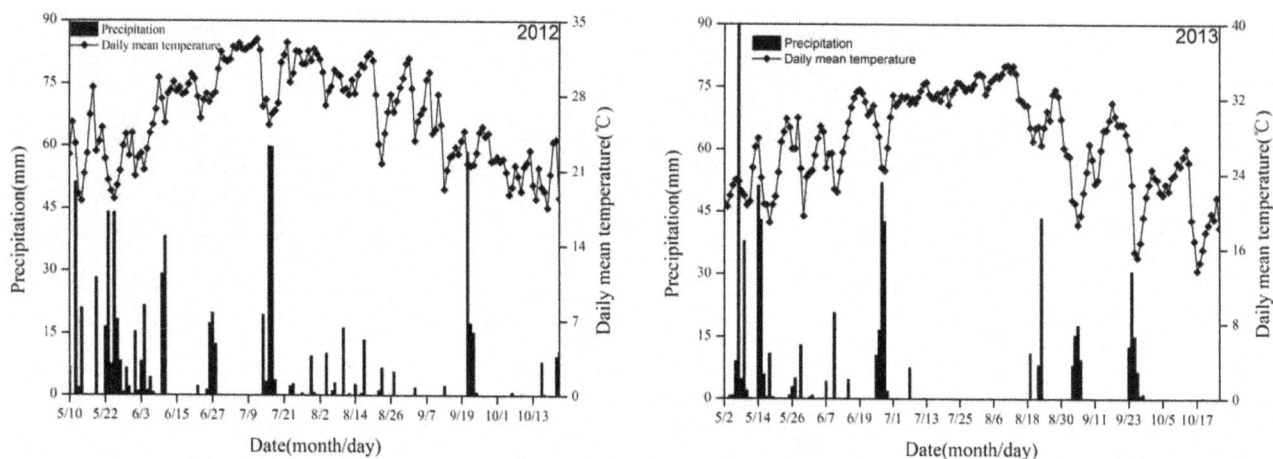

Figure 1. Daily precipitation and mean temperature at the study site in 2012 and 2013.

1.964 kg m^{-3} for N$_2$O); h is the headspace height of the chamber (m); and dc/dt is the slope of the curve of gas concentration variation with time.

The total emissions of CH$_4$ and N$_2$O were sequentially computed from the emissions between every 2 adjacent intervals of the measurements, based on a non–linear, least–squares method of analysis [34,35].

GWPs is defined as the cumulative radiative forcing both direct and indirect effects integrated over a period of time from the emission of a unit mass of gas relative to some reference gas. Carbon dioxide was chosen as this reference gas. The GWPs conversion parameters of CH$_4$ and N$_2$O (over 100 years) were adopted with 25 and 298 kg ha^{-1} CO$_2$-equivalent [5].

Statistical analysis

Data presented herein are means of 3 replicates in each treatment. All data were expressed as mean ± standard error. The data were analyzed as a randomized complete block, using the

PROC ANOVA procedure of SAS [36]. Mean values were compared using the least significant difference (LSD) test, and a probability value of 0.05 was considered to indicate statistical significance.

Results

Characteristics of CH$_4$ emission flux from early and late rice fields

In the early rice season, the curve of CH$_4$ flux was low when early rice was newly transplanted, but increased quickly until the first peak about 2 weeks after transplanting, and then dramatically declined to a low level with relative stability with the second small peak appeared at 36 and 35 d after transplanting in 2012 and 2013, respectively (Fig. 2). The gradual increase of CH$_4$ emission after transplanting resulted from the decomposition of organic matter and the growth of rice. The second peak was mainly because of the continuous decomposition of organic matter under

Table 1. Management practices of different cropping systems.

Crop	Date (month/day)		Field management
	2012	2013	
Early rice	4/12	4/5	Sowing and seedling raising
	5/9	5/1	Paddy tillage
	5/10	5/2	Transplanting (16 cm×16 cm)
	5/18	5/10	Urea were applied at 130.0 kg ha^{-1} for top-dressed at tillering
	6/7–6/15	5/27–6/5	Drained out water and dried the soil at maximum tillering stage
	6/16–7/13	6/6–7/13	Wetting–drying alternation irrigation
	7/18	7/18	Grains were harvested
Late rice	6/25	6/27	Sowing and seedling raising
	7/21	7/19	Paddy tillage (The rate of early rice straw returning was 4 500.0 kg ha^{-1})
	7/22	7/20	Transplanting (16 cm×16 cm)
	7/30	7/28	Urea were applied at 156.5 kg ha^{-1} for top-dressed at tillering
	8/20–8/27	8/16–8/26	Drained out water and dried the soil at maximal tillering stage
	8/28–10/17	8/27–10/19	Wetting–drying alternation irrigation
	10/22	10/25	Grains were harvested

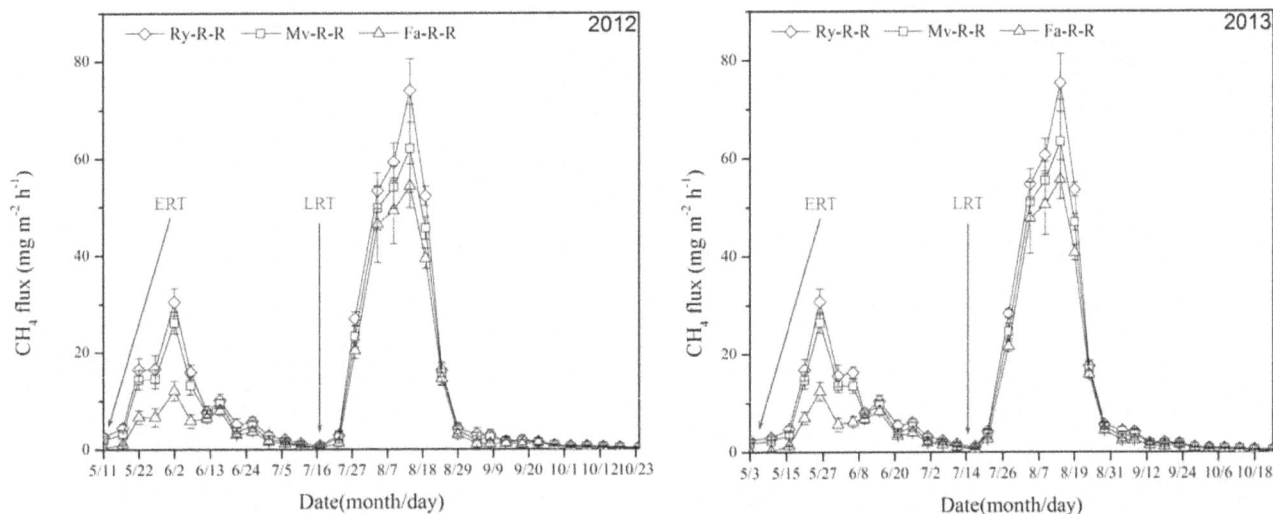

Figure 2. Effects of winter cover crops on CH$_4$ flux in early and late rice fields in 2012 and 2013. Ry-R-R: rice–rice–ryegrass cropping system; Mv-R-R: rice–rice–Chinese milk vetch cropping system; Fa-R-R: rice–rice cropping system with winter fallow. ERT: early rice transplanting; LRT: late rice transplanting. CH$_4$ emission rate is the mean of values measured within each treatment (n = 3).

high temperature. In the early rice season, the CH$_4$ flux values were significantly different among treatments with the order of Ry-R-R>Mv-R-R>Fa-R-R ($P<0.05$) (Fig. 2).

Methane emission in the late rice growing season mainly focused at tillering stage, and the peak value of CH$_4$ flux was observed at 23 and 24 d after transplanting in all treatments in 2012 and 2013, respectively. Then, the emission rate dramatically decreased to a low and stable level, especially from field drainage to harvest. The order of treatments in CH$_4$ emission was Ry-R-R>Mv-R-R>Fa-R-R (Fig. 2).

Characteristics of N$_2$O emission flux from early and late rice fields

The peak flux N$_2$O was emitted when the field was drained. Meanwhile, part of N$_2$O was emitted during wetting–drying alternation irrigation period. The first peak value of N$_2$O flux appeared at 7 and 15 d after transplanting in all treatments in 2012 and 2013, respectively, and then decreased. The order among treatments was Mv-R-R>Ry-R-R>Fa-R-R during the period from transplanting to field drainage, and Ry-R-R>Mv-R-R>Fa-R-R during wetting–drying alternation period. The N$_2$O flux in early rice paddy reached the highest peak at 32 and 35 d after transplanting in 2012 and 2013, respectively (Fig. 3).

In the late rice growing season, N$_2$O emission increased from field drainage to full heading stage, and mainly focused at booting stage. The order of N$_2$O emission fluxes among different treatments was Ry-R-R>Mv-R-R>Fa-R-R in the late rice growing season. In 2012, the average N$_2$O fluxes in the late rice growing season were 78.718 μg m^{-2} h^{-1} in Ry-R-R, 64.928 μg m^{-2} h^{-1} in Mv-R-R, and 32.275 μg m^{-2} h^{-1} in Fa-R-R. In 2013, the average N$_2$O fluxes in the late rice growing season were 81.453 μg m^{-2} h^{-1} in Ry-R-R, 67.662 μg m^{-2} h^{-1} in Mv-R-R, and 34.623 μg m^{-2} h^{-1} in Fa-R-R (Fig. 3).

Total CH$_4$ and N$_2$O emission from paddy fields in the growing durations of early and late rice

In the early rice growing season, the total CH$_4$ emissions of Ry-R-R and Mv-R-R were significantly higher than Fa-R-R ($P<0.05$), and the order of treatments was Ry-R-R>Mv-R-R>Fa-R-

R (Table 2). The straws of winter cover crops incorporated into soil provided favorable soil condition and sufficient substance to be decomposed in the early rice season; therefore, the CH$_4$ emission quantities in straw returning treatments were higher than Fa-R-R ($P<0.05$). In 2012, the total CH$_4$ emissions from paddy fields during late rice entire growing season were 35.673 g m^{-2} in Ry-R-R, 31.542 g m^{-2} in Mv-R-R, 27.874 g m^{-2} in Fa-R-R. In 2013, the total CH$_4$ emissions from paddy fields during late rice whole growing season were 38.606 g m^{-2} in Ry-R-R, 34.358 g m^{-2} in Mv-R-R, 30.550 g m^{-2} in Fa-R-R. The order of treatments in total CH$_4$ emission was Ry-R-R>Mv-R-R>Fa-R-R (Table 2).

Compared to Fa-R-R, the other treatments increased total N$_2$O emissions in the early rice growing season, and the N$_2$O emissions increased by 0.05 g m^{-2} (131.58%) in Ry-R-R and 0.063 g m^{-2} (165.79%) in Mv-R-R in 2012, and by 0.058 g m^{-2} (138.1%) in Ry-R-R and 0.068 g m^{-2} (161.90%) in Mv-R-R in 2013, respectively. Similar results were observed in the late rice growing season in 2012, the total N$_2$O emissions increased by 0.104 g m^{-2} (144.44%) in Ry-R-R and 0.073 g m^{-2} (101.39%) in Mv-R-R. And the total N$_2$O emissions increased by 0.108 g m^{-2} (135.00%) in Ry-R-R and 0.076 g m^{-2} (95.00%) in Mv-R-R in 2013 (Table 2).

The emissions of CH$_4$ and N$_2$O were closely related to farming system, soil type, climate, and field management practices. Ry-R-R and Mv-R-R had larger total CH$_4$ emissions than Fa-R-R in the double rice growing season ($P<0.05$). Ry-R-R had the largest total N$_2$O emissions in the double rice growing season with the quantities of 0.264 g m^{-2} in 2012, and 0.288 g m^{-2} in 2013, respectively (Table 3).

Global warming potentials of CH$_4$ and N$_2$O

GWPs is an indicator to reflect the relative radioactive effect of a greenhouse gas, and the GWPs of CO$_2$ is defined as 1. In this study, the GWPs of CH$_4$ and N$_2$O from double cropping paddy fields varied with different winter cover crops, and the trend showed as Ry-R-R>Mv-R-R>Fa-R-R. In 2012, Ry-R-R had the largest GWPs (13281.79 kg CO$_2$-eq ha^{-1}) of total CH$_4$ and N$_2$O from double cropping paddy fields, followed by Mv-R-R (11657.44 kg CO$_2$-eq ha^{-1}), and Fa-R-R had the lowest GWPs

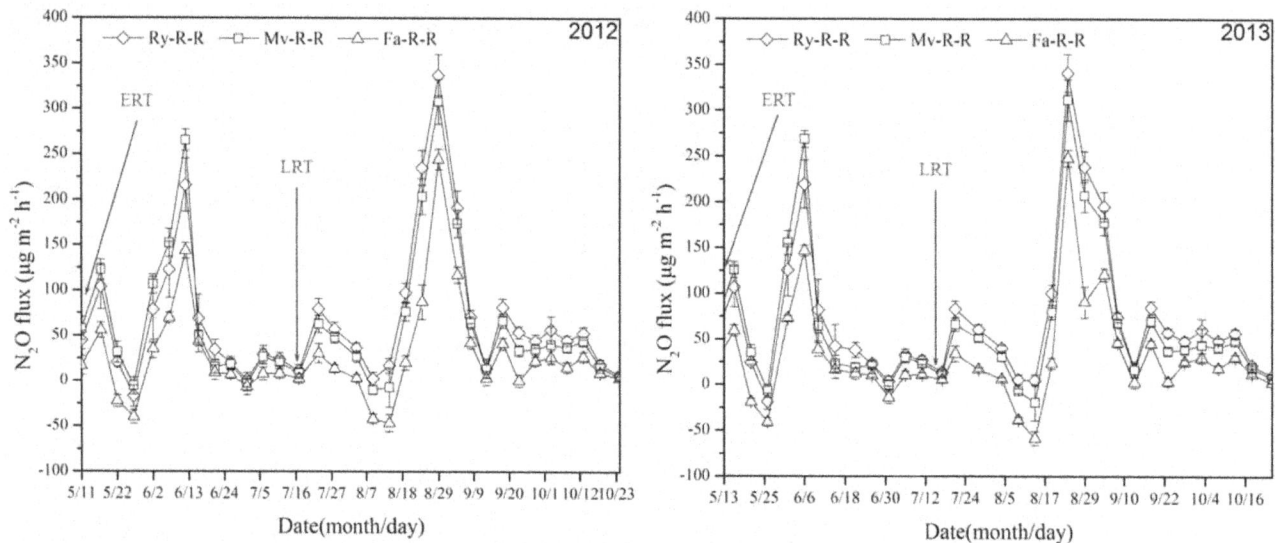

Figure 3. Effects of winter cover crops on N$_2$O flux in early and late rice fields in 2012 and 2013. Ry-R-R: rice–rice–ryegrass cropping system; Mv-R-R: rice–rice–Chinese milk vetch cropping system; Fa-R-R: rice–rice cropping system with winter fallow. ERT: early rice transplanting; LRT: late rice transplanting. N$_2$O emission rate is the mean of values measured within each treatment (n = 3).

of total CH$_4$ and N$_2$O (8993.12 kg CO$_2$–eq ha^{-1}). In 2013, Ry-R-R had the largest GWPs (14506.80 kg CO$_2$–eq ha^{-1}) of total CH$_4$ and N$_2$O from double cropping paddy fields, followed by Mv-R-R (12780.73 kg CO$_2$–eq ha^{-1}), and Fa-R-R had the lowest GWPs of total CH$_4$ and N$_2$O (9899.22 kg CO$_2$–eq ha^{-1}). According to GWPs, CH$_4$ from double cropping paddy fields had greater contribution to global warming than N$_2$O (Table 3).

Double rice grain yield of Mv-R-R was the highest, the lowest was Fa-R-R (Table 3). We also estimated per yield GWPs which was calculated as GWPs divided by rice grain yield. As is shown in Table 3, per yield GWPs of Ry-R-R was significantly higher than Mv-R-R and Fa-R-R ($P < 0.05$), and the lowest was Fa-R-R.

Discussion

CH$_4$ emission

Methane emission is complex processes including production, oxidation, and emission. Chidthaisong et al. [37] reported that the highest CH$_4$ peaks were observed at flowering and heading stages, which could be related to the development of intense reducing conditions in the rice rhizosphere. In this study, we found that CH$_4$ emission was low in paddy fields after transplanting during early rice–growing season, and increased with the decomposition of organic matters and growth of rice. In addition, CH$_4$ emission was influenced by soil temperature and soil redox potential (Eh). Yu et al. [38] reported that CH$_4$ emission showed an exponential decrease by an Eh increase. In this study, the CH$_4$ flux and total CH$_4$ emission from paddy fields during the early and late rice growing season were much larger in Ry-R-R and Mv-R-R compared to Fa-R-R, which was similar to the result by Lee et al. [22]. The reasons for above result may be: first, microbial activities were improved after returning straws of winter cover crops into the soil due to the supplements of carbon source and energy for microbial activities to accelerate consumption of soil oxygen and decrease of soil Eh; second, methanogens became active due to the large quantities of C source, which provided reactive substrate for CH$_4$ emission from paddy fields. In the early rice growing season, the order of CH$_4$ flux and total CH$_4$ emission from paddy fields varied among treatments, which were highly related to the

returning straw type, and straw decomposition rate. During the late rice growing season, the CH$_4$ emission increased gradually with the decomposition of organic matters and growth of rice after transplanting, and reach the peak value at tillering stage in all treatments. However, CH$_4$ emissions in both rice seasons were reduced in a large extent after field drying, because (1) soil aeration was improved during this period, and the activities of methanogens were restricted; and (2) the physiological activity of rice plant decreased, thereby limiting the ability for transportation and emission of CH$_4$ [39].

Although straw returning helps to maintain soil fertility and protect environment, but it enhances CH$_4$ emission simultaneously. Pandey et al. [12] showed that CH$_4$ emission was positively related to straw returning amount under permanent flooding condition, whereas N$_2$O emission had a reverse relationship with the amount of straw returning. In this study, we found that CH$_4$ flux in the late rice growing season was much higher than that in the early rice growing season, and peak appeared earlier. As straws of early rice (4500 kg ha^{-1}) returned to field before transplanting of late rice, the paddy soil of late rice was under anoxic condition after transplanting, which was favorable for CH$_4$ production and emission. Temperature was the major reason for the differences in the CH$_4$ emission pattern between the early and the late rice season. Soil temperature had a predictive functional relationship with CH$_4$ emission. Zhang et al. [40] reported that there was a strong positive correlation between CH$_4$ emission and soil temperature. In this experimental area, the late rice season was the hottest time in summer (Fig. 1). Therefore, high temperatures enhanced the decomposition of crop straws in the moist environment. In contrast to the warm temperatures of the late rice season, the air temperatures of the early rice season were lower, which resulted in slower crop straws decomposition and little CH$_4$–substrate. Hence, these differences in weather factors (e.g., temperature) resulted in the different characteristics of CH$_4$ between the early and the late rice seasons. However, there were significantly differences among treatments although they had similar trends. This indicated that CH$_4$ flux and emission from paddy fields were affected by different winter cover crops.

Table 2. Effects of winter cover crops on CH_4 and N_2O emission from rice fields during whole growing season of early and late rice (g m^{-2}).

Year	Treatment	CH_4			N_2O		
		Early rice	Late rice	Total	Early rice	Late rice	Total
2012	Ry-R-R	14.235±0.411a	35.673±1.030a	49.908±1.441a	0.088±0.003a	0.176±0.05a	0.264±0.008a
	Mv-R-R	12.092±0.349b	31.542±0.912b	43.634±1.260b	0.101±0.003b	0.145±0.04b	0.246±0.007a
	Fa-R-R	6.732±0.194c	27.874±0.805c	34.606±0.999c	0.038±0.001c	0.072±0.02c	0.110±0.003b
2013	Ry-R-R	15.906±0.459a	38.606±1.115a	54.512±1.574a	0.100±0.003b	0.188±0.006a	0.288±0.008a
	Mv-R-R	13.523±0.390b	34.358±0.992b	47.882±1.382b	0.110±0.003a	0.156±0.005b	0.266±0.008a
	Fa-R-R	7.535±0.218c	30.550±0.882c	38.085±1.099c	0.042±0.002c	0.080±0.002c	0.122±0.004b

Ry-R-R: rice–rice–ryegrass cropping system; Mv-R-R: rice–rice–Chinese milk vetch cropping system; Fa-R-R: rice–rice cropping system with winter fallow.
Values are presented as mean ± SE (n = 3). Means in each column with different letters are significantly different at the $P < 0.05$ level.

Table 3. Double rice grain yield, global warming potentials (GWPs) of CH_4 and N_2O and per yield GWPs from rice fields under different cropping patterns.

Year	Treatment	CH_4 emission (g m^{-2})	N_2O emission (g m^{-2})	GWPs of CH_4 (kg CO$_2$-eq ha^{-1})	GWPs of N_2O (kg CO$_2$-eq ha^{-1})	GWPs of CH_4 and N_2O (kg CO$_2$-eq ha^{-1})	Double rice grain yield (kg ha^{-1})	Per yield GWPs CO$_2$ (kg kg^{-1})
2012	Ry-R-R	49.908±1.441a	0.264±0.008a	12494.38±360.68a	787.41±22.73a	13281.79±383.41a	13800.23±398.38a	0.96±0.03a
	Mv-R-R	43.634±1.260b	0.246±0.007a	10923.85±315.34b	733.58±21.18a	11657.44±336.52b	15089.30±435.59a	0.77±0.02b
	Fa-R-R	34.606±0.999c	0.110±0.003b	8663.66±250.10c	329.46±9.51b	8993.12±259.61c	14359.00±414.51a	0.63±0.02c
2013	Ry-R-R	54.512±1.574a	0.288±0.008a	13646.99±393.95a	859.81±24.82a	14506.80±418.76a	14738.87±425.47a	0.98±0.03a
	Mv-R-R	47.882±1.382b	0.266±0.008a	11987.20±346.04b	793.53±22.91a	12780.73±368.95b	14896.57±430.03a	0.86±0.02b
	Fa-R-R	38.085±1.099c	0.122±0.004b	9534.57±275.24c	364.64±10.53b	9899.22±285.77c	13625.16±322.60a	0.73±0.02c

Ry-R-R: rice–rice–ryegrass cropping system; Mv-R-R: rice–rice–Chinese milk vetch cropping system; Fa-R-R: rice–rice cropping system with winter fallow.
Values are presented as mean ± SE (n = 3). Means in each column with different letters are significantly different at the $P < 0.05$ level.

N$_2$O emission

The emissions of N$_2$O are closely related to soil moisture, oxygen, temperature, content of soil organic matter and pH [4,11,17]. Great positive interaction has been reported between N$_2$O emission and green manure or chemical nitrogen fertilizer in early rice growing season [41]. In this study, we found that N$_2$O emission in the early rice growing season focused in the period of field drainage, and the Ry-R-R and Mv-R-R with winter cover crops had more N$_2$O emissions than Fa-R-R in both rice growing seasons (Fig. 3). N$_2$O emission from paddy field is promoted with the amount of straw returning via increasing soil denitrification, which provides the soil microbial substrates and energy for soil nitrification and denitrification process [42]. Different ranking of treatments in N$_2$O flux and total N$_2$O emission might be related to the decomposition rates of winter crop species during the rice growing season. In the late rice growing season, the total N$_2$O emissions of treatments Ry-R-R and Mv-R-R were significantly higher than Fa-R-R ($P<0.05$). This possibly results from that soil nitrification and denitrification process has been facilitated after the early rice straw returning through carbon and energy resource regulation (Table 1); a small amount of winter crop straw remains in the soil until the growing season of late rice; and tillage practice before late rice transplanting helps the incorporation of straws into soil, which may improve the soil nitrification and denitrification process.

Global warming potentials of CH$_4$ and N$_2$O

Global warming potential can be used as an index to estimate the potential effects of different greenhouse gases on the global climate system. Bhatia et al. [5] estimated that GWPs of rice–wheat system increased by 28% on full substitution of organic N by chemical N. Zhu et al. [43] reported that the highest GWPs was found in Chinese milk vetch incorporation in double cropping rice system, which was 21–325% higher than the other three treatments. In this study, the GWPs of CH$_4$, N$_2$O or both had different orders. For a comprehensive consideration, GWPs of both CH$_4$ and N$_2$O is more important to assess the effect of a farming system on climate warming. Therefore, it is necessary to make a combined estimate of global warming effects of CH$_4$ and N$_2$O emitted from each treatment. Thus, we introduced the GWPs and per yield GWPs into this study for global warming calculations. Although the global warming effect of N$_2$O is 12 times as large as that of CH$_4$, CH$_4$ emissions were nearly 370 times that of N$_2$O, resulting in the majority of GWPs originating

from CH$_4$ (Table 3). Therefore, it is certain that the GWPs and per yield GWPs values for Ry-R-R and Mv-R-R were larger than Fa-R-R ($P<0.05$), due to their greater CH$_4$ emissions. But the GWPs of CH$_4$ and N$_2$O and per yield GWPs of Mv-R-R was significantly lower than Ry-R-R ($P<0.05$). It should be mentioned that, the cultivation of ryegrass, Chinese milk vetch and its incorporation is a process involving C accumulation from the atmosphere to the soil, while the production of synthetic nitrogen fertilizer consumes fossil fuels that release C and contribute to greenhouse gas emissions. Therefore, we recommend Mv-R-R pattern in double cropping rice areas in the Middle and Lower reaches of Yangtze River in China, which correspond to Chinese milk vetch as winter cover crop + double rice.

Conclusions

The emissions of CH$_4$ and N$_2$O from double cropping paddy fields were significantly enhanced by returning different winter cover crops. The effects on CH$_4$ and N$_2$O fluxes and emissions were different among treatments, and the emission characteristics varied greatly between early and late rice growing season. The orders of treatments were Ry-R-R>Mv-R-R>Fa-R-R for total emissions of CH$_4$ and N$_2$O during double rice seasons, and Ry-R-R>Mv-R-R>Fa-R-R for GWPs of total CH$_4$ and N$_2$O from double cropping paddy fields. Compared with Ry-R-R, Mv-R-R and Fa-R-R reduced CH$_4$ emission during rice growing seasons. The GWPs (based on CH$_4$ emission) under Mv-R-R and Fa-R-R was significantly ($P<0.05$) lower than Ry-R-R. Although the cumulative N$_2$O emission under Ry-R-R and Mv-R-R were higher than that from Fa-R-R ($P<0.05$), GWPs of N$_2$O was relatively low compared to that of CH$_4$. The GWPs (based on CH$_4$ and N$_2$O) of Mv-R-R and Fa-R-R is lower than that of Ry-R-R ($P<0.05$). Meanwhile, the GWPs of CH$_4$ and N$_2$O and per yield GWPs of Mv-R-R was significantly lower than Ry-R-R ($P<0.05$). Thus, Mv-R-R is beneficial in GHG mitigation and it can be extended as an excellent cropping pattern in double rice cropped regions.

Author Contributions

Conceived and designed the experiments: XPX GLY. Performed the experiments: HMT. Analyzed the data: HMT WGT. Contributed reagents/materials/analysis tools: JMS KW WYL. Wrote the paper: HMT.

References

1. Levy PE, Mobbs DC, Jones SK, Milne R, Campbell C, et al. (2007) Simulation of fluxes of greenhouse gases from European grasslands using the DNDC model. Agric Ecosyst Environ 121: 186–192.

2. Saggar S, Hedley CB, Giltrap DL, Lambie SM (2007) Measured and modeled estimates of nitrous oxide emission and methane consumption from a sheep-grazed pasture. Agric Ecosyst Environ 122: 357–365.

3. Hernandez-Ramirez G, Brouder SM, Smith DR, Van Scoyoc GE (2009) Greenhouse gas fluxes in an eastern corn belt soil: Weather, nitrogen source, and rotation. J Environ Qual 38: 841–854.

4. Wassmann R, Neue HU, Ladha JK, Aulakh MS (2004) Mitigating greenhouse gas emissions from rice-wheat cropping systems in Asia. Environ Devel Sustain 6: 65–90.

5. Bhatia A, Pathak H, Jain N, Singh PK, Singh AK (2005) Global warming potential of manure amended soils under rice-wheat system in the Indo-Gangetic plains. Atmospheric Environ, 39(37): 6976–6984.

6. Verge XPC, Kimp CD, Desjardins RL (2007) Agricultural production, greenhouse gas emissions and mitigation potential. Agric Forest Meteorol 142: 255–269.

7. Lokupitiya E, Paustian K (2006) Agricultural soil greenhouse gas emissions: A review of national inventory methods. J Environ Qual 35: 1413–1427.

8. Verma A, Tyagi L, Yadav S, Singh SN (2006) Temporal changes in N$_2$O efflux from cropped and fallow agricultural fields. Agric Ecosyst Environ 116: 209–215.

9. Liu H, Zhao P, Lu P, Wang YS, Lin YB, et al. (2008) Greenhouse gas fluxes from soils of different land-use types in a hilly area of South China. Agric Ecosyst Environ 124: 125–135.

10. Tan Z, Liu S, Tieszen LL, Tachie-Obeng E (2009) Simulated dynamics of carbon stocks driven by changes in land use, management and climate in a tropical moist ecosystem of Ghana. Agric Ecosyst Environ 130: 171–176.

11. Kallenbach CM, Rolston DE, Horwath WR (2010) Cover cropping affects soil N$_2$O and CO$_2$ emissions differently depending on type of irrigation. Agric Ecosyst Environ 137: 251–260.

12. Pandey D, Agrawal M, Bohra JS (2012) Greenhouse gas emissions from rice crop with different tillage permutations in rice-wheat system. Agric Ecosyst Environ 159: 133–144.

13. Yagi K, Minami K (1990) Effect of organic matter application on methane emission from some Japanese paddy fields. Soil Sci Plant Nutr 36: 599–610.

14. Yagi K, Tsuruta H, Kanda KI, Minami K (1996) Effect of water management on methane emission from a Japanese rice paddy field: Automated methane monitoring. Global Biogeochem Cycles 10: 255–267.

15. Nishimura S, Sawamoto T, Akiyama H, Sudo S, Yagi K (2004) Methane and nitrous oxide emissions from a paddy field with Japanese conventional water

management and fertilizer application. Global Biogeochem Cycles 18, GB2017, doi:10.1029/2003GB002207

16. Al-Kaisi MM, Yin X (2005) Tillage and crop residue effects on soil carbon and carbon dioxide emission in corn-soybean rotations. J Environ Qual 34: 437–445.

17. Yao Z, Zheng X, Xie B, Mei B, Wang R, et al. (2009) Tillage and crop residue management significantly affects N-trace gas emissions during the non-rice season of a subtropical rice-wheat rotation. Soil Biol Biochem 41: 2131–2140.

18. Groot TT, VanBodegom PM, Harren FJM, Meijer HAJ (2003) Quantification of methane oxidation in the rice rhizosphere using ^{13}C-labelled methane. Biogeochemistry 64: 355–372.

19. Yue J, Shi Y, Liang W, Wu J, Wang CR, et al. (2005) Methane and nitrous oxide emissions from rice field and related microorganism in black soil, northeast China. Nutr Cy Agroecosyst 73: 293–301.

20. Sethunathan N, Kumaraswamy S, Rath AK, Ramakrishnan B, Satpathy SN, et al. (2000) Methane production, oxidation, and emission from Indian rice soils. Nutr Cy Agroecosyst 58: 377–388.

21. Ma J, Xu H, Yagi K, Cai ZC (2008) Methane emission from paddy soils as affected by wheat straw returning mode. Plant Soil 313: 167–174.

22. Lee CH, Park KD, Jung KY, Ali MA, Lee D, et al. (2010) Effect of Chinese milk vetch (*Astragalus sinicus* L.) as a green manure on rice productivity and methane emission in paddy soil. Agric Ecosyst Environ 138: 343–347.

23. Bouwman AF (1998) Nitrous oxides and tropical agriculture. Nature 392: 866–867.

24. Cai ZC, Lanughlin RJ, Stevens RJ (2001) Nitrous oxide and dinitrogen emissions from soil under different water regimes and straw amendment. Chemosphere 42: 113–121.

25. Rittera WF, Scarborough RW, Chirnside AEM (1998) Winter cover crops as a best management practice for reducing nitrogen leaching. J Contam Hydrol 34: 1–15.

26. Hermawan B, Bomke AA (1997) Effects of winter cover crops and successive spring tillage on soil aggregation. Soil Tillage Res 44: 109–120.

27. McCracken DV, Smith MS, Grove JH, MacKown CT, Blevins RL (1994) Nitrate leaching as influenced by cover cropping and nitrogen source. Soil Sci Soc Am J 58: 1476–1483.

28. Barnes JP, Putnam AR (1983) Rye residues contribute weed suppression in no-tillage cropping systems. J Chem Ecol 9: 1045–1057.

29. Powlson DS, Prookes PC, Christensen BT (1987) Measurement of soil microbial biomass provides an early indication of changes in total soil organic matter due to straw incorporation. Soil Biol Biochem 19(2): 159–164.

30. Mitchell JP, Shennan C, Singer MJ, Peters DW, Miller RO, et al. (2000) Impacts of gypsum and winter cover crops on soil physical properties and crop productivity when irrigated with saline water. Agr Water Manag 45: 55–71.

31. Chang HL, Ki DP, Ki YJ, Muhammad AA, Dokyoung L, et al. (2010) Effect of Chinese milk vetch (*Astragalus sinicus* L.) as a green manure on rice productivity and methane emission in paddy soil. Agr Ecosyst Environ 138: 343–347.

32. Salmeróna M, Isla R, Cavero J (2011) Effect of winter cover crop species and planting methods on maize yield and N availability under irrigated Mediterranean conditions. Field Crops Res 123: 89–99.

33. Zheng X, Wang M, Wang Y, Shen R, Li J, et al. (1998) Comparison of manual and automatic methods for measurement of methane emission from rice paddy fields. Adv Atmos Sci 15: 569–579.

34. Parashar DC, Gupta PK, Rai J, Sharma RC, Singh N (1993) Effect of soil temperature on methane emission from paddy field. Chemosphere 26: 247–250.

35. Singh JS, Singh S, Raghubanshi AS, Saranath S, Kashyap AK (1996) Methane flux from rice/wheat agroecosystem as affected by crop phenology, fertilization and water lever. Plant Soil 183: 323–327.

36. SAS Institute (2003) SAS Version 9.1.2 2002–2003. SAS Institute Inc., Cary, NC.

37. Chidthaisong A, Obata H, Watanabe I (1999) Methane formation and substrate utilization in anaerobic rice soils as affected by fertilization. Soil Biol Biochem 31: 135–143.

38. Yu K, Bohme F, Rinklebe J, Neue HU, DeLaune RD (2007) Major biogeochemical processes in soils-A microcosm incubation from reducing to oxidizing conditions. Soil Sci Soc Am J 71: 1406–1417.

39. Yang X, Shang Q, Wu P, Liu J, Shen Q, et al. (2010) Methane emissions from double rice agriculture under long-term fertilizing systems in Hunan, China. Agric Ecosyst Environ 137: 308–316.

40. Zhang HL, Bai XL, Xue JF, Chen ZD, Tang HM, et al. (2013) Emissions of CH$_4$ and N$_2$O under different tillage systems from double-cropped paddy fields in Southern China. PLoS ONE 8(6): e65277. doi:10.1371/journal.pone.0065277.

41. Petersen SO, Mutegi JK, Hansen EM, Munkholm LJ (2011) Tillage effects on N$_2$O emissions as influenced by a winter cover crop. Soil Biol Biochem 43: 1509–1517.

42. Huang Y, Zou JW, Zheng XH, Wang YS, Xu XK (2004) Nitrous oxide emissions as influenced by amendment of plant residues with different C: N ratios. Soil Biol Biochem 36: 973–981.

43. Zhu B, Yi LX, Hu YG, Zeng ZH, Tang HM, et al. (2012) Effects of Chinese Milk Vetch (*Astragalus sinicus* L.) residue incorporation on CH$_4$ and N$_2$O emission from a double-rice paddy soil. J Integrative Agric 11(9): 1537–1544.

18

The IQD Gene Family in Soybean: Structure, Phylogeny, Evolution and Expression

Lin Feng[1,9]**, Zhu Chen**[1,9]**, Hui Ma**[1]**, Xue Chen**[1]**, Yuan Li**[1]**, Yiyi Wang**[1]**, Yan Xiang**[1,2]*****

1 Laboratory of Modern Biotechnology, School of Forestry and Landscape Architecture, Anhui Agricultural University, Hefei, China, **2** Key Laboratory of Crop Biology of Anhui Agriculture University, Hefei, China

Abstract

Members of the plant-specific IQ67-domain (IQD) protein family are involved in plant development and the basal defense response. Although systematic characterization of this family has been carried out in *Arabidopsis*, tomato (*Solanum lycopersicum*), *Brachypodium distachyon* and rice (*Oryza sativa*), systematic analysis and expression profiling of this gene family in soybean (*Glycine max*) have not previously been reported. In this study, we identified and structurally characterized IQD genes in the soybean genome. A complete set of 67 soybean IQD genes (*GmIQD1–67*) was identified using Blast search tools, and the genes were clustered into four subfamilies (IQD I–IV) based on phylogeny. These soybean IQD genes are distributed unevenly across all 20 chromosomes, with 30 segmental duplication events, suggesting that segmental duplication has played a major role in the expansion of the soybean IQD gene family. Analysis of the Ka/Ks ratios showed that the duplicated genes of the GmIQD family primarily underwent purifying selection. Microsynteny was detected in most pairs: genes in clade 1–3 might be present in genome regions that were inverted, expanded or contracted after the divergence; most gene pairs in clade 4 showed high conservation with little rearrangement among these gene-residing regions. Of the soybean IQD genes examined, six were most highly expressed in young leaves, six in flowers, one in roots and two in nodules. Our qRT-PCR analysis of 24 soybean IQD III genes confirmed that these genes are regulated by MeJA stress. Our findings present a comprehensive overview of the soybean IQD gene family and provide insights into the evolution of this family. In addition, this work lays a solid foundation for further experiments aimed at determining the biological functions of soybean IQD genes in growth and development.

Editor: Marc Robinson-Rechavi, University of Lausanne, Switzerland

Funding: This work was supported by grants from the National Natural Science Foundation of China (No. 31370561),Specialized research Fund for the Doctoral Program of Higher Education(No.20133418110005), Anhui Provincial Natural Science Foundation (No. 1308085MC36)and Anhui Agricultural University disciplinary construction Foundation(No.XKTS2013001) The funders had no role in study design, data collection and analysis, decision to publish, or preparation of the manuscript.

Competing Interests: The authors have declared that no competing interests exist.

* Email: xiangyanahau@sina.com

⑨ These authors contributed equally to this work.

Introduction

Ca^{2+} is a pivotal cytosolic second messenger involved in many physiological processes such as plant growth [1], plant-pathogen interactions [2], photosynthetic electron transport and photophosphorylation [3], regulation of stomatal aperture [4], hormonal regulation [5] and so on. Plants produce calcium signals by adjusting cytoplasm Ca^{2+} levels at specific times, places and concentrations [6], responding to numerous extracellular stimuli including physical signals (light, temperature, gravity, etc.) and chemical signals (plant hormones, pathogenic bacteria inducing factors, etc.) [7].

The transmission of these intracellular calcium signals relies on the oscillation signal generated by voltage- and ligand-gated Ca^{2+}-permeable channels (influx) and by Ca^{2+}-ATPases and antiporters (efflux) to return to resting Ca^{2+} levels [8,9]. The conduction of calcium signals is also dependent on downstream Ca^{2+} sensors. These Ca^{2+} sensors detect changes in Ca^{2+} levels by binding to Ca^{2+} via domains such as EF hands, which undergo conformational changes [10]. Consequently, calcium signature information is decoded and relayed by these Ca^{2+} sensors [6,11–13].

To date, approximately four major classes of Ca^{2+} sensors have been identified in plants. Most of these sensors contain the Ca^{2+}-binding EF-hand motif, a conserved helix-loop-helix structure that can bind to a single Ca^{2+} ion [7]. The four major classes of Ca^{2+} sensors are as follows: class A: calmodulin (CaM), containing four EF-hand motifs; class B: calcineurin B-like (CBL) proteins, possessing three EF-hand motifs; class C: Ca^{2+}-dependent protein kinases (CDPK), containing four EF-hand motifs and a Ca^{2+}-dependent Ser/Thr protein kinase domain and class D: lacking EF-hand motifs [7,14–19].

Calmodulin (CaM) and calcineurin B-like (CBL) proteins, which lack catalytic activity, are sometimes referred to as "Ca^{2+} sensor relays" [15,19,20]. In contrast, CDPK proteins, which function as catalytic effectors, are referred to as "Ca^{2+} sensor responders" [18]. Among these Ca^{2+}-binding proteins, calmodulin is the most extensively studied Ca^{2+} sensor. Calmodulin is small, acid resistant, heat resistant and highly stable. This multifunctional protein is widespread in eukaryotic cells, highly conserved and has at least 30 multiple target proteins or enzymes [21–23].

To mediate intracellular calcium signaling pathways, Ca^{2+} sensor relays expose their negative hydrophobic surfaces after they undergo conformational changes induced by Ca^{2+} binding. As a result, the affinity between Ca^{2+} sensor relays and their effectors are enhanced, and the biochemical activities of target proteins are modulated by Ca^{2+} sensor relays [6,12,14,19].

In the final phase of the calcium signal transduction process, the target effectors respond to specific extracellular signals by regulating various cellular activities. Calmodulin interacts with numerous target proteins termed calmodulin-binding proteins (CaMBPs), mainly by recognizing and targeting calmodulin-binding domains (CaMBD; basic amphiphilic helices usually composed of 16–35 amino acid residues) in the CaMBPs via its negative hydrophobic pockets [12,22,24].

CaMBD amino acid sequences contain three CaMBD motifs that are grouped into two categories, including a Ca^{2+}-independent motif termed the IQ motif and two Ca^{2+}-dependent motifs referred to as the 1-5-10 motif and the 1-8-14 motif. The number and positions of these motifs in different CaMBPs are variable [25–27]. Due to the diversity of the motif arrangement, there are a variety of diverse CaMBPs with disparate functions, which are implicated in plant development, metabolic regulation, stress reactions, defense reactions, transcriptional regulation and so on [28,29].

Plant-specific IQ67 domain (IQD) protein families were first identified in *Arabidopsis* and rice by Abel et al. (2005) [30]. These proteins have two common features in their IQ67 domains (67 conserved amino acid residues) [31]. One feature is 1–3 copies IQ motifs are separated by 11 and 15 residues and overlapped certain regions with 1–4 copies 1-5-10 motif as well as 1-8-14 motif. The other hallmark is a highly conserved exon-intron boundary that interrupts codons 16 and 17 with a 0 phase intron [31–33]. To date, IQD gene families have been identified in four genomes (*Arabidopsis*, rice, tomato and *Brachypodium distachyon*), including approximately 30 IQD genes per species (33 in *Arabidopsis*, 29 in rice, 34 in tomato and 23 in *Brachypodium distachyon*), and the functions of several members of the IQD family have been reported [30,34,35]. Overexpression of *Arabidopsis IQD1* can mediate the accumulation of glucosinolate in response to insect herbivory [31]. *Arabidopsis IQD22* contributes to the negative feedback regulation of gibberellin (GA) [36]. The tomato *SUN* gene plays a role in elongating tomato fruit by increasing the vertical division of cells and reducing horizontal cell divisions [37–39].

Soybean serves as a major source of vegetable proteins and edible oil and own the ability to fix atmospheric nitrogen via its intimate symbiosis with microorganisms. This crucial leguminous seed crop has high economic and nutritional value [40,41], serving as a main food crop for humans and animals in many parts of the world. Nevertheless, soybean production is limited by many biotic stresses. For example, Asian soybean rust (ASR, caused by the fungus *Phakopsora pachyrhizi*) results in soybean yield losses ranging from 10 to 80% in various countries [42,43].

In this study, we identified and characterized 67 soybean IQD genes. Among these, we subjected 24 IQD III genes to qRT-PCR analysis to investigate their response to MeJA stress. We determined that all 24 soybean IQD genes are stress-responsive. Our findings lay the foundation of further investigations of the functions of these calmodulin target proteins in soybean.

Results

Identification and annotation of IQD genes in soybean

As described in previous studies, IQD proteins, which are plant-specific calcium-dependent calmodulin-binding proteins, contain 67 amino acid residues in their central regions referred to as the IQ67 domain, including three CaM recruitment motifs exhibiting unique repetitive patterns. The Ca^{2+}-independent IQ motif (IQxxxRGxxxR or its more relaxed version [ILV]QxxxRxxxx [R, K]) is present in 1–3 copies and overlaps with 1–4 copies of the Ca^{2+}-dependent 1-5-10 motif ([FILVW]x$_3$ [FILV]x$_4$ [FILVW]) and the 1-8-14 motif ([FILVW]x$_6$ [FAILVW]x$_5$ [FILVW]) by several conserved basic and hydrophobic amino acid residues flanking these motifs [30,34]. These features allow the IQ67 domain to fold into a basic amphiphilic helix structure, which enables these proteins to perform specific roles.

To conducted genome-wide identification of IQD gene families in soybean, we performed *Glycine max* genome BLASTP analysis. Through removing redundant sequences and pattern identificating, a total of 67 IQD genes were identified in the soybean genome, which is twice that of *Arabidopsis* (Table 1 and 2). We named these 67 IQD genes *GmIQD1* to *GmIQD67* according to their physical locations (from top to bottom) on chromosomes 1–20 (Table 1).

The physicochemical parameters of each gene were calculated using ExPASy. Although all of the GmIQD genes encode the conserved IQ67 domains (Figure S1), their sequences are highly diverse with respect to size (141–904 aa) and molecular mass (16.3–99.2 kDa; Table 1). Almost all soybean IQD proteins (97%) have relatively high isoelectric points (pI>7.0 with an average of 10.1), except for *GmIQD9* (pI 5.4) and *GmIQD17* (pI 5.7; Table 1). All soybean IQD proteins were submitted to TargetP and Wolf PSORT to predict their subcellular localizations. Wolf PSORT revealed that fifty-seven soybean IQD proteins are localized to the nucleus, nine to the chloroplast and one to the endoplasmic reticulum. TargetP analysis revealed that fifteen soybean IQD proteins are located in the mitochondria, five in the chloroplast, one in the secretory pathway and forty-six in other compartments (Table 1). The detailed parameters are provided in Table 1.

Phylogenetic and structural analyses of the soybean IQD genes

To infer the similarity and evolutionary ancestry of soybean IQD proteins, we constructed an unrooted phylogenetic tree of the 67 soybean protein sequences. The soybean IQD gene family was categorized into four major subfamilies (subfamily I, II, III and IV; Figure 1a) according to phylogenetic analysis of IQD genes in *Arabidopsis*, rice, tomato and *Brachypodium distachyon* [30,34,35]. Subfamily I was further divided into four subclasses (clade Ia, Ib, Ic and Id), and subfamily II and III were divided into two subclasses (clade IIa and IIb; clade IIIa and IIIb) based on bootstrap values, the existence and positions of introns and the presence of protein motifs flanking the IQ67 domain (Figure 1 and 2). Subfamily I (containing 27 members) is the largest group, followed by subfamily III (24) and subfamily IV (10). Subfamily II has the fewest IQD gene members (6). This distribution pattern is similar to that observed for IQD genes in *Arabidopsis* and rice (Table 2). The phylogenetic tree reveals that 62 of the 67 soybean IQD genes form 31 gene pairs with strong bootstrap values (Figure 1a).

To further examine the structural diversity of the IQD genes in soybean, we deduced the exon/intron organization of individual GmIQD genes (Figure 1b). A comparison of the 67 genomic loci

Table 1. List of 67 IQD genes identified in soybean, their sequence characteristics and subcellular localization.

Name	Gene Identifier	Chr.	Location coordinates (5'-3')	ORF length (bp)	Protein Length (aa.)	pI	Mol.Wt. (kD)	Exons	WoLF PSORT	TargetP
GmIQD1	Glyma01g01030	1	681417–683646	1263	420	10.3	46.6	3	N	M0.65/4
GmIQD2	Glyma01g05100	1	4750065–4755456	1692	563	9.7	61.7	6	N	C0.68/3
GmIQD3	Glyma01g42620	1	53843322–53846963	1191	396	10.4	44.4	4	N	M0.54/5
GmIQD4	Glyma02g00710	2	502944–506146	1254	417	9.6	46.8	3	N	?
GmIQD5	Glyma02g02370	2	1778568–1785636	1692	563	9.8	61.7	6	N	M0.43/5
GmIQD6	Glyma02g15590	2	14083370–14089609	1608	535	10.8	60.1	5	N	?
GmIQD7	Glyma03g33560	3	41092523–41096935	1434	477	10.0	53.3	5	N	?
GmIQD8	Glyma03g40630	3	46330165–46332185	1125	374	10.5	42.4	3	N	?
GmIQD9	Glyma04g02830	4	2030287–2036251	2715	904	5.4	99.2	6	N	?
GmIQD10	Glyma04g05520	4	4187757–4190317	1353	450	10.5	49.9	5	N	?
GmIQD11	Glyma04g23760	4	27192306–27195532	1353	450	9.8	50.8	5	N	?
GmIQD12	Glyma04g34150	4	40144241–40151603	1752	583	9.4	64.6	6	N	?
GmIQD13	Glyma04g41380	4	47220698–47225472	1392	463	9.6	51.7	4	N	?
GmIQD14	Glyma05g01240	5	785189–792757	1761	586	9.7	64.9	6	N	?
GmIQD15	Glyma05g03450	5	2638386–2641896	1338	445	10.0	48.9	4	N	?
GmIQD16	Glyma05g35920	5	39871246–39873985	1128	375	10.0	41.4	4	N	M0.81/3
GmIQD17	Glyma06g02841	6	1950849–1956820	2532	843	5.7	92.8	6	N	?
GmIQD18	Glyma06g05530	6	3957759–3960421	1353	450	10.7	49.8	5	E.R.	?
GmIQD19	Glyma06g13470	6	10606168–10611219	1341	446	9.7	50.1	4	N	?
GmIQD20	Glyma06g20341	6	16752231–16759304	1755	584	9.5	64.9	6	N	?
GmIQD21	Glyma07g01040	7	607467–610485	1302	433	10.0	47.9	5	N	?
GmIQD22	Glyma07g01760	7	1164144–1167157	1191	396	10.2	44.6	3	N	?
GmIQD23	Glyma07g05680	7	4335391–4339373	1641	546	10.3	61.2	5	N	?
GmIQD24	Glyma07g14910	7	14801071–14803234	1398	465	10.0	51.8	3	C	S0.90/1
GmIQD25	Glyma07g32531	7	37416802–37421879	873	290	10.6	32.8	5	N	?
GmIQD26	Glyma07g32860	7	37753882–37759623	1602	533	10.9	59.7	5	N	?
GmIQD27	Glyma08g03710	8	2630927–2633769	1311	436	10.2	48.2	3	N	M0.77/4
GmIQD28	Glyma08g20430	8	15453660–15456579	1266	421	10.4	46.4	5	N	M0.56/5
GmIQD29	Glyma08g21430	8	16271106–16273575	1209	402	10.3	45.2	3	N	?
GmIQD30	Glyma08g40880	8	40742659–40748073	1644	547	9.8	60.6	6	C	C0.75/3
GmIQD31	Glyma09g26630	9	33163730–33169453	1449	482	10.0	53.3	4	C	M0.51/4
GmIQD32	Glyma09g30780	9	37552192–37557238	1305	434	10.1	48.0	6	N	?
GmIQD33	Glyma09g35920	9	41794962–41798738	1407	468	9.9	52.6	5	N	M0.60/4

Table 1. Cont.

Name	Gene Identifier	Chr.	Location coordinates (5'-3')	ORF length (bp)	Protein Length (aa.)	pI	Mol.Wt. (kD)	Exons	WoLF PSORT	TargetP
GmIQD34	Glyma10g00630	10	386683–389158	1272	423	9.5	47.5	3	N	M0.33/5
GmIQD35	Glyma10g05720	10	4477640–4481520	1425	474	10.0	52.8	5	N	?
GmIQD36	Glyma10g35721	10	43974896–43978361	1452	483	10.6	53.0	5	N	?
GmIQD37	Glyma10g38310	10	46118444–46123432	1395	464	10.4	51.0	4	C	M0.54/5
GmIQD38	Glyma10g39030	10	46764292–46767407	1410	469	9.7	52.0	4	N	?
GmIQD39	Glyma11g20880	11	17714458–17717939	1374	457	10.0	51.7	5	N	M0.67/3
GmIQD40	Glyma12g01410	12	842971–846738	1383	460	10.0	51.8	5	N	M0.56/4
GmIQD41	Glyma12g31610	12	35181013–35188577	1269	422	9.9	46.5	6	N	?
GmIQD42	Glyma12g35711	12	38833825–38837834	885	294	9.8	34.2	5	N	?
GmIQD43	Glyma13g20070	13	23539750–23543840	1413	470	10.1	52.3	5	N	C0.34/5
GmIQD44	Glyma13g24070	13	27399608–27404534	774	257	10.5	29.3	4	N	?
GmIQD45	Glyma13g30590	13	33154582–33158861	900	299	10.4	33.6	5	N	?
GmIQD46	Glyma13g34700	13	36237460–36241896	1173	390	9.8	45.5	6	N	?
GmIQD47	Glyma13g38800	13	39521853–39528595	1278	425	9.9	47.1	6	N	?
GmIQD48	Glyma13g42440	13	42441870–42445047	1239	412	10.3	45.8	5	N	?
GmIQD49	Glyma13g43031	13	42796469–42804226	1143	380	10.2	43.4	3	N	?
GmIQD50	Glyma14g11050	14	9335703–9339095	1254	417	10.3	47.3	5	N	?
GmIQD51	Glyma14g25860	14	31470493–31475301	1377	458	10.0	51.3	4	N	?
GmIQD52	Glyma15g02370	15	1595640–1598698	1137	378	10.2	43.3	3	N	?
GmIQD53	Glyma15g02940	15	2051157–2053854	1251	416	10.3	45.9	5	C	?
GmIQD54	Glyma15g08660	15	6125483–6129362	927	308	10.3	34.7	5	N	?
GmIQD55	Glyma16g02240	16	1759053–1762330	1653	550	10.2	61.6	5	N	M0.37/5
GmIQD56	Glyma16g22935	16	26564269–26565120	426	141	11.1	16.3	2	C	M0.82/4
GmIQD57	Glyma16g32161	16	35337880–35343544	1434	477	10.0	52.8	4	C	M0.56/4
GmIQD58	Glyma17g10660	17	8002515–8009332	1767	588	9.5	65.0	6	N	?
GmIQD59	Glyma17g14000	17	10763173–10767584	1344	447	10.0	48.9	4	N	?
GmIQD60	Glyma17g23770	17	23932487–23938307	1386	461	10.4	50.7	5	N	?
GmIQD61	Glyma17g34520	17	38500561–38503843	1242	413	10.4	46.7	5	N	?
GmIQD62	Glyma18g16130	18	16440695–16446996	1644	547	9.7	60.3	6	N	C0.65/4
GmIQD63	Glyma19g36270	19	43610551–43615073	1434	477	10.0	53.3	5	N	?
GmIQD64	Glyma19g43300	19	48995941–48998264	1113	370	10.6	42.2	3	N	?
GmIQD65	Glyma20g28800	20	37708013–37709907	1434	477	9.8	52.7	3	N	C0.67/5
GmIQD66	Glyma20g29550	20	38392614–38397440	1371	456	10.5	50.3	4	C	?
GmIQD67	Glyma20g31810	20	40423269–40426995	1470	489	10.4	53.7	5	C	?

bp, base pair; aa, amino acids; kD, kilo Dalton.
WoLF PSORT predictions: N (nucleus), C (chloroplast), ER (endoplasmic reticulum).
TargetP predictions: C (chloroplast), M (mitochondrion), S (secretory pathway),? (any other location),? values indicate score (0.00–1.00) and reliability class (1–5; best class is 1).

Table 2. Number of IQD genes in the soybean, rice, *Arabidopsis*, tomato and *Brachypodium distachyon* genomes.

Subfamily	Soybean	Arabidopsis	Tomato	Rice	Brachypodium distachyon
I	27	13	15	11	9
II	6	4	6	1	2
III	24	10	10	10	9
IV	10	5	3	3	2
Outgroup		1		4	1
Total number	67	33	34	29	23

with corresponding cDNA sequences revealed that most of the gene models predicted by GSDS are correct, except for one pair of genes (*GmIQD9/-17*). Both *GmIQD9* and *GmIQD17* encode six exons, but GSDS predicted that these genes contain only five exons. This unconformity is caused by the missing annotation of the fifth intron by GSDS. The schematic structures reveal that the coding sequence of each IQD gene contains 2–6 translated exons (Figure 1b), which is similar to that reported in *Arabidopsis*, rice and *Brachypodium distachyon* [30,34]. More than three–fifths of the soybean IQD gene family (41 members) contain five or six protein-coding exons, and one gene (*GmIQD56*, encoding the smallest protein, comprising 141 aa) contains two exons (Figure 1b). Most closely related soybean IQD members in the same subfamilies share similar intron numbers and exon lengths. Soybean IQD genes in subfamily II and IV possess five and six exons, respectively. Most members in subfamily III contain five exons, except for *GmIQD44* (four exons) and *GmIQD32*, *GmIQD41*, *GmIQD46* and *GmIQD47* (six exons). Subfamily I genes harbor 2–5 exons. All introns of most IQD genes are in phase-0 (interrupting two triplet codons exactly); a phase-1 intron (separating the first and second nucleotides of a codon) was found in 15 remaining IQD genes, and no phase-2 intron (splitting the second and third nucleotides of a codon) was found (Figure 1b).

The exon/intron organization of 31 paralogous pairs that clustered together at the terminal branch of the phylogenetic tree was further examined to obtain traceable intron gain/loss information. Although twenty-seven paralogous pairs exhibited conserved exon/intron structures, four pairs (*GmIQD16/-27*, *GmIQD38/-65*, *GmIQD25/-44* and *GmIQD42/-46*) showed certain variations (Figure 1b). These differences may have been derived from single intron loss or gain events during the long evolutionary period. Based on analysis of the exon/intron organization of IQD genes from soybean, *Arabidopsis* [30], rice [30], and *Brachypodium distachyon* [34], we inferred that both *GmIQD16* and *GmIQD38* gained the third intron; *GmIQD46* gained the first intron while *GmIQD44* lost the first intron. The second or third exons in the central regions of most members encode amino acids 17–67 of the IQ67 domain, except for *GmIQD46* (the fourth exon) and *GmIQD56* (the C-terminal exon), with a conserved phase-0 intron separating codons 16 and 17 (Figure 1b and Figure S1).

A total of 67 IQD genes from soybean were subjected to analysis with MEME to reveal conserved domains or motifs shared among related proteins. We identified 10 conserved motifs (Figure 2 and Table S1). Each of the putative motifs was annotated by searching Pfam and SMART. Motif 1 was found to encode the IQ domain. Motif 2 and motif 7 were found to encode proteins of unknown function (DUF4005) and (DUF3982). While the other subfamily-specific motifs have not functional annotation. As expected, most of the closely related members had common motif compositions,

suggesting functional similarities among IQD proteins within the same subfamily (Figure 2). The most common motif is motif 1, found in all sixty-seven soybean IQD genes (Figure 2). Motif 8 is mainly present in subfamily I besides one of *GmIQD14* exists in subfamily IV. Subfamily III members contain motif 1, motif 10, motif 4 and motif 3 in order, except for *GmIQD28* lacking motif 10. Motif 7 is peculiar to subfamily IV. To some extent, these subfamily-specific motifs may contribute to the functional divergence of IQD genes in soybean. The detailed information is shown in table S1. To predict calmodulin-binding sites, we searched the Calmodulin Target Database, which provides various structural and biophysical parameters for the 67 soybean IQD protein sequences. This analysis predicted that all soybean IQD proteins contain multiple IQ motifs and 1–3 strings of high-scoring amino acid residues (Table 3). These IQ motifs and amino acid residues indicate the locations of putative calmodulin interaction sites. Among the 67 IQD protein sequences, the predicted calmodulin binding sites of 50 sequences overlap with the IQ67 domain (Figure 2).

Chromosomal location and gene duplication

The 67 soybean IQD genes were mapped to all 20 soybean chromosomes. The distribution of soybean IQD genes varies depending on the chromosome and appears to be unequal. Both chromosomes 11 and 18 contain only one soybean IQD gene, while chromosomes 13, which possesses seven IQD genes, has the highest number of IQD genes per chromosome. Although high densities of IQD genes were found on some chromosomal regions, for instance, the bottom of chromosome 13, these is no substantial clustering of soybean IQD genes on the map (Figure 3).

We investigated gene duplication events to further understand the expansion mechanism of the soybean IQD family. Except for three genes (*GmIQD11*, *-39* and *-51*) located outside of a duplicated block, 64 genes were mapped onto 48 related duplicated blocks (Figure 3 and Table S2). Among these, twenty-two block pairs retained thirty GmIQD gene pairs, whereas the remaining four duplication blocks harbor *GmIQD3*, *-32*, *-56* and *-60* respectively, but lack IQD sisters in their corresponding synteny blocks (Figure 3 and Table S2). Analysis of GmIQD paralogous pairs showed that one pair (*GmIQD11/-39*) appear to be closely related paralogs, sharing 91.2% identity (Table S3) as well as similar exon–intron organization. However, both of them exist outside of any duplicated blocks. Except for *GmIQD11/-39*, 30 out of 31 gene pairs have remained in conserved positions on segmental duplicated blocks, indicating that these genes were generated by segmental duplication. Furthermore, we analyzed the adjacent genes to determined whether tandem duplication has taken place. A pair of genes separated by three or fewer genes within a 100-kb region on a

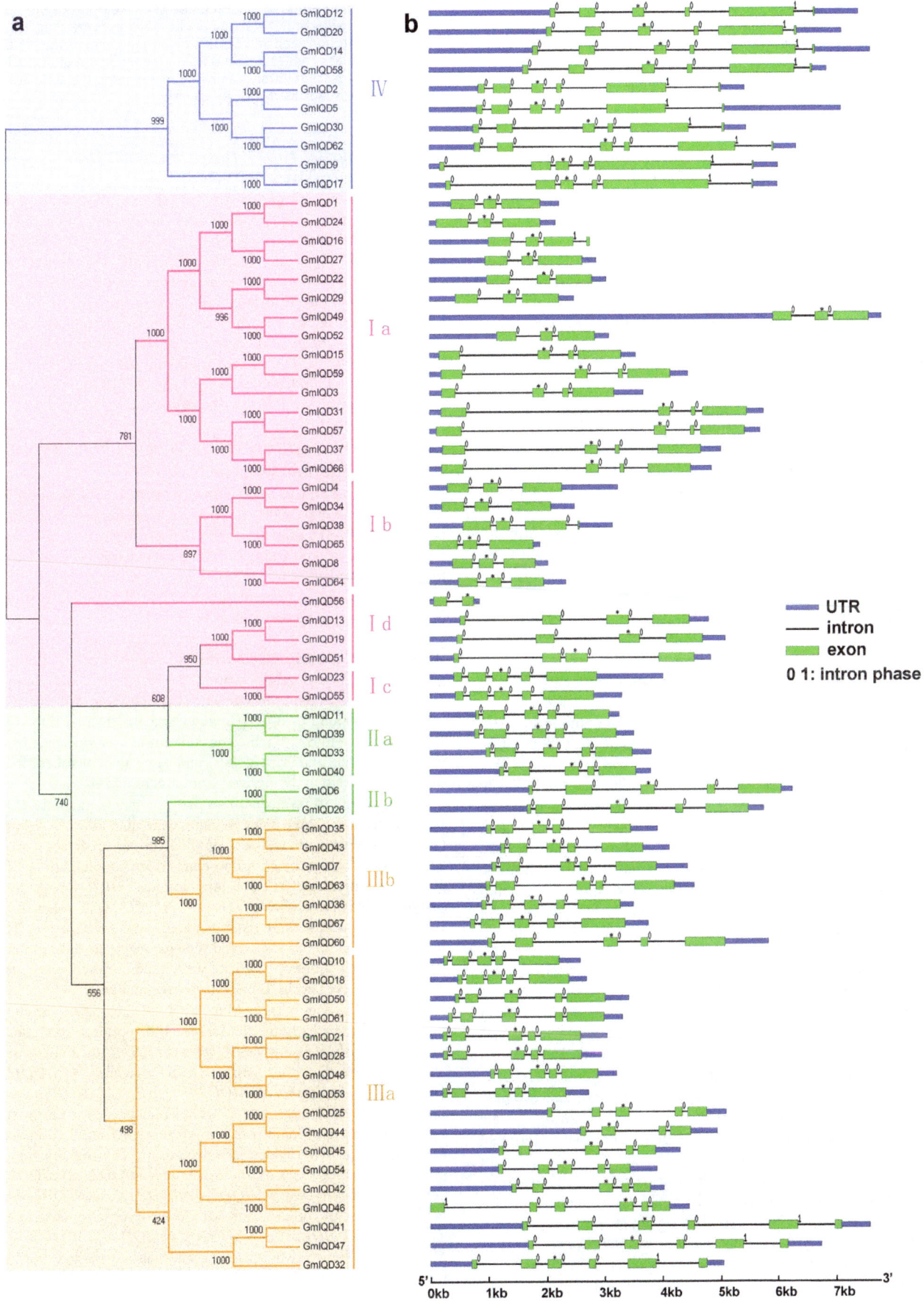

Figure 1. Phylogenetic relationships and exon/intron organization of soybean IQD genes. a: Unrooted tree generated with Clustal X2.0 using the full-length amino acid sequences of the 67 soybean IQD proteins by the Neighbor-Joining (NJ) method with 1,000 bootstrap replicates. Subfamilies and subclasses of IQD genes (I–IV) are highlighted with different colored backgrounds and vertical bars next to the gene names of the tree. **b:** Exon/intron organization of soybean IQD genes. Green boxes represent exons and black lines represent introns. Untranslated regions (UTRs) are indicated by blue boxes. Numbers 0 and 1 represent the splicing phases. The sizes of exons and introns can be estimated using the scale at the bottom. The exon encoding amino acids 17–67 of the IQ67 domain in each soybean gene is indicated with an asterisk.

chromosome may have resulted from tandem duplication. According to this criterion, no pair was found to be generated by tandem duplication. Therefore, segmental duplication appears to have played a crucial role in the expansion of the IQD gene family in soybean (Figure 3 and Table S2).

To explore the selective constraints on duplicated soybean IQD genes, we calculated the ratio of nonsynonymous versus synonymous substitutions (Ka/Ks) for each pair of duplicated IQD genes. In general, a ratio of 1 indicates that both genes are drifting neutrally; a Ka/Ks ratio >1 indicates accelerated evolution with positive selection, while a ratio <1 indicates functional constraint, with negative or purifying selection of the genes[44]. The Ka/Ks ratios from 31 soybean IQD paralogous pairs (Table 4) were less than 0.6. This result suggests that the soybean IQD gene family has evolved mainly under the influence of strong purifying selection pressure, with limited functional divergence occurring after segmental duplication. Duplication of these 31 paralogous pairs was estimated to have occurred between 6.39 to 17.94 Mya (Table 4), according to the divergence rate of 6.1×10^{-9} synonymous mutations per synonymous site per year, as previously proposed for soybean [45,46].

Comparative analysis of the IQD genes in soybean, *Arabidopsis*, rice, tomato and *Brachypodium distachyon*

The development of comparative genomics has enabled the analysis of the same protein families among different species. We constructed an NJ phylogenetic tree using 184 full-length protein sequence to reveal the evolutionary relationships among soybean, *Arabidopsis*, rice, tomato and *Brachypodium distachyon* IQD proteins [34]. In *Arabidopsis*, the IQD gene family is divided into four subfamilies, with *AtIQD33* (containing a C-terminally truncated IQ67 domain) as the outgroup. Therefore, based on their phylogenetic relationships, the combined phylogenetic tree can be divided into five distinct subfamilies (I to V; Figure 4) [30]. In general, IQD I genes comprise the largest subfamily in these plant species, except for *Brachypodium distachyon*, where both IQD I and III comprise the largest subfamilies. By contrast, IQD V genes comprise the smallest IQD subfamily (Figure 4, Table 2).

To illustrate the paralogous and orthologous relationships among IQD family members, the subfamilies were further divided into subgroups using previously defined clades from studies of *Arabidopsis*, rice and tomato IQDs, as shown in Figure 4. IQD subfamily I was divided into four subclasses, i.e., a, b, c and d, and clade b was further divided into two clades, b_1 and b_2. Because one of the IQD Ib clades only contains four IQD genes (*BdIQD11, BdIQD19, OsIQD19* and *OsIQD20*) from monocots, we assigned these four genes to the rice- and *Brachypodium distachyon*-specific Ib_2 clade. The clade containing the genes encoding C-terminal IQ67 domains was defined as Id. Notably, no members of *Brachypodium distachyon* were detected in this clade, suggesting that *Brachypodium distachyon* IQD family lost its members of this subgroup during the long period of evolution. Both IQD II and IQD III subfamilies were divided into two subclasses, a and b, which were designated as described by Zejun et al.(2013) and Abel et al. (2005) [35]. The C-terminally truncated IQ67 domain-

containing genes (*At IQD33, OsIQD28* and *BdIQD14*) comprise IQD V subfamily (Figure 4) [30,34].

The combined phylogenetic tree reveals that most genes in the IQD family, especially the duplicated genes, are contained in paralogous pairs in each species, which supports the occurrence of species-specific IQD gene duplication events. By contrast, we identified 20 pairs of orthologous genes from monocotyledons (rice and *Brachypodium distachyon*) distributed among all of the subfamilies. In addition, two pairs of orthologous genes from dicotyledons (soybean and tomato) stemming from subfamily I (*GmIQD56* and *SlSUN9*) and subfamily III (*GmIQD60* and *SlSUN13*) were found. And *AtIQD20* and *OsIQD26*, members of subfamily I, formed a pair of orthologous genes.

Conserved microsynteny of IQD III genes from soybean, *Arabidopsis* and tomato

The analysis of microsynteny provides valuable information for identifying gene expansion patterns and inferring gene orthology or paralogy. We combined genetic and phylogenetic analyses to perform microsynteny analysis of three dicotyledons, i.e., soybean, tomato and *Arabidopsis*.

To provide a basic framework for the identification of IQD III orthologous or paralogous genes, 44 IQD III genes, including 24 predicted soybean IQDs, 10 *Arabidopsis* IQDs and 10 tomato IQDs, were classified into four distinct clades, clade 1 (thirteen genes), clade 2 (five genes), clade 3 (eleven genes) and clade 4 (fifteen genes), based on phylogenetic analysis (Figure S2). Clade 1, 2 and 3 correspond to IQD IIIa and clade 4 corresponds to IQD IIIb (Figure 4 and S2). Each clade contains at least one gene from soybean, tomato and *Arabidopsis*, indicating that members from different species may be derived from a common ancestor.

Subsequently, to produce a comparative genetic map, 44 IQD III genes from the three dicot genomes were used as anchor genes. Conserved microsynteny was identified through reciprocal pairwise comparisons of the chromosomal regions containing IQD III genes. Microsynteny relationships among *AtIQD3, AtIQD4, AtIQD5, GmIQD32, GmIQD56* or *SlSUN13* with other IQD III members in these three dicot genomes were not observed. The map reveals that the 38 conserved syntenic segments diverged into four groups (Figure 5), which were anastomosed with the classification revealed by phylogenetic tree analysis.

In clade 1 (Figure 5a), Map a shows a higher level of microsynteny, with both the same and opposite directions. *SlSUN11/GmIQD21* and *GmIQD53/GmIQD48* exhibit remarkable opposite-direction microsynteny, while *GmIQD10/GmIQD18, GmIQD21/GmIQD28, AtIQD7/AtIQD8* and *AtIQD8/SlSUN11* are aligned with flanking gene pairs in the same order but discordant transcriptional orientation. In addition, genes in map a were divided into two groups (Figure 5a), i.e., one group with higher levels of microsynteny (*GmIQD21, GmIQD28, GmIQD53, GmIQD48, AtIQD7, AtIQD8* and *AtIQD8* and *SlSUN11*) and the other group with lower levels of microsynteny (*GmIQD61, GmIQD50, GmIQD10, GmIQD18, AtIQD6* and *SlSUN22*). These two groups were also detected in the phylogenetic tree of IQD III genes (Figure S2). In clade 2 (Figure 5b), two pairs from soybean and tomato, *GmIQD41/*

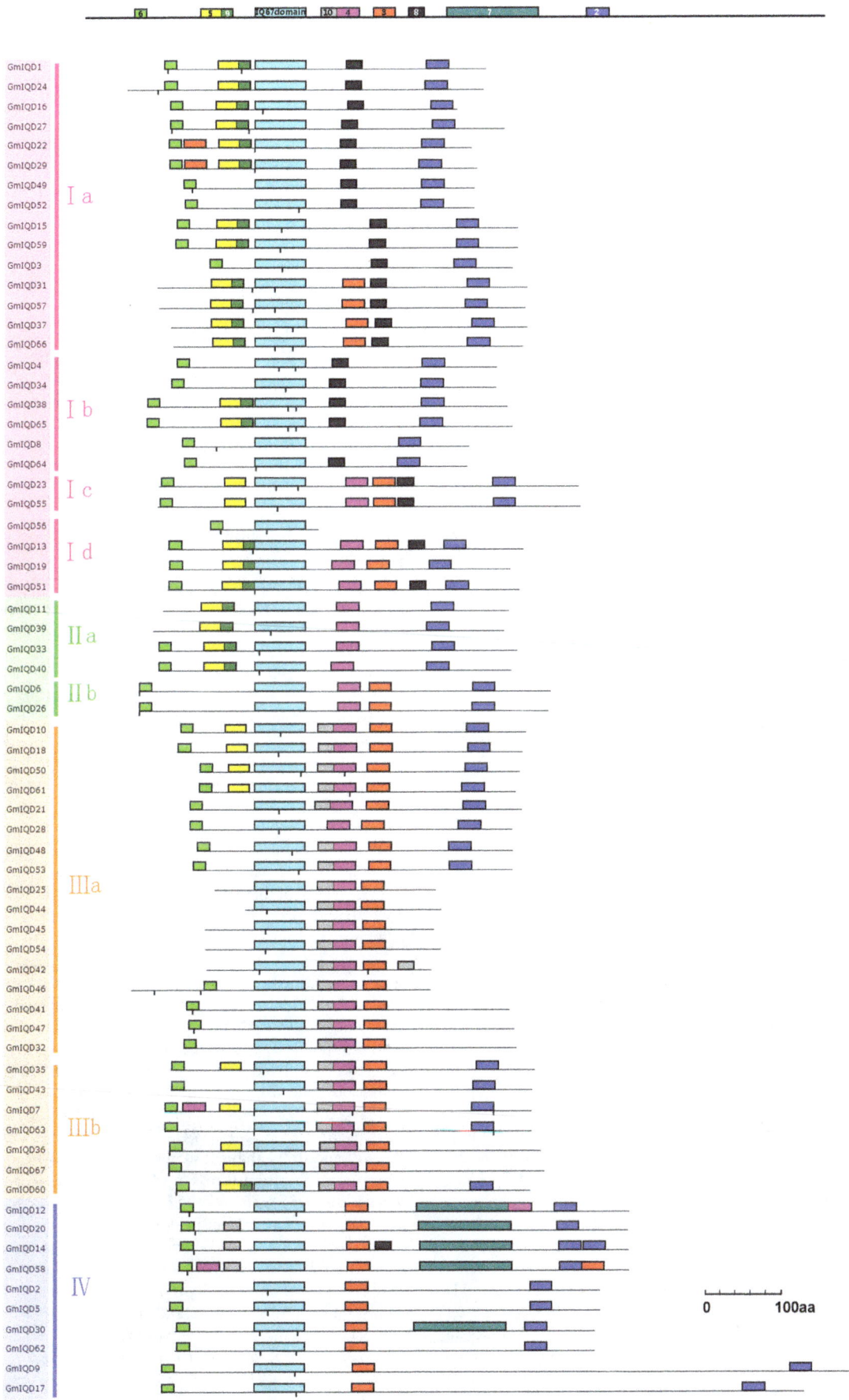

Figure 2. Motif patterns in 67 IQD proteins of soybean. The schematic soybean IQD proteins were aligned relative to the IQ67 domain (motif 1, light blue box). The lengths of the proteins and motifs can be estimated using the scale. Motifs shared by at least four soybean IQD proteins are depicted at the reference bar on top of each alignment. The positions of putative calmodulin-binding sites predicted by the Calmodulin Target Database are indicated by vertical tick marks below each protein model. Subfamilies and subclasses (I–IV) of IQD proteins are highlighted with colored backgrounds and colored vertical bars as in Figure 1a to the right of the gene names.

SlSUN31 and *GmIQD47/SlSUN31*, exhibited microsynteny. However, the predicted duplicated pair *GmIQD41/GmIQD47* had no detectable linkage with each other. High level of microsynteny exists in Clade 3, with most pairs in reverse order (Figure 5c), especially *GmIQD42/GmIQD46*, *GmIQD45/ GmIQD54* and *GmIQD25/GmIQD44*. *GmIQD45/GmIQD44* and *GmIQD44/SlSUN33* were identified as having same-direction microsynteny. In clade 4 (Figure 5d), we also observed a higher level of microsynteny. Except for *AtIQD2/SlSUN29*, *SlSUN30/SlSUN29* and *GmIQD67/GmIQD36*, which are aligned in the opposite direction, most gene pairs in this clade have successive collinearity in order and the same orientation.

Two regions are considered to have originated from a large-scale duplication event when five or more protein-coding gene pairs flanking the anchor point are ligatured with the best non-self match (E-value <1e−10) [47,48]. Applying this standard, except for the pair *GmIQD41/GmIQD47*, all soybean IQD III paralogous gene pairs were generated from a large-scale duplication event, which further supports the results of soybean gene duplication analysis (Figure 5, Table S4).

To estimate the degree of conserved gene content and order, the synteny quality was calculated [47]. The average synteny quality of IQD III genes from the three dicotyledons genomes was 18.41% (Table S5d). Due to the large number of syntenic genes between tomato and soybean, the synteny quality between these genomes is 26.39%; this value is higher than that observed in the Sl/At synteny blocks (16.68%). The lowest synteny quality (12.15%) was found between soybean and *Arabidopsis* (Table S5). Details of this comparative analysis are shown in Table S5.

Expression patterns of soybean IQD genes in various tissues

To gain insight into the expression patterns of soybean IQD genes in various tissues, we searched the RNA-Seq Atlas of *Glycine max*; this atlas provides high-resolution gene expression data from 14 diverse tissues, including aerial tissues (young leaf, flower, one-cm pod, pod-shell 10-DAF and pod-shell 14-DAF), underground tissues (root and nodule) and seed tissue at various stages of development (seed 10-DAF, seed 14-DAF, seed 21-DAF, seed 25-DAF, seed 28-DAF, seed 35-DAF and seed 42-DAF). Because the expression profiles of eight IQD genes (*GmIQD17, -20, -25, -36, -42, -49, -56, -57*) weren't obtained in the database, we only examined the expression patterns of fifty-nine IQD genes (Figure 6 and Table S6).

Most soybean IQD genes exhibit broad expression patterns (Figure 6). Forty-four soybean IQDs are expressed in all of the seven tissues (young leaves, flowers, one-cm pod, pod-shell, roots, nodules and seed). The heat map also revealed that the majority of GmIQDs showed preferential expression. Based on a hierarchical clustering analysis, fifty-nine IQD genes were mainly clustered into four groups (A–D) (Figure 6). Group A showed partial expression in young leaves, group B in roots, group C in nodules and group D in flowers. Eight GmIQDs (*GmIQD8, -13, -19, -29, -36, -52, -64* and *-67*) showed marked high transcript abundance profiles in only a single tissue. Among the fifty-nine soybean IQD genes examined, six showed the highest transcript accumulation in young leaves (*GmIQD13, -18, -19, -29, -50* and *-61*), six in flowers

(*GmIQD11, -15, -38, -52 -65* and *-67*), one in roots (*GmIQD26*) and two in nodules (*GmIQD8* and *-64*; Figure 6). Genes in different subfamilies have their primary abundant transcripts, for instance, GmIQD I in leaves, flowers and nodules, GmIQD II in flowers and roots, GmIQD III in young leaves, flowers and roots and GmIQD IV in roots and young leaves (Figure 6). These subfamily-specific tissue expression patterns may be closely related to gene functions. The expression patterns of the paralogous pairs were also revealed by heat maps; paralogous pairs with high sequence similarity have similar expression patterns. The best examples of this include *GmIQD8/-64* and *GmIQD6/-26*, which are strongly expressed in nodules and root respectively, with little or no expression in other tissues. Expression divergence was also found in paralogous pairs. For example, *GmIQD15* is highly expressed in flowers, while its paralog, *GmIQD59*, is highly expressed in nodules.

Examination of soybean IQD gene expression by qRT-PCR

Since soybean production is limited by stress, it is important to identify the master regulators of stress responses in soybean, as well as their regulatory pathways. According to microsynteny analysis, the high level of microsynteny indicates that IQD III genes existed before the divergence of the three dicotyledon genomes examined (soybean, tomato and *Arabidopsis*). In addition, IQD III genes in the same clade may share common ancestors and play similar roles in these species. *AtIQD1*, which belongs to the IQD III subfamily, plays a major role in the response to biotic stress, as it mediates the accumulation of glucosinolate in response to phytophagous insect attack. Jasmonic acid methyl ester (MeJA), the plant hormones and the signal molecules, widely exists in plants, which triggers expression of plants defense genes by exogenous applications and has similar effects with mechanical damage and insect herbivory [49,50]. Based on these, we subjected 24 members of the soybean IQD III subfamily to real-time quantitative PCR (qRT-PCR) analysis to examine their regulation by MeJA.

The qRT-PCR results show that all 24 genes are MeJA-responsive, but some differences were observed among these genes (Figure 7). Although 23 genes were upregulated by MeJA treatment, *GmIQD21* was obviously downregulated (<0.5 folds) at all time points. Eleven of the twenty-three upregulated GmIQD III genes exhibited minor changes in expression (relative expression scale from 0 to 5 and lower), including *GmIQD10, -18, -21, -25, -42, -47, -53, -61, -7, -36* and *-63*. By contrast, 12 genes (*GmIQD2, -32, -41, -44, -45, -46, -48, -50, -53, -54, -35, -43, -60* and *-67*) exhibited major changes in expression (relative expression scales from 0 to 5 up 0 to 80). The expression of six genes (*GmIQD35, -36, -47, -54, -63* and *-67*) peaked relatively early (at 1 h of treatment); *GmIQD54* and *-67* were strongly upregulated (more than 26-fold and 34-fold, respectively). Eight genes (*GmIQD10, -18, -28, -41, -42, -43, -53* and *-60*) were highly expressed at 4 h; *GmIQD28* and *-60* had the highest expression level more than 12-fold and *GmIQD41* had the highest expression level more than 28-fold. While seven genes (*GmIQD7, -25, -32, -44, -46, -48* and *-61*) exhibited the highest expression levels at 8 h; *GmIQD48* were strongly induced by more than 35-fold. Only one gene (*GmIQD45*) had the highest expression level

Table 3. Predicted calmodulin binding sites in soybean IQD proteins.

Group	Name	Gene Identifier	Predicted calmodulin binding sequence	
Ia	GmIQD1	Glyma01g01030	7-WVKSLFGIRREKEKKLN	100-V**AVVRLTSQGRGRTMFG**
	GmIQD3	Glyma01g42620	94-VRGH**IERKR**TAEW	
	GmIQD15	Glyma05g03450	136-LVRG**HIERKR**TAEWL	
	GmIQD16	Glyma05g35920	120-G**QERLAVVKIQT**FFR	
	GmIQD22	Glyma07g01760	109-FSGS**REKWA**AVKI	
	GmIQD24	Glyma07g14910	39-MGRATRW**VKSLFGIRKE**	
	GmIQD27	Glyma08g03710	2-GRAIRWLKGLFGIRTDRER	102-RDTTFGGAGQERL**AVVKI**
			164-LIRAQATVRSKKSRNEAHR	
	GmIQD29	Glyma08g21430	108-FSGS**REKWA**AVKI	
	GmIQD31	Glyma09g26630	123-RRVAEETTA**AAVKIQSAFR**	153-K**ALVKLQALVRGHIVRKQT**
	GmIQD37	Glyma10g38310	136-ALVKLQALVRGHIVRKQS	158-**RRMQTLVRLQAQARASRA**
	GmIQD49	Glyma13g43031	8-**LKG LLGKKKEKDYCGY**	
	GmIQD52	Glyma15g02370	148-**AQAVARSVRARRSM**	
	GmIQD57	Glyma16g32161	121-RVANETTA**AAVKIQSAFRG**	150-K **ALVKLQALVRGHIVRKQT**
	GmIQD59	Glyma17g14000	137-LVRG**HIERKR**TAEW L	
	GmIQD66	Glyma20g29550	133-LKALVKLQALVRGHIVRKQS	155-**RRMQTLVRLQAQARASRA**
Ib	GmIQD4	Glyma02g00710	133-LQALVRGHLVRKQ**A**RETL	155-AL**VIAQS**RARAQRA
	GmIQD8	Glyma03g40630	46-RR**WSFGKLTGAGH**KF	
	GmIQD34	Glyma10g00630	148-LVRK**QARETLRCIQA**LVIA	
	GmIQD38	Glyma10g39030	181-RKQ**AKATL**RC	193-ALVTAQ
	GmIQD64	Glyma19g43300	92-KDK**NKAATKIQA**SF	
	GmIQD65	Glyma20g28800	182-RKQ**AKATL**RC	194-AL**VTAQ**AR
Ic	GmIQD23	Glyma07g05680	152-LV**KLQALVRGHNVR**KQA	180-RVQARVLDQRIRSSL
	GmIQD55	Glyma16g02240	154-LV**KLQALVRGHN**VR	
Id	GmIQD13	Glyma04g41380	109-YGRQ**SKEERAAILIQ**SYYR	
	GmIQD19	Glyma06g13470	119-IL**IQSYYRGYL**ARRALRALKG	
	GmIQD51	Glyma14g25860	111-RQ**SKEERAATLIQ**SYYRGYLARRALRAL	
	GmIQD56	Glyma16g22935	13-RGRFLRSS	73-GHLARR**AYKALKSLVKLQA**LVR
IIa	GmIQD11	Glyma04g23760	119-K**IQESSAIKIQIAFRGY**L	
	GmIQD33	Glyma09g35920	125-IKESA**AAIKIQTAFRG**Y	
	GmIQD39	Glyma11g20880	132-KIQES**SAIKIQTAYRGY**LA	
	GmIQD40	Glyma12g01410	125-IKESA**AAIKIQTAFRG**Y	
IIb	GmIQD6	Glyma02g15590	1-**MGKKGSWFSAI**	
	GmIQD26	Glyma07g32860	1-**MGKKGSWFSAI**	
IIIa	GmIQD10	Glyma04g05520	131-VRG**RQVRKQAAVTLRCMQ**ALVRVQA	
	GmIQD18	Glyma06g05530	136-VRG**RQVRKQAAVTLRCMQ**ALVRVQAR	
	GmIQD21	Glyma07g01040	117-AIFR**GWQVRKQAAVTLR**CMQ	
	GmIQD25	Glyma07g32531	67-AYK**ARKYL**HRLR	
	GmIQD28	Glyma08g20430	117-AIFR**GWQVRKQAAVTLR**CMQ	
	GmIQD32	Glyma09g30780	205-RQEAAA**KRGRAMAY**AL	
	GmIQD41	Glyma12g31610	3-V**SGKWIKALVGLKKSEKP**G	90-R **EELAAIRIQTAFRGFLA**
			207-AKRERAMAYALSHQWQAG	
	GmIQD42	Glyma12g35711	68-AATRIQNAFRSFMARRTL	210-**LGKESWGWSWTERWVAAR**
	GmIQD44	Glyma13g24070	27-AYK**ARKYLH**RLRG	
	GmIQD45	Glyma13g30590	78-RAY**KARKAL**RRMKGFTK**LKIL**TEG	
	GmIQD46	Glyma13g34700	25-EIKHLIQRGWVV	90-LKRN**KRMGAK**KWF
	GmIQD47	Glyma13g38800	3-**VSGKWIKALVGLKKSEKP**	204-**AKRERAMAYALSHQWQAG**
	GmIQD48	Glyma13g42440	123-LRCMQA**LVRVQARVRA**R	
	GmIQD50	Glyma14g11050	126-VRVQARVRAR	187-GAF**K RERAIAYS**LA
	GmIQD53	Glyma15g02940	138-LRCMQA**LVRVQARVRA**R	

Table 3. Cont.

Group	Name	Gene Identifier	Predicted calmodulin binding sequence	
	GmIQD54	Glyma15g08660	78-RAY**KARKAL**RRMKGFTK**LKIL**TEG	
	GmIQD61	Glyma17g34520	197-EGA**F KR**ERAIAYSL	
IIIb	GmIQD7	Glyma03g33560	116-P**KDEVAAIKIQTAFRGY**L	227-LSKYEATTRRERALAYA
			427-NG**KAEKGSFGSAKKRL**SF	
	GmIQD35	Glyma10g05720	111-EEM**AAIRIQKAFRGY**LA	218-KLLSKYEASMRRERAMAYS
	GmIQD36	Glyma10g35721	1-**MGRKGGWFSAV**	292-**HASAKSVASQTMSV**
	GmIQD43	Glyma13g20070	126-LARR**ELRAL**RGLV	
	GmIQD60	Glyma17g23770	1-**MGKKGSWFSAV**	
	GmIQD63	Glyma19g36270	116-P**KDEVAAIKIQTAFRGY**L	227-LSKYEATMRRERALAYA
			427-NA**KAEKGSFGSAKKRL**SF	
	GmIQD67	Glyma20g31810	1-**MGRKGGWFSAV**	293-**HASAKSVASQTMSV**
IV	GmIQD2	Glyma01g05100	130-LARQ**TFK**KLEGV	175-RGYNVRRS
	GmIQD5	Glyma02g02370	130-LARR**TLQ**KLKGV	
	GmIQD9	Glyma04g02830	173-QAI**IKMQILVRAR**RAR	
	GmIQD12	Glyma04g34150	13-LFGKKSS**K SNI**SK	153-KLQALVRGGRIRQS
	GmIQD14	Glyma05g01240	19-SKS**NISKGRE**KLV	
	GmIQD17	Glyma06g02841	175-II**KMQILVRARRA**WQ	
	GmIQD20	Glyma06g20341	20-KS**NIS**KGRE	
	GmIQD30	Glyma08g40880	113-QAA**IRGYQ**ARG	163-LARGYKVRHS
	GmIQD58	Glyma17g10660	12-VL**FGKKSSKSNI**SK	
	GmIQD62	Glyma18g16130	115-IRGYQARGTFKTL	161-QA**LARGYKVRHS**DV

Predicted calmodulin binding sites obtained from the Calmodulin Target Database are shown for strings of amino acid residues with a score of at least 7. Residues with the highest score (9) are highlighted in bold. Numbers before strings indicate the location of the first amino acid residues of the strings in soybean IQD protein sequences.

at 12 h, with a relative expression level approaching 70-fold. We also compared the expression profiles of paralogous pairs. Most paralogs in a pair had different expression profiles. For example, the expression of *GmIQD28* peaked at 4 h while its sister gene, *GmIQD21*, was downregulated at all time points, suggesting that these genes may play diverse roles in the response to MeJA stress.

Discussion

Structural characteristics of IQD proteins

The plant-specific IQD gene family has previously been comprehensively analyzed in *Arabidopsis*, rice, tomato and *Brachypodium distachyon*, this gene family has not been previously identified and annotated in soybean. We identified and characterized 67 IQ67 domain-encoding genes in soybean using genome-wide analysis. The IQD gene family in soybean is by far the largest one compared to that in other plant species (33 in *Arabidopsis*, 29 in rice, 34 in tomato and 23 in *Brachypodium distachyon*). At ~1,150 Mbp, with ~46,400 predicted coding genes, soybean possesses 9.2-fold larger genome size and 1.75-fold higher gene count than *Arabidopsis*, which has a genome of 125 Mbp and ~26,500 coding genes [51]. Given the obvious difference in genome size and estimated gene count between soybean and *Arabidopsis*, the IQD genes in soybean seems to be highly expanded. The presence of twice as many of these genes in soybean versus *Arabidopsis* may be mainly due to the recent polyploidy event and segmental duplication events in soybean's evolutionary history. It can be speculated that the presence of more IQD genes in soybean genome may reflect the great needs

for these genes coding for calcium signal regulatory components with functions in plant development, defense response or others.

The common feature of IQ67 domain proteins is the arrangement of three IQ motifs separated by 11 and 15 intervening amino acid residues (Figure S1). To date, at least five protein families containing IQ motifs, which play a role in the calcium signaling pathway, have been identified in *Arabidopsis*. These protein families include the cyclic nucleotide gated channels family (CNGC), the IQ-Motif family (IQM), the CaM-binding transcriptional activator family (CAMTA), the myosin family and the IQD family, which contain one, one, two, five and up to three IQ motifs, respectively [52–54]. The unique spacing of IQ motifs and exon/intron organization of each family suggest that these IQD protein families represent separate classes of putative calmodulin targets. The calmodulin-interacting peptides in AtIQD20 and CNGC proteins, which were experimentally verified, were previously predicted using the algorithm provided by the Calmodulin Target Database successfully [30]. In the current study, using the Calmodulin Target Database, we detected calmodulin-binding sites in all soybean IQD proteins, which strongly suggests that all IQD proteins have the potential to interact with calmodulin (Figure 2 and Table 3). Three aspects of IQD proteins appear to underlie the mechanism of interaction between IQD proteins and calmodulin: the number and specific composition of the IQ, 1-5-10 and 1-8-14 motifs, the predicted calmodulin binding site and the overall tertiary structure of the IQD protein.

Of the 31 soybean IQD paralogs examined, 27 exhibit highly conserved exon-intron structures, which is consistent with the high

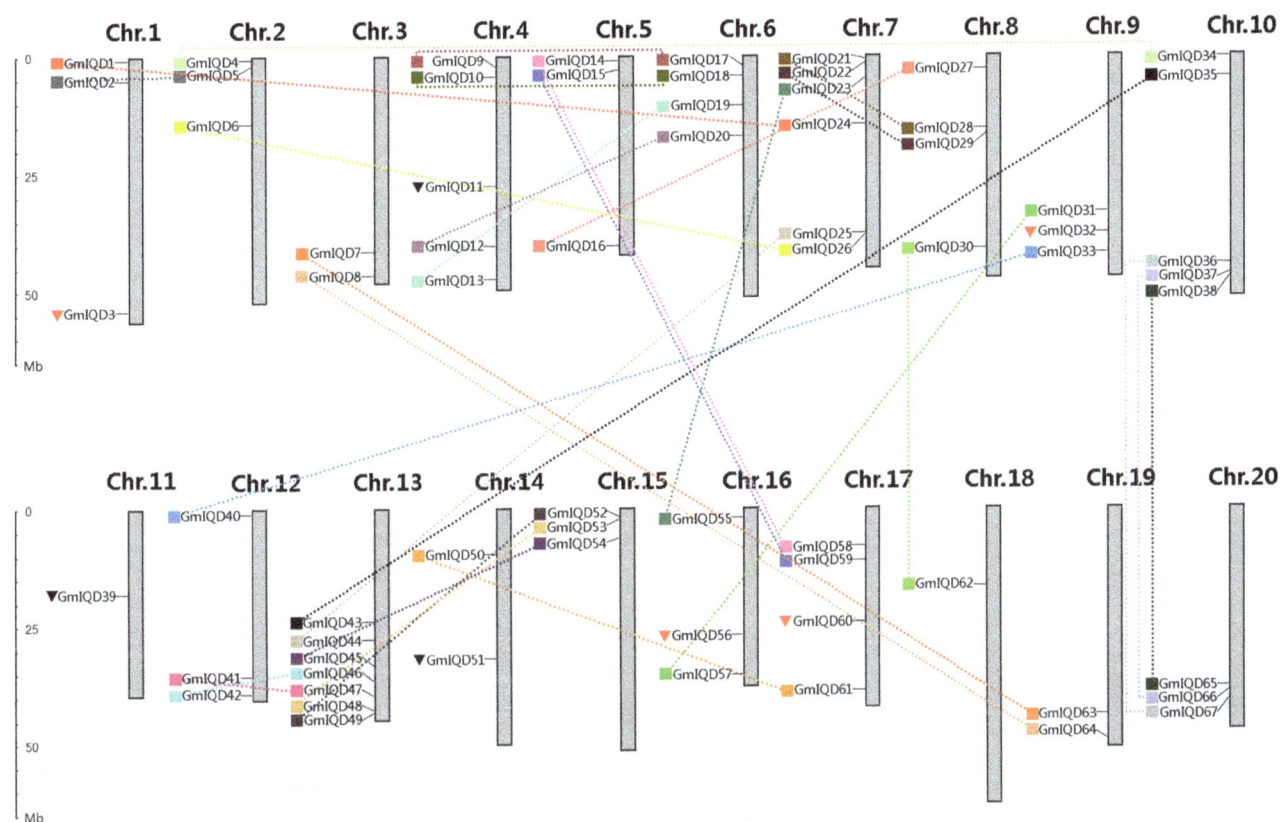

Figure 3. Chromosomal distribution and segmental duplication events for soybean IQD genes. The 67 IQD genes were mapped to the 20 soybean chromosomes. The duplicated paralogous pairs of IQD genes in the segmental duplicated blocks are indicated with small boxes of the same color and connected by dashed lines of the same color. Red triangles represent soybean IQD genes located on duplicated segments with the corresponding members lost. Black circles indicate soybean IQD genes not located in duplicated regions. Scale represents the length of the chromosome.

degree of position and phase conservation broadly found across angiosperms [41]. In addition, the sizes of related introns between paralogs are also highly conserved, indicating that few insertions and deletions have accumulated within introns over the past 13 million years [41]. Most introns in GmIQD genes are in phase-0. This strong bias for phase-0 introns in soybean IQD genes is also found in IQD genes of *Arabidopsis*, rice and *Brachypodium distachyon*. The strong bias for one intron phase class, along with the variation in the number of exons (two to six) and the sizes of encoded proteins, suggests that exon shuffling has played a prominent role during the evolution and diversification of IQD genes [30].

65 of 67 soybean IQD proteins have relatively high isoelectric points with an average of 10.1. It is very similar to *Arabidopsis* (10.3), rice (10.4) and *Brachypodium distachyon* (10.3) [30,34]. The extensive presence of the basic isoelectric point and high frequency of serine residues (*Arabidopsis*: ~11%, rice: ~11%, *Brachypodium distachyon*: ~11.5% and soybean: ~12%; Table S7) in IQDs suggest that the basic nature of IQDs is crucial to their biochemical functions [30,34]. The high isoelectric points are evocative of RNA-binding proteins although IQD proteins don't comprise currently known RNA-binding motifs. Fifty-seven soybean IQD proteins are localized to the nucleus, because of their high content of basic residues revealed by Wolf PSORT. TargetP analysis revealed that fifteen soybean IQD proteins are located in the mitochondria by identifying the presence of mitochondrial targeting peptide (mTP). The contradicting subcel-

lular localization predictions is due to the different algorithm used by Wolf PSORT and TargetP. Most soybean IQD protein members are likely to function in the nucleus, as nucleus specific Ca^{2+}-signatures are reported to generate in plant cells [55–57] and calmodulin and related Ca^{2+}sensor proteins may play a regulatory role in nuclear processes such as transcription [58,59]. Observably, *Arabidopsis IQD1* was reveraled to target to microtubules as well as the cell nucleus and nucleolus [32]. In vitro binding to single-stranded nucleic acids suggests *AtIQD1* and other IQD family members may control and fine-tune gene expression and protein sorting by facilitating cellular RNA localization [32].

Phylogenetic analysis and evolution of IQD family genes

IQD proteins are an ancient family of CaM/CML binding proteins that originated during the early evolution of land plants, as IQD genes are present in *Physcomitrella patens*. ESTs corresponding to IQD proteins for angiosperm species (*Arabidopsis*, rice, etc.) and at least nine homologous sequences in the gymnosperm pine (*Pinus* ssp.) corresponding to IQD proteins were identified suggesting that the IQD gene family originated not later than the split of gymnosperms and angiosperms about 300 Myr ago [30]. We performed a genome-wide comparison of plant IQD members from monocots (rice and *Brachypodium distachyon*) and eudicots (soybean, *Arabidopsis* and tomato) to explore how the IQD gene family has evolved. The plant IQD members from monocots (rice and *Brachypodium distachyon*) and eudicots (soybean, Arabidopsis and tomato) appear to be more closely

Table 4. Divergence between paralogous IQD gene pairs in soybean.

Group	No.	Paralogous pairs	Ka	Ks	Ka/Ks	Duplication date (MY)	Duplicate type
Ia	1	GmIQD1-GmIQD24	0.0474	0.0802	0.5914	6.57	S
	2	GmIQD16-GmIQD27	0.045	0.195	0.228	15.99	S
	3	GmIQD22-GmIQD29	0.029	0.108	0.267	8.83	S
	4	GmIQD49-GmIQD52	0.041	0.106	0.388	8.66	S
	5	GmIQD15-GmIQD59	0.041	0.158	0.258	12.96	S
	6	GmIQD31-GmIQD57	0.029	0.147	0.194	12.03	S
	7	GmIQD37-GmIQD66	0.030	0.116	0.260	9.48	S
Ib	8	GmIQD4-GmIQD34	0.044	0.124	0.356	10.18	S
	9	GmIQD38-GmIQD65	0.054	0.111	0.485	9.07	S
	10	GmIQD8-GmIQD64	0.039	0.134	0.293	11.00	S
Ic	11	GmIQD23-GmIQD55	0.017	0.086	0.193	7.08	S
Id	12	GmIQD13-GmIQD19	0.057	0.164	0.346	13.47	S
IIa	13	GmIQD11-GmIQD39	0.043	0.093	0.460	7.60	O
	14	GmIQD33-GmIQD40	0.022	0.094	0.238	7.70	S
IIb	15	GmIQD6-GmIQD26	0.022	0.091	0.245	7.49	S
IIIa	16	GmIQD10-GmIQD18	0.030	0.152	0.197	12.43	S
	17	GmIQD50-GmIQD61	0.031	0.162	0.189	13.30	S
	18	GmIQD21-GmIQD28	0.041	0.125	0.325	10.22	S
	19	GmIQD48-GmIQD53	0.029	0.111	0.262	9.08	S
	20	GmIQD25-GmIQD44	0.052	0.157	0.335	12.84	S
	21	GmIQD45-GmIQD54	0.034	0.105	0.325	8.57	S
	22	GmIQD42-GmIQD46	0.058	0.219	0.263	17.94	S
	23	GmIQD41-GmIQD47	0.037	0.095	0.387	7.78	S
IIIb	24	GmIQD35-GmIQD43	0.033	0.114	0.293	9.34	S
	25	GmIQD7-GmIQD63	0.024	0.093	0.253	7.61	S
	26	GmIQD36-GmIQD67	0.045	0.127	0.349	10.43	S
IV	27	GmIQD12-GmIQD20	0.054	0.134	0.400	10.99	S
	28	GmIQD14-GmIQD58	0.035	0.118	0.297	9.66	S
	29	GmIQD2-GmIQD5	0.059	0.109	0.537	8.93	S
	30	GmIQD30-GmIQD62	0.067	0.151	0.443	12.35	S
	31	GmIQD9-GmIQD17	0.028	0.078	0.363	6.39	S

S: segmental duplication, O: other events.

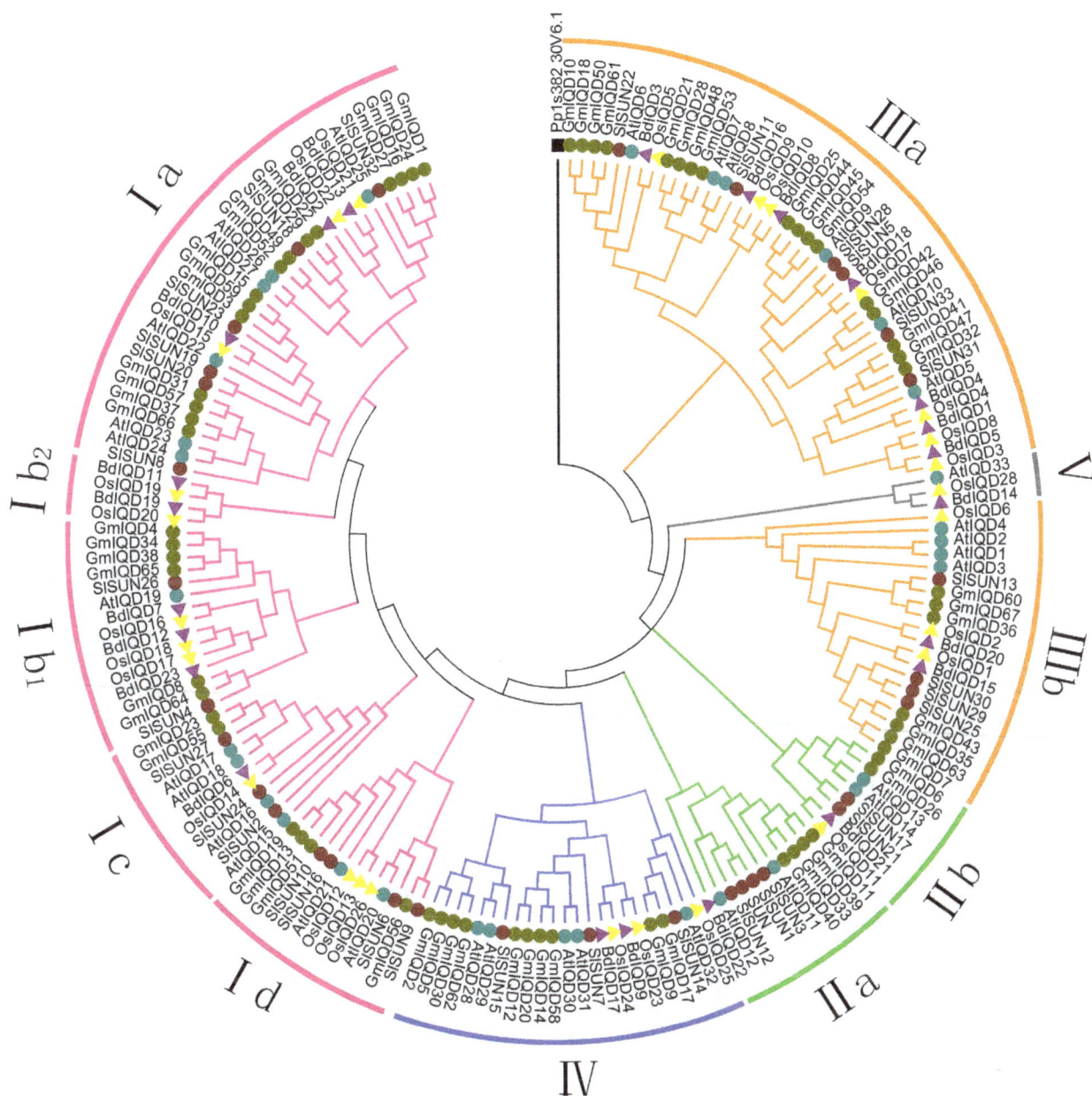

Figure 4. Phylogenetic tree of full-length IQD proteins from soybean, *Arabidopsis*, rice, tomato and *Brachypodium distachyon*. The tree was generated with Clustal X2.0 using the NJ method. Dicotyledons (soybean, tomato and *Arabidopsis*) IQD proteins are marked with colored dots. Monocotyledons (rice and *Brachypodium distachyon*) are marked with colored triangles. A moss IQD protein (Pp1s38230v6), used as the outgroup, is marked with a black box. Each IQD subfamily is indicated by a specific color.

related to each other than to IQD genes of the same species in a different subfamily. This alternating distribution of monocots and eudicots in all subfamilies suggests that an ancestral set of IQD genes have existed before the dicot–monocot split (Figure 4, Table 4). The presence of five distinct subfamilies of IQD genes and the presence of both monocots and eudicots containing members in all five subfamilies indicate IQD genes have diversified before the monocot–eudicot split (Figure 4). These subfamilies include 23 pairs of orthologous genes, suggesting that orthologous genes may have originated from a common ancestor (Figure 4). About half of the orthologous genes (10 pairs; *BdIQD1/OsIQD8*:N, *BdIQD5/OsIQD3*:N, *BdIQD8/OsIQD10*:C, *BdIQD9/OsIQD23*:C, *BdIQD11/OsIQD19*:N, *BdIQD14/*

OsIQD28:N, *BdIQD17/OsIQD24*:N *BdIQD18/OsIQD7*:N, *BdIQD20/OsIQD2*:N, *SlSUN13/GmIQD60*:N) have the same predicted subcellular localization suggesting that the encoded proteins may play similar roles in both species [30,34]. A total of 87% (20 pairs) of orthologous gene pairs from rice and *Brachypodium distachyon* are distributed in all subfamilies. However, only two pairs of orthologous genes from dicotyledons (soybean and tomato) are from subfamily I and III. This difference may be due to the fact that both rice and *Brachypodium distachyon* are in the grass family and are therefore more closely related than *Arabidopsis*, soybean and tomato, which belong to Cruciferae, Solanaceae and Leguminosae, respectively. The number of soybean genes in each subfamily is greater than that of the other four species examined

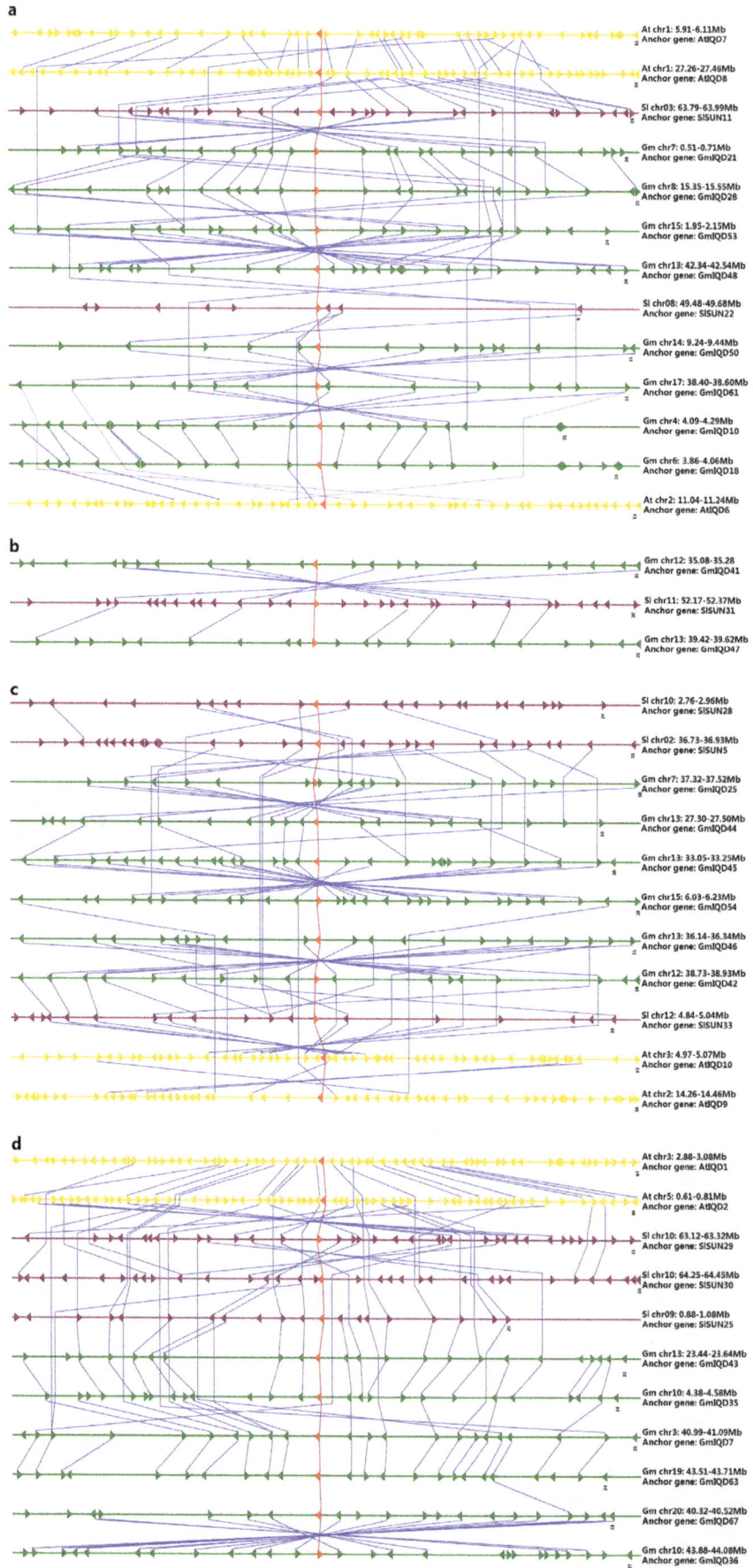

Figure 5. Microsynteny map of IQD III genes in soybean, *Arabidopsis* **and tomato. a, b, c** and **d:** four groups of syntenic segments. The relative positions of all flanking protein-coding genes were defined by the anchored IQD III genes (red). Colored horizontal lines represent chromosome segments of soybean (green), *Arabidopsis* (yellow) and tomato (purple). Triangles of the same color represent individual genes and their transcriptional orientations. The total number of genes on each segment is indicated to the right below the segment. Colored lines connect the conserved gene pairs among the segments (anchor gene pairs, red; others, blue).

suggesting that IQD counterparts in soybean may have undergone gene expansion.

The duplication of individual genes, chromosomal segments or entire genomes has been a major force in the evolution of plant genome structure and content during the process of genome evolution [60,61]. The soybean genome has undergone at least two round of duplication, resulting in the presence of significant features of remnants of a glycine-specific genome duplication that occurred ~13 Mya and fainter remnants of older polyploidies prior to the divergence of the papilionoids (58–60 Mya) that occurred ~58 Mya [41,51]. Thus, 75% of soybean genes are present in multiple copies [41]. Among 67 soybean IQD genes, *GmIQD3*, *-32*, *-56* and *-60* were found as single copies on duplication blocks. These results suggest that segmental duplication has occurred as a continuous process and dynamic changes may have occurred in a chromosomal segment that contained two ancestral IQD genes, leading to corresponding sister gene loss [62]. One paralogs (*GmIQD11/-39*) shares 91.2% identity and similar exon/intron organization, but exists outside of any duplicated blocks. this pair might have been produced by retrotransposition. A high proportion (approximately 96%) of soybean IQD genes reside preferentially in duplicated segments, suggesting that segmental duplications have played a prominent role in the expansion of the soybean IQD gene family. The duplicated IQD genes in soybean have been preferentially retained at the high rate of 92.5% (62/67), which is distinctly higher than the retention rate (67.3%) of duplicated paralogs in the 1.1-gigabase sequence of the soybean (cv. Williams 82) genome, in which 31,264 genes exist as 15,632 paralog pairs (out of the 46,430 predicted high-confidence genes that were duplicated and retained after the 13-Mya tetraploidy event) [63]. The higher retention rate corroborates previous findings that genes involved in signal transduction are preferentially retained following duplications[64]. Our calculation of the duplication dates of the 31 paralogous pairs revealed that all of the segmental duplication events in the soybean IQD family occurred during the recent whole genome duplication event.

During evolution, eukaryotic genomes have retained genes on corresponding chromosomes (synteny) and in corresponding orders (collinearity) to various degrees. Synteny broadly refers to parallels in gene arrangement in dissimilar genomes. Collinearity, a specific form of synteny, requires genes to occur in largely corresponding orders along the chromosomes of respective genomes. According to the microsynteny analysis, microsynteny relationships among *AtIQD3*, *AtIQD4* or *SlSUN13* with other IQD III members in these three dicot genomes were not observed indicating that either these genes are ancient genes without detectable linkage to other IQD genes or that they were formed through complete transposition and loss of their primogenitors. In addition, three different duplicated chromosomal segments (harboring *AtIQD5*, *GmIQD32* and *GmIQD6*) that lost their sister IQD genes lack detectable microsynteny relationships to all other IQD III genes in the soybean and *Arabidopsis* genome, respectively. In the four IQD III gene clades, genes from soybean, tomato and *Arabidopsis* exhibit high levels of microsynteny, which indicates the IQD III genes existed before the divergence of the three dicotyledons genomes (soybean, tomato and *Arabidopsis*).

Microsynteny was detected in most pairs, and alignment in clade 1–3 was discordant, suggesting that these genes may all be present in genome regions that were inverted, expanded or contracted after the divergence. Notably, most gene pairs in clade 4 have successive collinearity in order and the same orientation, which indicates high conservation among these IQD III gene-residing regions, with little rearrangement. The low (18.41%) synteny quality of IQD III genes from the three dicotyledon genomes (soybean, tomato and *Arabidopsis*) may have been due to the fact that these plants are not closely related; moreover, the gene density differs between *Arabidopsis* and the two other species. Significantly, the number of synteny blocks (31) within the soybean genome is much more than the number (3 or 4) of synteny blocks between tomato or *Arabidopsis* genomes, which suggests that soybean IQD III genes may have undergone large-scale duplication events and less rearrangement was followed (Figure 5 and Table S5b). The gene expansion pattern analysis of soybean paralogs indicates that most pairs were generated from a large-scale duplication, which supports the results of soybean gene duplication analysis, with the exception of *GmIQD41/-47*.

Organ- or tissue-specific expression of IQD genes and expression of GmIQD III genes under MeJA stress treatment

Organ- or tissue-specific expression patterns have been observed for quite a few members of the IQD family. However, the functions of soybean IQD genes remain unclear. We therefore performed a thorough analysis of the RNA-Seq Atlas to investigate organ- or tissue-specific expression of IQD genes and qRT-PCR to examine the expression of GmIQD III members under MeJA stress treatment.

The tissue expression data deficiency of eight soybean IQD genes potentially indicated that these are pseudogenes or express only at specific developmental stages or under special conditions. 65.7% soybean IQD genes constitutively express in all of the seven tissues suggesting that GmIQDs may play roles at multiple developmental stages. Eight GmIQD proteins peak in only one tissue indicating that these tissue-specific calmodulin target proteins may be limited to discrete cells or organs to regulate various cellular activities.

Except for group C comprised of genes from GmIQD I, group A, B and D comprise genes from four subfamilies indicating these soybean IQD genes exhibit similar transcript abundance profiles but are relatively phylogenetically distinct. The analysis indicated that only some members within the same phylogenetic subgroup share a similar expression profile in soybean organs/tissues during development, excluding *GmIQD6* and *GmIQD26* belong to GmIQD IIb. For instance, *GmIQD4*, *-34*, *-38* and *-65* belong to GmIQD Ib clustered in group with high expression in flowers suggesting their potential roles in flower formation. While the other two GmIQD Ib members (*GmIQD8* and *-64*) were detected in nodules indicating they may involve in fixing atmospheric nitrogen.

Members possessing similar sequences are clustered in the same subfamilies, which may have similar expression patterns or functions. In IQD subgroup Ia, *Arabidopsis IQD22* is involved in the negative feedback regulation of GA-responsive DELLA

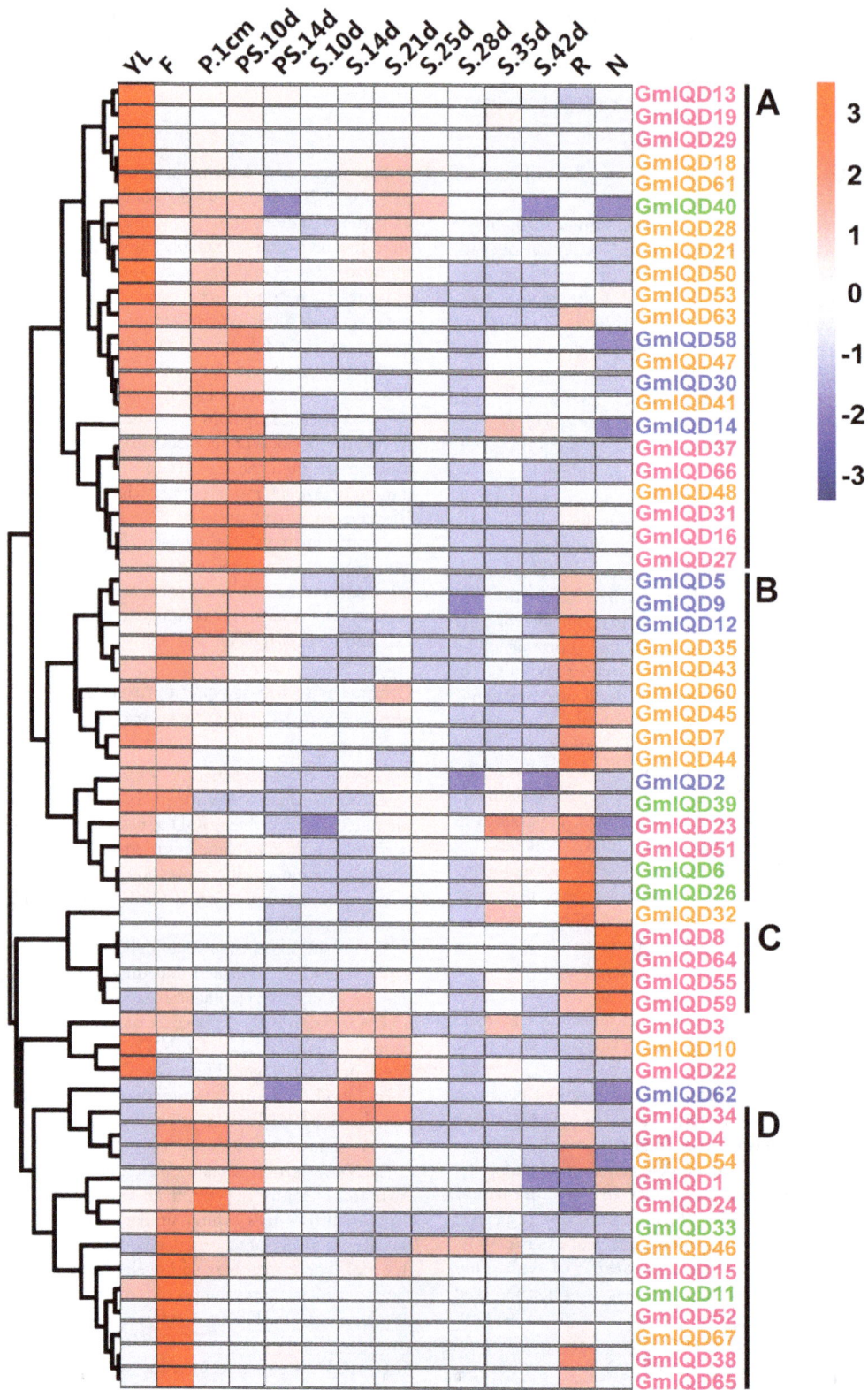

Figure 6. Hierarchical clustering of the expression profiles of soybean IQD genes in 14 tissues. RNA-seq relative expression data from 14 tissues were used to reconstruct the expression patterns of soybean genes. The raw data was normalized and retrieved from the online database http://soybase.org/soyseq/. The normal relative expression levels of 67 IQD genes are shown in Table S6. YL, young leaf; F, flower; P.1cm, one cm pod; PS.10d, pod shell 10 DAF; PS.14d, pod shell 14 DAF; S.10d, seed 10 DAF; S.14d, seed 14 DAF; S.21d, seed 21 DAF; S.25d, seed 25 DAF; S.28d, seed 28 DAF; S.35d, seed 35 DAF; S.42d, seed 42 DAF; R, root; N, nodule. Gene names in different subfamilies are highlighted with various colors. Genes clustered into four groups (A–D) are indicated by the black vertical bars.

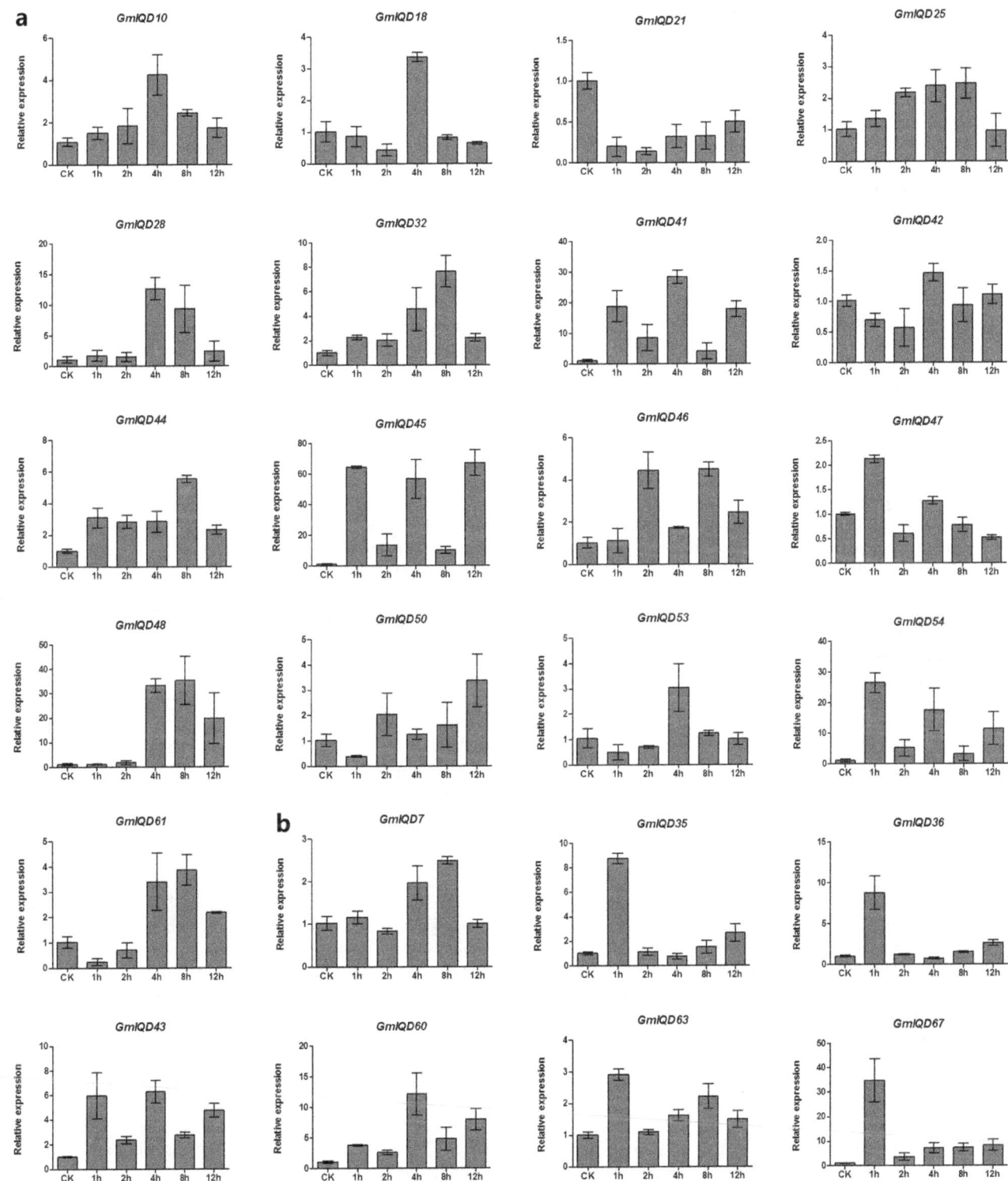

Figure 7. Expression patterns of 24 selected IQD III genes under MeJA stress using qRT-PCR. Relative expression levels of 24 IQD genes were examined by qRT-PCR and normalized to the expression of *CYP2*. Bars represent standard deviations (SD) of three biological replicates. Y-axes indicate the scale of the relative expression levels. X-axes show time courses of MeJA stress treatments for each gene. **a** and **b**: data for genes from IIIa and IIIb, respectively.

genes [36]. Subgroup Ia members of *Arabidopsis IQD26* has higher expression level in parts with divided vigorous growth and microtubule organization of leaves, root and flowers [65]. Eight of

the of the thirteen soybean IQD Ia members (*GmIQD3, -16, -22, -27, -29, -31 -37* and *-66*) have high expression in young fleaves. And six of them (*GmIQD15, -16, -27, -31 -37* and *-66*) have high expression in one-cm pod or pod-shell 10-DAF. *GmIQD22* have obviously higher expression at seed 21-DAF, the period of seed cell division (3, 4 weeks after flowering) [66]. These founding suggested that soybean IQD Ia members may function in transport of signaling molecules, nutrient transport and cell division.

Mapping and positional cloning of the *SUN* locus revealed that this member of the IQD II subfamily was generated by duplication of a 24.7-kb region carrying the tomato *IQD12* gene, a major gene involved in the control of fruit shape, particularly length, in tomato [35]. *SUN* is expressed the highest in hypocotyl and shoot apex. Overexpression of *SUN* causes root reduction when applied auxin and prostrate growth and twisted stems indicating that *SUN* can affect auxin transport or response [38]. *SlSUN1* shows slightly higher expression in the hypocotyl, flower at anthesis and fruit at 10 and 20 DPA [35]. *SlSUN12* and *SlSUN21* highly are expressed in the hypocotyls and root respectively [35]. *SlSUN17* evenly expressed in almost all tissues [35]. Soybean IQD II members showed the similar expression profile (*GmIQD11*: highest in flowers; *GmIQD33, -39* and *-40*: slight high in flowers and *GmIQD6* and *-26*: highest in roots) indicating that GmIQD II members may play similar role in plant development.

Arabidopsis IQD1 from the IQD III subfamily modulates the expression of several glucosinolate (GS) pathway genes, resulting in the alteration of glucosinolate content and composition to promote resistance to herbivory. *Arabidopsis IQD1*, the first functionally characterized IQD gene, is expressed in vascular tissues of hypocotyls, leaves, stems, flowers and roots, as revealed by histochemical analysis. Expression pattern analysis of soybean IQD genes revealed that genes from the soybean IQD III subfamily were mainly expressed in young leaves, flowers and roots (Figure 6). Jasmonic acid(JA) treatment leads to elevating levels of specific glucosinolate in *Arabidopsis* [67,68]. And overexpression of *AtIQD1* causes the accumulation of glucosinolates. However, *AtIQD1* expression is independent of JA, as steady-state *AtIQD1* mRNA expression levels are not appreciably altered when externally applied JA and are also unaffected in mutants defective in JA synthesis or signaling (JA *-jar1* and *fad3-2 fad7-2 fad8*) [31]. Indeed, *AtIQD1* increases resistance against herbivory by augmenting and fine-tuning glucosinolate accumulation [31]. Glucosinolates with important roles in plant defense and human nutrition are a small but diverse class of defense related secondary metabolites in cruciferous species such as *Brassica* crops and the *Arabidopsis thaliana* [69,70]. Obviously, soybean doesn't synthesize glucosinolates. Base on the microsynteny analysis of IQDIII members, we auspiciously found there exsited highly conserved microsynteny relationship between *AtIQD1* and soybean IQDIII members. And combined with phylogenetic analysis of IQDIII members, we speculated soybean IQDIII members might have the similar biological function with *AtIQD1* in defenses to insect herbivory.

Therefore, we performed qRT-PCR of 24 soybean IQD III with MeJA treatment to detect whether soybean IQD III genes defense to insect herbivory. Compared to *AtIQD1*, soybean IQD III genes exhibited different responses with the MeJA treatment. The qRT-PCR results showed that 23 of the 24 soybean IQD III subfamily genes were upregulated by MeJA treatment, except *GmIQD21*, seven genes (*GmIQD28, -41, -45, -48, -54, -60* and *-67*) were strongly induced by MeJA with relative expression more than 10-fold. *GmIQD45* even accumulated the highest transcripts approaching 70-fold at 12 h (Figure 7). Based on these, we speculate that IQD III genes in soybean may involve in defense to insect herbivory by JA pathway.

Orthologs may have equivalent functions, as they originated from a single ancestral gene in the last common ancestor of the species. Two pairs of orthologous genes (*SlSUN22/AtIQD6* and *SlSUN31/AtIQD5*) were found between tomato and *Arabidopsis* (Figure S2). Similar expression patterns of these two pairs in tomato and *Arabidopsis* have been reported; *SlSUN31* and *AtIQD5* are almost ubiquitously expressed, whereas *SlSUN22* and *AtIQD6* are highly expressed in young flower buds [30,34,35].

Duplicated genes may face three different fates: nonfunctionalization (one copy becomes silenced); neofunctionalization (one copy acquires a novel, beneficial function while the other copy retains the original function) or subfunctionalization (both copies become partially compromised by the accumulation of mutations) [45,71]. Paralogs originating from duplication within one organism may have more divergent functions. In the current study, several pairs of paralogs showed similar expression patterns, which suggests that they may share a common or similar function. For example, *GmIQD10/GmIQD18* were highly expressed in young leaves, and their expression peaked at 4 h in response to MeJA (Figure 7). Several pairs of paralogs have different expression patterns, suggesting that they play diverse roles in soybean development. For example, *GmIQD21/GmIQD28* are mainly expressed in young leaves. Upon MeJA treatment, *GmIQD28* was most highly expressed at 4 h while its sister gene *GmIQD21* was downregulated at all time points.

In conclusion, IQD proteins play fundamental roles in various plant developmental processes. Therefore, the systematic analysis of the soybean IQD gene family performed in the current study provides an important reference for further characterization of the biological functions of these proteins.

Materials and Methods

Identification of IQD family genes in soybean

To identify IQD proteins in soybean, the *Glycine max* genome database (release 1.0, http://www.phytozome.net/soybean.php) was searched using Basic Local Alignment Search Tool algorithms (BLASTP), with the published *Arabidopsis* IQD protein sequences and their IQ67 domain used as initial query sequences. Redundant sequences were then removed manually, and the Hidden Markov Model of the Pfam (http://pfam.sanger.ac.uk/search) [72] and SMART (http://smart.embl-heidelberg.de/) [73] databases were used to confirm each candidate sequence as a member of the IQD family [74]. A total of 33 *Arabidopsis*, one moss (*Physcomitrella patens*) and 23 *Brachypodium* IQD protein sequences were downloaded from Phytozome v9.0 (http://www.phytozome.net/), and 34 tomato IQD protein sequences were retrieved from the tomato WGS chromosomes (2.40; SL2.40) (SGNhttp://solgenomics.net). Finally, 27 rice IQD protein sequences were obtained from the TAIR database (http://rice.plantbiology.msu.edu). Accession numbers of published IQD proteins for *Arabidopsis*, rice, tomato, *Brachypodium* and moss were listed in Table S8.

Soybean IQD gene information, including the number of amino acids, ORF lengths and chromosome locations, was obtained from the Phytozome database. Physicochemical parameters including the molecular weight (kDa) and isoelectric point (pI) of each gene product were calculated using compute pI/Mw tool from ExPASy (http://www.expasy.org/tools/) and parameter (resolution) was set to average [75]. Subcellular localization was predicted using the TargetP 1.1 (http://www.cbs.dtu.dk/services/TargetP/) server and WoLF PSORT (http://wolfpsort.org/).

Multiple alignment and phylogenetic analysis of IQD family genes

Multiple sequence alignment of all predicted soybean IQD protein sequences was performed with Clustal X2.0 software using default parameters. Then, based on this alignment, phylogenetic trees were constructed using Clustal X2.0 with the Neighbor-Joining (NJ) method, and bootstrap analysis was conducted using 1,000 replicates [76]. An unrooted NJ tree of 184 the full-length IQD protein sequences from soybean, rice, *Arabidopsis*, tomato, *Brachypodium* was constructed using Clustal X2.0 with one moss IQD protein (Pp1s38230v6.1) as the outgroup. The GmIQD genes were classified according to their phylogenetic relationships with the corresponding *Arabidopsis* and rice IQD genes. For microsynteny analysis of IQD III genes across soybean, tomato and *Arabidopsis*, a phylogenetic tree was constructed using MEGA 5.0 with the NJ method and bootstrap analysis was conducted using 1,000 replicates.

Genomic structure

Exon and intron structures of individual soybean IQD genes were deduced using GSDS (Gene structure display server; http://gsds.cbi.pku.edu.cn/) via alignment of the cDNAs with their corresponding genomic DNA sequences [77].

Identification of conserved motifs and putative calmodulin-binding sites

Online MEME (Multiple Expectation Maximization for Motif Elicitation) (http://meme.sdsc.edu/meme4_4_0/intro.html) was performed to identify the conserved motif structures encoded by GmIQD genes. The parameters were as followings: number of repetitions - any, maximum number of motifs -10, and the optimum motif width was constrained to between 6 and 200 residues. In addition, each structural motif was annotated using the Pfam (http://pfam.sanger.ac.uk/search) and SMART (http://smart.embl-heidelberg.de/) tools. All IQD protein sequences were examined against the Calmodulin Target Database (http://calcium.uhnres.utoronto.ca/ctdb/ctdb/home.html) to predict putative calmodulin-binding sites.

Chromosomal location and gene duplication

The chromosomal location image of GmIQD genes was generated by MapInspect (http://www.plantbreeding.wur.nl/uk/software_mapinspect.html) according to chromosomal position information provided in the Phytozome database. To identify tandem and segmental duplications, two genes in the same species located in the same clade of the phylogenetic tree were defined as coparalogs. The SoyBase browser (http://soybase.org/gb2/gbrowse/gmax1.01/) [78] was queried to detect the segmental duplication coordinates of the target genes. Coparalogs were deemed to result from segmental duplication if they were located on duplicated chromosomal blocks [79]. Paralogs were deemed to be tandem duplicated genes if two genes were separated by five or fewer genes in a 100-kb region [80]. The local alignment of two protein sequences was calculated using the Smith-Waterman algorithm (http://www.ebi.ac.uk/Tools/psa/).

Calculation of Ka/Ks Values

Amino acid sequences of each paralog pair were first aligned using Clustal X2.0. Then, the multiple sequence alignments of proteins and the corresponding cDNA sequences were converted to codon alignments using PAL2NAL (http://www.bork.embl.de/pal2nal/) [81]. Finally, the resulting codon alignment was subjected to calculation of synonymous (Ks) and non-synonymous (Ka) substitution rates using the CODEML program of PAML [82].

Based on a rate of 6.1×10^{-9} substitutions per site per year, divergence time (T) was calculated using the Ks value with the formula: $T = Ks/(2 \times 6.1 \times 10^{-9}) \times 10^{-6}$ Mya [45].

Microsynteny analysis

For microsynteny analysis, IQD III genes from soybean, tomato and *Arabidopsis* as the anchors were localized to specific target genomic regions. Then, all protein-coding sequences of 100 kb flanking each anchor point were compared by pairwise BLASTP analysis. The syntenic blocks used to construct synteny analysis maps of the IQD III genes were obtained from the Plant Genome Duplication Database (http://chibba.agtec.uga.edu/duplication/), a web service providing synteny information in terms of collinearity between chromosomes [83–85]. A synteny block is defined as a region where three or more conserved homologs are located within a 100-kb DNA stretch in both genomes. Two regions were considered to have originated from a large-scale duplication event when five or more protein-coding gene pairs flanking the anchor point were ligatured with the best non-self match (E-value<1e−10) [48]. The relative syntenic quality in a region was calculated from the sum of the total number of genes in both conserved gene regions (excluding retroelements and transposons and collapsing tandem duplications) [47].

RNA-Seq atlas analysis

To acquire the tissue-specific transcript data, a list of 67 GmIQD gene names was entered to the RNA-Seq Atlas of *Glycine max* (http://soybase.org/soyseq/) [86]. The raw digital gene expression counts of the uniquely mappable reads were normalized using a slight variation of the reads/Kb/Million (RPKM) method and the normalized data was download from this database [86]. Hierarchical clustering analysis was conducted using clustering distance "correlation" (Pearson correlation) and the clustering method used "complete" (complete linkage method) in R [87]. A heat map was generated in R using the pheatmap function [87].

Plant growth and treatments

Soybean (*Glycine max* L.) Williams 82 was used in this study. Seedlings were grown in a growth chamber under the following conditions: temperature, 30°C; photoperiod, 12 h/12 h; photon flux density, 80 μmolm^{-2} s^{-1} and relative humidity, 50% [88]. For expression pattern analysis of soybean IQD genes under stress, four-week-old seedlings were treated with 100 μm MeJA in the growth chamber [89]. Jasmonic acid methyl ester (MeJA) (Sigma, 95%) was diluted 1:10 with 95% ethanol, followed by a further dilution with MilliQ water containing 0.1% Triton X-100, resulting in a final concentration of 100 μmol/L MeJA. Untreated seedlings were used as a control. Leaves of MeJA-treated plants were collected at 0, 1, 2, 4, 8 and 12 h. After collection, the samples were immediately frozen in liquid N_2 and stored at −80°C for RNA extraction. Three biological replicates were conducted per sample.

RNA extraction and qRT-PCR analysis

An RNAprep Pure Plant Kit (Tiangen) was used to isolate total RNA from each frozen sample. Possible contaminating genomic DNA was removed using DNaseI supplied in the kit. First-strand cDNA was synthesized from the RNA using a PrimeScript™ RT Master Mix Kit (TaKaRa) according to the manufacturer's instructions. Gene-specific primers for the 24 GmIQD genes were designed using Primer5.0 (Table S9). Primer specificity was first checked using the primer-BLAST tool available on the NCBI website. Subsequently, by performing analysis of melting curves and analysis of visualization of amplicon fragments, we found primers were gene-specific only when corresponding melting curves generated a single sharp peak and the primers demonstrated an electrophoresis pattern of a single amplicon with the correct predicted length. A housekeeping gene constitutively expressed in soybean, *CYP2* (cyclophilin) [46,90–93], was used as a reference for normalization. The qRT-PCR analysis was conducted on an ABI 7500 Real-Time PCR system (Applied Biosystems). The reactions were performed in a 20 µl volume containing 10 µl 2×SYBR® *Premix Ex Taq*™ II (TaKaRa), 6.0 µl ddH$_2$O, 0. 4 µl ROX Reference Dye II, 2.0 µl diluted cDNA and 0.8 µl of each gene-specific primer. The PCR conditions were as follows: Stage 1: 95°C for 30 s; stage 2: 40 cycles of 5 s at 95°C and 34 s at 60°C; stage 3: 95°C for 15 s, 60°C for 1 min, 95°C for 15 s. At stage 3, a melting curve was generated to estimate the specificity of the reactions. Three biological replicates were used per sample.

The relative expression levels were calculated as $2^{-\Delta\Delta CT}$ [$\Delta C_T = C_{T,\ Target} - C_{T,\ CYP2}$. $\Delta\Delta CT = \Delta C_{T,\ treatment} - \Delta C_{T,\ CK\ (0\ h)}$]. The relative expression levels ($2^{-\Delta\Delta CT,\ CK\ (0\ h)}$) in the untreated control plants were normalized to 1 as described previously [46,94,95]. If an efficiency of amplification was less than 2, the result was proofread. Statistical analyses were conducted using GraphPad Prism 5.01 software [96].

Supporting Information

Figure S1 Amino acid sequence alignments of IQ67 domains in soybean IQD protein sequences. The multiple alignment results indicate the highly conserved IQD domains among the 67 identified soybean IQD protein sequences. The positions of the conserved IQ calmodulin binding motifs are shown. Identical residues of proteins are marked with an asterisk. The consensus sequence at the bottom was constructed with greater than 50% conservation among the 67 soybean IQD proteins. Red arrow indicates the position of the conserved phase-0 intron, which divides codons 16 and 17 of the IQ67 domain.

Figure S2 Phylogenetic tree of full-length IQD III proteins from soybean, *Arabidopsis* and tomato. The tree was generated with MEGA 5.0 using the NJ method with 1,000 bootstrap replicates. Dicotyledon (soybean, tomato and *Arabidopsis*) IQD proteins are marked with colored dots. IQD III proteins from soybean, *Arabidopsis* and tomato were divided into four clades (1–4) presented by different color.

Table S1 Detailed information about the 10 motifs in soybean IQD proteins.

Table S2 Recent synteny blocks of soybean and soybean (13 Mya) genomes containing IQD genes.

Table S3 Pairwise identities between paralogous pairs of IQD genes from soybean.

Table S4 The synteny blocks used to construct micro-synteny map.

Table S5 Number of conserved gene pairs and synteny blocks and relative syntenic quality.

Table S6 Transcription of soybean IQD genes, as determined by RNA-seq analysis.

Table S7 Animo acid content of 67 soybean IQD proteins.

Table S8 Accession numbers of IQDs for *Arabidopsis thaliana*, rice, tomato, *Brachypodium distachyon* and moss.

Table S9 List of primer sequences used for qRT-PCR analysis of the 24 soybean IQD III genes.

Author Contributions

Conceived and designed the experiments: LF YX. Performed the experiments: LF XC. Analyzed the data: LF ZC. Contributed reagents/materials/analysis tools: LF XC YW. Wrote the paper: LF ZC YL.

References

1. Hepler PK, Vidali L, Cheung AY (2001) Polarized cell growth in higher plants. Annual review of cell and developmental biology 17: 159–187.
2. Du L, Ali GS, Simons KA, Hou J, Yang T, et al. (2009) Ca2+/calmodulin regulates salicylic-acid-mediated plant immunity. Nature 457: 1154–1158.
3. Harada A, Shimazaki K-i (2009) Measurement of changes in cytosolic Ca2+ in Arabidopsis guard cells and mesophyll cells in response to blue light. Plant and cell physiology 50: 360–373.
4. Ng C, Mcainsh MR, Gray JE, Hunt L, Leckie CP, et al. (2001) Calcium-based signalling systems in guard cells. New Phytologist 151: 109–120.
5. Reddy AS (2001) Calcium: silver bullet in signaling. Plant Science 160: 381–404.
6. Dodd AN, Kudla J, Sanders D (2010) The language of calcium signaling. Annu Rev Plant Biol 61: 593–620.
7. Day IS, Reddy VS, Shad Ali G, Reddy AS (2002) Analysis of EF-hand-containing proteins in Arabidopsis. Genome Biol 3: RESEARCH0056.
8. Evans NH, McAinsh MR, Hetherington AM (2001) Calcium oscillations in higher plants. Current opinion in plant biology 4: 415–420.
9. Harper JF (2001) Dissecting calcium oscillators in plant cells. Trends in plant science 6: 395–397.
10. Ali GS, Reddy VS, Lindgren PB, Jakobek JL, Reddy A (2003) Differential expression of genes encoding calmodulin-binding proteins in response to bacterial pathogens and inducers of defense responses. Plant Molecular Biology 51: 803–815.

11. Snedden WA, Fromm H (2001) Calmodulin as a versatile calcium signal transducer in plants. New Phytologist 151: 35–66.
12. Bouché N, Yellin A, Snedden WA, Fromm H (2005) Plant-specific calmodulin-binding proteins. Annu Rev Plant Biol 56: 435–466.
13. Luan S, Kudla J, Rodriguez-Concepcion M, Yalovsky S, Gruissem W (2002) Calmodulins and calcineurin B–like proteins calcium sensors for specific signal response coupling in plants. The Plant Cell Online 14: S389–S400.
14. Ranty B, Aldon D, Galaud J-P (2006) Plant calmodulins and calmodulin-related proteins: multifaceted relays to decode calcium signals. Plant Signal Behav 1: 96.
15. Zhang H, Yin W, Xia X (2008) Calcineurin B-Like family in Populus: comparative genome analysis and expression pattern under cold, drought and salt stress treatment. Plant Growth Regulation 56: 129–140.
16. Zhang C, Bian M, Yu H, Liu Q, Yang Z (2011) Identification of alkaline stress-responsive genes of CBL family in sweet sorghum (Sorghum bicolor L.). Plant Physiol Biochem 49: 1306–1312.
17. Yu Y, Xia X, Yin W, Zhang H (2007) Comparative genomic analysis of CIPK gene family in Arabidopsis and Populus. Plant Growth Regulation 52: 101–110.
18. Zuo R, Hu R, Chai G, Xu M, Qi G, et al. (2013) Genome-wide identification, classification, and expression analysis of CDPK and its closely related gene families in poplar (Populus trichocarpa). Mol Biol Rep 40: 2645–2662.
19. Reddy AS, Ali GS, Celesnik H, Day IS (2011) Coping with stresses: roles of calcium- and calcium/calmodulin-regulated gene expression. Plant Cell 23: 2010–2032.

20. Batistic O, Kudla J (2004) Integration and channeling of calcium signaling through the CBL calcium sensor/CIPK protein kinase network. Planta 219: 915–924.

21. Boonburapong B, Buaboocha T (2007) Genome-wide identification and analyses of the rice calmodulin and related potential calcium sensor proteins. BMC Plant Biol 7: 4.

22. Perochon A, Aldon D, Galaud J-P, Ranty B (2011) Calmodulin and calmodulin-like proteins in plant calcium signaling. Biochimie 93: 2048–2053.

23. DeFalco T, Bender K, Snedden W (2010) Breaking the code: Ca2+ sensors in plant signalling. Biochem J 425: 27–40.

24. Chang'en T, Yuping Z (2014) Research Progress in Plant IQ Motif-containing Calmodulin-binding Proteins. Chinese Bulletin of Botany 48: 447–460.

25. Fischer C, Kugler A, Hoth S, Dietrich P (2013) An IQ domain mediates the interaction with calmodulin in a plant cyclic-nucleotide-gated channel. Plant and cell physiology 54: 573–584.

26. Bähler M, Rhoads A (2002) Calmodulin signaling via the IQ motif. FEBS letters 513: 107–113.

27. Hoeflich KP, Ikura M (2002) Calmodulin in action: diversity in target recognition and activation mechanisms. Cell 108: 739–742.

28. Bhattacharya S, Bunick CG, Chazin WJ (2004) Target selectivity in EF-hand calcium binding proteins. Biochimica et Biophysica Acta (BBA)-Molecular Cell Research 1742: 69–79.

29. Clapperton JA, Martin SR, Smerdon SJ, Gamblin SJ, Bayley PM (2002) Structure of the complex of calmodulin with the target sequence of calmodulin-dependent protein kinase I: studies of the kinase activation mechanism. Biochemistry 41: 14669–14679.

30. Abel S, Savchenko T, Levy M (2005) Genome-wide comparative analysis of the IQD gene families in Arabidopsis thaliana and Oryza sativa. BMC Evol Biol 5: 72.

31. Levy M, Wang Q, Kaspi R, Parrella MP, Abel S (2005) Arabidopsis IQD1, a novel calmodulin-binding nuclear protein, stimulates glucosinolate accumulation and plant defense. Plant J 43: 79–96.

32. Burstenbinder K, Savchenko T, Muller J, Adamson AW, Stamm G, et al. (2012) Arabidopsis Calmodulin-binding Protein IQ67-Domain 1 Localizes to Microtubules and Interacts with Kinesin Light Chain-related Protein-1. Journal of Biological Chemistry 288: 1871–1882.

33. Abel S, Burstenbinder K, Muller J (2013) The emerging function of IQD proteins as scaffolds in cellular signaling and trafficking. Plant Signal Behav 8: e24369.

34. Filiz E, Tombuloglu H, Ozyigit II (2013) Genome-wide analysis of IQ67 domain (IQD) gene families in Brachypodium distachyon. Plant Omics 6: 425–432.

35. Huang Z, Van Houten J, Gonzalez G, Xiao H, van der Knaap E (2013) Genome-wide identification, phylogeny and expression analysis of SUN, OFP and YABBY gene family in tomato. Mol Genet Genomics 288: 111–129.

36. Zentella R, Zhang ZL, Park M, Thomas SG, Endo A, et al. (2007) Global analysis of della direct targets in early gibberellin signaling in Arabidopsis. Plant Cell 19: 3037–3057.

37. Xiao H, Jiang N, Schaffner E, Stockinger EJ, van der Knaap E (2008) A retrotransposon-mediated gene duplication underlies morphological variation of tomato fruit. Science 319: 1527–1530.

38. Wu S, Xiao H, Cabrera A, Meulia T, van der Knaap E (2011) SUN regulates vegetative and reproductive organ shape by changing cell division patterns. Plant Physiol 157: 1175–1186.

39. Kamenetzky L, Asis R, Bassi S, de Godoy F, Bermudez L, et al. (2010) Genomic analysis of wild tomato introgressions determining metabolism- and yield-associated traits. Plant Physiol 152: 1772–1786.

40. Soto-Valdez H, Colin-Chavez C, Peralta E. Fabrication and Properties of Antioxidant Polyethylene-based Films Containing Marigold (Tagetes erecta) Extract and Application on Soybean Oil Stability; 2012. DEStech Publications, Inc. pp. 206.

41. Schmutz J, Cannon SB, Schlueter J, Ma J, Mitros T, et al. (2010) Genome sequence of the palaeopolyploid soybean. Nature 463: 178–183.

42. Silva DC, Yamanaka N, Brogin RL, Arias CA, Nepomuceno AL, et al. (2008) Molecular mapping of two loci that confer resistance to Asian rust in soybean. Theor Appl Genet 117: 57–63.

43. Helfer S (2014) Rust fungi and global change. New Phytologist 201: 770–780.

44. Nekrutenko A, Makova KD, Li W-H (2002) The KA/KS ratio test for assessing the protein-coding potential of genomic regions: an empirical and simulation study. Genome research 12: 198–202.

45. Lynch M, Conery JS (2000) The evolutionary fate and consequences of duplicate genes. Science 290: 1151–1155.

46. Chen X, Chen Z, Zhao H, Zhao Y, Cheng B, et al. (2014) Genome-Wide Analysis of Soybean HD-Zip Gene Family and Expression Profiling under Salinity and Drought Treatments. PloS one 9: e87156.

47. Cannon SB, McCombie WR, Sato S, Tabata S, Denny R, et al. (2003) Evolution and microsynteny of the apyrase gene family in three legume genomes. Molecular Genetics and Genomics 270: 347–361.

48. Zhang X, Feng Y, Cheng H, Tian D, Yang S, et al. (2011) Relative evolutionary rates of NBS-encoding genes revealed by soybean segmental duplication. Molecular Genetics and Genomics 285: 79–90.

49. Creelman RA, Tierney ML, Mullet JE (1992) Jasmonic acid/methyl jasmonate accumulate in wounded soybean hypocotyls and modulate wound gene expression. Proceedings of the National Academy of Sciences 89: 4938–4941.

50. HU W-z, JIANG A-l, YANG H, LIU C-h, HE Y-b (2012) Effect of Jasmonic acid methyl ester treatment on the physiological and biochemical reactions of fresh-cut apple. Science and Technology of Food Industry 16: 086.

51. Cannon SB, Shoemaker RC (2012) Evolutionary and comparative analyses of the soybean genome. Breeding science 61.

52. Reddy A, Day IS (2001) Analysis of the myosins encoded in the recently completed Arabidopsis thaliana genome sequence. Genome Biol 2: 1–17.

53. Gao F, Han X, Wu J, Zheng S, Shang Z, et al. (2012) A heat-activated calcium-permeable channel–Arabidopsis cyclic nucleotide-gated ion channel 6–is involved in heat shock responses. The Plant Journal 70: 1056–1069.

54. Bürstenbinder K, Savchenko T, Müller J, Adamson AW, Stamm G, et al. (2013) Arabidopsis Calmodulin-binding Protein IQ67-Domain 1 Localizes to Microtubules and Interacts with Kinesin Light Chain-related Protein-1. Journal of Biological Chemistry 288: 1871–1882.

55. Pauly N, Knight MR, Thuleau P, van der Luit AH, Moreau M, et al. (2000) Cell signalling: control of free calcium in plant cell nuclei. Nature 405: 754–755.

56. Xiong TC, Jauneau A, Ranjeva R, Mazars C (2004) Isolated plant nuclei as mechanical and thermal sensors involved in calcium signalling. The Plant Journal 40: 12–21.

57. Lecourieux D, Ranjeva R, Pugin A (2006) Calcium in plant defence-signalling pathways. New Phytologist 171: 249–269.

58. Anandalakshmi R, Marathe R, Ge X, Herr J, Mau C, et al. (2000) A calmodulin-related protein that suppresses posttranscriptional gene silencing in plants. Science 290: 142–144.

59. Yoo JH, Park CY, Kim JC, Do Heo W, Cheong MS, et al. (2005) Direct interaction of a divergent CaM isoform and the transcription factor, MYB2, enhances salt tolerance in Arabidopsis. Journal of Biological Chemistry 280: 3697–3706.

60. Du J, Tian Z, Sui Y, Zhao M, Song Q, et al. (2012) Pericentromeric effects shape the patterns of divergence, retention, and expression of duplicated genes in the paleopolyploid soybean. The Plant Cell Online 24: 21–32.

61. Ohno S (1970) Evolution by gene duplication: London: George Alien & Unwin Ltd. Berlin, Heidelberg and New York: Springer-Verlag.

62. Schlueter JA, Dixon P, Granger C, Grant D, Clark L, et al. (2004) Mining EST databases to resolve evolutionary events in major crop species. Genome 47: 868–876.

63. Shoemaker RC, Schlueter J, Doyle JJ (2006) Paleopolyploidy and gene duplication in soybean and other legumes. Current opinion in plant biology 9: 104–109.

64. Dreze M, Carvunis A-R, Charloteaux B, Galli M, Pevzner SJ, et al. (2011) Evidence for network evolution in an Arabidopsis interactome map. Science 333: 601–607.

65. Su-Juan WH-YGZ-QWZ-JLZ-WWZ-JC (2008) Isolation and Characterization of Calmodulin-binding Protein AtIQD26 in Arabidopsis thaliana. Progress in Biochemistry and Biophysics 35: 703–711.

66. Hajduch M, Ganapathy A, Stein JW, Thelen JJ (2005) A systematic proteomic study of seed filling in soybean. Establishment of high-resolution two-dimensional reference maps, expression profiles, and an interactive proteome database. Plant Physiol 137: 1397–1419.

67. Brader G, Tas É, Palva ET (2001) Jasmonate-dependent induction of indole glucosinolates in arabidopsis by culture filtrates of the nonspecific pathogen-erwinia carotovora. Plant Physiology 126: 849–860.

68. Cipollini D, Enright S, Traw M, Bergelson J (2004) Salicylic acid inhibits jasmonic acid-induced resistance of Arabidopsis thaliana to Spodoptera exigua. Molecular Ecology 13: 1643–1653.

69. Fahey JW, Zalcmann AT, Talalay P (2001) The chemical diversity and distribution of glucosinolates and isothiocyanates among plants. Phytochemistry 56: 5–51.

70. Wittstock U, Halkier BA (2002) Glucosinolate research in the Arabidopsis era. Trends in plant science 7: 263–270.

71. Dittmar K, Liberles D (2011) Evolution after gene duplication: John Wiley & Sons.

72. Finn RD, Mistry J, Schuster-Böckler B, Griffiths-Jones S, Hollich V, et al. (2006) Pfam: clans, web tools and services. Nucleic acids research 34: D247–D251.

73. Letunic I, Copley RR, Schmidt S, Ciccarelli FD, Doerks T, et al. (2004) SMART 4.0: towards genomic data integration. Nucleic Acids Research 32: D142–D144.

74. Bateman A, Coin L, Durbin R, Finn RD, Hollich V, et al. (2004) The Pfam protein families database. Nucleic acids research 32: D138–D141.

75. Gasteiger E, Gattiker A, Hoogland C, Ivanyi I, Appel RD, et al. (2003) ExPASy: the proteomics server for in-depth protein knowledge and analysis. Nucleic acids research 31: 3784–3788.

76. Hu R, Qi G, Kong Y, Kong D, Gao Q, et al. (2010) Comprehensive analysis of NAC domain transcription factor gene family in Populus trichocarpa. BMC plant biology 10: 145.

77. Guo A-Y, Zhu Q-H, Chen X, Luo J-C (2007) GSDS: a gene structure display server]. Yi chuan = Hereditas/Zhongguo yi chuan xue hui bian ji 29: 1023.

78. Grant D, Nelson RT, Cannon SB, Shoemaker RC (2010) SoyBase, the USDA-ARS soybean genetics and genomics database. Nucleic acids research 38: D843–D846.

79. Wei F, Coe E, Nelson W, Bharti AK, Engler F, et al. (2007) Physical and genetic structure of the maize genome reflects its complex evolutionary history. PLoS Genetics 3: e123.

80. Wang L, Guo K, Li Y, Tu Y, Hu H, et al. (2010) Expression profiling and integrative analysis of the CESA/CSL superfamily in rice. BMC plant biology 10: 282.

81. Suyama M, Torrents D, Bork P (2006) PAL2NAL: robust conversion of protein sequence alignments into the corresponding codon alignments. Nucleic acids research 34: W609–W612.

82. Yang Z (2007) PAML 4: phylogenetic analysis by maximum likelihood. Molecular biology and evolution 24: 1586–1591.

83. Lee T-H, Tang H, Wang X, Paterson AH (2013) PGDD: a database of gene and genome duplication in plants. Nucleic acids research 41: D1152–D1158.

84. Tang H, Bowers JE, Wang X, Ming R, Alam M, et al. (2008) Synteny and collinearity in plant genomes. Science 320: 486–488.

85. Tang H, Wang X, Bowers JE, Ming R, Alam M, et al. (2008) Unraveling ancient hexaploidy through multiply-aligned angiosperm gene maps. Genome research 18: 1944–1954.

86. Severin AJ, Woody JL, Bolon Y-T, Joseph B, Diers BW, et al. (2010) RNA-Seq Atlas of Glycine max: a guide to the soybean transcriptome. BMC plant biology 10: 160.

87. Team RC (2012) R: A language and environment for statistical computing.

88. Hyun TK, Eom SH, Jeun YC, Han SH, Kim J-S (2013) Identification of glutamate decarboxylases as a γ-aminobutyric acid (GABA) biosynthetic enzyme in soybean. Industrial Crops and Products 49: 864–870.

89. Cheng Q, Zhang B, Zhuge Q, Zeng Y, Wang M, et al. (2006) Expression profiles of two novel lipoxygenase genes in *Populus deltoides*. Plant Science 170: 1027–1035.

90. Gutierrez N GMJ, Palomino C, et al. (2011) Assessment of candidate reference genes for expression studies in Vicia faba L. by real-time quantitative PCR. Molecular Breeding: 13–24.

91. Vívian de Jesus Miranda RRC, Antônio Américo Barbosa Viana , Osmundo Brilhante de Oliveira Neto , Regina Maria Dechechi Gomes Carneiro , Thales Lima Rocha , Maria Fatima Grossi de Sa , Rodrigo Rocha Fragoso (2013) Validation of reference genes aiming accurate normalization of qPCR data in soybean upon nematode parasitism and insect attack. BMC Research Notes 6.

92. Le DT, Nishiyama R, Watanabe Y, Mochida K, Yamaguchi-Shinozaki K, et al. (2011) Genome-wide expression profiling of soybean two-component system genes in soybean root and shoot tissues under dehydration stress. DNA Res 18: 17–29.

93. Jian B, Liu B, Bi Y, Hou W, Wu C, et al. (2008) Validation of internal control for gene expression study in soybean by quantitative real-time PCR. BMC Mol Biol 9: 59.

94. Livak KJ, Schmittgen TD (2001) Analysis of Relative Gene Expression Data Using Real-Time Quantitative PCR and the $2^{-\Delta\Delta CT}$ Method. methods 25: 402–408.

95. Peng X, Zhao Y, Cao J, Zhang W, Jiang H, et al. (2012) CCCH-type zinc finger family in maize: genome-wide identification, classification and expression profiling under abscisic acid and drought treatments. PLoS One 7: e40120.

96. Bryfczynski SP, Pargas RP (2009) GraphPad: a graph creation tool for CS2/CS7. ACM SIGCSE Bulletin 41: 389–389.

19

Evaluation of the Agronomic Performance of Atrazine-Tolerant Transgenic *japonica* Rice Parental Lines for Utilization in Hybrid Seed Production

Luhua Zhang[1,9], **Haiwei Chen**[1,9,¤], **Yanlan Li**[1], **Yanan Li**[1], **Shengjun Wang**[2], **Jinping Su**[2], **Xuejun Liu**[2], **Defu Chen**[1]*, **Xiwen Chen**[1]*

1 Laboratory of Molecular Genetics, College of Life Sciences, Nankai University, Tianjin, China, **2** Tianjin Crop Research Institute, Tianjin, China

Abstract

Currently, the purity of hybrid seed is a crucial limiting factor when developing hybrid *japonica* rice (*Oryza sativa* L.). To chemically control hybrid seed purity, we transferred an improved atrazine chlorohydrolase gene (*atzA*) from *Pseudomonas* ADP into hybrid *japonica* parental lines (two maintainers, one restorer), and Nipponbare, by using *Agrobacterium*-mediated transformation. We subsequently selected several transgenic lines from each genotype by using PCR, RT-PCR, and germination analysis. In the presence of the investigated atrazine concentrations, particularly 150 μM atrazine, almost all of the transgenic lines produced significantly larger seedlings, with similar or higher germination percentages, than did the respective controls. Although the seedlings of transgenic lines were taller and gained more root biomass compared to the respective control plants, their growth was nevertheless inhibited by atrazine treatment compared to that without treatment. When grown in soil containing 2 mg/kg or 5 mg/kg atrazine, the transgenic lines were taller, and had higher total chlorophyll contents than did the respective controls; moreover, three of the strongest transgenic lines completely recovered after 45 days of growth. After treatment with 2 mg/kg or 5 mg/kg of atrazine, the atrazine residue remaining in the soil was 2.9–7.0% or 0.8–8.7% respectively, for transgenic lines, and 44.0–59.2% or 28.1–30.8%, respectively, for control plants. Spraying plants at the vegetative growth stage with 0.15% atrazine effectively killed control plants, but not transgenic lines. Our results indicate that transgenic *atzA* rice plants show tolerance to atrazine, and may be used as parental lines in future hybrid seed production.

Editor: Jin-Song Zhang, Institute of Genetics and Developmental Biology, Chinese Academy of Sciences, China

Funding: This work was supported by the Key Project of Tianjin Science and Technology Support Program (11ZCGYNC01000), the Key Program of the Natural Science Foundation of Tianjin (12YFJZJC01700, 14JCZDJC34100), the grant of Natural Science Foundation of China (No. 31070273, 31070717) and the 111 Project (No. B08011). The funders had no role in study design, data collection and analysis, decision to publish, or preparation of the manuscript.

Competing Interests: The authors have declared that no competing interests exist.

* Email: chendefu@nankai.edu.cn (DC); xiwenchen@nankai.edu.cn (XC)

9 These authors contributed equally to this work.

¤ Current address: College of Life Sciences, Chifeng College, Chifeng, China

Introduction

Rice (*Oryza sativa*) is one of the most important staple food crops globally. According to the National Grain and Oil Information Center, the area of China planted with rice in 2012 was 3.0×10^7 hm^2, including 9.0×10^6 hm^2 of *japonica* rice [1]. *Japonica* rice is mainly planted in the northern region of the Qinling Mountains–Huai River, and its planted area has increased in recent years because of its high quality and good taste. *Japonica* rice production is currently dominated by conventional varieties, with hybrid rice accounting for only 3% of the cultivated area. On the other hand, *indica* hybrid rice represents 70–80% of the total planted area of *indica* rice [2]. Therefore, there is considerable potential for the development of *japonica* hybrid rice. An increase in the annual planted area of *japonica* hybrid rice from 3% to 50%, i.e., to reach 4.0×10^6 hm^2, is estimated to lead to the production of 3.5×10^9 kg of high-quality grain (www.cngrain.com/Publish/qita/200503/207290), thereby contributing considerably to meet consumer's demand for high-quality food both in China and globally.

The three-line system is a traditional and effective production method for hybrid *japonica* rice seed [3]. The most widely used male sterile line in the system is BT-type cytoplasmic male sterile (CMS). However, the panicle of this line is loosely enclosed when heading, and this appearance closely resembles that of the maintainer. This makes it difficult for farmers to distinguish the BT-CMS line when eliminating off-type plants [2]. Furthermore, the BT CMS line has good restorability, and may therefore be easily pollinated with exotic pollens that contaminated during mechanical harvesting and storage of seeds, and also with exotic pollens from other plants [4]. The use of contaminated CMS lines in seed production results in decreased hybrid seed purity. Therefore, off-type contamination must be eliminated as early as possible. This is largely a manual process and requires considerable labor input, particularly in Asia. On the one hand, the need for increased labor will increase the price of hybrid seeds, while on the other hand, the increase in manual procedures may lead to the production of false hybrids. Furthermore, as the Chinese economy develops, increasing numbers of young men are leaving their home towns to seek work in the cities, leaving the elderly and women to work on the farms. The transformation of heavy and complex farming to light and simple farming is therefore becoming increasingly important. Thus, ensuring hybrid seed purity and reducing labor costs are two key issues in hybrid *japonica* rice seed production.

Genetic engineering, especially herbicide resistance engineering, provides an efficient means of controlling purity in hybrid seed production. Yan first proposed a strategy of utilizing herbicide resistance genes to chemically control purity in hybrid seed production [5]. Since then, two-line hybrid rice production has been extensively investigated [6–12] and progress has recently been reviewed [13]. Additionally, some transgenic hybrid rice combinations have been used in field trials [7,8,10,12]. However, the research has mainly focused on the *bar* gene isolated from *Streptomyces hygroscopicus* [6,8,10,12]; other genes, such as the EPSPS (5-enolpyruvylshikimate-3-phosphate synthase) gene from *Agrobacterium* strain CP4, and the protoporphyrinogen oxidase gene from *Bacillus subtilis*, have rarely been investigated [11]. If the strategy is proven to be effective, chemical control of hybrid seed purity will be mainly dependent on an herbicide with a single mode of action, and this will hinder sustainable weed management.

Atrazine (6-chloro-N^2-ethyl-N^4-isopropyl-1,3,5-triazine-2,4-diamine) is a triazine herbicide, and is commonly used in maize, sorghum, and sugarcane fields [14]. By inhibiting electron transport to plastoquinone in the photosystem PSII, atrazine terminates photosynthesis and kills weeds [15]. Atrazine was once the most widely used herbicide worldwide, because of its low cost and high effectiveness [14]. We previously isolated an atrazine chlorohydrolase gene (*atzA*) from a soil bacterium *Pseudomonas* ADP [16], and modified this gene by using directed evolution, to improve the enzymatic activity [17]. In the present study, we transferred the improved *atzA* gene into breeding hybrid *japonica* parental lines. Our results indicate that the transgenic *atzA* rice lines show tolerance to atrazine, and may be used as parental lines to chemically improve seed purity in hybrid seed production.

Materials and Methods

Construction of plant expression vector

Ubiquitin promoter and an improved atrazine chlorohydrolase gene *atzA*-22-4 [17] were respectively amplified from pSTAR-LING, an RNAi intermediate vector for monocots (a kind gift from the Commonwealth Scientific and Industrial Research Organization, Australia), and *AtzA*-22-4 using primers SL-Ubi-F/SL-Ubi-R and atzA-TJ-F/atzA-TJ-R (Table 1). After respectively inserted into TaKaRa pMD19T-simple vector and confirmed by sequencing, the pAtzA-22-4-19T-simple was restricted with *Eco*RV to collect the 1.4 kb fragment, and then lignated with pUbi-19T-simple. The resulting plasmid was restricted with *Sac*I/*Spe*I to collect the 3.4 kb fragment, and then ligated with pCAMBIA1301. The recombinant plasmid p1301-ubi-22-14 was then introduced into *Agrobacterium tumefaciens* EHA105 by the freeze-thaw method [18].

Genetic transformation and plant regeneration

Oryza sativa L. Nipponbare and *japonica* hybrid rice parental lines in the three-line system, Jindao7 (maintainer), Jindao8 (maintainer) and Jinhui3 (restorer) were used for transformation. Mature seeds were dehulled, surface-sterilized and placed on NB medium (N6 macro elements, B5 micro elements and vitamins)

supplemented with 2 g/L proline, 3 mg/L 2, 4-D and 300 mg/L casein hydrolysate in dark at 28°C. After 2–3 weeks, the scutellum-derived calli were excised and subcultured every four weeks on the same medium but with 0.5 g/L proline, 2 mg/L 2, 4-D in dark at 28°C. The highly embryogenic compact calli (3–5 mm in diameter) that subcultured for less than five generations, were selected and co-cultivated with *A. tumefaciens* EHA105 harboring p1301C-ubi-22-14 on the co-cultivation medium (subculture medium but with 100 µM acetosyringone) for 3 days in dark at 28°C. Following that, the explants were then transferred into selection medium (subculture medium but with 50 mg/L hygromycin and 500 mg/L cefotaxime) in dark at 28°C for selection. After two cycles of selection, hygromycin-resistant calli were transferred onto pre-regeneration medium (NB medium with 0.5 g/L proline, 2 mg/L 6-BA, 1 mg/L NAA, 5 mg/L ABA, 300 mg/L casein hydrolysate and 50 mg/L hygromycin) for 14 to 21 days in dark at 28°C, then to regeneration medium (pre-regeneration medium but without 5 mg/L ABA) for 30 days under 54 µmol/m^2/s light at 28°C and finally to the rooting medium (MS medium with 1 mg/L IBA) for 15 days under light at 28°C.

Molecular analysis of transgenic lines

Genomic DNA was isolated from young leaves using a modified CTAB method [19]. PCR was performed to preliminarily select the transformed plants using primers atzA7/atzA8 [20] and SL-Ubi-C-F/SL-Ubi-C-R (Table 1). RT-PCR was performed to further confirm the expression of *atzA* in the PCR-positive transformed plants. Total RNA was extracted from young leaves using an RNAultra Extraction Kit (Qiagen). cDNA was synthesized using atzA8 and oligo(dT) as primers and SuperScript^{TM} II RNase H− (Invitrogen) as reverse transcriptase. RT-PCR was amplified using the cDNA as template and atzA7/atzA2 [20] as primers. β-actin (AB047313) was also amplified as the internal control using actin-R/actin-F as primers [21].

Germination and seedling growth in the presence of atrazine

Fifty seeds from each transgenic line and the respective control plant were directly sown on the surface of filter paper in plates containing 0, 75 or 150 µM atrazine. Seeds were placed in a growth chamber at 28°C with 16 h of 54 µmol/m^2/s light per day. Germination (based on radicles >2 mm) was recorded daily and the cumulative values at day 3 and day 7 were calculated to represent as germination potential and germination percentage. The shoot length, root length and their biomass were measured after 7 days. For seedling growth test, seven-day old seedlings germinated in absence of atrazine were placed on filter paper in pots, and incubated in Kimura B nutrition solution [22] containing 0, 75 or 150 µM atrazine in growth chamber described above. During the period, the nutrition solution containing the respective concentration of atrazine was added to keep the filter paper wet. The growth parameters as described above were measured after 10 days.

Soil-grown transgenic T_2 lines in the presence of atrazine

Two-week old seedlings of similar size that germinated in absence of atrazine were transplanted into pots containing 1.12 kg of soil with 0, 2, or 5 mg/kg of atrazine. Nine plants were planted in each pot, and irrigated with the same amount of water every day and with Kimura B nutrition solution [22] twice a week. Plants were incubated in the greenhouse in a 14-h light/10-h dark cycle (28/25°C) at 300 µmol m^{-2}s^{-1} light and 75% relative humidity. Plant growth and chlorophyll content were measured at

Table 1. Primers used in this study.

Primer	Sequence (5′→3′)
SL-Ubi-F	tgagctcctgcagtgcagcgtgacccggtcgt, with *Sac*I site
SL-Ubi-R	tgatatcctgcagaagtaacaccaaacaacag, with *Eco*RV site
atzA-TJ-F	tgatatcatgcaaacgctcagcatccag, with *Eco*RV site
atzA-TJ-R	tgatatcactagtctagaggctgcgccaagctg, with *Eco*RV and *Spe*I site
SL-Ubi-C-F	cccgccgtaataaatagacac
SL-Ubi-C-R	accacaccacatcatcacaac

15-day intervals. For chlorophyll content analysis, approximately 10 mg of leaves was extracted with 5 ml acetone and then quantified the absorbance at 663.6 nm and 646.6 nm, as described by Porra et al. [23]. Chlorophyll a $= 12.25 \times A_{663.6} - 2.55 \times A_{646.6}$, Chlorophyll b $= 20.31 \times A_{646.6} - 4.91 \times A_{663.6}$.

HPLC analysis of atrazine and its metabolite hydroxyatrazine in the plant and soil

Atrazine and hydroxyatrazine in the rice leaves and soil were also determined in the soil-grown experiment. One gram of leaves, or five grams of soil samples that collected from the soil layer mixture as thick as possible, were extracted with 5 mL dichloromethane. After 30 min of incubation at room temperature, the mixture was centrifuged at 15,294 g for 10 min. The supernatant was transferred to a new tube and filtered through a 0.2-μm filter and then subjected to CoM 6000 HPLC system analysis on an analytical C18 column (5 μm, 250 mm×4.6 mm) at 30°C with linear gradients phase as follows: 0 to 6 min, 10% to 25% acetonitrile; 6 min to 21 min, 25% to 65% acetonitrile; 21 min to 23 min, 65% to 100% acetonitrile; and 23 min to 25 min, 100% acetonitrile [24] at the wavelength of 228 nm.

Spraying atrazine to transgenic T2 lines

Two-week old seedlings from the germination experiment were also transplanted into pots with soil. Each pot contained 16 plants. After 30 days' growth, the last second leaves were taken and cut into 2–3 cm section, soaked in 0, 75 and 150 μM atrazine, and incubated at 25°C under light for 2 days. The color of the leaves was observed every day. After 40 days' growth, each pot of plants was sprayed with 20 ml of 0.15% atrazine solution.

Statistical analysis

All the experiments were performed for three times. Significant difference between the specific transgenic line and the respective control was performed using independent-samples t-test at 95% or 99% confidence with IBM SPSS Statistics 11.0. Values indicated by * or ** represented significantly difference at $P<0.05$ or $P<0.01$.

Results

Selection of transgenic *atzA* plants

PCR analysis revealed that 18 (100%) Nipponbare transformants, 15 out of 18 (83%) Jindao7 transformants, 14 out of 16 (88%) Jinhui1 transformants, and 17 (100%) Jindao8 transformants were positive. RT-PCR analysis further confirmed the expression of *atzA* in the PCR-positive lines (data not shown). Subsequently, the self-pollinated seeds (T$_1$ progenies) from each plant were germinated in the presence of hygromycin. Transgenic

lines that showed a segregation pattern of 3:1 resistant/sensitive in the germination test were selected to produce the T$_2$ generations. Individual plants, whose herbicide tolerance did not segregate in the germination test, were considered as homogenous lines and selected. For simplicity, in further research, we used only two or three independent transgenic T$_2$ lines for each genotype.

Germination of transgenic *atzA* lines in the presence of atrazine

To investigate the atrazine tolerance of transgenic lines during germination, we germinated seeds of three lines of each genotype and the respective controls in the presence of atrazine (Fig. 1A, Table 2). The germination potential of the wild types decreased as the atrazine concentration increased (58.3–98%, 22.4–64.2%, and 0%, respectively, in 0 μM, 75 μM, and 150 μM atrazine); on the other hand, the germination percentage significantly decreased only in presence of 150 μM atrazine (95.2–100%, 72.1–96.7%, and 0–42.2%, respectively, in 0 μM, 75 μM, and 150 μM atrazine). In the presence of the investigated atrazine concentrations, the highest germination potential was determined for Nipponbare, followed by Jinhui1, Jindao8, and Jindao7. In the presence of 0 μM or 75 μM atrazine, the germination potentials and percentages of almost all of the transgenic lines did not differ significantly from those of the respective controls. The exceptions were the germination potential for Jindao8 in the presence of 75 μM atrazine and the germination percentage for Jinhui1 in the presence of 75 μM atrazine. On the other hand, in the presence of 150 μM atrazine, all of the transgenic lines showed significantly higher germination potentials and germination percentages than did the respective controls. Moreover, two of the wild types (Nipponbare and Jinhui1) failed to germinate or only rarely germinated at day 3 or day 7. Interestingly, in the presence of 150 μM atrazine, the highest germination percentage was determined for Jindao7, followed by Jindao8, Jinhui1, and Nipponbare; on the other hand, Jindao7 showed the lowest germination potential. When germinated in the presence of atrazine, all of the transgenic lines produced larger seedlings (with taller shoots and longer roots) than did the respective control plants (Fig. 1A). However, the presence of atrazine significantly inhibited the growth of transgenic lines and wild types (Fig. 1A).

Seedling growth of transgenic *atzA* lines in the presence of atrazine

To investigate the tolerance of transgenic lines to atrazine, we transplanted 7-day old seedlings germinated in the absence of atrazine, to pots of soil containing different concentration of atrazine (Fig. 1B and Table 3). In the absence of atrazine, we determined no differences in shoot or root growth between any of the transgenic lines and the respective controls. However, in the

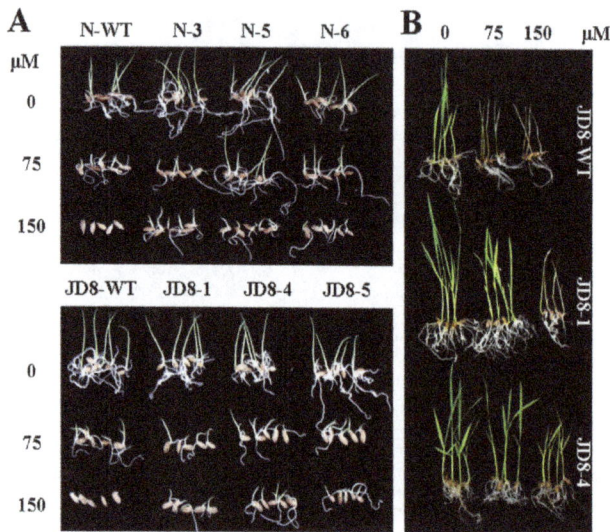

Figure 1. Germination and seedling growth of transgenic *atzA* **rice lines in the presence of atrazine. (A)** Representative images of seeds from the transgenic lines and the respective controls sown on plates containing 0, 75 or 150 μM atrazine for 7 days. **(B)** Representative images from seven-day old seedlings germinated in absence of atrazine were transplanted into pots containing 0, 75 or 150 μM atrazine for 10 days. For simplicity, only images of transgenic Jindao8 (JD8) and/or Nipponbare (N) were shown here.

Plants became yellowish, and subsequently rotted and died after 10 days (Fig. 1B).

Tolerance of soil-grown transgenic *atzA* lines to atrazine

To further investigate the tolerance of transgenic lines to atrazine, we transplanted 15-day-old seedlings germinated in the absence of atrazine, to pots of soil containing 0 mg/kg, 2 mg/kg, or 5 mg/kg of atrazine. We measured the plant height (as a non-destructive assay of plant growth) and chlorophyll content at 15-day intervals.

In the absence of atrazine, we determined no significant differences in plant height between the transgenic lines and the respective controls (Fig. 2A). In soil containing 2 mg/kg or 5 mg/kg of atrazine, all the transgenic lines were significantly taller than were the respective control plants. However, the presence of atrazine significantly inhibited the growth of transgenic lines and control plants (by 68.1–111.9% and 46.4–68.2%, respectively). In contrast to the control plants, all of the transgenic lines (except for the Jindao8) were slightly taller in soil containing 2 mg/kg of atrazine than in soil containing 5 mg/kg of atrazine. The growth of transgenic lines gradually recovered with an increase in the growth time. Further, three of the strongest transgenic lines (Jindao7–4, Jinhui1–5, and Jindao8–4) completely recovered after 45 days of growth (data not shown).

Atrazine may disrupt the electron flow of photosystem II and destroy photosynthetic pigments in plant leaves [15]. In the absence of atrazine, we observed no difference in chlorophyll content between transgenic lines and control plants (Fig. 2B). However, in soil containing 2 mg/kg or 5 mg/kg of atrazine, the transgenic lines retained more chlorophyll than did the respective control plants. With a few exceptions, the chlorophyll content of the transgenic lines decreased significantly in soil containing 5 mg/kg of atrazine, but not in soil containing 2 mg/kg of atrazine; on the other hand, the chlorophyll content of the control plants decreased significantly under both conditions. For each genotype grown in soil containing the same atrazine concentration, we observed no difference in chlorophyll content at different growth stages. After 45 days, the chlorophyll contents of the three

presence of 75 μM atrazine, all of the transgenic lines (except for Jindao8) were significantly taller than were the respective controls; further, all of the transgenic lines (except for Jinhui1) produced significantly more root biomass than did the respective controls. On the other hand, in the presence of 150 μM atrazine, shoot and root growth of almost all of the transgenic lines and the respective controls were significantly inhibited. The exceptions were the shoot growth of Nipponbare and the root growth of Jinhui1–5.

Table 2. Germination of transgenic *atzA* rice lines in the presence of atrazine.

Line	0 μM		75 μM		150 μM	
	3 d	7 d	3 d	7 d	3 d	7 d
N[a]-WT	98.0±2.0	100±0.0	54.0±1.6	94.5±1.4	0.0±0.0	0.0±0.0
N-3	98.0±1.6	94.2±0.0	60.0±2.2	98.2±1.1	64.3±1.6**	96.2±1.8**
N-5	96.0±2.0	100±0.0	64.0±1.3	92.1±1.5	62.3±2.6**	92.6±1.5**
JD7-WT	58.3±1.6	100±0.0	24.3±2.1	90.2±2.3	0.0±0.0	42.2±1.1
JD7-2	56.0±1.1	98.1±1.4	22.3±1.7	88.2±2.1	2.3±0.1**	88.5±1.6**
JD7-4	62.3±2.2	98.6±1.6	26.3±1.3	90.2±0.8	2.4±0.2**	86.5±1.3**
JH3-WT	86.3±2.0	95.2±2.8	64.2±2.2	72.1±1.1	0.0±0.0	4.5±0.0
JH-3	82.3±1.2	94.2±1.8	58.7±1.6	90.2±1.2*	32.3±0.9**	78.6±1.2**
JH3-5	88.3±3.2	92.2±1.3	66.3±2.7	94.1±1.0*	38.7±1.1**	80.5±1.2**
JD8-WT	82.0±4.2	98.7±0.8	22.4±1.6	96.7±1.0	0.0±0.0	26.5±1.3
JD8-1	78.3±2.5	99.2±0.8	48.3±1.1**	97.2±0.2	34.3±0.7**	94.5±2.8**
JD8-4	78.0±2.6	100±0.0	54.0±2.6**	98.3±1.0	35.2±1.6**	96.8±1.3**

Note: In each independent treatment 50 seeds were used for the experiment. Values are shown as mean ±SEM (n = 3). * or ** indicates significant difference (P<0.05) or highly significant difference (P<0.01) between the treatment and the control.
[a]N, JD7, JH3 and JD8 were the abbreviations of Nipponbare, Jindao7, Jindao8 and Jinhui3, respectively. For simplicity, only two transgenic lines for each genotype were shown here.

Table 3. Seedling growth performance of transgenic atzA rice lines in the presence of atrazine.

Line	0 μM		75 μM		150 μM	
	Shoot length (cm)	Root biomass (mg)	Shoot length (cm)	Root biomass (mg)	Shoot length (cm)	Root biomass (mg)
N [a]-WT	9.90±1.37	32.00±3.16	6.54±0.86 (66.0)[a]	21.00±2.55 (65.6)	5.07±1.17 (51.2)	21.75±3.77 (68.0)
N-3	10.55±1.49	30.25±3.50	9.56±2.08* (90.7)	29.50±3.14** (97.5)	7.55±1.20* (71.6)	21.50±6.40 (71.1)
N-5	8.78±2.42	32.75±4.50	8.94±1.33* (101.8)	29.85±1.83** (91.2)	9.39±0.51** (106.9)	21.00±2.71 (64.1)
JD7-WT	8.47±0.37	28.50±2.56	5.28±0.57 (62.4)	20.00±3.10 (70.2)	6.54±0.61 (77.2)	17.75±5.19 (62.3)
JD7-2	9.40±2.61	31.50±3.00	7.08±0.78* (75.3)	32.50±2.65** (103.2)	5.81±0.56 (61.8)	17.00±3.46 (54.0)
JD7-4	7.99±1.65	27.50±3.35	8.16±1.27* (102.2)	27.50±2.65** (100.0)	6.78±0.53 (84.8)	17.75±4.19 (64.5)
JH3-WT	9.83±1.25	24.00±3.07	5.80±0.43 (59.0)	21.75±3.36 (90.6)	4.62±0.42 (47.0)	13.25±6.24 (55.2)
JH-3	10.44±1.47	26.75±2.75	8.27±0.59* (79.2)	24.25±2.22* (90.7)	5.12±0.61 (49.0)	15.25±2.63 (57.0)
JH3-5	8.89±1.58	25.50±3.57	8.25±0.78* (92.8)	23.25±2.87* (83.8)	6.04±0.19* (67.9)	22.00±3.37** (86.3)
JD8-WT	7.88±0.93	26.75±3.55	7.20±1.85 (91.4)	21.75±2.87 (81.3)	6.43±0.35 (81.5)	12.25±3.46 (45.8)
JD8-1	8.11±1.51	27.50±2.15	7.18±0.40 (88.6)	27.50±1.52** (100.0)	5.65±1.12 (69.6)	11.50±1.91 (41.8)
JD8-4	7.90±0.72	27.75±2.53	7.64±0.69 (96.7)	27.75±1.79** (100.0)	5.64±0.78 (71.3)	11.75±2.99 (42.3)

Note: The parameters were recorded after 10 days. Values are shown as mean ± SEM ($n=3$). * or ** indicates significant difference ($P<0.05$) or highly significant difference ($P<0.01$) between the treatment and the control.
[a]N, JD7, JH3 and JD8 were the abbreviations of Nipponbare, Jindao7, Jindao8 and Jinhui3, respectively.
[b]Data in the bracket are the percentage of the treated values to the mock values.

strongest lines (Jindao7–4, Jinhui1–5, and Jindao8–4) were almost identical when grown in soil containing 5 mg/kg of atrazine and when grown in the absence of atrazine (data not shown).

Ability of soil-grown T$_2$ transgenic lines to degrade atrazine

To investigate the ability of the soil-grown transgenic plants to degrade atrazine, we determined the atrazine residue in the soil at 15-day intervals, after growth of seedlings in pots of soil containing 0 mg/kg, 2 mg/kg, or 5 mg/kg of atrazine. We observed that the atrazine residue in the soil decreased with an increase in growth time, for transgenic and control plants (Fig. 2C). However, the decrease was greater for transgenic lines than for control plants. After treatment with 2 mg/kg of atrazine, the atrazine residue remaining in the soil was 2.9–7.0% for transgenic lines, and 44.0–59.2% for control plants; after treatment with 5 mg/kg of atrazine, the atrazine residue remaining in the soil was 0.8–8.7% for transgenic lines and 28.1–30.8% for control plant. For each genotype at the same growth stage, the atrazine residue in the soil was higher after growth in 2 mg/kg atrazine than after growth in 5 mg/kg atrazine. We also observed that hydroxyatrazine, not atrazine (only trace) appeared in the leaves of the transgenic plants (Fig. S1), indicating that atrazine was metabolized rather than accumulated in the transgenic plants.

Utilization of transgenic plants in hybrid seed production

To investigate the tolerance of transgenic plants to the atrazine concentration used in weed control, we first examined the tolerance of leaf sections to atrazine solution. In the presence of 75 μM atrazine, the leaf sections of transgenic lines remained green, whereas those of control plants became bleached (Fig. 3A). In the presence of 150 μM atrazine, the leaf sections of transgenic lines became slightly bleached, but remained greener than those of control plants. We subsequently sprayed plants with 0.15% atrazine, and observed that wild-type plants became curled and withered after 6 days, whereas transgenic lines continued to grow well (Fig. 3B). Our results indicate that the transgenic rice plants

showed tolerance to the atrazine concentration used in weed control.

Discussion

Current research on the utilization of herbicide resistance genes in hybrid rice is focused on the two-line hybrid production system. Sterile lines are not stable in the two-line system, and therefore most japonica hybrid combinations are still produced by using three-line system. In the present study, we have developed atrazine-tolerant transgenic japonica rice parental lines with the potential to be used in future hybrid seed production.

Herbicide tolerance during germination is very important when utilizing transgenic atzA lines in hybrid seed production, because off-type plants should be eliminated as early as possible. Atrazine is a slightly water-soluble herbicide, with a saturated concentration of 153 μM. Kawahigashi [25] previously showed that an atrazine concentration of 100 μM did not affect the germination of rice plants. Therefore, we used two atrazine concentrations (75 μM and 150 μM) in the germination test, to evaluate the tolerance of transgenic rice lines. In contrast to Kawahigashi [25], we observed that the transgenic lines germinated well in the presence of atrazine whereas the respective controls did not. However, as the seedling growth of transgenic lines were also inhibited by atrazine, chemical control of contamination during germination may not be appropriate.

To evaluate the tolerance of seedlings to atrazine, we used 7-day-old seedlings germinated in the absence of atrazine as a starting material, to avoid cumulative effects derived from germination. We observed that almost all of the transgenic lines grew well in the presence of 75 μM atrazine with no difference from that in absence of atrazine. Therefore, 75 μM atrazine may be used for effective control of off-type plants during the early seedling growth stage. Our data also suggested spraying plants with 0.15% atrazine (equivalent to the standard dosage used in the field for weed control, i.e., 200–250 g 40% suspension concentrate in 30~50 kg of water) at the subsequently vegetative growth stage could serve as an alternative means of chemically controlling

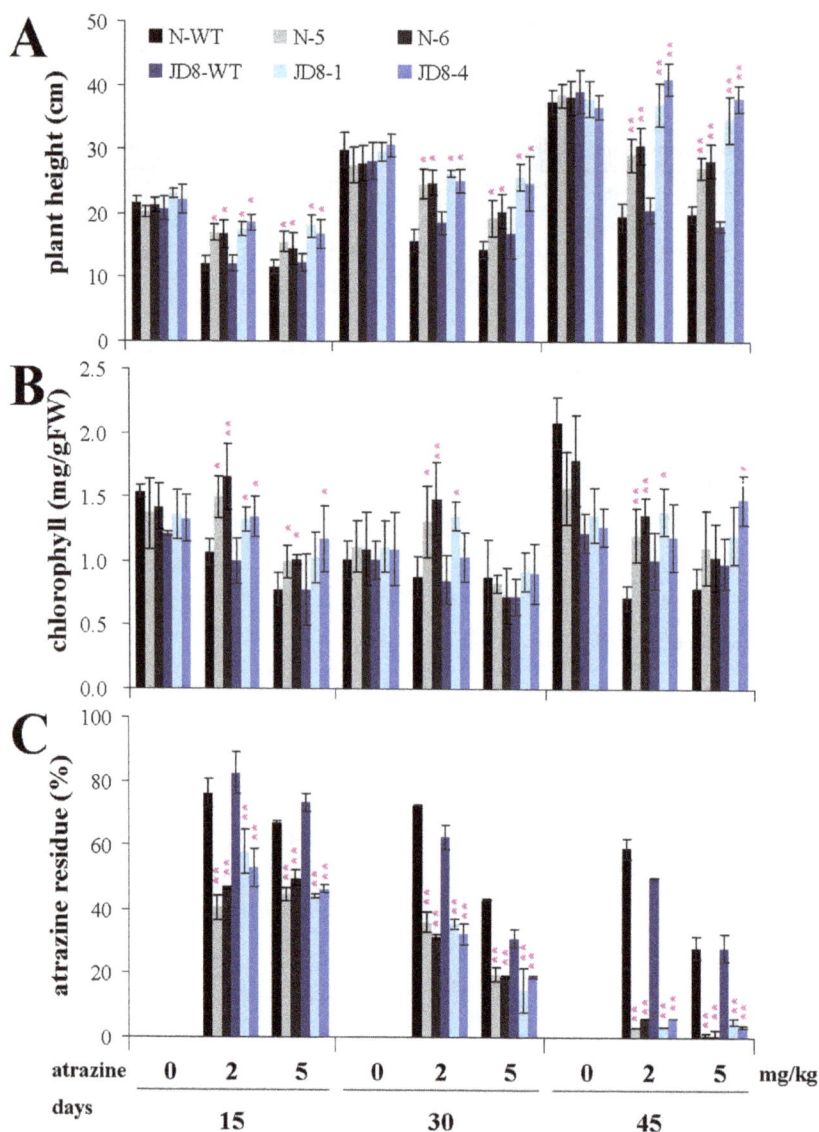

Figure 2. Growth and physiology of transgenic *atzA* rice lines in the presence of atrazine. Fifteen-day old seedlings germinated in absence of atrazine were transplanted into pots with 1.12 kg soil containing 0, 2 or 5 mg/kg of atrazine. (**A**) plant height; (**B**) chlorophyll content; and (**C**) atrazine residue were measured at 15-day intervals. Values are shown as mean ±SEM ($n = 3$). * or ** over the bar indicates significant difference ($P<0.05$) or highly significant difference ($P<0.01$) between the treatment and the control. For simplicity, only data of transgenic Nipponbare (N) and Jindao8 (JD8) were shown here.

off-type plants. However, we did not test the tolerance of transgenic *atzA* lines at the reproductive growth stage, which is important for the utilization in mechanical harvesting of hybrid seed production. Further studies are required to investigate whether the lethal dosage for untransformed parental lines is tolerated at the reproductive stage.

Atrazine degrades relatively slowly, with an average half-life of 4–57 weeks [26]. Therefore, application of this herbicide will inevitably lead to accumulation in the field, which may affect current crop growth or damage sensitive succession crops [27]. The transgenic rice lines were taller, and had higher chlorophyll contents than did the respective control plants when growing in the soil of 2 mg/kg or 5 mg/kg atrazine. Three of the transgenic lines (Jindao7–4, Jinhui1–5, and Jindao8–4) completely recovered after 45 days of growth in the presence of atrazine, suggesting that an atrazine concentration up to 5 mg/kg does not affect sustained

growth of transgenic lines. Our data also suggested that almost all of the atrazine was degraded after growth of the transgenic rice plants for 45 days. This was also verified by the appearance of hydroxyatrazine in the leaves of transgenic plants. Besides, atrazine metabolite hydroxyatrazine could be directly determined in transgenic plants. However, as the leave, stem and root of the plants all possibly metabolize atrazine [28], it would be more complicated if we want to quantitatively evaluate the atrazine metabolized by plants. That is why we chose to determine the atrazine residue in the soil after growth of the transgenic plants in this study.

Utilization of herbicide resistance genes in hybrid seed production may be achieved by introducing the genes into a maintainer or restorer line in the three-line system. When an herbicide-resistant maintainer is obtained, its corresponding CMS line, which possesses herbicide resistance, may be created by using

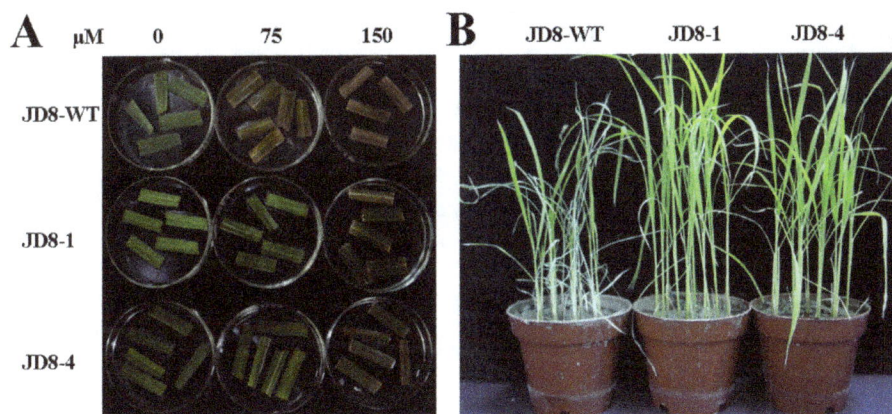

Figure 3. Tolerance of mature transgenic _atzA_ plants to atrazine. (A) Representative images of leave sections in the presence of atrazine. Fifteen-day old seedlings were transplanted into pots containing soil. After 30 days' growth, the last second leaves were cut into 2–3 cm section, and soaked in 0, 75 and 150 µM atrazine, and incubated at 25°C under light for 2 days. **(B)** Representative images of mature plants after sprayed with atrazine. Fifteen-day old seedlings were transplanted into pots containing soil. After 40 days' growth, each pot of transgenic lines (9 plants) was sprayed with 20 mL of 0.15% atrazine solution. For simplicity, only data of transgenic Jindao8 (JD8) was shown here.

a backcross. The resulting herbicide-resistant CMS breeder seeds may be used to cross with the original maintainer (herbicide sensitive), to multiply the herbicide-resistant foundation seeds. During this process, the selfed maintainer and off-type plants may easily be eliminated by herbicide spraying, thus ensuring the purity of CMS seeds. The herbicide-resistant CMS line may also be used in mechanical harvesting, by spraying herbicide to kill the pollen plants after pollination, and harvesting seeds from the surviving CMS plants. A similar result may be achieved when crossing the herbicide-resistant CMS line with the herbicide-sensitive restorer line. Alternatively, the herbicide-resistant restorer line may be crossed with the herbicide-sensitive sterile line during the hybrid seed production process, by spraying herbicide to kill the off-type plants. Further, the herbicide-resistant maintainer and restorer lines may be useful for eliminating off-type plants, and for weed management when growing hybrid rice.

Field trials with genetically modified rice are strictly controlled in China, and therefore we were unable to evaluate our herbicide-resistant hybrid rice combinations on a field scale. Further studies are required to verify the validity of utilizing atrazine-tolerant transgenic lines in the field, and to determine whether herbicide-resistant hybrid rice combinations retain their original superiority in terms of yield. Nevertheless, the development of atrazine-tolerant transgenic _japonica_ rice parental lines represents a valuable tool for future application in hybrid seed production.

Supporting Information

Figure S1   HPLC analysis for determination of atrazine and its metabolite hydroxyatrazine in leaves of transgenic and WT rice plants after grown in 5 mg/kg atrazine soil for 45 days. (A) Standard of hydroxyatrazine (I) and atrazine (II). **(B)** Wild type (WT). **(C–E)** Transgenic Nipponbare (N-6), Jindao7 (JD7-4) or Jindao8 (JD8-4), respectively.

Author Contributions

Conceived and designed the experiments: DFC XWC. Performed the experiments: LHZ HWC YLL YNL SJW JPS. Analyzed the data: LHZ DFC XWC. Contributed reagents/materials/analysis tools: SJW XJL DFC XWC. Wrote the paper: DFC XWC.

References

1. Peng C, Zhang H (2013) Analysis of paddy and rice markets at home and abroad in the first quarter of 2013 and its prospect. Agric Outlook (4): 4–9 (in Chinese, with English abstract).
2. Shi KB, Deng HS (2004) Chinese _japonica_ hybrid rice technology innovation seminar held in Sanya. Hybrid Rice 19: 76 (in Chinese).
3. Li S, Yang D, Zhu Y (2007) Characterization and use of male sterility in hybrid rice breeding. J Integ Plant Biol 49: 791–804.
4. Tang SZ, Zhang HG, Zhu ZB, Liu C, Li P, et al. (2010) Application of HL type male sterile cytoplasm in _japonica_ hybrid rice breeding. Chin J Rice Sci 24: 116–124 (in Chinese, with English abstract).
5. Yan W (2000) Crop heterosis and herbicide. US Patent 6066779.
6. Hu G, Xiao H, Yu Y, Zhu Z, Si H, et al. (2000) _Agrobacterium_-mediated transformation of the restorer lines of two-line hybrid rice with _bar_ gene. Chin J Appl Environ Biol 6: 511–515 (in Chinese, with English abstract).
7. Li Y, Xu Q, Duan F, Liu G, Yan W (2000) Breeding of herbicide-resistant hybrid rice combinations. Hybrid Rice 15(6): 9–11.
8. Xue S, Zhang W, Gong X, Shen L, Huang D, et al. (2001) Breeding of isotype restorer line with _bar_ gene from Miyang 46 and its combinations. J Zhejiang Agric Sci (4): 181–183 (in Chinese).
9. Ramesh S, Nagadhara D, Pasalu IC, Kumari AP, Sarma NP, et al. (2004) Development of stem borer resistant transgenic parental lines involved in the production of hybrid rice. J Biotechnol 111: 131–141.
10. Xiong X, Tang L, Deng X, Xiao G (2004) A preliminary report on the experiments of herbicide-resistant two-line hybrid rice Xiang 125S/Bar 68-1. Hybrid Rice 19(5): 41–43.
11. Wu F, Wang S, Li S, Zhang K, Li P (2006) Research progress on herbicide resistant transgenic rice and its safety issues. Mol Plant Breed 4: 846–852.
12. Xiao G, Yuan L, Sun S (2007) Strategy and utilization of a herbicide resistance gene in two-line hybrid rice. Mol Breed 20: 287–292.
13. Xiao G (2009) Recent advances in development of herbicide resistance transgenic hybrid rice in China. Rice Sci 16: 235–239.
14. Udiković-Kolić N, Scott C, Martin-Laurent F (2012) Evolution of atrazine-degrading capabilities in the environment. Appl Microbiol Biotechnol 96: 1175–1189.
15. Rutherford AW, Krieger-Liszkay A (2001) Herbicide-induced oxidative stress in photosystem II. Trends Biochem Sci 26: 648–653.
16. Cai B, Han Y, Liu B, Ren Y, Jiang S (2003) Isolation and characterization of an atrazine-degrading bacterium from industrial wastewater in China. Lett Appl Microbiol 36: 272–276.
17. Wang Y, Li X, Chen X, Chen D (2013) Directed evolution and characterization of atrazine chlorohydrolase variants with enhanced activity. Biochemistry (Moscow) 78: 1104–1111.
18. Höfgen R, Willmitzer L (1988) Storage of competent cells for _Agrobacterium_ transformation. Nucleic Acids Res 16: 9877.

19. Loh JP, Kiew R, Kee A, Gan LH, Gan YY (1999) Amplified fragment length polymorphism (AFLP) provides molecular markers for the identification of *Caladium bicolor* cultivars. Ann Bot 84: 155–161.

20. Wang H, Chen X. Xing X, Hao X, Chen D (2010) Transgenic tobacco plants expressing *atzA* exhibit resistance and strong ability to degrade atrazine. Plant Cell Rep 29: 1391–1399.

21. Chen D, Chen H, Zhang L, Shi X, Chen X (2014) Tocopherol-deficient rice plants display increased sensitivity to photooxidative stress. Planta 239: 1351–1362.

22. Ma J, Takahashi E (1990) Effect of silicon on the growth and phosphorus uptake of rice. Plant Soil 126: 115–119.

23. Porra RJ, Thompson WA, Kriedemann PE (1989) Determination of accurate extinction coefficients and simultaneous equations for assaying chlorophylls *a* and *b* extracted with four different solvents: verification of the concentration of chlorophyll standards by atomic absorption spectroscopy. Biochim Biophys Acta - Bioenergetics 975: 384–394.

24. de Souza ML, Sadowsky MJ, Wackett LP (1996). Atrazine chlorohydrolase from *Pseudomonas* sp. strain ADP: gene sequence, enzyme purification, and protein characterization. J Bacteriol 178: 4894–4900.

25. Kawahigashi H, Hirose S, Ohkawa H, Ohkawa Y (2007) Herbicide resistance of transgenic rice plants expressing human CYP1A1. Biotechnol Adv 25: 75–84.

26. Erickson LE, Lee KH, Sumner DD (1989) Degradation of atrazine and related *s*-triazines. Crit Rev Environ Control 19: 1–14.

27. Rhine ED, Fuhrmann JJ, Radosevich M (2003) Microbial community responses to atrazine exposure and nutrient availability: linking degradation capacity to community structure. Microbiol Ecol 46: 145–160.

28. Wang L, Samac DA, Shapir N, Wackett LP, Vance CP, et al. (2005) Biodegradation of atrazine in transgenic plants expressing a modified bacterial atrazine chlorohydrolase (*atzA*) gene. Plant Biotech J 3: 475–486.

PERMISSIONS

All chapters in this book were first published in PLOS ONE, by The Public Library of Science; hereby published with permission under the Creative Commons Attribution License or equivalent. Every chapter published in this book has been scrutinized by our experts. Their significance has been extensively debated. The topics covered herein carry significant findings which will fuel the growth of the discipline. They may even be implemented as practical applications or may be referred to as a beginning point for another development.

The contributors of this book come from diverse backgrounds, making this book a truly international effort. This book will bring forth new frontiers with its revolutionizing research information and detailed analysis of the nascent developments around the world.

We would like to thank all the contributing authors for lending their expertise to make the book truly unique. They have played a crucial role in the development of this book. Without their invaluable contributions this book wouldn't have been possible. They have made vital efforts to compile up to date information on the varied aspects of this subject to make this book a valuable addition to the collection of many professionals and students.

This book was conceptualized with the vision of imparting up-to-date information and advanced data in this field. To ensure the same, a matchless editorial board was set up. Every individual on the board went through rigorous rounds of assessment to prove their worth. After which they invested a large part of their time researching and compiling the most relevant data for our readers.

The editorial board has been involved in producing this book since its inception. They have spent rigorous hours researching and exploring the diverse topics which have resulted in the successful publishing of this book. They have passed on their knowledge of decades through this book. To expedite this challenging task, the publisher supported the team at every step. A small team of assistant editors was also appointed to further simplify the editing procedure and attain best results for the readers.

Apart from the editorial board, the designing team has also invested a significant amount of their time in understanding the subject and creating the most relevant covers. They scrutinized every image to scout for the most suitable representation of the subject and create an appropriate cover for the book.

The publishing team has been an ardent support to the editorial, designing and production team. Their endless efforts to recruit the best for this project, has resulted in the accomplishment of this book. They are a veteran in the field of academics and their pool of knowledge is as vast as their experience in printing. Their expertise and guidance has proved useful at every step. Their uncompromising quality standards have made this book an exceptional effort. Their encouragement from time to time has been an inspiration for everyone.

The publisher and the editorial board hope that this book will prove to be a valuable piece of knowledge for researchers, students, practitioners and scholars across the globe.

LIST OF CONTRIBUTORS

Qurban Ali Panhwar and Shamshuddin Jusop
Department of Land Management, Faculty of Agriculture, Universiti Putra Malaysia (UPM), Serdang, Selangor, Malaysia

Umme Aminun Naher and Md Abdul Latif
Institute of Tropical Agriculture, Universiti Putra Malaysia (UPM), Serdang, Selangor, Malaysia
Bangladesh Rice Research Institute, Gazipur, Bangladesh

Radziah Othman
Department of Land Management, Faculty of Agriculture, Universiti Putra Malaysia (UPM) Serdang, Selangor, Malaysia
Institute of Tropical Agriculture, Universiti Putra Malaysia (UPM), Serdang, Selangor, Malaysia

Mohd Razi Ismail
Institute of Tropical Agriculture, Universiti PutraMalaysia (UPM), Serdang, Selangor, Malaysia

Prasenjit Saha and Eduardo Blumwald
Department of Plant Sciences, University of California Davis, Davis, California, United States of America

Satoko Yoneyama, Masaru Sakurai, Koshi Nakamura, Yuko Morikawa and Hideaki Nakagawa
Department of Epidemiology and Public Health, Kanazawa Medical University, Ishikawa, Japan

Katsuyuki Miura
Department of Health Science, Shiga University of Medical Science, Otsu, Japan

Motoko Nakashima
Department of Community Health Nursing, School of Nursing, Kanazawa Medical University, Ishikawa, Japan

Katsushi Yoshita
Department of Food and Human Health Science Osaka City University, Graduate School of Human Life Science, Osaka, Japan

Masao Ishizaki
Department of Social and Environmental Medicine, Kanazawa Medical University, Ishikawa, Japan

Teruhiko Kido
School of Health Science, College of Medical, Pharmaceutical and Health Science, Kanazawa University, Kanazawa, Japan

Yuchi Naruse
Department of HumanScience and Fundamental Nursing, Toyama University School of Nursing, Toyama, Japan

Kazuhiro Nogawa and Yasushi Suwazono
Department of Occupational and Environmental Medicine, Graduate School of Medicine, Chiba University, Chiba, Japan

Satoshi Sasaki
Department of Social and Preventive Epidemiology, the University of Tokyo, Tokyo, Japan

Xiaochan He
State Key Laboratory Breeding Base for Zhejiang Sustainable Pest and Disease Control, Institute of Plant Protection and Microbiology, Zhejiang Academy of Agriculture Sciences, Hangzhou, China
Jinhua Research Academy of Agricultural Sciences, Jinhua, China

Xiaojun Zhou and Yujian Sun
Jinhua Research Academy of Agricultural Sciences, Jinhua, China

Guanchun Gao
School of Medicine Science, Jiaxing University, Jiaxing, China

Yajun Yang, Junce Tian, Zhongxian Lu, Xusong Zheng and Hongxing Xu
State Key Laboratory Breeding Base for Zhejiang Sustainable Pest and Disease Control, Institute of Plant Protection and Microbiology, Zhejiang Academy of Agriculture Sciences, Hangzhou, China

Koji Miyamoto and Hisakazu Yamane
Department of Biosciences, Teikyo University, Utsunomiya, Tochigi, Japan
Biotechnology Research Center, The University of Tokyo, Bunkyo-ku, Tokyo, Japan

Takashi Matsumoto
Genome Research Center, NODAI Research Institute, Tokyo University of Agriculture, Setagaya-ku, Tokyo, Japan

Atsushi Okada, Kohei Komiyama, Tetsuya Chujo, Hideaki Nojiri and Kazunori Okada
Biotechnology Research Center, The University of Tokyo, Bunkyo-ku, Tokyo, Japan

Hirofumi Yoshikawa
Genome Research Center, NODAI Research Institute, Tokyo University of Agriculture, Setagaya-ku, Tokyo, Japan
Department of Bioscience, Tokyo University of Agriculture, Setagaya-ku, Tokyo, Japan

Catarina Cardoso, Tatsiana Charnikhova, Muhammad Jamil and Maryam Amini
Laboratory of Plant Physiology, Wageningen University, Wageningen, the Netherlands

Pierre-Marc Delaux and Dominique Lauressergues
Laboratoire de Recherche en Sciences Végétales, Unité Mixte de Recherche
(UMR) 5546, Université de Toulouse, Castanet-Tolosan, France
Laboratoire de Recherche en Sciences Végétales, Unité Mixte de Recherche (UMR) 5546, Centre National de la Recherche Scientifique (CNRS), Castanet-Tolosan, France

Francel Verstappen and Harro Bouwmeester
Laboratory of Plant Physiology, Wageningen University, Wageningen, the Netherlands Centre for Biosystems Genomics, Wageningen, the Netherlands

Carolien Ruyter-Spira
Laboratory of Plant Physiology, Wageningen University, Wageningen, the Netherlands Bioscience, Plant Research International, Wageningen, the Netherlands

Hui Xia, Xiaoguo Zheng, Liang Chen, Huan Gao, Hua Yang, Ping Long, Jiajia Li and Lijun Luo
Shanghai Agrobiological Gene Center, Shanghai, China

Jun Rong
Center for Watershed Ecology, Institute of Life Science and Key Laboratory of Poyang Lake Environment and Resource Utilization, Ministry of Education, Nanchang University, Nanchang, China

Baorong Lu
Ministry of Education Key Laboratory for Biodiversity and Ecological Engineering, Fudan University, Shanghai, China

Chunyan Yu, Sha Su, Yichun Xu, Yongqin Zhao, An Yan, Linli Huang, Imran Ali and Yinbo Gan
Zhejiang Key Lab of Crop Germplasm, Department of Agronomy, College of Agriculture and Biotechnology, Zhejiang University, Hangzhou, China

Chao Chen, Shu Fu, FangSen Xue and LiuFeng Wang
Institute of Entomology, Jiangxi Agricultural University, Nanchang, Jiangxi Province, China

KeJian Lin
state Key Laboratory for Biology of Plant Diseases and Insect Pests, Institute of Plant Protection, Chinese Academy of Agricultural Sciences, Beijing, China

Shanshan Yang, Rebecca L. Murphy, Daryl T. Morishige and John E. Mullet
Department of Biochemistry and Biophysics, Texas A&M University, College Station, Texas, United States of America

Patricia E. Klein
Department of Horticultural Sciences and Institute for Plant Genomics and Biotechnology, Texas A&M University, College Station, Texas, United States of America

William L. Rooney
Department of Soil and Crop Sciences, Texas A&M University, College Station, Texas, United States of America

Uma M. Singh, Muktesh Chandra and Anil Kumar
Department of Molecular Biology and Genetic Engineering, Govind Ballabh Pant University of Agriculture and Technology, Pantnagar, Uttarakhand, India

Shailesh C. Shankhdhar
Department of Plant Physiology, Govind Ballabh
Pant University of Agriculture and Technology,
Pantnagar, Uttarakhand, India

**Xiuqin Zhao, Wensheng Wang, Fan Zhang, Jianli
Deng, Zhikang Li and Binying Fu**
Institute of Crop Sciences, National Key Facility for
Crop Gene Resources and Genetic Improvement,
Chinese Academy of Agricultural Sciences, Beijing,
China

Yuke Geng
Key Laboratory of Crop Gene Resources and
Germplasm Enhancement, Ministry of Agriculture/
Institute of Crop Science, Chinese Academy of
Agricultural Sciences, Beijing, China
College of Biological sciences, China Agricultural
University, Beijing

Binshuang Pang
Beijing Engineering and Technique Research Center
of Hybrid Wheat, Beijing
Academy of Agricultural and Forestry Sciences,
Beijing, China

Chenyang Hao, Xueyong Zhang and Tian Li
Key Laboratory of Crop Gene Resources and
Germplasm Enhancement, Ministry of Agriculture/
Institute of Crop Science, Chinese Academy of
Agricultural Sciences, Beijing, China

Saijun Tang
College of Biological sciences, China Agricultural
University, Beijing, China

**Emelie Lindquist, Mohamed Alezzawi and Henrik
Aronsson**
Department of Biological and Environmental
Sciences, University of Gothenburg, Gothenburg,
Sweden

**Tong Zhou, Linlin Du, Hui Feng, Lijiao Wang,
Ying Lan, Feng Sun, Lihui Wei, Yongjian Fan and
Yijun Zhou**
Institute of Plant Protection, Jiangsu Academy of
Agricultural Sciences, Nanjing, China

Cunyi Gao
Institute of Plant Protection, Jiangsu Academy of
Agricultural Sciences, Nanjing, China
College of Life Science, Nanjing Agricultural
University, Nanjing, China

Wenbiao Shen
College of Life Science, Nanjing Agricultural
University, Nanjing, China

**Syed Tahir Ata-Ul-Karim, Xia Yao, Xiaojun Liu,
Weixing Cao and Yan Zhu**
National Engineering and Technology Center for
Information Agriculture, Jiangsu Key Laboratory
for Information Agriculture, Nanjing Agricultural
University, Nanjing, Jiangsu, P. R. China

**Hai-Ming Tang, Xiao-Ping Xiao, Wen-Guang
Tang, Ke Wang, Ji-Min Sun, Wei-Yan Li and
Guang-Li Yang**
Hunan Soil and Fertilizer Institute, Changsha, PR
China

**Lin Feng, Zhu Chen, Hui Ma, Xue Chen, Yuan Li
and Yiyi Wang**
Laboratory of Modern Biotechnology, School
of Forestry and Landscape Architecture, Anhui
Agricultural University, Hefei, China

Yan Xiang
Laboratory of Modern Biotechnology, School
of Forestry and Landscape Architecture, Anhui
Agricultural University, Hefei, China
Key Laboratory of Crop Biology of Anhui
Agriculture University, Hefei, China

**Luhua Zhang, Haiwei Chen, Yanlan Li, Yanan Li,
Defu Chen and Xiwen Chen**
Laboratory of Molecular Genetics, College of Life
Sciences, Nankai University, Tianjin, China

Shengjun Wang, Jinping Su and Xuejun Liu
Tianjin Crop Research Institute, Tianjin, China

Index

www.ingramcontent.com/pod-product-compliance
Lightning Source LLC
Chambersburg PA
CBHW082050190326
41458CB00010B/3496